NESS/SCIENCE/TECHNOLOGY DIVISION
CHICAGO PUBLIC LIBRARY
400 SOUTH STATE STREET
CHICAGO, IL 60605

R06007 53049

D1786744

Chicago Public Library

REFERENCE

Form 178 rev. 11-00

FORM 125 M

NATURAL SCIENCES &
USEFUL ARTS DEPT.

Public Library

Received Sept 13, 1972

Essays in Evolution and Genetics in Honor of Theodosius Dobzhansky

Essays in
Evolution and Genetics

A Supplement to Evolutionary Biology

APPLETON-CENTURY-CROFTS
EDUCATIONAL DIVISION/MEREDITH CORPORATION
New York

in Honor of
Theodosius Dobzhansky

EDITORS:

MAX K. HECHT, *Queens College*
WILLIAM C. STEERE, *The New York Botanical Garden*

FOREWORD by L. C. DUNN, *Columbia University*

Copyright © 1970 by MEREDITH CORPORATION

All rights reserved. This book, or parts thereof, must not be used or reproduced in any manner without written permission. For information address the publisher, Appleton-Century-Crofts, Educational Division, Meredith Corporation, 440 Park Avenue South, New York, New York 10016.

710-1

Library of Congress Catalog Card Number: 74-105428

PRINTED IN THE UNITED STATES OF AMERICA
390-43040-4

To our colleague, Theodosius Dobzhansky, on the occasion of his seventieth birthday we dedicate with great pleasure this volume of articles by his friends and students

Contributors

FRANCISCO J. AYALA   *The Rockefeller University, New York*

JOHN BEARDMORE   *Department of Genetics, University College of Swansea, Glam., Wales*

DANKO BRNCIC   *Departamento de Genética, Escuela de Medicina, Universidad de Chile, Santiago, Chile*

HAMPTON L. CARSON   *Department of Biology, Washington University, St. Louis, Missouri*

E. R. CREED   *The Genetics Laboratories, Department of Zoology, The University, Oxford, England*

W. H. DOWDESWELL   *School of Education, Bath University of Technology, Bath, England*

LEE EHRMAN   *The Rockefeller University, New York*

VILHELMS EICHE   *Department of Forest Genetics, Royal College of Forestry, Stockholm, Sweden*

E. B. FORD   *The Genetics Laboratories, Department of Zoology, The University, Oxford, England*

MARCEL FLORKIN   *Department of Biochemistry, University of Liège, Liège, Belgium*

ÅKE GUSTAFSSON   *Institute of Genetics, Lund University, Lund, Sweden*

D. ELMO HARDY   *Department of Entomology, University of Hawaii, Honolulu, Hawaii*

HOWARD LEVENE Columbia University, New York

C. C. LI Graduate School of Public Health, University of Pittsburgh, Pittsburgh, Pennsylvania

K. G. McWHIRTER Department of Genetics, The University of Alberta, Edmonton, Canada

BERNARD RENSCH Zoologisches Institut, Wilhelms-Universität, Münster, West Germany

ROLLIN RICHMOND The Rockefeller University, New York

ROSE M. RIZKI The University of Michigan, Ann Arbor, Michigan

T. M. RIZKI The University of Michigan, Ann Arbor, Michigan

E. SCHOFFENIELS Department of Biochemistry, University of Liège, Liège, Belgium

GEORGE GAYLORD SIMPSON Harvard University and The University of Arizona, Tucson, Arizona

ELIOT B. SPIESS Department of Biological Sciences, University of Illinois at Chicago Circle, Chicago, Illinois

HERMAN T. SPIETH Department of Zoology, University of California, Davis, California

G. LEDYARD STEBBINS Department of Genetics, University of California, Davis, California

WILSON D. STONE late professor, Genetics Foundation, University of Texas, Austin, Texas

BRUCE WALLACE Cornell University, Ithaca, New York

M. J. D. WHITE Department of Genetics, University of Melbourne, Australia

Foreword

It is not often that one has the opportunity to send a public birthday greeting to a friend and colleague of many years, and to congratulate him on having reached the age of reason. In fact it happens only once, and comes then as a surprise. Surely it was only a few years ago that we sat together at an International Genetics Congress in Ithaca, and only yesterday that we became members of the same department. The eighth floor of Schermerhorn Hall had a north end where the flies were and a south end furnished with mice, and in between, a seminar room and laboratory. There the distances were short and the doors open and the coffee pot busy. But it now appears that yesterday has fallen thirty years behind and that we have grown up.

I find it interesting and appropriate that Dobzhansky's lifetime spans the period of maturation of the fields to which this volume is devoted. This is true in a chronological sense for his birth occurred in the same year, 1900, in which modern genetics began. The rediscovery of Mendel's principles and the interpretation of the nature of heredity and variation to which this event led were necessary prerequisites to the development of evolutionary biology as presented in this collection of essays.

Dobzhansky's career spans this field in another sense as well, for there are few areas of genetics and of evolutionary thought to which his work has not contributed. Most of these contributions have arisen from first-hand observation in nature and from experimental analysis in the laboratory. It was this new approach to problems of evolution, guided by theory elaborated from simple mendelian assumptions, that led to the rapid rise of population genetics. Dobzhansky and his collaborators and students played leading roles in that development, and influenced both the direction and momentum of growth of a central field of biology.

This volume is the work of his colleagues. One might appropriately call them his fellow travellers for they have been heeding the call of the same ideological drum. But there are others, not of his professional school, whose lives have been influenced by him. Some of these were students who followed careers not closely related to evolutionary biology and genetics but yet benefitted from his teaching in other ways. Others were members of the audiences he has addressed at meetings of scientists, philosophers, humanists or laymen. Surely their number is legion and surely few of them have forgotten the sharp, incisive discourses they heard from him. His lectures are generally not designed to persuade but to elucidate and clarity is usually the virtue achieved.

Finally there are those beyond the sound of his voice who have read the works in which he has expressed his views as to what science is about and what it is for. Many of them must have gained the impression from him that while science is certainly about the real world it serves the larger good of human beings who may by its influence become not merely reasonable but creative beings.

It is pleasant to observe that those who have written these articles in honor of a many-sided person themselves represent many aspects of evolutionary biology. The ballast of this book is provided by new observation, facts like those in which Dobzhansky has been interested, dealing with animals, plants and man, related through their bearing on general evolutionary theory, invigorated by a dash of philosophical reflection and speculation.

This variety should be relished by the man to whom the collection is dedicated and by many others interested in the fields which have been strongly influenced by him. As an old friend, I hope I may be allowed to express to him the thanks of those many of us who have benefitted from his work and scientific comradeship and to wish him well upon this anniversary and days of fruitful work and pleasant journeys to follow.

L. C. Dunn

Contents

Contributors vii

Foreword *by L. C. Dunn* ix

1 **Theodosius Dobzhansky Up to Now**   *Howard Levene, Lee Ehrman, and Rollin Richmond* 1
   Introduction 1
   Scientific Work 3
   A Collecting Expedition 7
   Acknowledgment 15
   Career Summary 16
   Theodosius Dobzhansky: Bibliography 1918-1969 18

2 **Uniformitarianism. An Inquiry into Principle, Theory, and Method in Geohistory and Biohistory**   *George Gaylord Simpson* 43
   Introduction 43
   Origins; Hutton 45
   Lyell; Conybeare 48
   Lyell; Darwin 53
   Since Darwin 55
   Immanence and Configuration 58
   Classification of Current Issues 60
   Naturalism 61
   Actualism 61
   Historicism 66
   Evolutionism 68
   Gradualism 72
   Historical Inferences 81
   Summary and Conclusion 87
   References 91

3 **Evolution of Matter and Consciousness and Its Relation to Panpsychistic Identism** *Bernard Rensch* 97
   Introduction 97
   Natural Laws and Their Epigenetic Manifestation 98
   Causal Analysis of Life Processes 101
   Factors and Laws of Organismic Evolution 104
   Psychogenesis and the Epistemological Analysis of Matter 109
   Physical Analysis of Matter 112
   The Psychophysical Substrate and the Problem of the Psychophysical Connection 113
   The Panpsychistic, Polynomistic, and Realistic Identism 117
   References 118

4 **Competition, Coexistence, and Evolution** *Francisco J. Ayala* 121
   Introduction 121
   The Ecological Niche 122
   The Concept of Competition 124
   The Principle of Competitive Exclusion 128
   The Logistic Theory of Population Growth 130
   Volterra's Equations for Competitive Exclusion 130
   Coexistence of Related Species in Nature 136
   Coexistence in the Laboratory 146
   The Role of Natural Selection 151
   Summary 154
   References 155

5 **Adapted Molecules** *Marcel Florkin and E. Schoffeniels* 159
   Adaptation at the Molecular Level 159
   Molecular Adaptation and Phylogeny 163
   Molecular Adaptation at the Population Level 166
   Conclusions 170
   References 171

6 **Variation and Evolution in Plants: Progress During the Past Twenty Years** *G. Ledyard Stebbins* 173
   Introduction 174
   Variation Patterns 174
   Examples of Variation Patterns 177
   The Basis of Individual Variation 180
   Natural Selection and Variation in Populations 182
   Genetic Systems as Factors in Evolution 185
   Isolation and the Origin of Species 187

Hybridization and its Effects	192
Polyploidy and Apomixis	193
Structural Hybridity and Chromosomal Evolution	197
Evolution Above the Species Level	200
References	203

7 Population Research in the Scandinavian Scots pine (*Pinus sylvestris L.*): Recent Experimentation *Vilhelms Eiche and Åke Gustafsson*   209

Introduction	209
Material and Methods	210
Climatic Conditions of the Experimental Plantations and the Natural Populations Involved	212
Cumulative Mortality and Survival	214
Cumulative Mortality and Survival in Selected Experimental Plantations	217
Single-Year Mortality	223
Height Growth	228
Conclusions and Summary	231
References	234

8 Heterozygosity and Genetic Polymorphism in Parthenogenetic Animals *M. J. D. White*   237

Introduction	237
Nature of Parthenogenesis	238
Parthenogenesis in Vertebrates	241
Parthenogenesis in Insects	248
Conclusions	256
References	259

9 Evolutionary Studies on *Maniola jurtina* (Lepidoptera, *Satyridae*): The "Boundary Phenomenon" in Southern England 1961 to 1968 *E. R. Creed, W. H. Dowdeswell, E. B. Ford, K. G. McWhirter*   263

Dedication	264
Studies of the Southern English and East Cornish Types up to 1960	268
1961 to 1968	270
Discussion	282
Summary	285
Acknowledgments	286
References	286

10	**The Genetic Basis of a Cell-Pattern Homology in** *Drosophila* **Species** T. M. Rizki and Rose M. Rizki	289
	Introduction	289
	Kynurenine Distribution in the Larval Fat Body	290
	Hybrid Autofluorescent Patterns	291
	Acknowledgments	297
	References	298
11	**Ecological Factors and the Variability of Gene-Pools in** *Drosophila* John Beardmore	299
	Introduction	299
	Relations Between Ecological Heterogeneity, Genetic Variability, and Fitness in *Drosophila* Populations	301
	Conclusions	312
	Summary	313
	Acknowledgments	313
	References	313
12	**Mating Propensity and Its Genetic Basis in** *Drosophila* Eliot B. Spiess	315
	Introduction	316
	Mating Propensity as a Genetic Character in *Drosophila*	318
	Chromosomal Polymorphism and Mating Propensity	343
	Factors Modifying Mating Propensity	356
	Summary	373
	Acknowledgments	375
	References	375
13	**Observations on the Microdispersion of** *Drosophila melanogaster* Bruce Wallace	381
	Introduction	381
	The Bryant Park Experiments	382
	Experimental Populations	386
	The "Tropical Rainforest" Greenhouse Experiments	393
	Venetian Experiments	395
	Discussion	396
	References	398
14	**Studies on the Evolutionary Biology of Chilean Species of** *Drosophila* Danko Brncic	401
	Introduction	401
	Widespread Species of *Drosophila*	405
	Endemic and Ecologically Restricted Species	411

	Endemic and Ecologically Versatile Species	418
	Isolation and Chromosomal Structure	428
	Acknowledgments	432
	References	432
15	**The Evolutionary Biology of the Hawaiian Drosophilidae**	
	Hampton L. Carson, D. Elmo Hardy, Herman T. Spieth, and Wilson S. Stone	437
	Introduction	438
	Habitat	439
	Drosophilid Fauna	450
	Biology and Behavior	469
	Chromosomal and Genetic Characteristics	492
	Evolution, Speciation, and Migration	520
	Summary	536
	Acknowledgments	540
	References	540
16	**Human Genetic Adaptation** *C. C. Li*	545
	Introduction	545
	Heritability	546
	Estimation of Heritability	550
	Random Mating	552
	The Markov Property	557
	Shall We Count the Living or the Dead?	560
	Simplest Selection Model	561
	Correlated Responses	564
	Genetic Improvement of Mankind	567
	Genetic Deterioration	568
	Decline of Intelligence	570
	Control of Human Evolution: Reproductive Specialization	574
	Genotype-Environment Interaction	575
	Summary and Conclusions	576
	References	577

Author Index — 579

Subject Index — 589

Essays in Evolution and Genetics in Honor of Theodosius Dobzhansky

1

Theodosius Dobzhansky Up to Now

HOWARD LEVENE
Columbia University, New York, N. Y.

LEE EHRMAN
The Rockefeller University, New York, N. Y.

ROLLIN RICHMOND
The Rockefeller University, New York, N. Y.

Introduction 1
Scientific Work 3
A Collecting Expedition 7
Acknowledgment 15
Career Summary 16
Theodosius Dobzhansky: Bibliography 1918-1969 18

Introduction

Theodosius Grigorievich Dobzhansky[1] (Феодосий Григорьевич Добржанский) was born on January 25, 1900, in the town of Nemirov,

---

[1] Sometimes spelled in the original form, Dobrzhansky.

which is located about 130 miles (200 km) southeast of Kiev in the Ukraine, and which at that time had a population of about 5,000. He was the only child of the former Sophia Voinarsky and Grigory Dobrzhansky, a teacher of mathematics. After his graduation from secondary school in Kiev, Dobzhansky majored in biology at the University of Kiev, from which he was graduated in 1921. He remained in Kiev at the Institute of Agriculture until 1924. He then moved to Leningrad, where he was a Lecturer in Genetics at the University from 1924 to 1927. While at this post, in 1926 he led an expedition to Central Asia to study domesticated animals.

In 1927 he received a Fellowship from the International Board of the Rockefeller Foundation to come to the United States, for two years, to work at Columbia University under the sponsorship of the future Nobel Laureate, Thomas Hunt Morgan. He accompanied Morgan when the latter moved to California Institute of Technology in 1928; he stayed on there becoming an Assistant Professor in 1929 and Professor of Genetics in 1936.

In 1940 he returned to Columbia University as Professor of Zoology, where he remained until 1962 when he moved to the Rockefeller Institute (now Rockefeller University).

On August 8, 1924, Dobzhansky wed Natalia Petrovna Sivertzev, a geneticist in her own right. Her "Uber den letalen Effekt einiger Gene bei Drosophila melanogaster," Arch. Entw. Mech. Org., 109:535-548, 1927, presented the initial demonstration that in crosses involving lethal genes the percentage of eggs deposited that hatch actually equals the percentage expected to be viable. Mrs. Dobzhansky's last publication (with Chana Malogolowkin-Cohen, Howard Levene, and Angela Solima Simmons) was "Inbreeding and the mutational and balanced loads in natural populations of *Drosophila willistoni*," Genetics, 50:1299-1311, 1964. It is our sad duty to add that Mrs. Dobzhansky died on February 22, 1969. The Dobzhanskys' only child, Sophia, is married to Professor Michael D. Coe of Yale University. Both Drs. Coe are anthropologists, and they have five children.

The scientific career of Theodosius Dobzhansky has extended over more than half a century from paper No. 1 in 1918; over 400 papers are listed in the accompanying bibliography. (The authors hope to continue this bibliography in later volumes of *Evolutionary Biology*.) Most readers will have seen only a small part of this output, and virtually all will recall mainly those facets of the work which are closest to their own special interests. Accordingly, it seems useful to begin this volume with a short overview of Dobzhansky's work to place it in better perspective. Rather than a strictly chronological review the authors have chosen for the most part to divide the work into broad categories, which are discussed separately. References are made mainly to papers that started a new line of work, and to recent ones that serve as points of entry to the whole.

Professor Dobzhansky indulging in one of his favorite activities (Raleigh, North Carolina, 1967). His lifelong passion for riding crystallized during an expedition he led to central Asia in 1926 to study domesticated animals. (Photograph by Dr. Lawrence E. Mettler.)

Scientific Work

Dobzhansky's early work in Russia from 1918 to 1927 was mainly on the *Coccinellidae* or ladybird beetles. The work started with an emphasis on systematics, but under the influence of his great teacher, Chetverikov, Dobzhansky's interest turned more to genetics and evolution, and his research soon branched into studies of geographic variation and of polymorphism, as well as such biological problems as migration and diapause. This work ended in a report on the origin of geographical races in this group (1928a). Thus this early work set the pattern for many of his later main interests. The interest in systematics has continued (see, e.g., 1960h, 1964g), and has included description of many new species (e.g., 1918a, 1944d, 1967y).

Also during this period he started his work on *Drosophila* genetics with an important paper on the manifold effects of genes (1927g); this study is still quoted as the first major investigation of this problem.

From 1927 to 1933, in Morgan's laboratory, Dobzhansky's main interest was in classical genetics. There was a series of papers on sex determination in *Drosophila,* inspired by C. B. Bridges, and a study of chromosomal aberrations, particularly a proof of the occurrence of translocations, which resulted in one of the first cytological proofs of the linear arrangement of genes on chromosomes, and the first cytological map (1929b). This work led to the hypothesis that the spindle-fiber attachment (now called the centrosome) was a permanent feature of the chromosome, that only chromosomes with exactly one of these attachments were viable, and to an explanation of the role of translocations in preventing crossing over (1930c, 1931c, 1933a). The work on position effects (1936b) came a little later.

In 1933(c) Dobzhansky's first paper on *Drosophila pseudoobscura* was published, one of a series on the reproductive isolation between "races" A and B. (These were later raised into the first pair of "sibling species" [morphologically indistinguishable but reproductively isolated species] in an important monograph written with Epling [19441].) In 1936 came the first paper on inversions in these species—the brilliant work with Sturtevant (1936h, 1938f), which showed how a phylogeny of inversions could be built up.

Several additional general papers on evolution were capped by the Jesup Lectures of 1936, delivered at Columbia University and resulting in the first edition of his most famous technical book, *Genetics and the Origin of Species* (1937a). This was a masterly synthesis of a century's work on field data, of the results of 35 years' laboratory study of genetics, of the theoretical work of Fisher, Haldane, Wright, and his old teacher Chetverikov, and of the experimental studies of Dobzhansky and a few other workers. With this book a new branch of scientific investigation came of age, and many workers were induced to enter it. The book is one of the few seminal publications of its generation, and looking back it seems amazing that it could have been written so soon after the field was opened up. A second edition became necessary in 1941; the third (1951), issued only 14 years after the first, showed tremendous progress, due in good part to the studies of Dobzhansky and his former colleagues and students. He is now (1969) at work on a completely new version with the tentative title *The Genetics of the Evolutionary Process*.

As late as the second edition of the work, the observation that differing frequencies of third chromosome inversions in *D. pseudoobscura* occur at different times during the season at a collecting station was ascribed to genetic drift. Continuing observation, however, showed a seasonal pattern, consistent from year to year, reported in 1943a, and shown in laboratory populations to be due to natural selection (1946b). Studies on seasonal and secular changes in *D. pseudoobscura* have continued ever since (e.g., 1952a, 19661), giving a picture of evolution in action. It should be noted that up to 1943 it was generally believed that the action of natural selection was so slow that no visible change could be detected in a lifetime, except perhaps when man had radically changed the environment. The discovery of strong selection actually in process, later confirmed in many other cases, has radically changed our views on the nature of evolution.

Studies of the incidence of lethal genes in natural populations have been continuous since 1941f; of heterosis, since 1942g. Recent papers are 1960m, 1963r, s, t. These studies have led to an elaborate theory of genetic population structure, developed partly in collaboration with Dobzhansky's former student, Bruce Wallace of Cornell University, with whom he coauthored *Radiation, Genes and Man* (1959p). While the classical theory considers a sexually reproducing species to be mainly homozygous for "normal," "wild type" genes, (except for deleterious recessives maintained by mutation), the Dobzhansky-Wallace theory holds that a sexually reproducing species contains a large store of genetic diversity, largely maintained by heterozygous advantage. Alleles are selected to be "good mixers,"—good in heterozygous condition with the other alleles found at other loci. The population as a whole is integrated, with a coadapted genotype. Hybrids with other populations having different sets of genes may show hybrid vigor in the F_1 but a breakdown of the coadapted system by recombination in the F_2, with a decline in fitness. If the hybrid population is maintained for a longer time, a new, and possibly even superior, coadapted system may be built up by natural selection. Experimental papers dealing with the problem include 1953g, 1954e, 1958j, 1960u, 1962j, 1963t, 1965b, h, 1966f, 1968t. A summary paper of earlier work is 1957a. An important aspect of this "balance" theory is that individuals that are very homozygous show inferior homeostasis, so that even if selected to have high viability in one environment, they tend to be poor in others—they are "narrow specialists" compared to those of more hetorozygous individuals (see, e.g., 1955j).

The presence of genes deleterious when homozygous is a "genetic load." A question has arisen between Dobzhansky and his followers and Muller and Crow and their followers as to whether this load is mainly mutational, or is balanced and due to heterosis. For a summary of Dobzhansky's position, and the evidence on which it is based, see 1963s, t, 1964f. An essential question at issue is the prevalence and importance of epistatic interactions between genes on the same or different chromosomes. Important evidence is provided in a series of papers on increase of variability by recombination (e.g., 1960u), and in other papers including 1965b, h.

The extremely important implications of sexual isolation for the process of speciation were presented in a series beginning with 1939g. In one direction this led to the remarkable picture of incipient speciation presented by the superspecies *D. paulistorum* in Latin America (summarized in 1963c, 1964g, 1967i, 1969h). "The spontaneous origin of an incipient species in the *Drosophila paulistorum* species complex" (1966d) led to reports of the "laboratory creation of a new species" in *The New York Times*, March 13, 1967, and *Time*, April 7, 1967, and new fame for its "creator."

In another direction such papers as 1945c, d, concerned more with the behavioral aspects of sexual isolation, led to a more general interest in behavior and to work still in progress on selection for geotaxis and phototaxis. This work included experiments that exemplify, admittedly in much oversimplified form, a model of a human population with upward and downward mobility from a small elite whose talents are determined in part genetically. The results are interesting and to some extent unexpected, but space precludes their discussion. (See 1962i, 1965i, 1967j, and especially 1969f.)

Other lines of work to be mentioned briefly are the study of the socalled "founder effect" or genetic drift in the broad sense, connected with the unpredictability of certain experiments (1953g, 1962g) and the unpredictability of evolution in general; differences in population structure in marginal and central populations of a species and between rare and common species (1957e, 1963r); and the problem of adaptedness and how to measure it aside from the usual measures of intrapopulational Darwinian fitness (numerous studies of monomorphic and polymorphic populations of *D. pseudoobscura* [e.g., 1961l], and the use of innate capacity to reproduce [1963v, 1964i].) It should also be mentioned that Dobzhansky is a good cytologist, and has done substantial cytogenetic work on ordinary and on giant salivary gland chromosomes (e.g. 1929a, 1930f, 1936a, 1944d, 1962f), aside from his constant use of salivary chromosomes as a tool.

In 1943, a desire to compare population structure and evolutionary processes in the relatively constant environment of a tropical rain forest, as contrasted with the changing season in all places previously studied, led to the first of Dobzhansky's many trips to Brazil, which inspired an active group of young Brazilian geneticists. Aside from numerous papers on *Drosophila*, his great interest in the tropics resulted in papers on evolution in the tropics (1950d) and on species diversity in tropical forest trees (1950q, 1953h).

Throughout his career, Dobzhansky has shown an interest in the broader philosophical implications of his work. With the rise of Lysenko in Russia, and of Nazi ideas of race, he wrote frequent reviews of works in these fields in an attempt to arouse his fellow scientists to these dangers. He was led gradually into a strong interest in the genetics and evolution of man and wrote frequently on these subjects, although except for a brief period he did no experimental research in human genetics. This interest led ultimately to the giving of the Silliman Lectures at Yale University, resulting in *Mankind Evolving* (1962a)—a masterful book on the modern genetical view of man and his evolution, and a synthesis of genetics, physical, and cultural anthropology, as remarkable in its way as was *Genetics and the Origin of Species* in its. In matters of general ethical concern *Mankind*

Evolving was foreshadowed by *The Biological Basis of Human Freedom* (1956a), and followed by *The Biology of Ultimate Concern* (1967f).

Other papers of broader scope are exemplified by comments on the Pope's discussion of genetics (1953d), chance and antichance in evolution (1963d), the philosophical basis of Darwinism (1968e), "Genetics and equality" (1962b), and defenses of organismic versus molecular biology in his Presidential Address to the American Society of Zoologists (1964a).

A Collecting Expedition

Like most prominent scientists today, Professor Dobzhansky is constantly on the move attending congresses, giving lectures, visiting colleagues, and very occasionally attending a committee meeting. Unlike many who travel a lot, Dobzhansky derives pleasure from air travel per se and indeed looks forward to such flights. Since he has such wide interests ranging from art and architecture, to people and the way they live, to all varieties of natural settings, almost any destination becomes interesting in its own right. But he delights most in his collecting trips.

One of us (R. R.) accompanied him on a recent trip to Colombia. A description of this trip may give some interesting sidelights on Dobzhansky's more relaxed moments.

You do not have to know Professor Dobzhansky very long before he insists that you call him Doby. He is often referred to as "The Professor" around the lab and by his former students. Far from being formal this is used as a familiar yet respectful form of address.

Our trip to South America took us to the southeastern provinces of Columbia, where we collected in the forests around Leticia, Mitu, and Valparaiso. The purpose of this trip was to collect *Drosophila* species of the *willistoni* group, since a large portion of the Professor's scientific contributions of the last decade or so have dealt with the evolution of these species.

Doby is by nature a very economical person, and this characteristic extends to his scientific work. He believes firmly in doing experiments in the simplest and most expeditious way, and is loath to abandon any effective procedure in favor of a supposedly "better" and more up-to-date one, and he has often expressed his disdain of "gadgeteers." It is thus not surprising that equipment taken on field trips is kept to the irreducible minimum: a dissecting microscope, vials to hold flies, needles, and collecting nets. Empty vials are carried in cardboard boxes, tied with string, which in turn serve as shipping containers when the flies are entrusted to an airline pilot for the trip back to the laboratory. Naturally the simplicity of equipment reduces baggage charges on long flights and is a blessing for movement in the field. Over the years, the simple preparations have been reduced to a routine, and we were amazed at the lack of fuss involved for a month-long expedition.

Doby's arrival in a South American country always creates excitement in its scientific community and he apparently enjoys the attention. Usually he is requested to give at least one lecture, which is always read in Spanish, having been previously translated by a Latin colleague. The translation is carefully checked and the difficult passages are mused over and frequently altered, for Doby speaks, reads, and understands Spanish well (in addition to at least six other languages), although he claims to lack the ability to write it acceptably.

His presence at the podium sets an unforgettable picture. He is white-haired, balding, and somewhat heavy, but is lively and full of energy. The next sensation can only be described as one of surprise, for his voice is high-pitched and has a raucous tonal quality which seems to be entirely foreign to the body from which it issues. Doby, however, uses the characteristics of his voice to underline the important points of his lectures. When lecturing formally, he rarely seems completely at ease. He usually reads

Passengers and baggage ride in the same compartment on a DC-3 headed for Mitu, Colombia; February, 1968.

from a text and is somewhat subdued, although still managing to inject the necessary emphases. During the discussion, however, or on less formal occasions, his animation increases, and he seems almost to bounce about the platform. At such a time, the most important points seem to come like explosions, and no listener is permitted to overlook the essential points. Even on formal occasions he often uses a query to inject humor into the proceedings. Since for so many years he has spoken a language that is not his mother tongue, whenever he speaks he articulates each word carefully and as a result is one of the easiest speakers to understand.

Long-distance travel in Latin America is primarily by air, which enables one to see the topography of large areas. Propeller-driven aircraft are widely used for intracontinental flights, and their slow speeds and relatively low altitudes make observation of the tropical flora and the beautiful peaks and crags of the three cordilleras of the Andes an exciting show. Even nonbiologists must be struck by the vast extent of the Amazonian tropical forest. While observing it, Doby seems to feel a sense of wonder and curiosity, which he exhibits with more than usual fervor. Map in hand, he watches the panorama go by not only for its natural beauty but with a professional eye as well, commenting on the suitability of the terrain as a *Drosophila* habitat. He inevitably files away not only possible future collecting sites but material that will aid in constructing a general picture of the geographical distribution of the flies.

The forest, which is so easily visible from an airplane, seems almost to have vanished when we emerge on the ground. The *Drosophila* we seek cannot compete with what The Professor characteristically calls "garbage species," which are, of course, commensal with the source of garbage. The more "noble" species, *Drosophila paulistorum* and its relatives, are restricted to good secondary or primary forest, which we must find if we are to be successful.

Local travel in the Amazon basin is almost exclusively by water. A practical means of reaching the virgin forest surrounding an outpost is to make a suitable arrangement with one of the local inhabitants who knows the labyrinth of waterways feeding into the Amazon, and who is wealthy enough to own a motorboat. Such a man is found and he promises to take us the next morning at eight o'clock. With his usual pessimism, Doby is sceptical of the early hour and predicts that we shall not leave on time. The result is as predicted, but while the rest of us feel frustrated, he spends the morning hours writing (in Russian) in his ever-present journal and composing travel letters. (These could by now be combined into a short book relating his thoughts on many aspects of the people, the flora, and the general nature of the tropics.) While many of us seem content to spend a five-minute wait doing essentially nothing, he refuses to do so.

He cannot abide wasting time, which he considers the most precious commodity in the world. Time must always be employed pleasantly or usefully. Life is short, there is much to be done, let us get on with it. "Waste of time" is his most devastating criticism of anything. His ability to utilize snatches of time is almost as important for his unusual productivity as his boundless energy and ability to work long hours, seven days a week, for long stretches of time. On the other hand, his enjoyment of music, art, architecture, scenery, good food, and good friends is just as intense.

Our boatman finally appears, without an explanation, and even though Doby is as frustrated as the rest of us by now, he points out that a leisurely, unhurried approach to life is highly adaptive in this climate. Our destination has been well worth the wait, for a view (this time looking up) of a virgin rain forest awaits us.

Outside the forest the sun is intense, but once one enters, the intensity of the light falls dramatically. To describe a tropical forest is nearly an impossibility, but in one of the letters that he writes home to be circulated among his friends, Doby wrote a description of a tropical jungle on the opposite side of the Andes. We can do no better than to quote from it here to show his feeling for tropical nature.

"We descended to a narrow plateau with numerous small coffee plantations and cattle ranches, and finally plunged into the canyon of the river Anchicaya. This proved to be an unforgettable place. A deep canyon with a cascading river on the bottom; the road hacked out of the steep slope, mostly high above the river; half a hundred waterfalls, some of them so close to the road that the car is showered in passing; and before all else—a riot of vegetation from the tops of the mountains down to the water's edge."

"I do not recall having ever seen high mountains more thoroughly clothed with vegetation. The gentler parts of the slopes support a rain forest, with masses of epiphytes and lianas on every branch and forming suspension bridges between the crowns of neighboring trees; where the slope is too steep or the rock too bare of soil there is a tangled mass of smaller trees, bushes, and vines, with many species of palms spreading their feathery crowns over the rest; even the apparently vertical cliffs are all green, with some special vegetation peculiar to them. Only the streams and the places where the waterfalls are beating against the rock interrupt this continuous cover of vegetation, but a few feet or even inches away from the cascading water there is a rich green mat of plant life. Flowers are rather few—one sees an occasional flower-covered tree somewhere on the slope, and not numerous but quite diversified flowers hanging over the cliffs along the road, among them a beautiful and large white orchid with a lavender-colored throat, a species which I do not remember having seen either in nature or in orchidaria."

The large trees and graceful lianas of a forest near Valparaiso, Colombia catch Doby's eye. February, 1968.

"The superabundance of water which renders all this riot of vegetation possible makes walking or climbing exceedingly difficult, as I found out by experience while attempting to place the baits for *Drosophila* collecting. The surface of the rock is slippery like wet soap, the soil a sticky mud, the fallen leaves a spongy mass filled with water. Moreover, especially as one descends towards the sea level, it gets hot. Not so hot while you are sitting quietly with the wind produced by the motion of the car blowing in your face; but any kind of physical exertion, even such a modest one as swinging a net for Drosophila capture, makes one covered with sweat. One gets to understand just why it is that so many inhabitants of wet tropics are 'lazy,' and why slavery was almost widespread and lasted longest in hot lands. Being 'lazy' is here just plain common sense."

"Now, to many people places like the canyon of the Anchicaya river would be just 'green Hell' and places to be avoided. I do not agree with this; my feeling is exactly the opposite—such places are fascinating and exhilarating. The heat and the sweating are surely uncomfortable; but to watch the superabundance of life pouring out of every square foot of soil, and even out of an apparently naked rock, is a sight which I never

Even distinguished professors get hot while collecting, and dress accordingly. Doby examines the catch in the forest shade near Mitu, Colombia; February, 1968.

get tired watching. It is, I feel, symbolic of life's mastery over death. It is an apotheosis of life. It is not something for a biologist only; it touches some chord very deep inside the human nature."

These words reveal part of the fascination which the rain forest holds for Doby. The chance to collect *Drosophila* seems almost to become of secondary importance when he finds himself in the forest. He points again and again to the huge size of the trees, the graceful spiraling lianas climbing high into the canopy, and the fluttering of the large irridescent blue Morpho butterflies which dart in and out of the patches of sunlight. The forest, the beauty of the life forms created by natural selection, is a true esthetic adventure for this man. He welcomes the presence of Morphos even though they may become nuisances by congregating about *Drosophila* traps in large numbers. After reaching the last of our *Drosophila* traps, he would often stand for 15 minutes or so looking up into the canopy or pointing out especially graceful lianas to his companions and deriving obvious pleasure from their sense of wonder of the place and of having introduced them to it.

Doby never loses the opportunity to teach, and a field excursion provides many chances. The scientific names of trees and insects and their principal characteristics are often pointed out, and The Professor does not forget

Collecting Drosophila in the tropical rain forests. Leticia, Colombia; February, 1968.

which ones he showed you. Tomorrow, you will be asked to supply the name and the little bit of natural history which went with it. If you do not remember, Doby's remark is usually a taunting, "Shame on you," and a curt repetition of the information follows.

The presence of "Norteamericanos" with nets and strange-looking instruments like binocular microscopes aroused the curiosity of a few of the local natives. Doby took in hand the education of our audience. One by one each onlooker was instructed to look through the microscope and examine the fly. The first response was a startled look followed by a series of chuckles. The Professor has described another occasion when the audience was much larger and individual instruction was impossible, but the reactions of the people were the same.

"My activities naturally excited much interest, and the table on which I set my very simple laboratory became surrounded by a circle of onlookers. As a matter of self-protection, I pretended to be a normal Gringo, who neither speaks nor understands a word of Spanish. My companion rose however to the occasion, and delivered regular lectures, describing the strange business of this curious Gringo who has traveled all the way from Nueva York to Buenaventura for the sole purpose of collecting the 'moscas de fruta.' The attitude of the audience was interesting to watch. Colombian truckdrivers are at least as roughneck an outfit as their opposite numbers elsewhere, and their educational level is probably lower. Nevertheless, the word 'science' has magic in it. The Gringo may be a bit crazy, but he evidently possesses a power which ordinary people do not possess. He can put a *Drosophila* to sleep, make it as big as a cow by his microscope, and then wake it up in a little glass tube with a clean white cotton plug and some grey jelly on the bottom. These tubes will then be sent as far as Nueva York, for some esoteric purpose known as 'estudio cientifico' (scientific research). The Gringo has a lot of knowledge, and his knowledge makes him a superior person. The prestige and authority of science have reached even the Buenaventura road."

Our attempts to study the distribution of *D. paulistorum* took us to the city of Florencia in southcentral Colombia. This small city would be undistinguished were it not for the fact that it lies in the shadow of one of the greatest mountain chains in the world. The city is constructed around a Catholic monastery which provides religious and educational opportunities for the surrounding region. Doby's exploration of the city naturally led him to the center, where a large and well-worn church is the dominant structure. Here he began a casual conversation with a priest who stood nearby, and returned sometime later to our hotel obviously very deeply impressed by the commitment of this priest to a life far from his home and away from the luxuries of European civilization. Doby does not question or judge the man's motives for his choice; he only appreciates that the choice has been made.

The Professor very carefully regulates the baggage allowance of his companions during collecting trips and extends his prohibitions to permitting only two books per person. As a result, two long and at the same time interesting works must be found which will fill the stretches between collecting forays. Doby usually brings only one treatise, which requires slow and careful reading. Having spent several days laboring over his book, he inquired of the rest of us what we were reading. One of us had a copy of Dostoevski's *The Brothers Karamazov*. This finding pleased Doby, and after dinner that evening he started a discussion of the book and its author and gave his interpretations of the great passages it contains. He had, of course, read it in Russian many years ago, but had never read an English translation. His demanding work was quickly forgotten, and he spent several days engrossed in Dostoevski. Some time later, we were discussing Dostoevski again, and with a hint of a proud smile he asked if we knew that Professor Dobzhansky was remotely related to this great novelist? With modesty, but with recognition of his obvious talent, he told us that his present literary abilities might be traced to the presence of a few Dostoevskian genes in his genotype.

Doby's collecting trips last, for the most part, about a month or so. As is true of the rest of us, he ends them with mixed feelings. After all, the trip was undertaken to collect flies and once the material has been obtained and sent North, his curiosity as to what has been found becomes almost unendurable. On returning to New York, he will burst into the laboratory, after a brief rest or none at all, demanding details on the classification of the flies and the results of the first experiments. On the other hand, he leaves the beauty of tropical nature with regret, for it not only rewards an esthetic desire, but seems to afford a feeling of peace not often met with in today's world. His admiration of the tropical forest and its fauna is almost childlike in nature, as if his experiences there recalled a boyhood dream which he could periodically renew by seeing these wonderful phenomena.

During his stay, Doby frequently acquires some typical products of the region and one or more bottles of rum or some other local specialty. These are proudly shown and served in the following months, and he relives the trip as he tells of their origin, the delights of the excursion, and the opportunity once again to visit and enjoy tropical nature.

Acknowledgment

In acknowledgment, the authors would especially like to thank Mr. Andrea Palestrina for his great assistance, particularly in preparing the bibliography.

Career Summary

Curriculum Vitae

January 25, 1900	Born, Nemirov, Russia
1921	Graduate
	University of Kiev
1921–1924	Assistant Professor of Zoology
	Polytechnic Institute, Kiev
1924–1927	Lecturer in Genetics
	(colleague of Prof. J. Philipchenko)
	University of Leningrad
1927–1929	Fellow
	International Education Board
	The Rockefeller Foundation
	with Professor T. H. Morgan, at Columbia University, 1927–1928; at California Institute of Technology, 1928–1929
1929–1936	Assistant Professor of Genetics
	California Institute of Technology
1936–1940	Professor of Genetics
	California Institute of Technology
1940–1958	Professor of Zoology
	Columbia University
1958–1962	daCosta Professor of Zoology
	Columbia University
1962–	Professor
	The Rockefeller University
	[The Rockefeller Institute until 1965]

Societies

President	Genetics Society of America, 1941
”	American Society of Naturalists, 1950
”	Society for the Study of Evolution, 1951
”	American Society of Zoologists, 1963
”	American Teilhard de Chardin Association, 1969
Member	National Academy of Sciences (U.S.A.)
”	American Philosophical Society, Councilor 1969–
Foreign Member	Royal Swedish Academy of Sciences
”	Royal Danish Academy of Sciences
”	Brazilian Academy of Sciences
”	Academia Leopoldina
”	Accademia Nazionale dei Lincei
”	Royal Society of London
Honorary Member	Genetics Society of Japan

Doctor of Science (Honorary)

Universidad de São Paulo, Brazil (1943)
College of Wooster, Ohio, U. S. A. (1945)
University of Munster, Germany (1958)
University of Montreal, Canada (1958)
University of Chicago, Illinois, U. S. A. (1959)
University of Sydney, Australia (1960)
Columbia University, New York, U. S. A. (1964)
Oxford University, England (1964)
University of Louvain, Belgium (1965)
Clarkson College of Technology, New York, U. S. A. (1965)
Kalamazoo College, Michigan, U. S. A. (1966)
University of Michigan, Ann Arbor, U. S. A. (1966)
Syracuse University, New York, U. S. A. (1967)
University of Padua, Italy (1968)
University of California at Berkeley, U S. A. (1968)
Northwestern University, Illinois, U. S. A. (1968)

Awards

Daniel Giraud Elliot Medal
National Academy of Sciences, 1946
(Second edition of *Genetics and The Origin of Species*)

Kimber Genetics Award
National Academy of Sciences, 1958

Darwin Medal
Academia Leopoldina,
1959

Anisfield-Wolf Award
(*Mankind Evolving*)
1963

Pierre Lecomte du Nouy Award
(*Mankind Evolving*)
1963

The National Medal of Science
The President of The United States,
1964

Addison Emery Verrill Medal
Peabody Museum of Natural History
Yale University, 1966

Gold Medal Award for Distinguished Achievement in Science
The American Museum of Natural History, 1969

Theodosius Dobzhansky: Bibliography, 1918-1969

1918. Description of a new species of the genus *Coccinella* from the neighbourhood of Kiev. (Russian.) *In* Materials for the Fauna of Southwestern Russia, 2:46-47.

1922a. Über Massenauftreten und Wanderungen der *Coccinellidae*. (Russian.) Izvestiya Otdela prikladnoĭ entomologii, Sel'sko-khozyaĭstvennyĭ uchenyĭ komitet [Report of the Bureau of Applied Entomology, Agriculture Science Committee], 2:103-124.

1922b. Über die imaginale Diapause bei den *Coccinellidae*. (Russian.) Izvestiya Otdela prikladnoĭ entomologii, Sel'sko-khozyaĭstvennyĭ uchenyĭ komitet [Report of the Bureau of Applied Entomology, Agriculture Science Committee], 2:229-234.

1923. Tres novae *Coccinellidarum* species e fauna Rossiae Asiaticae (*Coleoptera*). (Latin and Russian.) Russkoe Entomologicheskoe Obozrenie [Revue Russe d'Entomologie], 18:99-102. (With A. Semenov-Tian-Shanskij.)

1924a. Die geographische und individuelle Variabilität von *Harmonia axyridis* Pallas in ihren Wechselbeziehungen. Biol. Zentralbl. 44:401-421.

1924b. Die weiblichen Generationsorgane der Coccinelliden als Artmerkmal betrachtet (*Coleoptera*). Entomol. Mitteil., 13:18-27.

1924c. Beitrag zur Kenntnis des weiblichen Geschlechtsapparates der Coccinelliden. Z. wiss. Insektenbiol. 19:98-100.

1924d. Die geographische und individuelle Variabilität von *Adalia bipunctata* L. und *Adalia decempunctata* L. (Coleoptera, Coccinellidae). (Russian with German synopsis.) Russkoe Entomologicheskoe Obozrenie [Revue Russe d'Entomologie], 18:201-212.

1924e. Zur Erforschung der Anschwemmungsfauna des Dnjeprs bei Kiev. (Russian with German synopsis.) Russkiĭ Gidrobiologicheskiĭ Zhurnal [Russian Hydrobiological Journal], 3:221-223.

1924f. Über den Bau des Geschlechtsapparats einiger Mutanten von *Drosophila melanogaster* Meig. Z. Indukt. Abstamm. Vererbungsl., 34:245-248.

1925a Zur Kenntnis der Gattung *Coccinella* auct. Zool. Anz., 62:241-249.

1925b. Die paläarktischen Arten der Gattung *Coccinula Dobzhansky*. Zool. Anz., 64:277-284.

1925c. Über das Massenauftreten einiger Coccinelliden im Gebirge Turkestans. Z. wiss. Insektenbiol., 20:249-256.

1926a. Die paläarktischen Arten der Gattung *Coccinella* L. (German with Russian synopsis.) Russkoe Entomologicheskoe Obozrenie [Revue Russe d'Entomologie], 20:16-32.

1926b. Über die Morphologie und systematische Stellung einiger Gattungen der Coccinellidae (tribus Hippodamiina). Zool. Anz., 69:200-208.

1926c. Reproductive organs of lady-bird beetles (*Coccinellidae*) as a species and a group character. (Russian.) Izv. Akad. Nauk SSSR 1926, 1385-1394.

1926d. Reproductive organs of lady-bird beetles (*Coccinellidae*) as a species and a group character. (Russian.) Izv. Akad. Nauk SSSR 1926, 1556-1586.

1927a. Die Coccinelliden Heptapotamiens (Semiretshje). (Russian.) Russkoe Entomologicheskoe Obozrenie [Revue Russe d'Entomologie], 21:43-52.

1927b. Neue und wenig bekannte Coccinelliden. (German.) Russkoe Entomologicheskoe Obozrenie [Revue Russe d'Entomologie], 21:212-217.
1927c. Zwei neue *Pharoscymnus*-Arten nebst einem Beitrag zur Kenntnis der Morphologie der *Coelopterina* (Coleoptera, Coccinellidae). (German.) Russkoe Entomologicheskoe Obozrenie [Revue Russe d'Entomologie], 21:240-244.
1927d. Materials for a fauna of Coccinellidae (Coleoptera) of the Jakutsk Province. (Russian.)
1927e. Horses of the nomadic population of Semiretshje. (Russian.) Materialy Osobogo Komiteta po Issledovanii v Soivznykh i Avtonomnykh Respublik, Akad. Nauk SSSR [Materials of the Special Committeee for the Study of Allied and Autonomous Republics], 8:16-131. Leningrad.
1927f. Zur Kenntnis der Vererbung der Farbe und Zeichnung beim kirghisischen Pferd. (Russian with German synopsis.) Izvestiya Byuro po Genetike i Evgenike [Bulletin of the Bureau of Genetics and Eugenics], 5:79-108.
1927g. Studies on the manifold effect of certain genes in *Drosophila melanogaster*. Z. Indukt. Abstamm. Vererbungsl., 43:330-388.
1927h. Die geographische Variabilität von *Coccinella septempunctata* L. Biol. Zentralbl., 47:556-569. (With N. P. Sivertzev-Dobzhansky.)
1927i. Die Larve von *Sylphopsyllus desmanae* Ols., Parasit der Moschusratte, als Kriterium seiner genetischen Beziehungen und seiner systematischen Stellung. (Russian with German synopsis.) Russkoe Entomologicheskoe Obozrenie [Revue Russe d'Entomologie], 21:8-16. (With A. Semenov-Tian-Shanskij.)
1928a. The origin of geographical varieties in Coccinellidae. Trans. Fourth Int. Cong. Entomology, Ithaca, 1928, 2:563.
1928b. Horses of the nomadic population of the Semipalatinsk Province. (Russian.) Materialy Osobogo Komiteta po Issledovanii v Soivznykh i Autonomnykh Respublik, Akad. Nauk SSSR [Materials of the Special Committeee for the Study of Allied and Autonomous Republics Academy of Sciences USSR] (Leningrad), 18:22-183. (With B. P. Vojtiazky.)
1928c. A review of maral breeding in southern Altai. (Russian.) Materialy Osobogo Komiteta po Issledovanii v Soivznykh i Avtonomnykh Respublik, Akad. Nauk SSSR [Materials of the Special Committee for the Study of Allied and Autonomous Republics, Academy of Science USSR] (Leningrad), 18:184-241.
1928d. The effect of temperature on the viability of superfemales in *Drosophila melanogaster*. Proc. Nat. Acad. Sci. (U.S.A.), 14:671-675.
1928e. Reproductive systems of triploid intersexes in *Drosophila melanogaster*. Amer. Natural., 62:425-434. (With C. B. Bridges.)
1929a. The influence of the quantity and quality of chromosomal material on the size of the cells in *Drosophila melanogaster*. Z. wiss. Biol. Abt. D. Wilhelm Roux' Arch. Entwicklungsmech. Organismen, 115:363-379.
1929b. Genetical and cytological proof of translocations involving the third and the fourth chromosomes of *Drosophila melanogaster*. Biol. Zentralbl., 49:408-419.
1929c. A homozygous translocation in *Drosophila melanogaster*. Proc. Nat. Acad. Sci. (U.S.A.), 15:633-638.

1930a. The manifold effects of the genes Stubble and Stubbloid in *Drosophila melanogaster*. Z. Indukt. Abstamm. Vererbungsl., 54:427-457.
1930b. Genetical and environmental factors influencing the type of intersexes in *Drosophila melanogaster*. Amer. Natural., 64:261-271.
1930c. Translocations involving the third and the fourth chromosomes of *Drosophila melanogaster*. Genetics, 15:347-399.
1930d. Time of development of the different sexual forms in *Drosophila melanogaster*. Biol. Bull., 59:128-133.
1930e. Studies on the intersexes and supersexes in *Drosophila melanogaster*. Bulletin of the Bureau of Genetics (Leningrad), 8:91-158.
1930f. Cytological map of the second chromosome of *Drosophila melanogaster*. Biol. Zentralbl., 50:671-685.
1930g. Reciprocal translocations in *Drosophila* and their bearing on *Oenothera* cytology and genetics. Proc. Nat. Acad. Sci. (U.S.A.), 16:533-536. (With A. H. Sturtevant.)
1931a. Interaction between female and male parts in gynandromorphs of *Drosophila simulans*. Z. wiss. Biol. Abt. D. Wilhelm Roux' Arch. Entwicklungsmech. Organismen, 123:719-746.
1931b. Translocations involving the second and the fourth chromosomes of *Drosophila melanogaster*. Genetics, 16:629-658.
1931c. The decrease of crossing-over observed in translocations and its probable explanation. Amer. Natural., 65:214-232.
1931d. The North American beetles of the genus *Coccinella*. Proc. U. S. Nat. Mus., 80:1-32.
1931e. Evidence for multiple sex factors in the X-chromosome of *Drosophila melanogaster*. Proc. Nat. Acad. Sci. (U.S.A.), 17:513-518. (With J. Schultz.)
1931f. Contributions to the genetics of certain chromosome anomalies in *Drosophila melanogaster*. II. Translocations between the second and third chromosomes of *Drosophila* and their bearing on *Oenothera* problems. Carnegie Institution of Washington Publ. No. 421, pp. 29-59. (With A. H. Sturtevant.)
1932a. Deletion of a section of the X-chromosome of *Drosophila melanogaster*. Bull. Lab. Genet. (Leningrad), 9:193-216.
1932b. Studies on chromosome conjugation. I. Translocations involving the second and the Y-chromosomes in *Drosophila melanogaster*. Z. Indukt. Abstamm. Vererbungsl., 60:235-286.
1932c. The Baroid mutation in *Drosophila melanogaster*. Genetics, 17:369-392.
1932d. Cytological map of the X-chromosome of *Drosophila melanogaster*. Biol. Zentralbl., 52:493-509.
1932e. Contribution à la connaissance des Coccinellides (Coleoptera) de la Yakoutie. (Russian.) Trudy Zoologicheskogo Instituta Akad. Nauk SSSR [Proceedings of the Zoological Institute of the Academy of Sciences USSR], 1:483-486.
1932f. Changes in dominance of genes lying in duplicating fragments of chromosomes. Proc. Sixth Int. Cong. Genetics, Ithaca, 2, 45-46. (With A. H. Sturtevant.)
1933a. Studies on chromosome conjugation. II. The relation between crossing-over and disjunction of chromosomes. Z. Indukt. Abstamm. Vererbungsl., 64:269-309.

1933b. Geographical variation in lady-beetles. Amer. Natural., 67:97-126.
1933c. On the sterility of the interracial hybrids in *Drosophila pseudoobscura*. Proc. Nat. Acad. Sci. (U.S.A.), 19:397-403.
1933d. Role of the autosomes in the *Drosophila pseudoobscura* hybrids. Proc. Nat. Acad. Sci. (U.S.A.), 19:950-953.
1933e. Intersterile races of *Drosophila pseudoobscura* Frol. Biol. Zentralbl., 53:314-330. (With R. D. Boche.)
1933g. Genes that effect early developmental stages of *Drosophila melanogaster*. Z. wiss. Biol. Abt. D. Wilhelm Roux' Arch. Entwicklungsmech. Organismen, 130:109-130. (With F. N. Duncan.)
1933f. The mutant "Proboscipedia" in *Drosophila melanogaster*—a case of hereditary homoosis. Z. wiss. Biol. Abt. D. Wilhelm Roux' Arch. Entwicklungsmech. Organismen, 127:575-590. (With C. B. Bridges.)
1933h. Deficiency and duplication for the gene "bobbed" in *Drosophila melanogaster*. Genetics, 18:173-192. (With N. P. Sivertzev-Dobzhansky.)
1933i. Triploid hybrids between *Drosophila melanogaster* and *Drosophila simulans*. J. Exp. Zool., 65:73-82. (With J. Schultz.)
1934a. Are racial and specific characters non-Mendelian? J. Mammal., 15:1-3.
1934b. Studies on hybrid sterility. I. Spermatogenesis in pure and hybrid *Drosophila pseudoobscura*. Z. Zellforsch., 21:169-223.
1934c. Studies on chromosome conjugation. III. Behavior of duplicating fragments. Z. Indukt. Abstamm. Vererbungsl., 68:134-162.
1934d. Obzor Iavleniĭ Perestroĭki Khromosomal'nogo Apparata. [Survey of phenomena of the reconstruction of the chromosomal apparatus]. Trudy po prikladnoĭ botanike genetike i selektsii, Seriia 2 Bull. Appl. Bot., Genet. Select., 6:147-171.
1934e. Sex in *Drosophila* and other organisms. Amer. Natural., 68:190-191. (With J. Schultz.)
1934f. The distribution of sex-factors in the X-chromosome of *Drosophila melanogaster*. J. Genet., 28:349-386. (With J. Schultz.)
1934g. The relation of a dominant eye color in *Drosophila melanogaster* to the associated chromosome rearrangement. Genetics, 19:344-364. (With J. Schultz.)
1935a. Maternal effect as a cause of the difference between the reciprocal crosses in *Drosophila pseudoobscura*. Proc. Nat. Acad. Sci. (U.S.A.), 21:443-446.
1935b. A critique of the species concept in biology. Philos. Sci., 2:344-355.
1935c. The Y-chromosomes of *Drosophila pseudoobscura*. Genetics, 20:366-376.
1935d. *Drosophila miranda*, a new species. Genetics, 20:377-391.
1935e. Some remarks on R. Goldschmidt's critique of the hypothesis of multiple sex-genes. J. Genet., 31:155-156.
1935f. A unique chromosome group in a species of *Drosophila*. Amer. Natural., 69:62.
1935g. A list of Coccinellidae of British Columbia. J. New York Entomol. Soc., 43:331-336.
1935h. Oxygen consumption of *Drosophila* pupae. II. *Drosophila pseudoobscura*. Z. Vergleich. Physiol., 22:473-478. (With D. F. Poulson.)
1935i. Fecundity in *Drosophila pseudoobscura* at different temperatures. J. Exp. Zool., 71:449-464. (With D. F. Poulson.)

1935j. Further data on maternal effects in *Drosophila pseudoobscura* hybrids. Proc. Nat. Acad. Sci. (U.S.A.), 21:566-570. (With A. H. Sturtevant.)
1936a. The persistence of the chromosome pattern in successive cell divisions in *Drosophila pseudoobscura*. J. Exp. Zool., 74:119-135.
1936b. Position effects of genes. Biol. Rev., 11:364-384.
1936c. Studies on hybrid sterility. II. Localization of sterility factors in *Drosophila pseudoobscura* hybrids. Genetics, 21:113-135.
1936d. L'effet de Position et la Théorie de l'Hérédité. 37 pp. (*Exposés des Génétique*, II, publié sous la direction de Boris Ephrussi.) Paris, Hermann et Cie.
1936e. Induced chromosomal aberrations in animals. *In* Duggar, B. M., ed. Biological Effects of Radiation: Mechanism and Measurement of Radiation, Applications in Biology, Photochemical Reactions, Effects of Radiant Energy on Organisms and Organic Products. New York, McGraw-Hill Book Company, Vol. 2, pp. 1167-1208.
1936f. Studies on hybrid sterility. IV. Transplanted testes in *Drosophila pseudoobscura*. Genetics, 21:832-840. (With G. W. Beadle.)
1936g. Studies on hybrid sterility. III. A comparison of the gene arrangement in two species, *Drosophila pseudoobscura* and *Drosophila miranda*. Z. Indukt. Abstamm. Vererbungsl., 72:88-114. (With C. C. Tan.)
1936h. Inversions in the third chromosome of wild races of *Drosophila pseudoobscura*, and their use in the study of the history of the species. Proc. Nat. Acad. Sci. (U.S.A.), 22:448-450. (With A. H. Sturtevant.)
1936i. Geographical distribution and cytology of "sex-ratio" in *Drosophila pseudoobscura* and related species. Genetics, 21:473-490. (With A. H. Sturtevant.)
1936j. Observations on the species related to *Drosophila affinis*, with descriptions of seven new forms. Amer. Natural., 70:174-184.
1937a. Genetics and the Origin of Species. (Columbia Biological Series, no. 11.) New York, Columbia University Press, xvi, 364 pp. (2nd revised edition, xvii, 446 pp., 1941i. 3rd ed., 1951i. Japanese transl., Tokyo, 1953. Spanish transl. F. Cordon, Genetica y el Origen de los Especes. 391 pp. Madrid, Revista de Occidente., 1955. Paperback ed., 1964q.)
1937b. Further data on *Drosophila miranda* and its hybrids with *Drosophila pseudoobscura*. J. Genet., 34:135-151.
1937c. Further data on the variation of the Y-chromosome in *Drosophila pseudoobscura*. Genetics, 22:340-346.
1937d. Genetic nature of species differences. Amer. Natural., 71:404-420.
1937e. What is a species? Scientia (Bologna), 61:280-286.
1938a. The raw materials of evolution. Carnegie Institution of Washington Suppl. Publ. No. 38. (Repr. from Scientific Monthly, 46:445-449.)
1938b. Genetic structure of natural populations. Carnegie Institution of Washington Year Book No. 37, pp. 323-325.
1938c. Genetics of natural populations. I. Chromosome variation in populations of *Drosophila pseudoobscura* inhabiting isolated mountain ranges. Genetics, 23:239-251. (With M. L. Queal.)
1938d. Genetics of natural populations. II. Genic variation in populations

of *Drosophila pseudoobscura* inhabiting isolated mountain ranges. Genetics, 23:463-484. (With M. L. Queal.)

1938e. A possible method for locating favorable genes in maize. J. Amer. Soc. Agron., 30:668-675. (With M. M. Rhoades.)

1938f. Inversions in the chromosomes of *Drosophila pseudoobscura*. Genetics, 23:28-64. (With A. H. Sturtevant.)

1939a. Studies on the genetic structure of natural populations. Carnegie Institution of Washington Year Book No. 38, pp. 287-289.

1939b. Genetics of natural populations. IV. Mexican and Guatemalan populations of *Drosophila pseudoobscura*. Genetics, 24:390-412.

1939c. Fatti e problemi della condizione "rapporto-sessi" [sex-ratio] in *Drosophila*. Scientia Genetica (Torino), 1:67-75.

1939d. Microgeographic variation in *Drosophila pseudoobscura*. Proc. Nat. Acad. Sci. (U.S.A.), 24:311-314.

1939e. Experimental studies on genetics of free-living populations of *Drosophila*. Biol. Rev. 14:339-386.

1939f. La composición genética de las poblaciones de *Drosophila pseudoobscura* que habitan México y Guatemala. Rev. Soc. Mexicana Hist. Nat., 1:15-17.

1939g. An experimental study of sexual isolation in *Drosophila*. Biol. Zentralbl., 58:589-607. (With P. C. Koller.)

1939h. Sexual isolation between two species of *Drosophila*—a study of the origin of an isolating mechanism. Genetics, 24:97-98. (With P. C. Koller.)

1939i. Structure and variation of the chromosomes in *Drosophila azteca*. J. Hered., 30:3-19. (With D. Sokolov.)

1939j. Morphological differences between the "races" of *Drosophila pseudoobscura*. Amer. Natural., 73:5-25. (With K. Mather.)

1940a. Studies on the genetic structure of natural populations. Carnegie Institution of Washington Year Book No. 39, pp. 244-247.

1940b. Speciation as a stage in evolutionary divergence. Amer. Natural., 74:312-321.

1941a. Studies on the genetic structure of natural populations. Carnegie Institution of Washington Year Book No. 40, pp. 271-276.

1941b. Discovery of a predicted gene arrangement in *Drosophila azteca*. Proc. Nat. Acad. Sci. (U.S.A.), 27:47-50.

1941c. The race concept in biology. Sci. Monthly, 52:161-165.

1941d. Chromosomal differences between races and species of *Drosophila*. University of Pennsylvania Bicentennial Conference in Cytology, Genetics and Evolution. Philadelphia, University of Pennsylvania Press, pp. 47-57.

1941e. On the genetic structure of natural populations of *Drosophila*. Proc. Seventh Int. Genetical Cong., Cambridge, Cambridge University Press, pp. 104-108.

1941f. Genetics of natural populations. V. Relations between mutation rate and accumulation of lethals in populations of *Drosophila pseudoobscura*. Genetics, 26:23-51. (With S. Wright.)

1941g. Intersexes in *Drosophila pseudoobscura*. Proc. Nat. Acad. Sci. (U.S.A.), 27:556-562. (With B. Spassky.)

1941h. Beetles of the genus *Hyperaspis* inhabiting the United States. Smithsonian Institution Publ. No. 3642. (Sm. Musc. Coll. 101:1-94.) 94 pp.

1941i. Genetics and the Origin of Species. 2nd revised ed. New York, Columbia University Press, xvii, 446 pp. (See 1937a.)
1942a. Studies on the genetic structure of natural populations. Carnegie Institution of Washington Year Book No. 41, 228-234.
1942b. Races and methods of their study. Trans. New York Acad. Sci., 2: 115-123.
1942c. Biological Symposia, Vol. 6. Th. Dobzhansky, ed. Pennsylvania, The Jaques Cattell Press, xii, 355 pp.
1942d. Biological adaptation. Sci. Monthly, 55:391-402.
1942e. Genetics of natural populations. VI. Microgeographic races in *Linanthus parryae*. Genetics, 27:317-332. (With C. Epling.)
1942f. Genetics of natural populations. VII. The allelism of lethals in the third chromosome of *Drosophila pseudoobscura*. Genetics, 27: 363-394. (With S. Wright and W. Hovanitz.)
1942g. Genetics of natural populations. VII. Concealed variability in the second and fourth chromosomes of *Drosophila pseudoobscura* and its bearing on the problem of heterosis. Genetics, 27:464-490. (With A. M. Holz and B. Spassky.)
1942h. Temperature and "sex-ratio" in *Drosophila pseudoobscura*. Proc. Nat. Acad. Sci. (U.S.A.), 28:45-48. (With C. D. Darlington.)
1942i. Darwin and our intellectual heritage. Science, 95:303-304.
1943a. Genetics of natural populations. IX. Temporal changes in the composition of populations of *Drosophila pseudoobscura*. Genetics, 28:162-186.
1943b. Genetics and human affairs. Teaching Biologist, 12:97-106.
1943c. O gen como unidade auto-reproductora da fisiologia celular. Revista de Agricultura [Piracicaba], 18:387-396. Cochabamba, Bolivia.
1943d. Heterosis. Revista de Agricultura [Piracicaba], 18:397-398. Cochabamba, Bolivia.
1943e. The species concept. Revista de Agricultura [Piracicaba], 18:441-442. Cochabamba, Bolivia.
1943f. Chromosomal aberrations in Brazilian *Drosophila ananassae*. Proc. Nat. Acad. Sci. (U.S.A.), 29:301-305. (With A. Dreyfus.)
1943g. A re-examination of the problem of manifold effects of genes in *Drosophila melanogaster*. Genetics, 28:295-303. (With A. M. Holz.)
1943h. Chromosome complements of some south-Brazilian species of *Drosophila*. Proc. Nat. Acad. Sci. (U.S.A.), 29:368-375. (With C. Pavan.)
1943i. Studies on Brazilian species of *Drosophila*. Bol. Fac. Filos. Ciencia Letras da Univ. São Paulo (Biologia Geral), 36:7-72. (With C. Pavan.)
1943j. Genetics of natural populations. X. Dispersion rates in *Drosophila pseudoobscura*. Genetics, 28:304-340. (With S. Wright.)
1944a. Genetic structure of natural populations. Carnegie Institution of Washington Year Book No. 43, pp. 120-127.
1944b. Distribution of Heterochromatin in the chromosomes of *Drosophila pallidipennis*. Amer. Natural., 78:193-213.
1944c. Rules of geographic variation. Science, 99:127-128.
1944d. Chromosomal races in *Drosophila pseudoobscura* and *Drosophila persimilis*. Carnegie Institution of Washington Publ. No. 554, pp. 47-144.

1944e. On species and races of living and fossil man. Amer. J. Phys. Anthrop., 2:251-265.
1944f. What is heredity? Science, 100-406. (Repr., *In* Amer. Biology Teacher, 7:127-128. 1945.)
1944g. Experiments on sexual isolation in *Drosophila*. I. Geographic strains of *Drosophila willistoni*. Proc. Nat. Acad. Sci. (U.S.A.), 30:238-244. (With E. Mayr.)
1944h. Mecanismo da Evolucão e Origem das Especies. Bol. No. 2 dos Cursos de Aperfeiçoamento e Especilisação do Ministério da Agricultura. 111 pp. (Trans. by dra. Rosina de Barros) Published by Centro Nacional de Pesquisas Agronomicas do Rio de Janeiro.
1944i. Experiments on sexual isolation in *Drosophila*. II. Geographic strains of *Drosophila prosaltans*. Proc. Nat. Acad. Sci. (U.S.A.), 30:340-345. (With G. Streisinger.)
1944j. Genetics of natural populations. XI. Manifestation of genetic variants in *Drosophila pseudoobscura* in different environments. Genetics, 29:270-290. (With B. Spassky.)
1944k. Experiments on sexual isolation in *Drosophila*. III. Geographic strains of *Drosophila sturtevanti*. Proc. Nat. Acad. Sci. (U.S.A.), 30:335-339.
1944l. Taxonomy, geographic distribution, and ecology of *Drosophila pseudoobscura* and its relatives. Carnegie Institution of Washington Publ. No. 554, pp. 1-46. (With C. Epling.)
1944m. *Drosophila* mutants. Science, 100:52. [Review]
1944n. Genes and the Man. Science, 100:103. [Review]
1944o. Evolution, creation, and science. Amer. Natural., 79:73-75. [Review]
1945a. Genetic structure of natural populations. Carnegie Institution of Washington Year Book No. 44, pp. 127-134.
1945b. Directly observable genetic changes in populations of *Drosophila pseudoobscura*. Biometrics Bull., 1:7-8.
1945c. Experiments on sexual isolation in *Drosophila*. IV. Modification of the degree of isolation between *Drosophila pseudoobscura* and *Drosophila persimilis* and of sexual preferences in *Drosophila prosaltans*. Proc. Nat. Acad. Sci. (U.S.A.), 31:75-82. (With E. Mayr.)
1945d. Experiments on sexual isolation in *Drosophila*. V. The effect of varying proportions of *Drosophila pseudoobscura* and *Drosophila persimilis* on the frequency of insemination in mixed populations. Proc. Nat. Acad. Sci. (U.S.A.), 31:274-281. (With H. Levene.)
1945e. Incipient reproductive isolation between two subspecies of *Drosophila pallidipennis*. Genetics, 30:429-438. (With J. T. Patterson.)
1945f. The science of man in the world crisis. Amer. J. Phys. Anthrop., 3:105-106. [Review]
1945g. An outline of politico-genetics. Science, 102:234-236. [Review] (Also, Review of General Semantics, 3:68-70.)
1946a. Genetic structure of natural populations. Carnegie Institution of Washington Year Book No. 45, pp. 162-171.
1946b. Genetics of natural populations. XII. Experimental reproduction of some of the changes caused by natural selection in certain populations of *Drosophila pseudoobscura*. Genetics, 31:125-156. (With S. Wright.)
1946c. Complete reproductive isolation between two morphologically similar species of *Drosophila*. Ecology, 27:205-211.

1946d. Heredity and Its Variability, by Trofim Denisovich Lysenko. Trans. by Th. Dobzhansky. New York, King's Crown Press, 65 pp.
1946e. Heredity, Race, and Society. New York, Penguin Books, Inc. 115 pp. (With L. C. Dunn.) (Revised and enlarged ed. 1952, New York, The New American Library, Inc. 144 pp. 3rd ed., 1956, New York, Mentor Books.) (Herança, raça e sociedade. 1952. Transl. by O. Frota-Pessoa. Rio de Janeiro, Livraria-editodia da Casa do Estudante do Brasil. 138 pp.) (Herança, raça e sociedade. 1962. Transl. by M. I. Rocha e Silva. São Paulo, Livraria Pioneira.) (El veratha was el sulala wa el mugtammaa. 1956. Arabic transl. by Al Din Farrag. 192 pp.))Hérédité, race, et société. 1964. Transl. by R. Graulich. Bruxelles, Charles Dessart.) (Arv, rase og samfunn. 1965. Transl. by M. Havtum. Oslo, Cappelens Forlag.) (Arv, ras och samhälle. 1956. Transl. by I. Kiellander. Ystad (Sweden), Kooperative Föbundets Bokförlag.)
1946f. Experiments on sexual isolation in *Drosophila*. VIII. Influence of light on the mating behavior of *Drosophila subobscura, Drosophila persimilis,* and *Drosophila pseudoobscura*. Proc. Nat. Acad. Sci. (U.S.A.), 32:226-234. (With B. Wallace.)
1946g. Genetics of natural populations. XIII. Recombination and variability in populations of *Drosophila pseudoobscura*. Genetics, 31:269-290.
1946h. Lysenko's "genetics". J. Hered., 37:5-9. [Review]
1946i. The new genetics in the Soviet Union. Amer. Natural., 80:649-651. [Review]
1947a. Genetic structure of natural populations. Carnegie Institution of Washington Year Book No. 46, pp. 155-165.
1947b. Adaptive changes induced by natural selection in wild populations of *Drosophila*. Evolution, 1:1-16.
1947c. Effectiveness of intraspecific and interspecific matings in *Drosophila pseudoobscura* and *Drosophila persimilis*. Amer. Natural., 81:66-72.
1947d. A directional change in the genetic constitution of a natural population of *Drosophila pseudoobscura*. Heredity (London), 1:53-64.
1947e. N. I. Vavilov, a martyr of genetics. J. Hered., 38:226-232.
1947f. Natural selection and the mental capacities of mankind. Science, 105: 587-590. (With M. F. Ashley Montagu.)
1947g. Evolutionary changes in laboratory cultures of *Drosophila pseudoobscura*. Evolution, 1:191-216. (With B. Spassky.)
1947h. Genetics of natural populations. XIV. A response of certain gene arrangements in the third chromosome of *Drosophila pseudoobscura* to natural selection. Genetics, 32:142-160.
1947i. Genetics of natural populations. XV. Rate of diffusion of a mutant gene through a population of *Drosophila pseudoobscura*. Genetics, 32:303-324. (With S. Wright.)
1947j. Cytology of evolution and evolution of cytology. J. Hered., 38:21-22. [Review]
1948a. Genetic structure of natural populations. Carnegie Institution of Washington Year Book No. 47, pp. 193-203.
1948b. Genetics of natural populations. XVI. Altitudinal and seasonal changes produced by natural selection in certain populations of

Drosophila pseudoobscura and Drosophila persimilis. Genetics, 33:158-176.
1948c. Chromosomal variation in populations of Drosophila pseudoobscura which inhabit northern Mexico. Amer. Natural., 82:97-106.
1948d. Genetics of natural populations. XVII. Proof of operation of natural selection in wild populations of Drosophila pseudoobscura. Genetics, 33:537-547. (With H. Levene.)
1948e. Genetics of natural populations. XVIII. Experiments on chromosomes of Drosophila pseudoobscura from different geographic regions. Genetics, 33:588-602.
1948f. The suppression of crossing over in inversion heterozygotes of Drosophila pseudoobscura. Proc. Nat. Acad. Sci. (U.S.A.), 34:137-141. (With C. Epling.)
1948g. Cashiering Homo sapiens—a geneticist's view. J. Hered., 29:141-142. [Review]
1948h. Darwin's finches and evolution. Ecology, 29:219-220. [Review]
1948i. Morphogenesis and adaptation. Sci. Monthly, 67:308-310. [Review]
1949a. Genetic structure of natural populations. Carnegie Institution of Washington Year Book No. 48, pp. 201-212.
1949b. Observations and experiments on natural selection in Drosophila. Proc. Eighth Int. Cong. Genetics, pp. 210-224. Lund, Berlingska Boktryckeriet.
1949c. The suppression of a science. Bull. Atomic Scientists, 5:144-146.
1949d. Toward a modern synthesis. Evolution, 3:376-377. [Review]
1949e. On some of the problems of population genetics and evolution. In Symposium on Ecological and Genetic Factors in Animal Speciation. Ric. Sci. (Roma), Suppl., 19:11-17.
1949f. Review and perspectives of the Symposium on Ecological and Genetic Factors of Speciation and Evolution. Ric. Sci. (Roma), Suppl., 19:128-134.
1949g. Conceito fundamentais de genetica. Agronomía (Rio de Janeiro), 8:253-258, 8:341-357.
1949h. [Foreward] In Schumalhausen, I. I. Transl. by I. Dordick, Dobzhansky, Th., ed. Factors of Evolution. Philadelphia, The Blakiston Co.
1949i. Principles of Genetics. 4th ed. New York, McGraw-Hill Book Company. 505 pp. (With E. W. Sinnott and L. C. Dunn.) (5th ed., 1958. New York, McGraw-Hill Book Company, xiv, 459 pp.) (Principios de genetica. Spanish transl. by Antonio Prevosti. 1961. Barcelona, Omega. 581 pp.) (Principi di Genetica. 1965. Italian trans. by B. Battaglia and G. Colombo. Padova, Piccin Editore.)
1949j. The willistoni group of sibling species of Drosophila. Evolution, 3:300-314. (With H. Burla, A. B. daCunha, A. R. Cordeiro, Ch. Malogolowkin and C. Pavan.)
1949k. Marxist biology, French style. J. Hered., 40:78-79. [Review]
1950a. The genetic basis of evolution. Sci. Amer., 182:32-41.
1950b. Heredity, environment, and evolution. Science, 11:161-166.
1950c. Genetics for the elite. Sci. Monthly, 70:133. [Review]
1950d. Evolution in the tropics. Amer. Sci., 38:209-221.
1950e. An appraisal of Charles Darwin's historical role. J. Hered., 41:40-41. [Review]

1950f. Nature and origins of races. Encyclopedia Americana, Vol. 23, pp. 107-111. New York, Americana Corp.
1950g. Genetics of natural populations. XIX. Origin of heterosis through natural selection in populations of *Drosophila pseudoobscura*. Genetics, 35:288-302.
1950h. The genetic nature of differences among men. *In* Persons, S., ed. Evolutionary Thought in America. New Haven, Yale University Press.
1950i. The chromosomes of *Drosophila willistoni*. J. Hered., 41:156-158.
1950j. Genetics. Sci. Amer., 183:55-58.
1950k. Mendelian populations and their evolution. Amer. Natural., 84:401-418. (Repr. *In* Dunn, L. C., ed. 1951. Genetics in the Twentieth Century. New York, The Macmillan Company. pp. 573-589.)
1950l. The science of ecology today. Quart. Rev. Biol., 25:408-409. [Review]
1950m. Local and seasonal variations in relative frequencies of species of *Drosophila* in Brazil. J. Anim. Ecol., 19:1-14. (With C. Pavan.)
1950n. A comparative study of chromosomal polymorphism in sibling species of the *willistoni* group of *Drosophila*. Amer. Natural., 84:229-246. (With H. Burla and A. B. daCunha.)
1950o. Comparative genetics of *Drosophila prosaltans*. Heredity (London), 4:189-200. (With B. Spassky and S. Zimmering.)
1950p. Comparative genetics of *Drosophila willistoni*. Heredity (London), 4:201-215. (With B. Spassky.)
1950q. Some attempts to estimate species diversity and population density of trees in Amazonian forests. Bot. Gaz., 111:413-524. (With G. Black and C. Pavan.)
1950r. Diurnal behavior of some neotropical species of *Drosophila*. Ecology, 31:36-43. (With C. Pavan and H. Burla.)
1950s. Adaptive chromosomal polymorphism in *Drosophila willistoni*. Evolution, 4:212-235. (With A. B. daCunha and H. Burla.)
1950t. Population density and dispersal rates. Ecology, 31:393-404. (With H. Burla, A. B. daCunha, A. G. L. Cavalcanti, and C. Pavan.)
1950u. Principles of Human Genetics. Amer. J. Phys. Anthrop., 8:1-4. [Review]
1950v. Death of a science in Russia—Heredity, East and West. J. Hist. Med. Allied Sci., 3:339-342. [Reviews]
1951a. Human diversity and adaptation. Sympos. Quant. Biol., 15:385-400.
1951b. Race and humanity. Science, 113:264-266. [Reviews]
1951c. Evolution in process. Sci. Monthly, 62:403-404.
1951d. Human races in the light of genetics. Int. Social Sci. Bull. (UNESCO), 3:660-663.
1951e. Experiments on sexual isolation in *Drosophila*. X. Reproductive isolation between *Drosophila pseudoobscura* and *Drosophila persimilis* under natural and under laboratory conditions. Proc. Nat. Acad. Sci. (U.S.A.), 37:792-796.
1951f. Development of heterosis through natural selection in experimental populations of *Drosophila pseudoobscura*. Amer. Natural., 85:247-264. (With H. Levene.)
1951g. On food preferences of sympatric species of *Drosophila*. Evolution, 5:97-101. (With A. B. daCunha and A. Sokoloff.)
1951h. Concealed genic variability in Brazilian populations of *Drosophila*

willistoni. Genetics, 36:13-30. (With C. Pavan, A. R. Cordeiro, N. P. Dobzhansky, C. Malogolowkin, B. Spassky, and M. Wedel.)

1951i. Genetics and the Origin of Species. 3rd ed. New York, Columbia University Press. (See 1937a.)

1952a. Genetics of natural populations. XX. Changes induced by drought in *Drosophila pseudoobscura* and *Drosophila persimilis*. Evolution, 6:234-243.

1952b. Adaptedness of individuals and populations. Amer. Natural., 86: 121-122.

1952c. Experimental evolution in *Drosophila*. Texas J. Sci., 4:545-550.

1952d. André Dreyfus e a escola brasileira de biologia geral. Ciência e Cultura (São Paulo), 4:166-169.

1952e. Lysenko's "michurinist" genetics. Bull. Atomic Scientists, 8:40-44.

1952f. Heredity and environment. Main Current in Modern Thought, 8: 108-109.

1952g. A comparative study of mutation rates in two ecologically diverse species of *Drosophila*. Genetics, 37:650-664. (With B. Spassky and N. Spassky.)

1952h. Two recent versions of eugenics. Amer. Natural., 86:61-62. [Review]

1953a. Natural hybrids of two species of *Arctostaphylos* in the Yosemite region of California. Heredity (London), 7:73-79.

1953b. Some new trends in population genetics and in evolutionary studies. International Union of Biological Sciences Symposium on Genetics of Population Structure. Pavia. pp. 95-97.

1953c. Russian genetics. *In* Soviet Science. Washington, Amer. Assoc. Adv. Sci., pp. 1-7.

1953d. A comment on the discussion of genetics by His Holiness, Pius XII. Science, 118:561-563.

1953e. Le naufrage de la biologie en Russie. Preuves (Paris), Aug.-Sept., pp. 92-98. (Repr. March 1954. *In* Science et Liberté, 37:48-54. (Das Schicksal der biologischen Wissenschaft in Russland. *In* Wissenschaft und Freiheit. Berlin, Grunewald. pp. 222-235.)

1953f. Evolution in *Drosophila*. Evolution, 7:92-93. [Review]

1953g. Indeterminate outcome of certain experiments on *Drosophila* populations. Evolution, 7:198-210. (With O. Pavlovsky.)

1953h. An estimate of the number of species of trees in an Amazonian forest community. Bot. Gaz., 114:467-477. With J. Murca Pires and G. A. Black.)

1953i. Genetics of natural populations. XXI. Concealed variability in two sympatric species of *Drosophila*. Genetics, 38:471-484. (With B. Spassky.)

1953j. The genetics of homeostasis in *Drosophila*. Proc. Nat. Acad. Sci., (U.S.A.), 39:162-171. (With B. Wallace.)

1953k. The theory of the gene. Amer. Natural., 87:119-123. [Review]

1953l. Evolution in action. Amer. J. Phys. Anthrop., 11:605-607. [Review]

1953m. Lysenko progresses backwards. J. Hered., 44:20-22. [Review]

1954a. Evolution as a creative process. *In* Atti IX Cong. Int. Genetica, Part 1. Caryologia, 6, Suppl.: 435-449.

1954b. An ethical problem for scientists in a divided world. Science, 119:908-909.

1954c. Animal breeding under Lysenko. Amer. Natural., 88: 165-167. [Review]

1954d. Animal species and their evolution. Heredity (London), 8:286-287. [Review]
1954e. Combining ability of certain chromosomes in *Drosophila willistoni*, and invalidation of the "wild-type" concept. Amer. Natural., 88:75-86. (With A. R. Cordeiro.)
1954f. A further study of chromosomal polymorphism in *Drosophila willistoni* and its relation to the environment. Evolution, 8:119-134. (With A. B. daCunha.)
1954g. Excretion in human urine of an unknown amino acid derived from dates. Nature (London), 174:533. (With S. Gartler.)
1954h. Strangler trees. Sci. Amer., 190:78-80. (With J. Murça Pires.)
1954i. Interaction of the adaptive values in polymorphic experimental populations of *Drosophila pseudoobscura*. Evolution, 8:335-349. (With H. Levene and O. Pavlovsky.)
1954j. Environmental modification of heterosis in *Drosophila pseudoobscura*. Proc. Nat. Acad. Sci. (U.S.A.), 40:407-415. (With N. P. Spassky.)
1954k. Genetics of natural populations. XXII. A comparison of the concealed variability in *Drosophila prosaltans* with that in other species. Genetics, 39:472-487. (With B. Spassky.)
1954l. Rates of spontaneous mutation in the second chromosomes of the sibling species, *Drosophila pseudoobscura* and *Drosophila persimilis*. Genetics, 39:899-907. (With B. Spassky and N. Spassky.)
1954m. The problem of adaptive differences in human populations. Amer. J. Human Genet., 6:199-207. (With B. Wallace.)
1954n. On the nature of species in the USSR. Syst. Zool. 3:66-68.
1954o. The facts of life. Amer. J. Phys. Anthrop., 12:619-623. [Review]
1954p. Symposia of the society for experimental biology. VII. Science, 119:165-166.
1954q. Ideologie und Forschung in der Sowjetischen Naturwissenschaft. Science, 119:545-546. [Review]
1954r. The facts of life. Ann. Hum. Genet., 19:75-77. (With L. S. Penrose.) [Review]
1955a. Evolution, Genetics and Man. New York, John Wiley & Sons, Inc. 398 pp. (Die Entwicklung zum Menschen, Evolution Abstammung und Verebung ein Abriss. 1958. Transl. by F. Schwanitz. Hamburg, Paul Parey.) (Evolutie en erfelijkheid. 406 pp. Dutch transl. by C. E. B. Bremekamp. 1961. Utrecht, Aula-Boeken. 428 pp.)
1955b. A review of some fundamental concepts and problems of population genetics. Sympos. Quant. Biol., 20:1-15.
1955c. The crisis in Soviet biology. *In* Simmons, E. J., ed. Continuity and Change in Russian and Soviet Thought. Cambridge, Harvard University Press. pp. 329-346.
1955d. Genetic homeostasis. Animal cytology and evolution. Atlas of men. Evolution, 9:100-104. [Reviews]
1955e. Differentiation of nutritional preferences in Brazilian species of *Drosophila*. Ecology, 36:34-39. (With A. B. daCunha.)
1955f. An extreme case of heterosis in a Central American population of *Drosophila tropicalis*. Proc. Nat. Acad. Sci., (U.S.A.), 41:289-295. (With O. Pavlovsky.)
1955g. Chromatographic studies on urinary excretion patterns in monozygotic and dizygotic twins. I. Methods and analysis. Amer. J. Hum.

Genet., 7:93-107. (With H. K. Berry, S. M. Gartler, H. Levene, and R. H. Osborne.)

1955h. Chromatographic studies on urinary excretion patterns in monozygotic and dizygotic twins. II. Heritability of the excretion rates of certain susbstances. Amer. J. Hum. Genet., 7:108-121. (With S. M. Gartler and H. K. Berry.)

1955i. Genetics of natural populations. XXIII. Biological role of deleterious recessives in populations of *Drosophila pseudoobscura.* Genetics, 40:781-796. (With O. Pavlovsky, B. Spassky, and N. Spassky.)

1955j. Genetics of natural populations. XXIV. Developmental homeostasis in natural populations of *Drosophila pseudoobscura.* Genetics, 40: 797-808. (With H. Levene.)

1955k. Evolution as a process. Science, 121:162-163. [Review]

1955l. A return to reason. Science, 121:188-189. [Review]

1956a. The Biological Basis of Human Freedom. New York, Columbia University Press, vi, 139 pp. (Paperback ed., 1960.) (Las Bases Biologicas de la Libertad Human. 1957. Transl. by D. Brncic. Buenos Aires, El Ateneo. 120 pp.) (Le Basi Biologiche della Liberta Umana. 1960. Milano, Feltrinelli. 146 pp.)

1956b. The genetic basis of systematic categories. In Biological Systematics, 16th Ann. Biol. Colloq. Corvallis, Oregon State College. pp. 37-42.

1956c. Inside human nature. *In* White, L. ed., Frontiers of Knowledge. New York, Harper & Row, Publishers. pp. 1-15.

1956d. What is an adaptive trait? Amer. Natural., 90:337-347.

1956e. Genetics of natural populations. XXV. Genetic changes in populations of *Drosophila pseudoobscura* and *Drosophila persimilis* in some localities in California. Evolution, 10:82-92.

1956f. A evoluçao humana. Rev. Antropol. (São Paulo), 4:97-102.

1956g. Balanced polymorphism in *Drosophila* and in *Homo.* Biologica (Santiago, Chile), 22:7-10.

1956h. Does natural selection continue to operate in modern mankind? Amer. Anthrop., 58:591-604. Repr., Ann. Rep. Smithsonian Inst., 1958, pp. 359-374. (With G. Allen.)

1956i. A chromatographic investigation of urinary amino acids in the great apes. Amer. J. Phys. Anthrop., 14:41-57. (With I. L. Firschein and S. M. Gartler.)

1956j. Studies on the ecology of *Drosophila* in the Yosemite region of California. I. The occurrence of species of *Drosophila* in different life zones and at different seasons. Ecology, 37:526-533. (With D. M. Cooper.)

1956k. Studies on the ecology of *Drosophila* in the Yosemite region of California. IV. Differential attraction of species of *Drosophila* to different species of yeasts. Ecology, 37:544-550. (With D. M. Cooper, H. J. Phaff, E. P. Knapp, and H. L. Carson.)

1956l. Heredity, race, and society. 3rd ed. New York, Mentor Books. (With L. C. Dunn.) (See 1946e.)

1957a. Mendelian populations as genetic systems. Sympos. Quant. Biol., 22:385-393.

1957b. The biological concept of heredity as applied to man. *In* Milbank Memorial Fund Conference. (Repr., as The concept of heredity as it applies to man. *Columbia University Forum,* 1:24-27, 1957.)

(Also *In* Fried, M. H., ed. Readings in Anthropology, 1959. Vol. 1, pp. 89-95.)

1957c. Genetic loads in natural populations. Science, 126.3266:191-194.

1957d. On methods of evolutionary biology and anthropology. I. Biology. Amer. Sci., 45:381-392.

1957e. Genetics of natural populations. XXVI. Chromosomal variability in island and continental populations of *Drosophila willistoni* from Central America and the West Indies. Evolution, 11:280-293.

1957f. The X-chromosome in the larval salivary glands of hybrids of *Drosophila insularis* and *Drosophila tropicalis*. Chromosoma, 8:691-698.

1957g. What is environment? Amer. Natural., 91:269-271.

1957h. The southernmost drosophilids. Amer. Natural., 91:127-128. (With D. Brncic.)

1957i. *Drosophila insularis,* a new sibling species of the *willistoni* group. Univ. Texas Publ. 5721, pp. 39-47. (With L. Ehrman and O. Pavlovsky.)

1957j. An experimental study of interaction between genetic drift and natural selection. Evolution, 11:311-319. (With O. Pavlovsky.)

1957k. Heterosis and elimination of weak homozygotes in natural populations of three related species of *Drosophila.* Proc. Nat. Acad. Sci. (U.S.A.), 43:226-234.

1958a. Species after Darwin. *In* Barnett, S. A., ed. Century of Darwin. London, Heinemann. pp. 19-55.

1958b. The causes of evolution. *In* Huxley, J. S., ed. A Book that Shook the World. Pittsburgh, University of Pittsburgh Press. pp. 12-29.

1958c. Evolution at work. Science, 127:1091-1098. (Repr. *In* Howells, W., ed., Ideas on Human Evolution. 1952. Cambridge, Harvard University Press. pp. 19-35.

1958d. Genetics of homeostasis and senility. Ann. New York Acad. Sci., 71:1234-1241.

1958e. Lysenko at bay. J. Hered., 49:15-17.

1958f. Genetics of natural populations. XXVII. The genetic changes in populations of *Drosophila pseudoobscura* in the American Southwest. Evolution, 12:385-401.

1958g. Interracial hybridization and breakdown of coadapted gene complexes in *Drosophila paulistorum* and *Drosophila willistoni.* Proc. Nat. Acad. Sci. (U.S.A.), 44:622-629. (With O. Pavlovsky.)

1958h. Adaptive polymorphism and developmental homeostasis in populations of *Drosophila pseudoobscura*. Proc. 10th Int. Cong. Genetics, Toronto, University of Toronto Press. Vol. 2, pp. 15-16. (With J. A. Beardmore and O. Pavlovsky.)

1958i. New evidence of heterosis in naturally occurring inversion heterozygotes in *Drosophila pseudoobscura*. Heredity (London), 12:37-49. (With H. Levene.)

1958j. Dependence of adaptive values of certain genotypes in *Drosophila pseudoobscura* on the composition of the gene pool. Evolution, 12:18-23. (With H. Levene and O. Pavlovsky.)

1958k. Release of genetic variability through recombination. I. *Drosophila pseudoobscura*. Genetics, 43:844-867. (With H. Levene, B. Spassky, and N. Spassky.)

1958l. Principles of genetics. 5th ed. New York, McGraw-Hill Book Company. 459 pp. (With E. W. Sinnott and L. C. Dunn.) (See 1949i.)

1959a. Changes in inversion frequencies in California populations of *Drosophila pseudoobscura* since 1941. London, Proc. 15th Int. Cong. Zool. pp. 169-170.
1959b. Genetics and the destiny of man. Proc. 10th Int. Cong. Genetics, 1:468-474. (Repr. 1959, Yellow Springs, Ohio, Antioch Rev., 19: 57-68.)
1959c. Evolution of genes and genes in evolution. Sympos. Quant. Biol., 24:15-30.
1959d. On selection of gene systems in natural populations. *In* Harrison, G. A. et al., eds. Sympos. Soc. Study Human Biol. Vol. 2, Natural Selection in Human Populations, New York, Pergamon Press, Inc.
1959e. Variation and evolution. Proc. Amer. Philos. Soc., 103:252-263.
1959f. Human nature as a product of evolution. *In* Maslow, A., ed. New Knowledge in Human Values. New York, Harper & Row, Publishers. pp. 75-85.
1959g. [Foreword] *In* Osborne, H., and De George, F. V., eds. Genetic Basis of Morphological Variation. Cambridge, Harvard University Press.
1959h. Genetics and the "average man." Challenge, 8:38-47.
1959i. Blyth, Darwin, and natural selection. Amer. Natural., 93:204-206.
1959j. Evolution, Marxian biology, and the social scene. Science, 129:1479-1480.
1959k. The evolution of living things. Science, 130:785. [Review]
1959l. Genetics of natural populations. XXVIII. Supplementary data on the chromosomal polymorphism in *Drosophila willistoni* in its relation to the environment. Evolution, 13:389-404. (With A. B. daCunha, O. Pavlovsky, and B. Spassky.)
1959m. Release of genetic variability through recombination. III. *Drosophila prosaltans*. Genetics, 44:75-92. (With H. Levene, B. Spassky, and N. Spassky.)
1959n. Possible genetic difference between the head louse and the body louse (*Pediculus humanus* L.) Amer. Natural. 93347-353. (With H. Levene.)
1959o. *Drosophila paulistorum*, a cluster of species in statu nascendi. Proc. Nat. Acad. Sci. (U.S.A.), 45:419-428. (With B. Spassky.)
1959p Radiation, Genes, and Man. New York, Holt, Rinehart & Winston, Inc. 205 pp. (London, Methuen. 1960.) (Adapted by Frank N. Paparello. 1963. New York Holt, Rinehart & Winston, Inc. 203 pp.) (Människan och strålningsriskerna. 1961. Swedish transl. Hans I. Gedin. Stockholm, Aluds/Bonniers. 428 pp.) (Arv, og stråling. 1962. Danish transl., D. J. Adler. København, Gyldendals Uglebøger. 203 pp. (With B. Wallace.)
1959q. [Foreword] *In*: Rensch, B. Evolution at the Species Level. New York, Columbia University Press.
1960a. The present evolution of man. Sci. Amer., 203:206-217.
1960b. Die Ursachen der Evolution. *In* Heberer, G. and Schwanitz, F. Hundert Jahre Evolutionsforschun. Stuttgart, Fischer Verlag. pp. 32-44.
1960c. Evolution und Unwelt. *In* Heberer, G. and Schwanitz, F. Hundert Jahre Evolutionsforschung. Stuttgart, Fischer Verlag. pp. 81-98.
1960d. Evolucion y genetica. Rev. Univ. Madrid, 8:165-186.
1960e. Bearing of evolutionary studies of *Drosophila* on understanding of human evolution. Scientia (Bologna), 54:1-4.

1960f. Evolutionism and man's hope. Sewanee Review, 68:274-288.
1960g. Evolution and environment. *In* Tax, S., ed. Evolution After Darwin. Chicago, University of Chicago Press. pp. 403-408. (La evolucion y el medio ambiente. 1962. La Educacion (Washington, Pan-American Union), 27:6-31.)
1960h. La specie. Un secolo dopo Darwin. Accad. Naz. Lincei, 357:41-52.
1960i. L'uomo. Prodotto singolare dell-evoluzione biologica. Accad. Naz. Lincei, 357:323-331.
1960j. One hundred years of Darwinian evolution. J. Sci. Indus. Res. (New Delhi), 19A:120-125.
1960k. Man consorting with things eternal. *In* Shapley, H., ed. Science Ponders Religion. New York, Appleton-Century-Crofts. pp. 117-135.
1960l. Bridging the gap between race and species. Proc. Linnean Soc. New South Wales, 85:322-327.
1960m. Genetics of natural populations. XXIX. The magnitude of the genetic load in populations of *Drosphilia pseudoobscura*. Genetics, 45:723-740. (With B. Spassky, N. Spassky, O. Pavlovsky, M. G. Krimbas, and C. Krimbas.)
1960n. Genetics of natural populations. XXX. Is the genetic load in *Drosophila pseudoobscura* mutational or balanced? Genetics, 45:741-753. (With C. Krimbas and M. G. Krimbas.)
1960o. Individuality, gene recombination, and non-repeatability of evolution. Austral. J. Sci., 23:71-74.
1960p. Darwin's biological work. J. Genet., 57:166-168.
1960q. Darwin's place in history. Heredity, (London) 14:450-451. [Review]
1960r. The radiation hazard. (Transcription of a panel meeting moderated by J. E. Rall.) Bull. New York Acad. Med., 36:804-827.
1960s. An attempt to compare the fitness of polymorphic and monomorphic experimental populations of *Drosophila pseudoobscura*. Heredity, 14:19-33. (With J. A. Beardmore and O. Pavlovsky.)
1960t. How stable is balanced polymorphism? Proc. Nat. Acad. Sci. (U.S.A.), 46:41-47. (With O. Pavlovsky.)
1960u. Release of genetic variability through recombination. V. Breakup of synthetic lethals by crossing over in *Drosophilia pseudoobscura*. Zool. Jahr. Abt. Syst., 88:57-66. (With B. Spassky.)
1960v. Eugenics in New Guinea. Science, 132:77.
1961a. Genetics. *In* Gray, P., ed. The Encyclopedia of the Biological Sciences. New York, Reinhold Publishing Corp. pp. 428-433.
1961b. Man and natural selection. Amer. Sci., 49:285-299. Repr. *In* Brode, W. R., ed. Science in Progress. 1963. New Haven, Yale University Press. pp. 131-154.
1961c. On the dynamics of chromosomal polymorphism in *Drosophila*. Roy. Entom. Soc. London, Sympos. 1:30-42.
1961d. Human races. Amer. J. Human Genet., 13: 349-350. [Review]
1961e. Adaptation in man and animals: a synthesis. Ann. New York Acad. Sci., 91:634-636.
1961f. Taxonomy, molecular biology, and the peck order. Evolution, 15:263-264. [Review]
1961g. A bogus "science" of race prejudice. J. Hered., 52:189-190. [Review]
1961h. Principles of human genetics. Science, 133:270-271. [Review]
1961i. The ethical animal. Science, 133:323-324. [Review]
1961j. Implications of evolution. Science, 133:752. [Review]

1961k. Soviet Marxism and natural science. Science, 133:1762-1763. [Review]
1961l. A further study of fitness of chromosomally polymorphic and monomorphic populations of Drosophila pseudoobscura. Heredity (London), 16:169-179. (With O. Pavlovsky.)
1961m. The evolutionary status of Drosophila serrata. Evolution, 15:461-467. (With W. B. Mather.)
1961n. Genetics, the core science of biology. Science, 134:2091-2092. [Review]
1962a. Mankind Evolving: the Evolution of the Human Species. (Silliman Memorial Lectures, 1959.) New Haven, Yale University Press. xiii, 381 pp. (Paperback ed., 1962.) (Dynamik der menschlichen Evolution. 1965. S. Fischer Verlag.) (De Biologische en culturele Evolutie van de Mens. 1965. Utrecht-Antwerpen, Anla-Boeken.) (L.Evoluzione della Specie Umana. 1965. Giulio Einaudi, Torino.) (L'homme en Evolution. 1966. Transl. by George and Simone Pasteur. Paris, Flammarion.)
1962b. Genetics and equality. Equality of opportunity makes the genetic diversity among men meaningful. Science, 137.3524:112-115. Washington D.C. (Repr. 1964. Yearbook Phys. Anthrop., 10, pp. 1-4. New York, American Journal, a Quarterly Review of Contemporary Thought, 3:372-379, 1964. American Review (New Delhi), 1964, pp. 79-86.
1962c. Genetics, society and evolution. Bull. New York Acad. Med., 38:451-459.
1962d. Evolutionary biology and modern culture. Rockefeller Inst. Quart., 6:1-9.
1962e. Rigid vs. Flexible chromosomal polymorphisms in Drosophila. Amer. Natural., 96:321-328.
1962f. A comparative study of the chromosomes in the incipient species of the Drosophila paulistorum complex. Chromosoma, 13:196-218. (With O. Pavlovsky.)
1962g. Genetic drift and natural selection in experimental populations of Drosophila pseudoobscura. Proc. Nat. Acad. Sci. (U.S.A.), 48:148-156. (With N. P. Spassky.)
1962h. Ein Beitrag zur genetischen Basis der Quanten-Evolution. In Drescher, W., ed. Evolution und Hominisation. Stuttgart, Gustav Fischer Verlag. pp. 64-73.
1962i. Selection for geotaxis in monomorphic and polymorphic populations of Drosophila pseudoobscura. Proc. Nat. Acad. Sci. (U.S.A.), 48. 10: 1704-1712. (With B. Spassky.)
1962j. Experimental proof of balanced genetic loads in Drosophila. Genetics, 47:1027-1042. (With B. Wallace.)
1962k. Two new species of Drosophila from New Guinea. (Diptera: Drosophilidae.) Pacific Insects (Honolulu), 4:245-249. (With W. B. Mather.)
1962l. Problemy radiacionnoi genetiki. Science, 137:276. [Review]
1963a. Mating preferences and sexual isolation within and between the incipient species of Drosophila paulistorum. Amer. Midl. Natural., 68:67-82. (With G. Carmody, A. Diaz Collazo, L. Ehrman, I. S. Jaffrey, S. Kimball, S. Obrebski, S. Silagi, T. Tidwell, and R. Ullrich.)

1963b. Biological evolution in island populations. *In* Fosberg, F. R., ed. Tenth Pacific Science Congress: Man's place in the Island Ecosystem. Honolulu, Bishop Museum Press. pp. 65-74.
1963c. Species in *Drosophila*. Proc. Linnean Soc. London, 174:1-12.
1963d. Scientific explanation: Chance and antichance in organic evolution. *In* Baumrin, B., ed. Philosophy of Science, Vol. 1. New York, Interscience. pp. 209-222.
1963e. Heredity in man. *In* Levin, L., ed. Grolier's Book of Popular Science. New York, Grolier, Incorporated. pp. 333-347.
1963f. Mankind evolving: a rejoinder. Perspect. Biol. Med., 6:275-276.
1963g. Anthropology and the natural sciences: the problem of human evolution. Current Athrop., 4:138-148.
1963h. Evolution—Organic and superorganic. Rockefeller Inst. Rev., 1:1-9. (Repr. Bull. Atomic Scientists, 5:4-8. 1964.)
1963i. Geographic and microgeographic races. Current Anthrop., 4:196-197.
1963j. Evolutionary and population genetics. Science, 142:1131-1135. (Repr. *In* Geerts, S. J., Genetics Today. Proc. 11th Int. Cong. Genetics, New York, Pergamon Press, Inc., Vol. 2, pp. 71-89. Birth Defects Reprint Series. 1966. New York. The National Foundation.)
1963k. Genetics of natural populations. XXXIII. A progress report on genetic changes in populations of *Drosophila pseudoobscura* and *Drosophila persimilis* in a locality in California. Evolution, 17:333-339.
1963l. Relative fitness of geographic races of *Drosophila serrata*. Evolution, 17:72-83. (With L. C. Birch, P. O. Elliot, and R. C. Lewontin.)
1963m. Genetic entities in hominid evolution. *In* Washburn, S. L., ed. Classification and Human Evolution. Chicago, Aldine Publishing Company.. pp. 347-362.
1963n. Cultural direction of human evolution: a summation. Human Biol., 35:311-316.
1963o. Evolutionism and man's hope. Lectures in Biological Sciences. University of Tennessee Chapter of Sigma XI. pp. 97-110.
1963p. Genetics of race equality. Eugenics Quart., 10:151-160. (Repr. 1966. *In* Muscatine, C., and Griffin, M., eds. The Borzoi College Reader, New York, Alfred A. Knopf, Inc.)
1963q. The nature of heredity. Ann. Rep. Proc. Papua and New Guinea Sci. Soc. p. 2.
1963r. Genetics of natural populations. XXXI. Genetics of an isolated marginal population of *Drosophila pseudoobscura*. Genetics, 48:91-103. (With A. S. Hunter, O. Pavlovsky, B. Spassky and B. Wallace.)
1963s. Genetics of natural populations. XXXII. Inbreeding and the mutational and balanced genetic loads in natural populations of *Drosophila pseudoobscura*. Genetics, 48:361-373. (With B. Spassky and T. Tidwell.)
1963t. Genetics of natural populations. XXXIV. Adaptive norm, genetic load and genetic elite in *Drosophila pseudoobscura*. Genetics, 48:1467-1485. (With B. Spassky.)
1963u. The origin of races. Sci. Amer., 208:169-172. [Review]
1963v. Relative fitness of geographic races of *Drosophila serrata*. Evolution, 17:72-83. (With L. C. Birch, P. O. Elliott, and R. C. Lewontin.)
1964a. Biology, molecular and organismic. Amer. Zool., 4:443-452. (Repr., The Graduate Journal, 7:11-15. 1965.)

1964b. Human genetics: an outsider's view. Sympos. Quant. Biol., 29:1-7.
1964c. Introduction to the Third Scientific Session, Centennial National Academy of Science. Proc. Nat. Acad. Sci. (U.S.A.), 51:907-908.
1964d. Heredity and the Nature of Man. New York, Harcourt, Brace & World, Inc., x, 179 pp. (*Vererbung und Menschenbild.* 1966. Transl. H. Heberer. München, Nymphenburger.)
1964e. The Mendel centennial. Rockefeller Inst. Rev., 2:1-6.
1964f. How do the genetic loads affect the fitness of their carriers in *Drosophila* populations? Amer. Natural., 98:151-166.
1964g. The superspecies *Drosophila paulistorum*. Proc. Nat. Acad. Sci. (U.S.A.), 51:3-9. (With L. Ehrman, O. Pavlovsky, and B. Spassky.)
1964h. Genetics of natural populations. XXXV. A progress report on genetic changes in populations of *Drosophila pseudoobscura* in the American Southwest. Evolution, 18:164-176. (With W. W. Anderson, O. Pavlovsky, B. Spassky, and C. J. Wills.)
1964i. The capacity for increase in chromosomally polymorphic and monomorphic populations of *Drosophila pseudoobscura*. Heredity, 19:597-614. (With R. C. Lewontin and O. Pavlovsky.)
1964j. Cultural direction of human evolution: a summation. *In* Garn, S., ed. Culture and the Direction of Human Evolution. Detroit, Wayne State University Press. pp. 93-98.
1964k. More on the mutation rates and on heterozygous effects of lethals in *Drosophila*. Amer. Natural., 98:449.
1964l. Eugenics: Hereditarian attitudes in American thought. Political Sci. Quart., 79:621-623. [Review]
1964m. The human species. An introduction to physical anthropology. Eugen. Quart., 11:121-122. [Review]
1964n. Oxford school of genetics. Science, 45:258-259. [Review]
1964o. Where is evolution taking us? A review of G. G. Simpson's "This View of Life." The New York Times, March 29, 1964.
1964p. In science, too, for the want of a nail a kingdom can be lost. A review of H. Selye's "From Dream to Discovery." The New York Times, May, 10, 1964.
1964q. Genetics and the origin of species. (Paperback printing of 3rd ed.) New York, Columbia University Press. 364 pp. (See 1937a.)
1964r. Evolutionary processes. Evolution, 18:139. [Review]
1965a. Genetic diversity and fitness. *In* Geerts, S. J., ed. Genetics Today. Proc. 11th Int. Cong. Genetics, 3:541-552. New York, Pergamon Press, Inc.
1965b. Bichromosomal synthetic semilethals in *Drosophila pseudoobscura*. Proc. Nat. Acad. Sci. (U.S.A.), 53:482-486. (With B. Spassky and W. Anderson.)
1965c. "Wild" and "domestic" species of *Drosophila*. *In* Baker, H. G. and Stebbins, G. L., eds. The Genetics of Colonizing Species. New York, Academic Press, Inc. pp. 533-546.
1965d. Mendelism, Darwinism, and evolutionism. Proc. Amer. Philos. Soc., 109:205-215.
1965e. On possible evolutionary consequences of different settlement patterns. Ekistics (Athens), 20:182-185.
1965f. Religion, death, and evolutionary adaptation. *In* Spiro, M. E., ed. Context and Meaning in Cultural Anthropology. New York, The Free Press. pp. 61-73.

1965g. Evolution and transcendence. Main Currents in Modern Thought, 22:3-9.
1965h. Genetics of natural populations. XXXVI. Epistatic interactions of the components of the genetic load in *Drosophila pseudoobscura*. Genetics, 52:653-664. (With B. Spassky and W. W. Anderson.)
1965i. Sexual selection, geotaxis, and chromosomal polymorphism in experimental populations of *Drosophila pseudoobscura*. Evolution, 19: 337-346. (With L. Ehrman, B. Spassky, and O. Pavlovsky.)
1966a. On types, genotypes, and the genetic diversity in populations. (In Japanese.) Iden (Tokyo), 19:30-33.
1966b. Determinism and indeterminism in biological evolution. *In* Smith, V. E., ed. Philosophical Problems of Biology. New York, St. John's University Press. pp. 55-66.
1966c. A geneticist's view of human equality. The Pharos of Alpha Omega Alpha, 29:12-16.
1966d. Spontaneous origin of an incipient species in the *Drosophila paulistorum* complex. Proc. Nat. Acad. Sci. (U.S.A.), 55:727-733. (With O. Pavlovsky.)
1966e. The code was broken. New York Times Book Review, April 17, 1966, and The University of Chicago Magazine, June, 1966. [Review]
1966f. Genetics of natural populations. XXXVII. The coadapted system of chromosomal variants in a population of *Drosophila pseudoobscura*. Genetics, 53:843-854. (With O. Pavlovsky.)
1966g. The Genetic Effects of Radiation. Tennessee, United States Atomic Energy Commission, Division of Technical Information. 49 pp. (With I. Asimov.)
1966h. Are naturalists old-fashioned? Amer. Natural., 100:541-550. (Expanded and modified versions, 1966j, 1968e)
1966i. Sind alle Menschen gleich erschaffen? Naturwissenschaft und Medizin, 3:3-13.
1966j. Kartezjanskie i Darwinowskie aspekty biologii. Problemy (Warsaw), No. 10, 590-604. (Portuguese version, Consideraçoes sobre a biologia Cartesiana e Darwiniana 1967. Ciência e Cultura São Paolo, vol. 19 no. 1. English version, On Cartesian and Darwinian aspects of biology. The Graduate Journal, 8:99-117. 1968.) (Cf. 1966h.)
1966k. An essay on religion, death, and evolutionary adaptation. Zygon, 1: 317-331.
1966l. Genetics of natural populations. XXXVIII. Continuity and change in populations of *Drosophila pseudoobscura* in the western United States. Evolution, 20:418-427. (With W. W. Anderson and O. Pavlovsky.)
1966m. The human revolution. Amer. Anthropol., 68:1084-1085. [Review]
1966n. L'Evolution des populations naturelles et experimentales de Drosophiles. Bull. Soc. Zool. France, 91:305-320.
1967a. On diversity and equality. Columbia University Forum, 10:5-6.
1967b. Changing Man. Science, 155:409-415.
1967c. Sergei Sergeevich Tshetverikov 1880-1959. Genetics, 55:1-3.
1967d. Genetic polymorphism. Eugen. Quart., 14:80-81. [Review]
1967e. Milislav Demerec. Amer. Philos. Soc. Year Book, 1966:115-121.
1967f. The Biology of Ultimate Concern. New York, The New American Library, Inc. 152 pp.

1967g. On the relationship of structural heterozygosity in the X- and third chromosomes of Drosophila pseudoobscura. Amer. Natural., 101: 89-92. (With W. W. Anderson and C. D. Kastritsis.)
1967h. Evolutionary Biology. Dobzhansky, Th., Hecht, M. K., and Steere, W. C., eds. New York, Appleton-Century-Crofts. Vol. 1, 444 pp.
1967i. Experiments on the incipient species of the Drosophila paulistorum complex. Genetics, 55:141-156. (With O. Pavlovsky.)
1967j. Responses of various strains of Drosophila pseudoobscura and Drosophila persimilis to light and to gravity. Amer. Natural., 101: 59-63. (With B. Spassky.)
1967k. Of flies and men. Amer. Psychol., 22:41-48. (Variant in Rockefeller Univ. Rev., Nov.-Dec. 1966, 14-22.)
1967l. Selection and inversion polymorphism in experimental populations of Drosphila pseudoobscura initiated with the chromosomal constitutions in natural populations. Evolution, 21:664-671. (With W. W. Anderson and C. D. Kastritsis.)
1967m. Genetic diversity and diversity of environment. Proc. 5th Berkeley Sympos. Math. Statistics and Probability. Berkeley, University of California Press. pp. 295-304.
1967n. On genetic aspects of human evolution. Proc. 3rd Int. Cong. Hum. Genet., Baltimore, The Johns Hopkins Press. pp. 361-365.
1967o. On types, genotypes and the genetic diversity in populations. In Spuhler, J. N., ed. Genetic Diversity and Human Behavior. Chicago, Aldine Publishing Company. pp. 1-18.
1967p. Creative evolution. Diogenes, 58:62-74. (French version, L'évolution créatrice. Diogène, 58:64-80.
1967q. Etude génétique des réactions de Drosophiles a la lumière et a la pesanteur. Ann. Biol. Clin. (Paris), 6:483-497.
1967r. Evolution: Implications for religion. Christian Century, 19:936-941.
1967s. The origin of genetics. A Mendel source book. BioScience [Review], (August), pp. 577-578.
1967t. Looking back at Mendel's discovery. Science, 156:1588-1589. [Review]
1967u. Cambiar al hombre. Razon y Fabula, 3:49-65. Extracts under the title "Changer l'homme." Centre d'Étude de Conséquence Générales de Grandes Techniques Nouvelles, 6:21-23.
1967v. Repeated mating and sperm mixing in Drosophila pseudoobscura. Amer. Natural., 101:527-533. (With O. Pavlovsky.)
1967w. An experiment on migration and simultaneous selection for several traits in Drosophila pseudoobscura. Genetics, 55:723-734. (With B. Spassky.)
1967x. Effects of selection and migration on geotactic and phototactic behaviour of Drosophila I. Proc. Roy. Soc. [Biol.], 168:24-47. (With B. Spassky.)
1967y. Drosophila pavlovskiana, a race or a species? Amer. Midl. Natural., 78:244-247. (With C. D. Kastritsis.)
1967z. Human values in an evolutionary world. In Hall, C. P., ed. Human Values and Advancing Technology. New York, Friendship Press. pp. 49-67.
1968a. Genetics of Natural Populations. XXXIX. A test of the possible influence of two insecticides on the chromosomal polymorphism in

	Drosophila pseudoobscura. Genetics, 58:423-434. (With W. W. Anderson, C. Oshima, T. Watanabe, and O. Pavlovsky.)
1968b.	Elementos de Genetica. Quart. Rev. Biol., 42:538. [Review]
1968c.	How has mankind originated? Man, 3:136-138. [Review]
1968d.	Evolution and behaviour. International Encyclopedia of the Social Sciences, 5:234-238.
1968e.	On some fundamental concepts of Darwinian biology. *In* Dobzhansky, Th., Hecht, M. K., and Steere, W. C., eds. Evolutionary Biology, Vol. 2. New York, Appleton-Century-Crofts. pp. 1-34.
1968f.	H. Bentley Glass, President Elect. Science, 159:750-751.
1968g.	On genetics and politics. Social Education, 32:142-146.
1968h.	Pierre Teilhard de Chardin as a scientist. *In* Chardin, Teilhard, ed. Letters to Two Friends. New York, The New American Library, Inc. pp. 219-227.
1968i.	Foreword. *In* Osborn, F., ed. The Future of Human Heredity. New York, Weybright & Talley, Inc. pp. v-vii.
1968j.	Human population cytogenetics. Amer. Sci. 56:162A. [Review]
1968k.	On diversity and equality. *In* Spackman, P., and Ambrose, L., eds. The Columbia University Forum Anthology. New York, Columbia University Press. Also *In* Mead, M., Dobzhansky, Th., Tobach, E., and Light, R. E., eds. Science and the Concept of Race. New York, Columbia University Press. pp. 77-79.
1968l.	Ethological isolation between sympatric and allopatric species of the *obscura* group of *Drosophila*. Anim. Behav., 16:79-87. (With L. Ehrman and P. A. Kastritsis.)
1968m.	Darwin versus Copernicus. *In* Rothblatt, B., ed. Changing Perspectives on Man. Chicago, University of Chicago Press. pp. 175-190.
1968n.	Teilhard de Chardin and the orientation of evolution. Zygon, 3:242-258.
1968o.	On genetics, sociology, and politics. Perspect. Biol. Med., 11:544-554.
1968p.	Ein Betrag zur genetischen Basis der Quanten-Evolution. *In* Kurth, G., ed. Evolution und Hominisation. Stuttgart, Gustav Fischer Verlag. pp. 32-42.
1968q.	Genetic differences between people cannot be ignored. Sci. Res. July 22, 1968. pp. 32-33.
1968r.	More bogus "science" of race prejudice. J. Hered. 59:102-104.
1968s.	Revival of genetics in the U.S.S.R. Quart. Rev. Biol., 43:56-59.
1968t.	Genetics of natural populations. XL. Heterotic and deleterious effects of recessive lethals in populations of *Drosophila pseudoobscura*. Genetics, 59:411-425. (With B. Spassky.)
1968u.	Chromosomal polymorphism in populations of *Drosophila immigrans* on the island of Maui, Hawaii. Texas, University of Texas Publ., 6818:381-386. (With R. Richmond.)
1968v.	Genetics of insect vectors of disease. Science, 160:1438-1439. [Review]
1969a.	The Pattern of Human Evolution. *In* Roslansky, J., ed. The Uniqueness of Man. New York, Appleton-Century-Crofts. pp. 41-70. (German transl. Die Struktur der menschlichen Evolution. Lutherische Rundschau, 19:135-158.)
1969b.	Gene. Encyclopedia Britannica, Vol. 10, pp. 65-80. (With R. Richmond.)

1969c. Heredity. Encyclopedia Britannica, Vol. 11, pp. 419-428.
1969d. [Foreword] *In* Berg. L. S. Nomogenesis or Evolution Determined by Law. (Paperback) Cambridge, Massachusetts Institute of Technology Press.
1969e. Artificial and natural selection for two behavioral traits in *Drosophila pseudoobscura*. Proc. Nat. Acad. Sci (U.S.A.), 62:75-80. (With B. Spassky.)
1969f. Effects of selection and migration on geotactic and phototactic behavior of *Drosophila*. II. Proc. Roy. Soc. [Biol.], 173:191-207. (With B. Spassky and J. Sved.)
1969g. The Parable of the Beast. Quart. Rev. Biol., 44:213-214. [Review]
1969h. Transitional populations of *Drosophila paulistorum*. Evolution, 23: 482-492. (With L. Ehrman and O. Pavlovsky.)
1969i. The rise and fall of T. D. Lysenko. Science, 164:1507-1509. [Review]
1969j. Evolution of mankind in the light of population genetics. Proc. 12th Int. Cong. Genetics, Tokyo. Sci. Counc. Jap., 3:281-292.

2

Uniformitarianism. An Inquiry into Principle, Theory, and Method in Geohistory and Biohistory

GEORGE GAYLORD SIMPSON

Harvard University and the University of Arizona

Introduction	43
Origins; Hutton	45
Lyell; Conybeare	48
Lyell; Darwin	53
Since Darwin	55
Immanence and Configuration	58
Classification of Current Issues	60
Naturalism	61
Actualism	61
Historicism	66
Evolutionism	68
Gradualism	72
Historical Inferences	81
Summary and Conclusion	87
References	91

Introduction

Many geologists have been selfconsciously perturbed by the fact that their science is extensively descriptive, and by the possibility that it may have little theoretical structure of its own. It has been supposed that its nondescriptive content may be largely, or even wholly, the application of the principles of other sciences to the special objects and events that geologists only describe. (For example, see Bradley, 1963; Kitts, 1963a.)

If asked to specify a principle that is more than a simple generalization of observations and that is strictly geological, geologists may come up with the *principle of superposition*. That is stated in a standard textbook (Dunbar and Rodgers, 1957, p. 110) as follows: "In a sequence of layered rocks, any layer is older than the layer next above." As a matter of fact, that statement is frequently untrue, as the quoted authors know and go on to explain, in part. It is true if restated in some such form as this: "If sediments are deposited in sequence each above the last and not thereafter changed in attitude, lower beds are older than higher beds." Although invariably true, the statement in that form seems both trivial and tautological. Nevertheless it was not obvious to early observers; its notice was an essential element in the rise of a science of geology; it is implicated in a truly basic and strictly geological concept, that of temporal succession of rocks and consequently of geological history; and it specifies one of a rather large number of methods for investigating rock sequence and earth history. That interpretation may seem so obvious as to have no alternatives and to need no statement, but there are alternatives, now almost but not quite universally discarded. For example, two once popular alternatives are: that the superposed rocks were simply created in that form; or that they were deposited all at once in a deluge. Regardless of whether one wants to call the temporal interpretation of superposition a principle or not, it is not really trivial either in the history or in the present practice of geology.

Something still more general lies back of the criterion (let us call it that) of superposition. It is not, in fact, observational and it is one of many things that show that geology is not basically more observational than other sciences, all of which do also depend on observation, and more. The events of deposition of a sequence of rocks, the forces involved, and the segment of time represented are not observed or observable. They are inferred on the basis of similar but quite distinct events, forces, and lapses of time that can be observed at present. The connection involves at least two still more general principles: that the properties of matter and energy have been the same in the past as in the present, and that no additional properties should be postulated unnecessarily. Those are the closely associated principles of uniformity and of simplicity. The latter principle, sometimes called *Ockham's* (or *Occam's*) *razor,* is not particularly associated with geology, although as pertinent there as anywhere. The former principle, which might be but is not called *Hutton's razor,* has been historically connected with geology, but the validity of that association has been challenged.

The principle of uniformity or the doctrine or school of uniformitarianism can be conventionally, although not precisely, dated from 1795. Ever since then it has been under constant attack. A principle that has so often been belabored and left for dead on the field and that still arises from the ground to fight again has, at least, extraordinary vitality. Discussion is as lively now as ever, including discussion by those who consider the principle outmoded, unnecessary, or simply wrong. Opinions are almost as numerous as those who express them. For some, uniformitarianism is the most important geological principle, and they may add that it is geology's major contribution to science. For others, regardless of source or value, it is not to be considered geological but a requisite of science in general, and they may add that even as such it has become merely banal. In the latter case, it appears that what interest it has is entirely historical and that it may be, and should be ignored at present. (Those who continue to write about uniformitarianism in that vein evidently would apply the conclusion rather to others than to themselves.) A currently popular opinion is that uniformitarianism comprises two quite distinct principles. Those who hold that opinion add that one of the principles is false, and the other is superfluous.

I hope to show that from time to time, and even to the present, the term "uniformitarianism" has covered not only two principles but many. Some are indeed of only historical interest, but some are highly pertinent to modern geology or science in general. Some are definitely geological, and some are not. Some are false, some are still debatable, and some have been firmly established. Some are superfluous or banal; some are indispensable. Interest in why uniformitarianism will not die, and hopes that even now its protean nature can be further clarified, are the excuse for adding to its studies.

Origins; Hutton

The history of uniformitarianism has been discussed in detail and at length, for example by (in alphabetical order): Adams (1938), Bailey (1967), Cannon (1960), Gillispie (1951), Hooykaas (1956, 1959, 1963), Hubbert (1967), Rudwick (1967), and Wilson (1967). (Full citations are given in the appended bibliography, and the works there cited in turn cite many others.) In spite of those excellent studies, the definitive history has yet to be written. For present purposes many details are unnecessary, but enough historical consideration is needed to determine what has been understood as uniformitarianism, to show how the subject has become so confused, and to suggest a conceivably better analysis.

The conventional starting date, 1795, is that of publication of the two volumes of *Theory of the Earth* by James Hutton although the first part of this had already appeared in a journal (Hutton, 1788) and a still earlier abstract had been printed (1785). A third volume was written, but part of it was lost and the remainder was not published until 1899, more than a century after Hutton's death. (A summary of all three volumes, much more readable than they are themselves, is given by Bailey, 1967.) Inevitably Hutton built on the basis of still earlier work (e.g., Toulmin, 1783) and some of his points had been anticipated by others, not only British but also Italian and Russian. However, it was Hutton's publication of 1795 which led toward modern geology through the lineage of Playfair (1802) and Lyell (1830-1833).

Although Hutton is now generally considered the founder of uniformitarianism, that was only one aspect of his work and not, at the time, the most important one. His theory also involved a number of other issues, all controversial at that time and some still so. Thus from the start uniformitarianism was intricately interwoven with varied principles and opinions with which its relationships even now have not been sufficiently analyzed. The core of the Huttonian theory was a geological cycle in which parts of the earth were supposed to be uplifted by crustal forces, essentially heat, and then worn down by erosion, especially by rivers. The crustal movements were postulated as catastrophic; they were literally called "catastrophes" by Hutton. Erosion, especially as seen in the formation of river valleys, was inferred by an admirable combination of observation and reasoning to be gradual. There later arose a debate between catastrophists and gradualists, and gradualism came to be associated with uniformitarianism. In fact Hutton was *both* a catastrophist and a gradualist, and so are most geologists to this day. Some geological processes *are* more rapid and more radical and others *are* slower and less disruptive. Hutton's belief that regional uplift belongs entirely in the first class was an oversimplification, but not a bad one in the state of knowledge of that time.

A subsidiary point that nevertheless loomed large for Hutton was his belief that the consolidation, or lithification, of sediments took place by heat and partial fusion following deep immersion in the crust. Although that can occur by what were later distinguished as metamorphic processes, as a general thesis Hutton's view was simply wrong. That is no longer an issue today, and it has had little involvement in uniformitarianism.

Another major point was Hutton's conclusion that rocks now called igneous are in fact igneous, having solidified from a hot, molten state. A related conclusion was that such rocks have commonly been injected into or below sediments, and are then younger than those sediments. Here essentially began the Vulcanist-Neptunist or Huttonian-Wernerian battle, which raged for decades after Hutton's death. As is well known, a basic difference between the schools (although by no means the only one) was the Wernerian, Neptunist view that rocks in general and granite and basalt in particular were precipitated from the waters of a worldwide primeval sea. The Vulcanist side of the argument became associated with uniformitarianism, but this would seem to be largely because late 18th and early 19th century uniformitarians (by any usual criterion) were generally also Vulcanists. There is some connection with what later came to be called actualism and is a basic element in uniformitarianism. Vulcanism does jibe better with the actual (in the sense of "present") state of the world, where lava is seen in molten state and seas are not universal, but presently observed processes of nature do not inevitably exclude the possibility of Neptunism. In fact, of course, the Neptunists were simply wrong on other grounds, and the bearing of their views on uniformity is not crucial.

Still another Huttonian theoretical element has been called providentialism. From his first brief extract onward, Hutton repeatedly insisted on a final cause, on the cycles of uplift, erosion and the rest having for their *purpose* the maintenance of a habitable and living world, occupied by plants, by animals, and preeminently by man. A final cause implies a first cause and a first cause implies a creation and a Creator. Those implications were accepted by Hutton, but he saw no reason to dwell on them. About as far as he went was to say that the purposeful orderliness of nature is "not unworthy of Divine wisdom." That must be viewed in the light of Hutton's conclusion, still often quoted today and considered scandalous in his day, that "we find no vestige of a beginning." Hutton, carefully pious, did not mean to deny that God created the universe. He simply considered it largely irrelevant to his inquiry. What he did imply by seeing neither beginning nor end is a theory of cyclic change without a secular trend. He could not identify a beginning in the record of the rocks, but only an unending repetition of change ever returning on itself, and he predicted the continuation of that process into an indefinite future. Here, then, is indeed a doctrine of uniformity, of a historical steady state overall, although it describes a dynamic, cyclic equlibrium and not a static one.

From the point of view of later geological theory and of then current theological objections, an important corollary of Hutton's "no vestige of a beginning" is its allowance for an indefinitely long span of geological time. In Hutton's day the creation of the world was generally dated at 4004 B.C., as fixed by Bishop Usher in his *Annales Veteris et Novi Testamenti* (1650-1654). Obviously that would not allow time for the erosional formation of valleys by the rivers flowing in them. Therefore that aspect, at least, of uniformitarianism, actualism, and gradualism demanded unspecified but great spans of time. Although by 1859 Usher's chronology had been abandoned by competent scientists, the same issue arose in regard to Darwin's gradualist theory of evolution. The debate was still being carried on by William Thomson (who by then had become Lord Kelvin) after Darwin and more than a century after Hutton (Kelvin, 1899; see also Hubbert, 1967). Thomson was willing to grant millions of years rather than Usher's thousands, but Darwin and other geologists were quite correctly unwilling to settle for mere millions. (See e.g., Thomson, 1862, and Darwin, 1959, Peckham reference XIV, 55.1:*f*; Darwin referred to Thomson = Kelvin as Thompson, and did not cite him until the 6th edition of the *Origin* although Thomson has published pertinent objections well before the 4th edition.)

Given the system of the earth, which, however or whenever it came to be, had been cycling for uncountable aeons, Hutton was very definite that its operation excludes the preternatural. That aspect of what later came to be called uniformitarianism in a broad sense was particularly objectionable to theologians and also to pious scientists in Hutton's day. Its formal exclusion of preternatural or miraculous intervention in history is still not universally accepted (see Hooykaas, 1959). As noted above, Hutton's theory included catastrophic events, but he considered them to be naturalistic and actualistic, that is excluding the miraculous or preternatural and involving only second causes, defined as forces now extant in nature. There was, however, still a general belief in the biblical deluge preternaturally caused. Especially at the hands of Cuvier (principal publication on this subject, 1812) and mostly after Hutton's time, that was expanded into a theory of successive revolutions, usually considered nonnaturalistic and nonactualistic. That particular version of catastrophism came to be called simply catastrophism, and even today the whole complex of issues is generally oversimplified and confused as a simple contrast of uniformitarianism and catastrophism. The confusion has been compounded by application of the term "neocatastrophism" to modern *naturalistic* theories analogous to revolutionism at least to the extent that they involve supposedly worldwide episodes of heightened tectonic or evolutionary activity. (See Schindewolf, 1963, and for a more moderate naturalistic revolutionist view of evolution see Newell, 1967; this is further discussed below.)

There is, finally but perhaps most fundamentally of all, throughout Hutton's work the conception that the rocks of the earth's crust constitute a historical record and that they can be interpreted as such.

In summary, Hutton's theory was far from simple but constituted a complex of stands taken on a variety of then, and to some extent still, controversial questions. It involved interlocking principles, not always clearly distinguished by Hutton himself and yet multiple. It is not clear in Hutton's works, and it is not always clear today, which of these can be, should be, or have been included in uniformitarianism, a term not proposed until long after Hutton's death.

Lyell; Conybeare

Although historians have lately agreed in ascribing the origin of uniformitarianism to Hutton, the body of theory later loosely understood under that name is both inchoate and somewhat incoherent in Hutton's own work. It was carried further by Playfair (1802) but became fully developed and consistently expressed only in Lyell (1830-1833), with whom modern geology can be said to begin even though he of course had numerous forerunners, Hutton among them. It was to Lyell's work that the term "uniformitarianism" was first applied, probably initially by Whewell (1832, 1837) in a derogatory sense (see Gould, 1965; Wilson, 1967).

The stark essential of Lyellian uniformitarianism appeared in his earliest subtitle (1830): "An attempt to explain the former changes of the earth's surface by reference to causes now in operation." When Lyell spoke of uniformity between past and present geological forces, he meant more than constancy of applicable physical laws or invariance of immanent characteristics of the universe. He did mean that, but he also meant that present causes are sufficient to explain past physical changes. He hedged a bit on biotic changes.) Thus he espoused Hutton's naturalism against the preternaturalism of catastrophists contemporaneous with Lyell. Furthermore he considered it probable that the average intensity of geological forces, "the energy of a cause," has tended to remain constant.

The last point was sharply attacked in Lyell's day and now, long after most of the arguments about immanent constancy and naturalism have died down, uniformity of intensity is still being argued in various guises. It is true that Lyell's original statements (1830–1833) on that principle would probably not be endorsed in detail by anyone today; in fact, they were later modified by Lyell himself. Nevertheless much of the criticism of Lyell and of uniformitarianism through the years (e.g., by Geikie, 1897) and even more recently (e.g., Gillispie, 1951) involves some misunderstanding of Lyell's views and intentions. Lyell of course knew from the start that the action of geologic forces varies greatly from time to time and place to place and in fact that catastrophes occur. That is true in present times, and therefore the conclusion that it was true at earlier geological times may be an application, not a contradiction, of actualism. Lyell's basic issue with the catastrophists was not on that point but on his firm rejection of preternaturalism. In that he was clearly right at least heuristically (and also, I believe, in principle) because the future progress of geology demanded that then still unpopular postulate.

Lyell did hold that the average intensity of geological forces, taking the earth as a whole, has tended to be approximately constant. His contemporaneous critics and later detractors have correctly insisted that this is not precisely true: intensities of such forces as mountain building, vulcanism, and glaciation, averaged for the whole earth, have been more intense at various times in the past than at others or than at present. But here Lyell's basic intention and more important point was to oppose the view, generally held by contemporaneous catastrophists and some others, that there has been a radical, unidirectional, secular reduction in the intensity of geological forces. That view was wrong, and Lyell's opposition to it was necessary and, again, highly heuristic, at least. It is related to the argument about geological time, for unless geological forces were formerly much more intense and have undergone secular reduction a short time scale cannot be accepted. In fact this was relevant to the argument for constancy of kinds of forces (actualism of Hooykaas and others, methodological uniformitarianism of Gould; see later pages) for it could be, and it was then argued that radical changes in intensities and their secular reduction implied inconstancy of forces and hence also preternaturalism.

In a broader, less literal, but essential sense, moreover, Lyell was right about this element of uniformitarianism. There have been revolution-like and catastrophe-like tectonic, volcanic, glacial, and other episodes, but they occur both early and late in geological time. They do not have a definite secular trend, and in the recorded parts of earth history there is no evidence of overall decrease (or increase) in energy flux (see Hubbert, 1967). These considerations led Lyell to conclude, much as Hutton had, that geological history has been in a dynamic equilibrium or steady state for which a beginning cannot be determined or an end predicted. Here there really is a serious weakness in Lyellian uniformitarianism, not only because later knowledge does enable a beginning to be inferred but also because of anomalies involved in this nonhistorical attitude toward history. That, too, will be more fully discussed here.

Major issues raised by Lyell can be specified in the light of criticism at that time. For present purposes that can be done in reference to a recently published (by Rudwick, 1967) letter to Lyell from Conybeare, one of Lyell's most friendly but decisive and well-informed critics. The date of Conybeare's letter is 1841, and it relates to the sixth edition (1840) of Lyell's *Principles*. By that time Lyell had modified his own views, as is indicated by the interesting weakening of the subtitle, quoted above for the first edition. In the sixth it had become: "The modern changes of the earth and its inhabitants considered as illustrative of geology." As will be further noted below, Lyell later retreated still further from the strictest or primitive uniformitarianism partly under the influence of Darwin.

It is not surprising or at this point even particularly interesting that Conybeare, a clergyman but also a geologist of first caliber, was a providentialist to the point of taking for granted that view and its argument from design. It is surprising that Conybeare was willing to accept a time scale of almost any length "short of the infinite." In fact he speaks of "Quadrillions of years" between the "oldest Cambrian and the Newest Pliocene," which is vastly too much. He did not grasp how damaging a long time scale is to the whole catastrophist argument. Gradualistic implications largely escaped him, and much of his letter, as of earlier published criticisms, consequently seems merely mistaken or irrelevant and need not now greatly concern us. Most noteworthy is his continued rejection of the Huttonian theory that rivers have formed the valleys they drain. For Conybeare, "In every single instance the evidence [has] proved this hypothesis to be totally inadequate and contradictory to the whole phaenomena." He even supposed, incorrectly, that Lyell had somewhat come around to his point of view in this respect.

On that subject Conybeare was a catastrophist and diluvialist of then typical stripe. He ascribed major erosional and depositional phenomena to periodic great floods. At that time (1840–1841) recognition of Pleistocene continental glaciation was just coming before geologists through the work of Agassiz, Buckland, and others. This shook the diluvialists because it otherwise explained phenomena they had ascribed to the Noachian deluge and upset then current historicism (see below) because the coming and going of an ice age contradicted the idea of a strictly unidirectional trend in earth history. On Lyell's ideas of climatic change, Conybeare wrote, "Almost thou persuadest me"—but he made it clear that "almost" was far from "quite." In general, the glacial theory exemplifies the uncertain status of both uniformitarianism and catastrophism at the time. Uniformitarians, in their role as actualists, could point to present glaciers as demonstrating the natural forces involved in Pleistocene continental glaciation. On the other hand, catastrophists in the role of (not necessarily unidirectional) historicists could point out that the Pleistocene phenomena exceeded the present ones in intensity or, at least, extent by several orders of magnitude.

The most forceful parts of Conybeare's arguments were historicist. He expounded the thesis of secular, unidirectional trends in earth history in opposition to the Hutton-Lyell "no vestige of a beginning,—no prospect of an end" (Hutton) or, in later terminology, dynamic steady state theory. Conybeare agreed "in believing [in] the absolute uniformity of the laws of nature and general physical causes"; in order words this clergyman was not a preternaturalist as regards the postcreational course of geological history. He agreed that earth history is to be read from the rocks in terms of second causes. But he believed that the causes have acted at different intensities at different times and—this is a main point—that the intensity has in general decreased. He argued for such a trend in physical geology, involving vulcanism (both volcanic and plutonic), tectonics, and sedimentation. He further and most interestingly argued not exactly for decreasing force but for a definite trend in biotic change, the sequence of fossils in the rocks, as a particular example (in reverse order): Man-Mammalia-Saurians-Cartilaginous fish. That is, in Conybeare's term, a converging series, or (in the opposite direction) a progressive one. Therefore it is "perfectly fatal" to Lyell's "system of a continually recurring series of identical terms." "The terminal point cannot be conceived indefinitely or even very remotely distant." Since Conybeare was ready to concede quadrillions (and British quadrillions at that —$n10^{24}$) of years for the ages of rocks then already known to be fossiliferous, his idea of "remotely distant" must have been rather extreme, but the point as opposed to "no vestige of a beginning" is clear enough.

Although Lyell had by this time (6th edition, 1840) conceded that nature might produce new species at successive intervals, he still denied the existence of directionalism or progression in organic as in inorganic history. He explained away the apparent progression on the grounds that the later, "higher," or "more perfect" organisms existed in earlier times but just had not been found. As evidence he pointed to the fact that mammals had been found in the Stonesfield "slate" (a limestone), Jurassic in age, hence long antecedent to the recognized Age of Mammals and contemporaneous with the saurians. That discovery had been made in 1764, but it was not published until 1824, only a few years before the first edition of Lyell's *Principles*. (For that segment of history see Simpson, 1928.)

Conybeare's riposte merits quotation: ". . . You surely cannot consider the exception of the wretched little marsupials of Stonesfield to counterbalance the general bearing of the whole evidence—for all that it would lead to is only this, that in the secondary [Mesozoic] strata a clan of Vertebrata intermediate in their plan between true Mammalia and the lower classes first shewed themselves." Although Stonesfield mammals are not marsupials, the essence of Conybeare's remark is perfectly correct.

Lyell did make one exception to the system of unbounded existence or continual recurrence: he acknowledged that man appeared late in geological history and is neither an original (from the first creation) nor (like groups periodically recreated) a recurrent phenomenon. Conybeare and others reproved him for this inconsistency, although it is not true that it vitiates the Lyellian system as a whole.

As Conybeare's commentator Rudwick (1967) points out, the truly basic disagreement between the archetypal uniformitarian Lyell and the nominal catastrophist Conybeare was not in fact on the subject of catastrophism as usually understood, but on that of a steady state versus a historical model of earth and life history. It complicates matters still more that the system of Cuvier, the truly archetypal catastrophist, in fact was closer to a dynamic steady state than to a really historical model. His catastrophes (like Hutton's for that matter) were cyclic, not secular in trend; he believed that changes in the fossil record were due to migration, not to appearance of really new species; and he opposed the *scala naturae* concept of progression in the animal kingdom and its history. Conybeare's concept was equally nonevolutionary, but it was based on the (supposed) *scala naturae,* which was later taken over, not altogether happily, into evolutionary thought. (On the *scala naturae* see Lovejoy, 1936.)

It has been noticed before, especially by Cannon (1960), that in several respects it was a version of catastrophism and not uniformitarianism that was more nearly appropriate and favorable for development of a theory of organic evolution. Even now this seems not to have been generally understood or sufficiently studied. Yet it is clear that the Hutton-Lyell steady state model excludes the *possibility* of evolutionary interpretation, while the historical model of Conybeare and others is just as consistent with an evolutionary explanation as with their nonevolutionary views. In fact it is more consistent with the former.

Lyell; Darwin

Darwin often referred to his admiration of Lyell and indebtedness to him, and this has been noted in every biography of Darwin and most commentaries on his theories. Nevertheless the nature of the intellectual relationship and the role of uniformitarianism in it have often been misunderstood. As a recent example, Hubbert (1967), in an otherwise impeccable essay on some aspects of uniformitarianism, has written: "In effect, what Darwin did was to develop paleontology from the point where Lyell had left it. By the rejection of Lyell's residual supernaturalism where organisms were concerned, and by an extension of Lyell's Principle of Uniformity to the plant and animal kingdoms, the theory of evolution was an almost inevitable consequence."

Even the preceding brief exposition here suffices to show that this interpretation is untenable. These points cannot be fully discussed here, and further study of them would be profitable, but a few citations from Darwin's own work may clarify them sufficiently for present purposes.

In the *Autobiography* (Lady Barlow's edition, cited here as Darwin, 1958; see p. 101) Darwin wrote, "The science of Geology is enormously indebted to Lyell—more so, as I believe, than to any other man who ever lived." Darwin learned and applied *geological* concepts from Lyell's *Principles* during the voyage of the Beagle. Darwin's first extensive, important research and reputation as a scientist were in geology, and his direct indebtedness here to Lyell is clear and was acknowledged. That was useful background, but it has no direct bearing on the origin of his evolutionary views. Later in the *Autobiography* (p. 119) Darwin wrote, "After my return to England, it appeared to me that by following the example of Lyell in Geology, and by collecting all facts that bore in any way on the variation of animals and plants under domestication and nature, some light might perhaps be thrown on the whole subject [of the explanation of adaptations]." Lyell's contribution here acknowledged is to *method*. The applications of this method to similar subject matter as regards the history of life by the two men led not only to different but also to opposite conclusions.

There, again, the contribution of Lyell, or of uniformitarianism, to Darwin's evolutionary views, as distinct from his way of reaching them, was nil. Incidentally, here and elsewhere Darwin indicates greater application of a purely inductive method, at least in principle, than is generally accepted by recent historians, philosophers, and methodologists of science. (That will be mentioned again in a different connection.)

In the *Origin* (Peckham's variorum edition, cited here as Darwin, 1959) there are 17 references to Lyell in the (variorum) index, but most of these have no direct bearing on the present subject and one (VII.382.65.0.50.51-63:*f*, by Peckham's reference system) relates not to support but to an objection by Lyell. One, however, is relevant (Morse's IV.124-125). There Darwin points to the then (1859 and subsequently) almost general acceptance of Lyell's views on the gradual erosional excavation of valleys and concludes: ". . . . So will natural selection, if it be a true principle, banish the belief of the continued creation of new organic beings, or of any great and sudden modification in their structure." That is an appeal to gradualism, not any other aspect of uniformitarianism. It is not clear and would be most interesting to know to what extent Darwin derived the idea of gradualism from Lyell or how much he only seized upon this as a support or parallel for his original views in a different field. Darwin also insisted on a long time scale and cited Lyell in support (Peckham reference IX.31-33). That is a requisite for gradualism, and on this point there is reason to believe that Darwin's thought was indeed strongly influenced by Lyell if not entirely derived from him. Incidentally, it is curious that although that passage was indexed in the first edition of the *Origin* and was retained in all later editions with only insignificant modification, the index reference disappeared and has even been missed by the variorum editor.

As far as I have learned, those are the most direct contributions of Lyellian uniformitarianism to Darwinian evolutionism. As is well known, Lyell was a determined opponent of organic evolution until long after the first publication of the *Principles* (1830). However, in the first (1859) edition of the *Origin* Darwin could already say that he had "reason to believe that one great authority, Sir Charles Lyell, from further reflexion entertains grave doubts" as to the immutability of species. In his fourth edition (1866) Darwin could say that Lyell "now almost gives up this view," and in the fifth (1869) triumphantly that "Sir Charles Lyell now gives the support of his high authority" to the mutability of species, as we have seen a *non-* and even *anti-*uniformitarian concept according to the original Lyellian system. Thus instead of Lyellian uniformitarianism leading toward Darwinian evolutionism, Darwinian evolutionism led Lyell away from what had been a major part of his uniformitarian concept. (It is interesting that Lyell, who had agreed that scientists should be destroyed at age 60 because thereafter they could not entertain new ideas, was 62 in 1859, when he had grave doubts, was 69 when he had almost given up his view of 39 years before, and was 72 when he had definitely subscribed to the radically different view that he had fought as false and subversive for most of his life. It is true that a younger man might have taken less than ten years to change his mind, but many younger men did not change their minds at all. Richard Owen and Louis Agassiz, the most eminent lifelong scientific opponents of Darwin, were only 55 and 52, respectively, in 1859, little older than Darwin himself.)

Since Darwin

As has now been indicated, by the later part of the 19th century numerous different concepts and theories had become related in some manner and degree with uniformitarianism in the broadest sense. Still others have arisen or, at least, have become more prominent since then.

One aspect of increasing importance is involved in the subdisciplines named "paleobiology" and "paleoecology" in the present century, although they were already being considered in a primitive way by Hutton and even before him. Both are basically applications of the aphorism that the present is the key to the past. That uniformitarian principle has been criticized as embracing so much that it does not denote anything definite. Yet in these connections it does indicate definite procedures and principles, used by historical geologists and paleontologists every day and taken for granted by them. Thus a major treatise on paleoecology (Ladd, 1957) applies uniformitarian methods and reasoning throughout without any of the contributors specifying them as such. Yet it is not true that what can be taken for granted requires no explicit statement. Statement is necessary for comprehension of the nature and philosophy of a science, and also on occasion for the practical reason that too much or (more rarely) too little may be taken for granted.

The very first requirement for a science of paleontology was an application of the principle that the present is a key to the past: the long-delayed general realization that a fossil resembling part of an organism now alive was, in fact, part of a living organism at some time in the past. The interpretation has come to involve not only the taxonomic affinities of past and present organisms, but also the physiology, behavior, and environmental relationships of past organisms. As a simple example, an extinct animal with skeleton and teeth generally like a recent dog is reasonably assumed to have been warm-blooded (more correctly, endothermic), with a coat of hair, living on land, and eating mainly vertebrate prey that it actively hunted. Although not really a pioneering production, Abel's *Palaeobiologie* (1911) was a milestone in the development of that aspect of paleontology. It is now a normal part of almost any paleontological study and of what is coming to be called general paleontology (exemplified by Simpson, 1953a; Brouwer, 1959).

The broader subject of paleoecology, equally based on application of the principle that the present is the key to the past, has developed largely out of paleobiological studies. Like neoecology, it involves communities and environments, hence the paleobiology of numerous associated fossil species and also clues to physical environments both from those species and from uniformitarian studies of the rocks in which they occur.

It is interesting that one of the earliest paleobiological-paleoecological problems is still discussed and that the name of uniformitarianism is, one might say, taken in vain in this connection. It was known at least as early as 1692 that extinct elephants (which, like most extinct species of elephants, are usually called mammoths) are found in Siberia and that a few of them were frozen before they had completely decayed. As living elephants occur only in warm temperate to tropical regions, it was an obvious but crude paleobiological inference that these extinct elephants also lived in warm climates. Their preservation by freezing thus seemed inexplicable in any but catastrophic terms: it was supposed either that they had really lived to the south and that their remains had been washed to Siberia by a (or *the*) flood, or else that the climate of Siberia, warm when elephants lived there, had catastrophically turned glacially cold. The latter idea was already espoused by the archcatastrophist Cuvier and has ever since been involved in arguments against various concepts of uniformitarianism. In our own day it has been enthusiastically supported and highly unusual mechanisms for it have been hypothesized by some writers who, it is fair to say, are not considered as high scientific authorities by the generality of professional geologists, notably Velikovsky (1955) and Hapgood and Campbell (1958).

Farrand (1961) has reviewed some of the now very extensive literature on the Siberian elephants. Its own characteristics show that this particular species (*Mammuthus primigenius*) was adapted to a cold climate, and the whole ecology as shown in fauna, flora, and matrix is also that of a cold climate, consistent with the presence of permafrost while the elephants were alive in that region. For those and other reasons too extensive for restatement here, Farrand concluded that the freezing of Siberian elephants, to the slight extent of its established occurrence, was an event expectable in their situation according to "uniformitarian concepts." Lippman (1962) sharply attacked Farrand's view, which he characterized as "gradualism," and supported Cuvierian catastrophism as the only reasonable explanation, "possibly by means of the mechanism suggested by Hapgood" cited above).

Farrand (1962) replied that Lippman "has apparently confused [gradualism] with uniformitarianism. Uniformitarianism ('the present is a key to the past') is the geologist's concept that processes that acted on the earth in the past are the same processes that are operating today, on the same scale and at approximately the same rates." Farrand added that catastrophes do happen today and that the sudden death and early freezing of a Siberian elephant was indeed catastrophic, but "such catastrophes are in accord with the doctrine of uniformitarianism," while catastrophes of the sort and on the scale envisioned by Lippman ([after Cuvier, Hapgood, and others]) are not.

That example is instructive especially because it is a debate involving uniformitarianism that has gone on literally for centuries and still continues in the 1960's. Few will disagree that Farrand's interpretation of the facts was right in the dialogue cited, and yet his concept of uniformitarianism is still equivocal and calls for further attention.

Some of the concepts that we have seen arising within the general sphere of geological uniformitarianism have also become involved in some of the broadest theoretical and philosophical questions of modern science. That is particularly true of the debates between the steady-state and big-bang models of cosmology. They carry aspects of the Lyellian debate about the earth out into the universe as a whole, although I do not happen to have seen this conceptual relationship noted before in so many words. The steady-state cosmological model is congruent with, or at least analogous to, the Hutton-Lyell steady-state geological model, both with "no vestige of a beginning." There is also a distinct but perhaps less close analogy between Lyell's last efforts to save his steady-state model of biohistory by continual spontaneous creation of species and the maintenance of a steady-state in an expanding universe by continual spontaneous creation of matter. Both are open to the objection that if preternatural causation of continuous "creation" is rejected, it must be considered uncaused, a concept that many of us find almost as uncongenial as that of preternatural causation.

On the other hand big-bang cosmology is a historical model analogous to that of the Conybearean geological catastrophists, and indeed has a beginning that can fairly be considered the greatgrandfather of all catastrophes! However, big-bang cosmology does not involve a Cuvierian succession of revolutions. After initial "creation" it follows a course more Huttonian in aspect and yet more directional than strictly Huttonian geohistory. Both in geology and in cosmology the historical model now dominates theoretical opinion, but in both cases with certain modifications and reservations. There is also a parallel between Lyell's view that, although marked local and temporary fluctuations occur, the worldwide average of geological forces is relatively uniform and the cosmological view that although local and temporal concentrations of matter and energy occur, the universe in the large is fairly homogeneous. (A convenient review of the present state of cosmology is given by McCrea, 1968.)

Immanence and Configuration

Thus various aspects of uniformitarianism, most of them originating in the field of geology, continue to ramify through that and other sciences. It has of course long and often been noticed that concepts embraced by, or related to, uniformitarianism are multiple and as a consequence are often ambiguously stated or wrongly understood. Important efforts to clarify the situation and provide more restricted definitions have lately been made successively and in part independently by Hooykaas (1956, 1959, 1963), Visotskii (1961), and Gould (1965). All three would subsume the principles or issues involved in uniformitarianism under two headings. One, called "actualism" by Hooykaas and Visotskii and "methodological uniformitarianism" by Gould, is the proposition that natural laws are invariable (e.g., Gould) and that those now ("actually," in European usage) observable are sufficient (e.g., Hooykaas). The second, called simply, "uniformitarianism" by Hooykaas and Visotskii and "substantive uniformitarnianism" by Gould, is the proposition of geohistorical uniformity in the intensities and rates of natural processes and in material conditions.

Those dichotomies of uniformitarianism, in the classical sense, can considerably clarify various issues. For instance, the dispute between Farrand and Lippman about the Siberian elephants would have been more nearly clear and decisive if Farrand (1961) could have been explicit that he was relying on actualism or methodological uniformitarianism and had not gone on (1962) to confuse the issue by defining his "uniformitarianism" as what Gould later labeled "substantive uniformitarianism," which for Farrand is an extreme form of Lyellian uniformitarianism, perhaps more extreme than Lyell's. And Lippman's objection would have been more pertinent and cogent if he had been clear that what he was opposing was *that* aspect of uniformitarianism and not actualism or methodological uniformitarianism. In fact he considered the issue to be one of gradualism versus catastrophism, an issue not raised by Farrand and not directly involved in the Visotskii-Hooykaas-Gould dichotomy.

The example just given illustrates how the dichotomy can be useful, but also one reason why it is insufficient. There are a number of issues, such as that of gradualism-catastrophism, historically important and still debated in one form or another that are not taken into account. An example even more important is the complex of issues involving historicism, retrodiction, and explanation, here to be referred to later. Another source of possible continuing confusion is that the two sides of the dichotomy as sometimes presented are not clear-cut. Moreover, each of the alternatives offered is still somewhat ambiguous. For instance the invariability of natural laws is indecisive about such basic problems as their sufficiency or as to whether the actions of all historically relevant laws are currently ("actually") observable. The definitions of uniformitarianism by Hooykaas and of substantive uniformitarianism by Gould, further, raise what is to some extent, at least, a *Scheinproblem,* because they are more rigid or extreme than the views of Hutton, Lyell, or most of their followers.

Because Gould's own extreme statement of it "is false and stifling to hypothesis formation," Gould proposes that this concept of uniformitarianism should simply be abandoned. Nevertheless there are aspects of uniformitarianism that as between "methodological" and "substantive" sides must be placed in the latter but that are real and current problems: especially matters of historicity, directionalism, revolutionism, and the like. Gould then proceeds also to demolish "methodological uniformitariansm" on grounds that it "amounts to an affirmation of induction and simplicity." He concludes that it is "subsumed in the simple statement: 'geology is a science.'" Therefore uniformitarianism in this sense is said to have no other than historical interest, and even that is so only in connection with banishing the supernatural from geology, and Gould would have the term dropped from current use. But here again there are methods and principles that must fall under, or at least are closely related to, "actualism" or "methodological uniformitarianism," that are current as to method and involve current problems, that are especially characteristic of geology even if not wholly confined to it, and that are not applicable to or do not involve all "the empirical sciences together." That is most obviously true of the methods and problems of retrodiction and historical explanation, which are peculiarly geological (including biohistorical) and which certainly are not now anachronistic, as Gould also recognizes.

"The present is the key to the past," although so broad a statement as to require analysis and specification, is still highly meaningful and is a basic principle especially arising from and important within studies of geohistory and biohistory.

The conclusion here is not that the dichotomy of actualism versus uniformitarianism or methodological versus substantive uniformitarianism is useless, unimportant, or passé, but quite the contrary, that with some restatement it is even more useful, fundamental, and correct than its proponents have claimed, or admitted. Most of the issues that have arisen in the long years of discussion of uniformitarianism do indeed fall into two classes that differ in an essential way. One has to do with the inherent properties of the universe (it is too restrictive and debatable to call those properties "laws"), that is, with what is *immanent* in it. The other has to do with the *configurations* that have arisen, and continue to arise, in historical sequence and in accordance with those immanent properties. Actualism and methodological uniformitarianism (as a general principle) belong to the former class, and uniformitarianism *sensu* Hooykaas and substantive uniformitarianism to the latter. Methodology in the more usual sense of the word does not strictly enter into either class, because its application to problems relevant to uniformitarianism involves the relationship between the two classes, that is, between immanence and configuration in geohistory and biohistory. (For further discussion of this broader and, I believe, more meaningful dichotomy see Simpson, 1963 and 1964.)

Classification of Current Issues

Burning issues to the founders of uniformitarianism, and not then always clearly distinguished from it, are now of no relevance in this connection. Thus we have seen that two of Hutton's major concerns were Vulcanism, in opposition to Neptunism, and a theory of lithification. We now consider that on the former subject he was mostly but not entirely right and on the latter mostly but not entirely wrong. We no longer argue those points, at least not in anything like Hutton's terms, and we see no special relationship between them and uniformitarianism. It is not surprising that some issues alive nearly two centuries ago are now dead. It is surprising that so many are not.

What remains to be done in the present essay is to specify and classify the principles and issues that do still have some interest within the general topic of uniformitarianism and to consider the present status of each. The following broad classification is proposed:

A. Concerned mostly with immanent properties
  1. Naturalism and its alternatives
  2. Actualism in its various different aspects or senses and its alternatives
B. Concerned mostly with configurations
  3. Historicism with its concomitants, aspects, and alternatives
  4. Evolutionism or biohistoricism, as a special, exceptionally important aspect of the foregoing
  5. Gradualism, catastrophism, neocatastrophism, revolutionism
C. Concerned mostly with special methodology
  6. Historical inference, retroduction, extrapolation, and related principles

The first words of those numbered topics will be used below as labels although the contents of each are more diverse.

Naturalism

Naturalism is a basic postulate of science as now almost always construed, a necessity of method and procedure in science regardless of what theological or philosophical stand may be taken on it. If only on heuristic grounds, scientific explanation must not invoke the supernatural, nonnatural, noumenal, or any other preternatural factor. Although Hutton was not the first to banish preternaturalism from consideration of the earth's history, we have seen that he did so firmly, and that was perhaps his greatest contribution. He sharply distinguished First Cause from second causes. He was a providentialist in that he considered the First Cause as ordaining a terrestrial system the final cause of which is the benefit of its inhabitants, especially man, but he believed that the operation of that system, once it had been caused, was by entirely rational second causes, with no preternatural intervention.

That is still a widely accepted view, and it contradicts no scientific canons as long as investigation and explanation in science are confined to naturalistic second causes. Many scientists as a matter of individual faith or opinion reject providentialism or final causes and believe that the First Cause (whatever they may call it) is literally ineffable and incomprehensible, not susceptible either to discussion or to investigation. Others of course follow a simple axiomatic naturalism, without concern for first, second, or final causes. For almost all present scientists naturalism, in reference to their work *as* scientists, is no longer a really active issue. It is a basis for uniformitarianism but not special to the latter.

Actualism

The term "actualism" is widely used in the present connection and is used here, but it is ambiguous, particularly in English, unless given special definition. It has been applied to the philosophical doctrine that the essence of the existent is activity, and it is so defined in some English dictionaries. Not in customary English but in the Romance languages, derivatives of Late Latin *actualis,* itself from classical *actus,* "motion," have come to mean not "actual" in the English sense (that is, "real" or "existent") but "present" *"now* existing": French *actuel,* Spanish *actual,* Portuguese *atual,* and Italian *attuale* all have that sense, and cognates appear in Germanic and Slavic languages as technical or adopted foreign terms. In those languages their variants of the term "actualism" refer to a doctrine or ism having to do with the present, the currently existent, and the term is now also used in English in that originally non-English sense.

The significance of the term here is that what exists now is postulated as having also existed in the past. It thus refers to the principle that the present is the key to the past and in loose usage is virtually a synonym of uniformitarianism in a broad sense. However, as has been noted, Hooykaas, Visotskii, and others have contrasted it with uniformitarianism by confining the latter term to *configurational* aspects of what is present ("actual") and using actualism to refer only to what is *immanent*. In that usage, actualism is the postulate or principle that the so-called laws of nature have been and are unchanging. That is made clearer and some serious difficulties are avoided if, as heretofore in the present study, for "laws" one substitutes "properties" or "inherent characteristics," that is, all immanent aspects of matter and energy as distinct from their position, arrangement, and activity at any one time.

Even in that restricted sense actualism involves more than one aspect or issue. In the usual connection with naturalism, an assumption is made that the immanent characteristics of the universe are now observable, a proposition not self-evident or provable as a complete generalization but obviously true in part, at least. It is almost always implied, although rarely stated, that actualism involves not only that present immanent characteristics have all existed throughout the past (always excepting First Cause or, if one likes, big-bang) but also that past immanent characteristics all exist (with the same exception) and probably are all observable at present. The latter distinct principle might be but, as far as I know has not been, called preteritism (*praeteritus*, "past"). The two principles are complementary but not necessarily equivalent. There is a good reason for preferring actualism to preteritism: science is necessarily based on the observable; the present is observable; the past is not.

Actualism in the full sense of the preceding paragraph is not an obvious *a priori* necessity, for conflicting principles readily can be and in fact have been proposed; but neither is it an arbitrary axiom. There is a large amount of observational evidence bearing on it and agreeing with it, even though in the nature of things its absolute, complete validity cannot be proved. Geologists and paleontologists have now accumulated a truly vast number of observations of recent configurations that have been visibly affected by immanent characteristics over periods up to more than three billion years. These are all consistent with actualism. That is the source and principal support of the canon of actualism, and it is generally taken to justify the acceptance of actualism where relevant in other sciences as well.

Further support, extensive in both space and time, is found in astronomy. Here, too, only what is now present (the actual in that sense) on or near the earth is observable, but the radiation observed here and now contains information about the far distant places and long past times whence and when it emanated. Much of that information is configurational; for example electromagnetic radiation tells us where various other bodies were relative to the present position of the earth when the emanations occurred, and if the red shift is correctly interpreted as a Doppler effect it tells us how rapidly that body was then receding relative to our present position. However, that radiation also contains information about immanent chemical and physical characteristics, and cosmic-ray or particle emissions bring us matter from points extremely distant in place and time. None of that information gives any indication of nonuniformity. It is all consistent with the view that electrons, protons, atoms, and so on, are identical everywhere and at all times and that "the universe follows the same physical laws throughout" (McCrea, 1968). That information is far less extensive than has been obtained by geology here on earth, but it is confirmatory.

McCrea (1968), just quoted as to the apparent bearing of the astronomical evidence on actualism (a term not used by him), nevertheless considers it naive to imply "that the universe suddenly came into existence and found a complete system of physical laws waiting to be obeyed." He adds, however, "Less crudely, according to this view, the notion that a changing universe should change in accordance with unchanging laws is regarded as acceptable." That is an excellent statement of the view of the geohistorian or cosmologist who is a uniformitarian as regards the immanent characteristics of the universe; in other words is an actualist as here defined, but who is not a uniformitarian as regards configurations, that is, not a uniformitarian *sensu* Hooykaas or a substantive uniformitarian *sensu* Gould. McCrea himself raises doubts as to whether the laws of physics as known here and now are applicable to the whole universe and whether they would permit prediction rather than only description of its behavior, hence (I would assume) whether they would permit retrodiction of history. (McCrea's discussion is here oversimplified and his paper is already an oversimplification of extended discussions there cited.) However, as there is no counterindication in any observations within our scope, most of us will be willing to accept actualism as a working principle, at least.

Actualism, or uniformitarianism in this sense, as considered up to this point is a *historical* principle. It refers to uniformity in all four dimensions of space and time. In fact it involves another premise which is rarely stated and is usually confused with historical actualism, from which it is nevertheless distinct: that the natural universe, as regards its immanent characteristics, is a single, consistent system at any one time, in other words that it makes what we humans consider as sense. That is probably the most fundamental or, at least, most necessary of all scientific principles, and, by the way, I think it is the only one we may really owe to the Greeks.

It is that aspect of uniformity and not actualism as a four-dimensional concept which is relevant to the nonhistorical aspects of science. Replications of physical and chemical observations over the few years that these have been made in rigorous fashion have been identical within the limits of experimental error. This comfortably confirms the assumption that here and now we are in a system that does make sense for practical purposes, at least. It also provides the basis for the application of actualistic principles: it establishes a set of probable immanent characteristics that can then be taken as also having four-dimensional uniformity and hence used for the interpretation of history. It does not, in itself, exclude contrary, non-actualistic postulates. For example: changes of immanent characteristics might be too slow to be detected in only a few centuries; quite different, additional immanent characteristics that existed in the past might not now exist and hence would not be detectable by nonhistorical methods; a switch from one consistent system to another might be undetectable because observations would continue to be consistent within each currently existing system; immanent characteristics of the present system might have arisen in the course of history and hence not have occurred at earlier times.

We thus see that actualism as here defined is in fact a principle special to geohistory, biohistory, and astrohistory and of limited application to nonhistorical aspects of science. Within its sphere, the question arises whether changes in immanent characteristics of the universe (or our segment of it) would be detectable by historical methods, whether actualism can be taken as a testable hypothesis. There is no possibility of testing every possible contradiction of the hypothesis, but that is a general disability of scientific theory-making. It might, for example, be difficult or impossible to distinguish a past small change in the immanent law of gravitation from a purely configurational change in the masses and distances involved. However, a major change in the immanent, such as abrogation of the inverse-squares law, would almost certainly be detectable. Many observations do indeed raise the chance of contradicting the hypothesis, and as none of them do so, considerable confidence is justified.

Further attention must now be given to the view that uniformity of the immanent, actualism, or methodological uniformitarianism is equivalent to the principle of simplicity, the principle of induction, or both. It would then have no significance as a distinct principle but would be a tautological description of science in general. "The Principle of Uniformity dissolves into a principle of simplicity that is not peculiar to geology but pervades all science and even daily life" (Goodman, 1967a). "The assumption of spatial and temporal invariance of natural laws is by no means unique to geology since it amounts to a warrant for inductive inference which, as Bacon showed nearly four hundred years ago, is the basic mode of reasoning in empirical science.... Methodological uniformitarianism amounts to an affirmation of induction and simplicity. But since those principles belong to the modern definition of empirical science in general, uniformitarianism is

subsumed in the simple statement: 'geology is a science' " (Gould, 1965).
"Simplicity is by no means a simple notion, and what is simplest in one sense may not be simplest in another. There is no single and unique ideal of simplicity" (Beck, 1953). The whole subject is in danger of slipping off into a semasiological morass (see, further, Goodman, 1967a, 1967b). By the complex criterion of such recent discussions, it is not clear that actualism is, in fact, in accordance with the principle of simplicity. It would be possible to agree with McCrea (1968) that it is more natural (simpler?) "to expect that, if the universe changes in the large, then its laws might also change in a way that could not be predicted," or with Russell (1953 [1929]) that "there must, at every moment, be laws hitherto unbroken which are now broken for the first time." However, it is possible and, in my opinion, desirable to adhere to a more naive concept of simplicity, nearer to the archaic canon of William of Ockham, and to postulate that immanent features of the universe now existing are valid in all dimensions until or unless contrary observations are made—which has not occurred. That is support for accepting actualism as a working principle, at least. I fail to see how it can be considered that such a principle "dissolves into" a quite different principle that is advanced as a possible criterion for accepting it. Indeed, in this context we would have no reason for traffic with the principle of simplicity if it were not for its possible but not ncessary relevance to actualism. Which principle dissolves into the other may become a mere semantic quibble; I submit that neither does.

In the relationship of actualism, or some other aspect of uniformity, to induction, the situation is reversed. The validity of induction depends on uniformity. Among the many attempts to justify induction is the postulate that the future will resemble the past, which is left-handed actualism or what I have here called preteritism. Black (1967) has recently provided a summary of this and other proposed justifications for induction and has rejected them all in spite of recognition that induction is to some extent, at least, justified by its works. It is used as a method in science and does work, but it works precisely to the extent that there really is uniformity in each particular case, a uniformity, however, that is as often contingent as immanent and as often statistical as absolute. Faith in induction arises from the experience of uniformity, which is its basis (e.g., Pap, 1953 [1949]).

Thus if we admit induction as a valid principle, it depends on a prior admission of a distinct principle of uniformity of one sort or another, and the latter cannot in any real sense be considered as the derivative or secondary principle of the two. It should also be noted that it is not now universally or even usually believed that induction is the basic mode of reasoning in science. Medawar (1967), for example, has reviewed the development of deductive reasoning and explanation in 19th century England and, as a devoted follower of Popper (e.g., 1935), has proclaimed that, "Induction is a myth." Perhaps even more generally accepted is the deductive system of Hempel and Oppenheim (1953 [1948]). In its original form that system is, indeed, inappropriate for historical explanation and hence for actualism, but Beckner (1967) has demonstrated that it can be modified, rather radically to be sure, in such a way as to apply to historical explanation without calling in induction. Thus induction cannot now be taken as definitive of empirical science, and in the opinion of some scientists and philosophers induction is not necessary or not even appropriate in science. In any case, it is quite distinct from actualism.

Confidence in a consistent system of the universe, as it now exists, is indeed a basic norm for almost all science. Actualism, as here more precisely defined, arose mainly from the study of geology and is specially involved in the historical aspects of geology. To that extent, it may properly be considered a geological principle. It does also apply to historical aspects of other sciences, especially biology and astronomy. Its relevance to the largely nonhistorical aspects of geology and to chemistry and physics as a whole is restricted.

Historicism

The term historicism is here used, with some stretching, as a tag for various principles and problems that arise from consideration of the configurations of the earth and the observable universe in relationship to time. Directly under this tag only some of the oldest and most general issues will be discussed. Evolutionism and gradualism, also aspects of historicism, will be considered subsequently, as they are exceptionally important and have their own special problems and principles.

The idea of a literally static, that is, unchanging earth has long been a *Scheinproblem*. From Hutton or before to today it has never been supported by anyone, but it has been repeatedly, brilliantly attacked; in still another figure, it has been a well-battered windmill for generations of Quixotes. The great virtue of the Hooykaas-Visotskii-Gould dichotomy of uniformitarianism is that it removes actualism from the arena of those foolish attacks. It has also somewhat, at least, clarified the usual but false alternatives of uniformitarianism *versus* catastrophism.

Real issues of historicism indicated in the preceding part of the present essay have often been neglected. Alternative views about the sequence of configurations include:

Steady-state models:
A cyclic steady state, with important, even catastrophic, changes in time but nevertheless with more or less regular return to essentially the same configurations (Hutton and followers).

A statistical steady state, also with important changes but these so localized and so distributed as to maintain a more-or-less constant average in space and time (Lyell and followers).
Historical model:
An irreversible sequence changing in a constant direction (Conybeare and many others).

Those three classic views have been stated in their extreme forms. It is not surprising that none can now be accepted in so extreme a form and that the present consensus includes features of all three.

As noted above, the classical form of the historical model, and the one that Lyell opposed, was extreme and involved directionalism or progressionism in the sense that the intensity of geological processes was supposed to have decreased continuously throughout earth history. As geologists and paleontologists are among those who seek to "prove all things" but do not invariably manage to "hold fast that which is good" (I Thessalonians, 21), it is not surprising that exactly the opposite form of directionalism has also been proposed, and that in fairly recent times. Schuchert (especially 1931) and others have concluded that geologic processes have accelerated in the course of geologic time. The same view as regards biological evolution has been even more widely advanced, for example in considerable but largely fallacious detail by Meyer (1954). The question of directionalism in biohistory will be discussed as an aspect of evolutionism. Directionalism of geohistory in either of its extreme forms is not now tenable. It is now clear that such processes as orogeny, vulcanism, and glaciation have varied greatly from time to time and place to place. At particular places and times in the past they have been both more and less active than at present. There is no evident regular progression either of decrease or of increase in their force. To that extent, Lyell's contention of configurational uniformity is confirmed.

On the other hand, it is no longer possible to accept the Hutton-Lyell model that involved no essential difference between early and late states of the earth: "No vestige of a beginning,—no prospect of an end" (Hutton, 1795). Hubbert (1967), in one of the clearest discussions of this aspect of directionalism (not his term), has pointed out that this is primarily a question of thermodynamics. When it was realized that geological processes involve the degradation of energy (increase of entropy, we say now), the principal source of that energy was supposed to be the heat of an originally molten earth (Thomson, 1862; Kelvin [=Thomson], 1899). Other known sources included solar radiation (both directly and as converted into winds and river flow), tides, and perhaps gravitation in other forms (e.g., compaction of the earth, impact of meteorites).

It was correctly maintained that secular cooling of the earth, with no internal generation of further heat, would not suffice to support observed geological expenditures of energy for more than a few hundred million years, at most. That was the basis for the geologically short time scale. In the present connection, it also seemed to indicate that geohistory must be directional and that its energy must have decreased rather steadily.

It now seems improbable that the earth was originally molten, and in any case that source would be entirely inadequate to balance the energy budget of geohistory on the much longer time scale now known to be more nearly correct. The source of energy missing in Thomson's calculations is now known to be radioactivity. But this, too, is a wasting resource. All energy transactions for the earth increase entropy and there is no possible way of reversing the thermodynamic flow, for the same reasons that a perpetual motion machine is impossible. Therefore Thomson was quite right in indicating an overall directionalism in earth history, although, in ignorance of radioactivity, he was wrong about the time scale. "Because of its involvement in thermodynamically irreversible processes, the earth history, despite the long time scale, can only be in the long run a unidirectional progression from some initial state characterized by a large store of available energy to a later state in which this energy has been discharged from the earth" (Hubbert, 1967). The sources of solar and tidal energy also have unidirectional progression. Thus there is (rather more than) a vestige of a beginning and there is a (clear) prospect of an end. The beginning can now be dated, as far as concerns this planet, as more or less 4½ billion (4.5×10^9) years ago, an approximation almost certainly of the right order of magnitude, at least. The date of the end cannot now be approximated, but it will surely come.

Although geological energy flow has a beginning and an end, it has been roughly constant or, at least, as previously indicated, has shown no secular trend over long periods of time. "Since the beginning of the Paleozoic era [about 6×10^8 years before present] . . . the average rate of thermally induced diastrophism and vulcanism need not have declined perceptibly," as Hubbert's (1967) conservative phrase puts it. However, this degree of configurational uniformitarianism does not preclude marked, even catastrophic, local and temporal geological changes, and furthermore this geohistorical conclusion does not necessarily apply to biohistory.

Evolutionism

An idea lately popular in some circles is that there is a grand process of evolution that is unitary through the whole sweep of things from the origins and transmutations of the elements through the histories of galaxies, stars, and planets to the origins of life and its progress to man and the other humanoids of the universe. Insofar as those things did occur, which is probable for all but not known for some of them, there is such a process. Its name is history. That it is a single process, following a predictable path, under the same principles and forces throughout is not science but inordinate popularization (e.g., Jastrow, 1967), forced analogism (e.g., Shklovskii and Sagan, 1966), or mushy mysticism (e.g., Teilhard, 1959). (The three books cited as examples—not horrible ones—are especially interesting together, as all deal in complete independence and in highly diverse ways with many of the same aspects of astrohistory, geohistory, and biohistory; they do not treat uniformitarianism in any more explicit way.) Transition was gradual and division is arbitrary, but evolution of organisms really has proceeded by quite different ways from changes in the inorganic—ways not contradictory, not even new in their basis in the immanent; just different. Use of the word "evolution" for inorganic changes tends to be misleading even if clearly defined, as it seldom is. In any modern usage, the term should imply historicity—change from one element to another in a physics laboratory is not evolution in any acceptable sense of the word.

For clarity in the present study evolution is confined to biohistory. Evolutionism refers to a historical model or theory of life as changing directionally and irreversibly in the course of descent. An alternative historical model is that of successive extinctions and creations, with the created species progressively different. It will be recalled that this was Conybeare's view, and it was widely shared in that period.

Three main steady-state models of biohistory have been proposed, two creationist and one, oddly enough, evolutionist. The only one to have wide acceptance—unfortunately it is still widely accepted—is that the present species of organisms or, in one variant, genera, were created once and for all and have not since changed significantly. The other steady-state creationist view, adopted for a time by Lyell before he became an evolutionist but never widely accepted, is that there have been repeated creations but within such repetitive taxonomic scope as essentially to maintain the status quo.

The evolutionary steady-state theory, as far as I know never adopted by anyone else, certainly not by the Neo-Lamarckians, was due to Lamarck (1809). According to him, organisms form a continuum, apart from a few nonessential perturbations (seized on by Neo-Lamarckians as the whole idea). Evolution occurs in the sense that the mass of beings constantly moves upward in the continuum from, so to speak, amoeba to man, yet essentially nothing becomes extinct and nothing new arises; amoebas and men continue as the flux passes from one to the other. Evolution simply flows like the water in a river along an established bed, with no new headwaters and no new mouth. (Simpson, 1961, 1964.)

Choice among those models has been so decisively settled as to require no discussion; not even designation is needed here. There are, however, two points special to the present topic that call for some notice: irreversibility and directionalism in evolutionism.

In biohistorical studies irreversibility usually refers to the generalization that in any given line of descent later populations do not return exactly to the condition of an ancestral population. This is commonly called "Dollo's law," after Dollo's statement of it, first in 1893. In fact, he was anticipated by Scott (1891), among others. Restated more rigorously as a generalization, more or less as in the first sentence of this paragraph, exceptions still cannot be completely ruled out *a priori* or on acceptable theoretical grounds, but they are extremely improbable and none are definitely known. In broader view, that is one side of a historical principle of the (usual) irrevocability of evolution, discussed, for example, in Simpson (1953b, especially pp. 310-312; see also Simpson, 1964, Chapter 9). The point to be made here is that when one designates evolutionism as a directional and irreversible model of biohistory, its irreversibility, although rooted in part in the phylogenetic generalization defined above, is a broader phenomenon of history in general.

In any historical model, as opposed to a steady-state model in which maintenance of or return to a given state is postulated, there is a difference between any earlier and any later state in the system as a whole. In the present application the system is that of all organic beings on the earth throughout time. In this model that system as a whole changes irreversibly through time, in addition to or regardless of what happens to any single element in it, simply because time itself is irreversible. The observational data show that this model is unquestionably the one to adopt. The known fossil record shows beyond any doubt that throughout its span, now over 3×10^9 years, change has been continuous, no fauna or flora of any one time being identical with any of a previous time. It may be questioned whether this would hold if the time intervals considered were made infinitesimally short. The answer is that by extrapolation it could be assumed to hold, but that the question has no real scientific meaning because the pertinent observations cannot be made.

A second point about evolutionism in the context of uniformitarianism follows from that just made. As regards directionalism there is a difference between biohistory and geohistory. It was previously noted that although geohistory is directional overall, there are long spans of time for which definitely directional change, as distinct from fluctuation or cycling, is not conspicuous. That is particularly true of the time from Cambrian to Recent, which is just the time for which both geohistory and biohistory are best known. During that time biohistory has been conspicuously directional in some respects. Thus we have a directional biohistory going on within a geohistorical scene that is, at most, distinctly less directional. It was this difference, hardly expectable *a priori*, which confused Lyell and long made him an antievolutionist. Again we see that evolutionism is not an outcome or an application but in this respect a contradiction of Lyellian uniformitarianism. (Many geologists now maintain that some configurational geological changes were occurring in an irreversible directional way from Cambrian to Recent, but this modification does not essentially invalidate the preceding remarks.)

For that general conclusion it does not matter whether biohistory is unidirectional or multidirectional, nor does it matter just what the direction or directions have been. Adequate discussion of those extremely complicated topics is impossible here, but brief notice is of interest as bearing on questions about directionalism as an aspect of general configurational uniformitarianism.

It must be assumed that earliest organisms were of few kinds and comparatively simple structure. In spite of the great inadequacy of the Precambrian fossil record, it does support that inference. The earliest known fossils, believed to be more than 3×10^9 years old, are indeed little varied and simple (Schopf and Barghoorn, 1967). Although biotas were not static, organisms thereafter as seen in the scant record did not become either highly varied or notably complex for a period of well over 2×10^9 years (e.g., Cloud, 1968). Marked increase in variety and organization first appears in fossils of probable but somewhat dubious late Precambrian age (e.g., Glaessner, 1962) and becomes strong and indubitable with the beginning of the Cambrian, about 6×10^8 years before present.

The number of now living species of organisms is not known, but it is probably in the millions, and they are stunning in their diversity. Some, such as ourselves, are vastly more complicated that any known or at all likely in the Precambrian. Thus we can make several definite directional statements about biohistory:

> The number of kinds of organisms has increased enormously.
> Their distinctions in structure, function, and ecology have become much greater.
> The structural complexity of some of them, or the average complexity for the whole, has strongly increased.

Those overall directional generalizations do not apply to all phases or all subdivisions of biohistory. In later phases of the history it is not clear or probable that the number of organic taxa in any one environment has increased significantly once adaptations to that environment had been well established. Increases since some time in the early Paleozoic seem to have resulted mainly from incursions into new environments (notably those of the land in the later Paleozoic) and from increased, usually local, variety in the kinds of existing environments. It follows from the second generalization above that the kinds of evolutionary changes have been far from unidirectional. They have been extremely multidirectional, and the concept of a mainstream or central line of evolutionary development is artificial. It is also noteworthy that within simple superspecific taxa reduction in number of species has been common. Indeed it is the usual outcome for taxa of high rank and ancient origin.

As regards complexity, evolution from unicellular organisms to multicellular organisms with organ differentiation certainly represents an increase in complexity. But once the latter stage was fully established, supposed further increase in complexity becomes largely a matter of definition, semantics, subjective opinion, or *ad hoc* criteria. It is questionable whether *Homo sapiens* is more complex than an Ordovician vertebrate by an objective, structural measure. In any case, increase in complexity has not been a universal characteristic of evolution. In some respects and in some taxa a definable decrease in complexity has occurred. There are also organisms now living that closely resemble the oldest known fossils in structure and that are unlikely ever to have had more complex phases in their ancestry.

It is probable that rates of evolution in the late Precambrian and early Paleozoic were more rapid, on an average, than in the early Precambrian. Since the early Paleozoic rates have varied greatly from one group to another and for particular groups at different times. Many examples both of acceleration and of deceleration are known. There is no clear indication of overall increase or decrease in average rates since the early Paleozoic. Meyer (1954), for example, has produced various graphs showing great acceleration in evolutionary rates since the early Paleozoic, but his placing of points along the time scale is arbitrary. Placing of the same data by equally or more defensible criteria eliminates regular acceleration. Meyer and others have evidently been influenced by the real acceleration of human technology (e.g., graphs in Meyer, 1954, Figs. 5, 6), but the analogy with rates of organic evolution is—obviously, one would suppose—false. Incidentally, all Meyer's graphs show rates becoming infinite at a finite time not now far in the future.

Supposed identifications of overall trends in later phases of biohistory are thus dubious, at best. Nevertheless all phases of biohistory, and perhaps especially these late phases, are directional in a more general sense. There is constant, irreversible change of biotas, which neither remain static nor fluctuate about a constant mean. (For more extended discussion and other references see Simpson, 1967.)

Gradualism

As usually conceived, the issue here is between gradualism, the doctrine that processes, originally geological processes, occur at slow rates and in small increments; and catastrophism, the doctrine that they occur at fast rates or instantaneously and in large increments. As noted in the preceding historical review, that issue was long confused by association of naturalism with the former, and preternaturalism with the latter view. Preternaturalism is now excluded from scientific consideration, but that does not settle the matter because there is no reason *a priori* why catastrophism should not be naturalistic.

Indeed the issue, if expressed in such simple and categorical terms, is ambiguous to the point of being either false or meaningless. "Slow," "fast," "small," and "large" are relative terms. How slow must a process be and how small its increments in order to be gradual rather than catastrophic? How fast and how large to be catastrophic rather than gradualistic? Moreover, it is obvious to any reasonably extensive observation that both gradual and catastrophic events do occur in nature, as was of course well known to the earliest proponents of uniformitarianism. To a present-day geohistorian or biohistorian these questions become meaningful only if they refer to specific processes or events in terms of probable ranges of rates in increments. For example, Hutton and all the uniformitarians were certainly right that most valleys have been formed by erosive action of their rivers at rates and with increments that all catastrophists would consider noncatastrophic. Nevertheless, rates and increments within any one valley differed at different times, and those of different valleys have differed still more greatly both at the same and at different times. Moreover there are valleys, such as the east African rifts, that were not primarily caused by erosion and that did sometimes have relatively large and exceedingly rapid increments, even though the rate of formation averaged over their whole history may have been moderate.

The possibility for any major developments of geohistory or biohistory to occur gradually depended on expansion of the time scale from the prescientific approximation of about 6×10^3 by biblical exegesis and even beyond the probable 1×10^8 and utmost limit of about 5×10^8 allowed by Thomson on thermodynamic estimates while physicists were still ignorant of radioactivity. That is no longer a question now that rocks have been dated with fair accuracy at about 3.5×10^9 years and reasonable evidence suggests about 4.5×10^9 years for the age of our planetary system. The possible lower limit for the rate of any historical process is very low indeed. There is no general theoretical upper limit, and investigations of rates for particular processes and events may follow the evidence without *a priori* restrictions.

(Thomson's arguments for a short time scale were previously mentioned [Thomson, 1862; Kelvin, 1899]. Recognition of a longer scale made possible by radioactivity may be dated from Rutherford [1904], although he had less definite predecessors and more definite followers. It was later found that radioactivity not only made the long time scale possible but also made its measurement possible. The substitution of the long for the short scale and the origins of radiometric dating are interestingly treated by

Badash [1968]. Results have been summarized by Kulp [1961], and various dating methods are conveniently discussed by Faul [1966]. See also Hubbert [1967].)

The present inquiry is not concerned with particular investigations on geological time except for the bearing of their accumulated results on more general questions related to uniformitarianism. In the field of geohistory, it is now clear that great, sudden, and worldwide catastrophes on the diluvialist-Neptunist-Cuvierian model have not occurred. Nevertheless it is clear that the intensities of all geological processes as seen in any one region have differed greatly at different times. Apparently or hypothetically contemporaneous maxima of such processes, especially tectonics or orogeny, within western Europe were early taken as division points in geological time. Division into Primary, Transition, Secondary, Tertiary, and Quaternary, names dating mostly from the 18th century, had that basis. A different major division into "Palæozoic, Mesozoic, and Kainozoic" (current American spellings: Paleozoic, Mesozoic, Cenozoic) was proposed in 1840 by John Phillips, who was a nephew of William Smith, "the father of geology." That marked a shift from a geohistorical to a biohistorical basis, to be discussed later. In consequence, the names Primary, Transition, and Secondary are no longer used, and Tertiary and Quaternary, although still accepted by the U. S. Geological Survey and other conservatives, are obsolescent. (For discussion of geological time classification with citations and quotations of original definitions of era, period, and epoch names see Wilmarth, 1925.)

During the 19th and the first half of the 20th centuries, a general belief developed that the major tectonic episodes first identified in Europe, along with other supposedly associated major geological and biological events, were worldwide. What were believed to be geologically brief and worldwide events of that nature came to be called revolutions. That view has been anathematized as neocatastrophism, but in fact geological revolutions have never been ascribed either the scope or the intensity of the events postulated in classical catastrophism, and of course no preternatural element was seen in them. This school or doctrine is more clearly designated as revolutionist in, of course, a geological sense. Although there were a few dissenters, it had become the consensus among geologists by 1948, when it was vigorously assailed by Gilluly (1949) in one of the extraordinarily rare presidential addresses (this one to the Geological Society of America) that was both original and influential. He demonstrated that tectonic and other events have occurred locally or regionally in a virtually random manner and that the supposed worldwide and synchronous association of the most intense of them in revolutions is largely spurious.

Revolutionism in geohistory has not been completely abandoned as a working principle or eliminated from all textbooks, but it is no longer the consensus. The term "revolution" is still in wide use, but revolutions are now usually named and understood as regional, not worldwide, episodes. Thus it is still common to refer to a Laramide revolution as occurring at the Mesozoic-Cenozoic boundary in the Rocky Mountain region of North America. However, studies made without that *parti pris* have shown that this was not a single, short tectonic episode but several over a long period of time, that their acme was not at what is accepted on other grounds as the Mesozoic-Cenozoic boundary, and that there is no reliable correlation with regional "revolutions" elsewhere (e.g., Eardley, 1951; L. S. Russell, 1951). More recently Gilluly (1967) has reviewed data for western United States from the Cambrian onward. (Precambrian data are still too scanty for significance, and really adequate data start with the Devonian.) His tabulations show that there is little evident synchronism at any time within this limited region (Montana-Wyoming-Colorado-New Mexico and hence west to the Pacific) and that there was no time when tectonic activity was not going on somewhere in that region.

For our present subject, the great interest of all this is that we seem finally to be confirming and returning to essentially the Lyellian concept of configurational (or substantive) uniformitarianism for the Paleozoic and Cenozoic, at least. (Lyell's data for western Europe covered approximately the same time range as Gilluly's for western United States.) This aspect of uniformitarianism is not a "law" or principle but is a generalization from observations. The indication is that revolutionism in its usual application to geohistory is dubious, at best.

That applies *a fortiori* to theories popular until recently and still not wholly abandoned that geologic activities have been rhythmic as a result of a cycle or of two or more superimposed cycles of constant period. A small classic by Holmes (1927) reviewed that and other theories of geologic time current rather more than a generation ago. The cyclic theory reached a culmination in a book by Umbgrove (1942; still pre-Gilluly!) appropriately titled *The Pulse of the Earth*. In essence, these are also uniformitarian dynamic theories which are steady-state rather than historical as they envision oscillation about a mean rather than directional change. This is somewhat Huttonian, but Hutton was less insistent on rhythmicity. They are even more susceptible to criticism that is revolutionism, which need not involve fixed periodism, and their postulate of such periodism rests on an even more slender basis of acceptable observation. There has been little serious discussion of a general rhythmic theory of the earth in the last quarter-century.

So much for those aspects of geohistory. Because evolution occurs in the setting of geohistory, it seems probable *a priori* that geological events have affected biohistory. That is unquestionably true as regards biogeography and many adaptive sequences. For example, the making and breaking of connections between continents and between oceans has greatly affected the distributions of organisms, with profound repercussions in the extinctions of some and progressive changes of others of the organisms involved. Adaptations to the special conditions of epicontinental seas or of cordilleras obviously have occurred when and where such seas spread and such cordilleras arose in the course of geohistory. Nothing that follows contradicts the importance of such interactions between geohistory and biohistory, but they are not in question here except as some of them bear on questions of gradualism and evolutionism. (On modern concepts and some conflicts about historical biogeography see, for example, Darlington, 1957, 1965; Simpson, 1965.)

Early in the 19th century it was already becoming evident that major changes in biotas had occurred around the times that we now designate as the Precambrian-Paleozoic, Paleozoic-Mesozoic, and Mesozoic-Cenozoic transitions. As noted above, those terms in–zoic (i.e., "pertaining to animals") followed faunal criteria, thus shifting the basis for major time divisions from geohistorical to biohistorical. There was, indeed, an assumption of synchronism between major geohistorical and biohistorical episodes as in the original historical models of the catastrophists. Diluvial catastrophes were believed to coincide with great biotic changes and to have some causal relationship to them. Strictly biogeographic and creationist interpretations of faunal change were replaced by evolutionism, and diluvial-catastrophist interpretations of geohistory were replaced largely by revolutionism. The theory of synchronism and causal relationship between major biohistorical and geohistorical events, however, continued to be supported by most geologists until relatively recently and still has advocates.

This subject required reconsideration as doubts arose about the revolutionist interpretation of geohistory and as more was learned about biohistory. Compilations of increasingly numerous and reliable data have shown that originations of new organic taxa and extinctions of old have occurred continuously throughout the time of most nearly adequate fossil record (Cambrian-Recent), at least, but that there have been peaks of maximal first and last appearances. For raw data on animals see Moore (1953 seq.), Piveteau (1952 seq.), and Romer (1966); for compilation and discussion see especially Henbest (1952a), Newell (1952, 1962, 1966), Schindewolf (1950, 1954, 1963), and Simpson (1952, 1953b, 1967). Unfortunately there are no fully comparable compilations for plants, and the detailed data are too scattered for citation here. Their general trend is indicated in Arnold (1947), Delevoryas (1962), Emberger (1944) and Mägdefrau (1967).

The sequence of terrestrial faunas across the conventionally agreed Mesozoic-Cenozoic boundary can now be more nearly followed in the

North American high plains and Rocky Mountain region than anywhere else. There is a test of the supposed synchronism of a tectonic (orogenic or diastrophic) revolution and a major faunal change. As previously noted, the supposed tectonic event, the "Laramide revolution," was not brief or clearly definable, and as far as a climax can be specified it evidently occurred millions of years after the major faunal change. The example is typical. Later study has confirmed and made more positive the moderate conclusion voiced by Henbest in 1949, in published version (1952b): "Clear, simple connections of evolutionary history with so-called diastrophic rhythms are not supported by unambiguous evidence." The symposium then led by Henbest was held at the annual meeting of the Geological Society of America following that at which Gilluly gave his epochal presidential address (published in 1949), previously mentioned. The weight of the evidence was quite clear; for example two other participants, Cooper and Williams (in Henbest, 1952a), showed "that the history of the brachiopods is characterized by evolutionary bursts and that these are distributed in time serially and without very clear relation to the geologic periods." Newell showed that among invertebrates "there is no evidence that there is increased evolutionary activity during diastrophic disturbances," and from a study of all the vertebrates I concluded that "little support is found . . . for the theory of simultaneous, worldwide physical and biological climaxes at the period and era boundaries."

Nevertheless, there have been times of particularly marked biotic change over much, or, apparently, all of the earth. Besides the three previously mentioned, one can be seen sometime in late Cambrian to early Ordovician, one in or shortly after the Devonian, one at or about the conventional Triassic-Jurassic boundary and one around the Pleistocene-Recent transition. The times are necessarily stated broadly because these are not sudden episodes pinpointed in geological time. Most of them involve high extinction rates followed by high origination rates. However, the first that can be definitely recognized (late Precambrian-early Cambrian) had high origination rates not, as far as known, preceded by high extinction rates, and the last (Pleistocene-Recent) had high extinction rates not yet, at least, followed by a clear rise in origination rates.

Schindewolf (1963) dislikes the label "neocatastrophist" and prefers to speak of these events as "anastrophes," mainly because he wants to emphasize origination (Greek *ana-*, "upward" or "anew") over extinction. Nevertheless his view (of course not preternatural) is catastrophic. He considers both extinctions and originations in these episodes as occurring suddenly, essentially synchronously over short periods of time, as a result of some factor intermittent at long intervals in geologic time and distinct from any causes of less pronounced continual extinction and origination between the major events. A suggestion as to cause, already made by Schindewolf in 1950 and recently independently advanced by others (see discussion in Simpson, 1968), is bursts of radiation from explosions of supernovae.

The observed data are quite conclusively opposed to that (or any other) form of neocatastrophism. The extinctions and, to still greater degree, the originations shown by Schindewolf (and, to be sure, many others) as if they were instantaneous and absolutely simultaneous show in actual observation as slow changes, ebbs and flows over geologically long periods of time. Just one example may be given to substantiate that criticism: Schindewolf (1963, Fig. 3) shows ten groups (all but one are orders) of mammals as appearing simultaneously exactly at the beginning of the Cenozoic. In fact the first occurrences of these mammals in the known fossil record are spread over a span of about 15 million years and only a single one (Taeniodonta) of the ten groups does appear in the record just at the conventional beginning of the Cenozoic.

The data (in works previously cited) show that extinctions and originations have been going on at all times for which the record is reasonably good and strongly suggest that that has been true ever since life originated. Peaks or accelerations and decelerations for different groups are not notably synchronized through most of the record. The multiple peaks associated in the major episodes are composed of individual peaks spread over considerable spans of time and each probably long drawn out. In short, there is a high probability that these events are gradualistic and not catastrophic in any usual sense of those words. The larger events are not rhythmic, but seem to be distributed more or less at random in geological time. It is possible, although not altogether probable, that the heaping up of rise and fall, or usually in reverse order, fall and rise, of various groups in the major episodes is largely coincidental. It is unlikely that it involves factors quite absent in intervening periods when extinction and origination were also going on. Nevertheless coincidences of accelerations and decelerations do occur, and it must be concluded that some causal factors have been more intense at certain times than at others.

Since the groups involved in the major, more or less revolutionary episodes are highly varied in structure, physiology, and ecology, it seems unlikely that the intensified factors are the same for all of them. Newell (1967) has argued for a "general explanation" of biotic revolutions, as he calls them, citing the principle of simplicity in support. It would not necessarily be contradictory but would indeed be in line with Newell's own discussion of the main proximate factors intensified in an episode of widespread extinction were different for different groups but themselves were related to epeirogenic movements, oscillations in sea level perhaps of only a few feet. Such movements could have quite different effects in different environments, for instance in broadest scale on land and in the sea, but everywhere there would be more or less radical and potentially deadly environmental changes of one sort or another. In any case the emptying of environmental niches and the appearance of new ones is in itself adequate to lead to an eventual increase in evolutionary rates, especially rates of taxonomic origination. No other factor, such as increased mutation rates from radiation or other causes, is required, and here it seems well to follow simplicity by not postulating what is not required.

There is an enormous and, on the whole, diverting literature on causes of extinction, some of it seriously argued from evidence, much more of it speculation, and no inconsiderable amount well beyond the fringe of lunacy. It need not be cited or reviewed here, where the real point is simply that a modified, relatively mild and gradualistic form of revolutionism is in accord with our present knowledge of biohistory, but that neocatastrophism is not. Probably this provisionally accepted model should be labeled neorevolutionism or biorevolutionism, because besides being gradualistic it negates or, at most, regards as nonproven the association of biohistorical with geohistorical revolutions.

Another aspect of gradualism has become involved in theories of evolutionary processes, as distinct from their results in the course of biohistory. In this field Darwin was a gradualist. He was aware of mutations with clearly discontinuous phenotypic effects, "sports" in the terms of his day, but he believed that evolution proceeded in the main by natural selection acting on "slight modifications," especially as more complex structures and characters are involved. "If it could be demonstrated that any complex organ existed, which could not possibly have been formed by numerous, successive, slight modifications, my theory would absolutely break down" (Darwin, 1959 [1859], Peckham reference VI 137). Terms such as "transitional gradations," "insensible steps," "finely graduated steps," and the like appear repeatedly in *The Origin of Species*. In the early 20th century pioneer geneticists challenged that view, believing that genetic variation is inherently discontinuous, that mutations usually have marked somatic effects, and that selection did not act by successive accumulation of slight variations but only by eliminating certain mutants.

Schindewolf, a paleontologist, developed on that basis a theory that organic taxa, at all levels of the taxonomic hierarchy, appear abruptly as such by single mutations, the higher or lower taxonomic levels being determined by the greater or lesser somatic effects of each mutation (Schindewolf, 1936, 1950a). This is related to Schindewolf's version of catastrophism, previously discussed, as it would make plausible the sudden rise of new high level taxa as part of the postulated catastrophes or, as Schindewolf would put it in this connection, anastrophes. Schindewolf's original evolutionary theory was a brilliant but, as it turns out, premature effort at synthesis of genetics and paleontology. In earlier studies I devoted some attention to its refutation (Simpson, 1944, 1953b). As far as I know, the Schindewolf theory is not now supported by any other paleontologist or biologist, and it need not be further discussed here. (It has more recently been supported by a philosopher, Grene, 1958, not on evidence or logic but because it better fitted a philosophical preconception; her views were sufficiently refuted by two biologists, Bock and von Wahlert, 1963, and also need no discussion here.)

It is now known that mutations, broadly speaking, an ultimate basis of variation, are indeed discontinuous. That requires some modification of Darwinian gradualism, but not abandonment of its essentials. The somatic effects of mutations vary from great to barely preceptible or, quite likely, to imperceptible by usual methods of observation. The probabilities that a mutation will survive or eventually spread in the course of evolution tend to vary inversely with the extent of its somatic effects. Most mutations with large effects are lethal at an early stage for the individual in which they occur and hence have zero probability of spreading. Mutations with small effects do have some probability of spreading and as a rule the chances are better the smaller the effect. Thus the usual, although not quite the only, materials for evolution are indeed "slight modifications" at the somatic level, as Darwin's acumen perceived. Moreover, despite the fact that a mutation is a discrete, discontinuous event at the cellular, chromosome, or gene level, its effects are modified by interactions in the whole genetic system of an individual (oddly enough, there is no generally accepted term for that important concept). They are also modified by varying environmental factors. The results are that for many mutations, the somatic effects in different individuals vary in an essentially continuous manner. Even an expression that is a marked modification in some individuals may be only the extreme of what is a gradual sequence in the population.

There is, finally, now another sense in which evolution is known to be usually gradualistic. The instantaneous origin of a new species by a single genetic event can occur but is unusual. It is practically confined to cases of increase in individual chromosome numbers happening to produce a system both viable and capable of reproduction but not capable of backbreeding into the parental population. In usual, or one might even say "normal," cases distinct evolutionary change involves the increase or decrease of proportions of genetic factors in whole populations, and that is a gradual process occurring in successions of generations. The prevailing modern theories of evolution are essentially, although not dogmatically, gradualistic. Stepanov (1959) as plaintively cited by Schindewolf (1963) has claimed that the evolutionary theory supported, in Stepanov's exemplification, by Newell, or by me is anti-Darwinian because it is not gradualist. That is an absurd misinterpretation of our position and that of the majority of modern evolutionists.

(For entrance into the vast literature on topics quickly skimmed here, see J. D. Watson, 1965, and Beadle and Beadle, 1966, at the molecular and cellular level; Dobzhansky, 1951, 1962, at the individual and populational level; and Dobzhansky, 1962, Grant, 1963, Mayr, 1963, and Simpson, 1964, 1967, at the level of general evolutionary theory.)

Historical Inferences

The present is the key to the past in more senses than one. What we know (or theorize) about the immanent characteristics of the universe is derived from observation of the present. If actualism is accepted as a working principle, it is then assumed that throughout time those characteristics have been the same as now, no more and no less. We also observe present configurations and from them infer configurations that preceded them. The principle of actualism is essential for such inferences, and that is its main interest. This, like other aspects of uniformitarianism, arose from the desire to read the history of the earth, and it has become an essential part of the methodology of all historical science.

Actualism is necessary for successful research into history, but it is rarely sufficient. Also involved are generalizations that are about configurations rather than immanence. If a tree is observed, it is a reasonable historical inference that it grew from a seed. That is an application of a generalization probable for the given case even though many exceptions are possible. (For example, tree ferns do not have seeds, and many trees can be grown from cuttings.) It is not a law of nature or an immanent feature of the universe that trees develop from seeds. Therefore this is not strictly an application of actualism, as previously defined, but it is an example of a necessary method in historical research. The method is similar when observations of erosional initiation of gullies and deepening of valleys are related to the classical problem of the causes of valleys.

In practice this aspect of uniformitarianism might be considered the most important of all because it is a major, while not the only, reliance for reconstruction of geohistory and biohistory. It might be called *procedural uniformitarianism*. It was in part this that Gould (1965) evidently had in mind when he applied the term methodological uniformitarianism to what, following Hooykaas, is here called actualism. However, actualism is a principle; its application, which is a method, is distinct from the principle itself. Further, the methods involved in historical inference are multiple, and what is here called procedural uniformitarianism is not altogether a simple application of actualism.

It has been claimed that growth of a tree or of a valley could in principle be wholly reduced to a microphenomenal basis and hence further to a strict interaction of immanence and configuration. In fact this has not been done at all in the example of a tree and only partially in the example of a valley. From the point of view of the historian (in any field of history) reduction to microphenomena and immanence would certainly be interesting and might be useful, but in the present state of the historical sciences that is never fully possible and it is rarely necessary. Reconstruction of the macrophenomena of geohistory and biohistory can be not only most practically but also most satisfactorily done by macromethods and macroprinciples. Explanation in this field is also fully practical only at the macrolevel. It is doubtful whether complete reduction to the microsciences of atomic physics and molecular chemisty is possible even in principle. (On the latter point, see for example Simpson, 1964, chapter 6; at a C.U.E.B.S. conference in June, 1968, Michael Scriven gave a paper, afterward distributed in manuscript form but not yet published, expounding the distinction between macro- and microsciences and the justification and need for macromethodology, macrologic, and macrophilosophy.)

Much of procedural uniformitarianism in geohistory and biohistory could be called detectival or of the "trout-in-the-milk school." A trout in that situation implies dilution; a tree implies reproduction; and a valley implies erosion. It is mostly this sort of interpretive procedure that may be called inductive, because it depends on generalization from a limited number of prior observations. As previously noted, the role of induction in science has been sharply questioned. That point need not be further discussed here, but it is noteworthy that induction in a strict and usual sense, or what is sometimes called enumerative induction, does not suffice in the examples given and seldom suffices for historical inference. Strict induction involves subjective formation of a class of observations and inference that some additional thing or process will have the characteristics of that class. That is involved in such examples as valley erosion, but there is also involved a gross extrapolation from the small to the large and the short to the long, from gullying in brief to valley formation in millennia or more. Inference from valley glaciation to an ice age or from annual changes in chromosome frequencies of *Drosophila* to the evolution of vertebrates involves so much extrapolation and so little strict induction that it can hardly be said to exemplify the latter.

Another aspect of historical inference is documentation. A document in itself is not necessarily historical. A fossil or a rock out of context is a document subject to description, analysis, and classification. It becomes historical only when fitted into a temporal sequence. The usual evidence consists of configurations formed in sequence and still, in the present that is here again the key to the past, retaining sequential relationships. In both geohistory and biohistory the classic and still the dominant method involves the principle of superposition, with which this essay began. It is supplemented in many ways, especially by the biohistorical principle that similar biotas are usually approximately contemporaneous. It does not follow that dissimilar biotas are of different ages. However, once different biotas have been established as sequential, their usefulness for geohistory and biohistory is supreme. Sequential evidence relevant to previous examples includes rings in trees and terraces in river valleys. There are many other kinds of sequential evidence (see, for example, Zeuner, 1950).

Other methods are not sequential in the same sense but depend on changes in configuration that are correlated with lapse of time. They thus afford, within widely variant limits of confidence, estimates as to when the changes began and thus dates to which historical configurations may be tied. Their accuracy depends, among other things, on how well their rates can be determined at the present time and on the relative constancy of that rate in the past. As regards configurational changes, we have seen that uniformity of the latter kind is often quite unreliable. Such early attempts as estimation of the age of the oceans by their saltiness gave grossly incorrect results, largely because determination of the present rate of increase was inaccurate and postulation of past constancy was false. At present, radiometric methods are incomparably the most important (reviewed in Faul, 1966). This is more directly and immediately an application of actualism than the historical methods previously exemplified, because it depends more directly and immediately on the immanent characteristics of matter, as seen in radioactive elements, and on postulation that those characteristics have been the same throughout measurable time.

Another extremely important historical method, also embodying an element of uniformity, is the comparative method. This depends on the fact that present configurations involving similar processes may represent different historical stages in the operations of those processes. That results when the pertinent processes began earlier in one case than in another, or when they acted more rapidly in one than in another, or both. A rill in a barnyard and the Grand Canyon represent, in the main, stages of valley erosion that began some millions of years apart. They can be connected because every intermediate configuration is now observable and can be interpreted as involving (among other factors for which allowance can be made) different periods of time since erosion began. This comparative method validates the previously mentioned extrapolation from the small amount of erosion now observable in action to the vast unobservable amount in the past of the Grand Canyon.

The comparative method is equally or even more essential in biohistory. For example, the reconstruction of phylogeny, an important although far from the only aspect of biohistory, necessarily brings in comparison of recent organisms. Different characteristics of organisms often evolve at different rates. If in one organism a given character has evolved more slowly than in a related organism, that particular character will be more primitive, that is, nearer the ancestral historical antecedent, in the former. It rarely happens that one organism is more primitive than a reasonably close relative in *all* respects. It also rarely happens that any organism has a character (above the molecular level, at least) *exactly* as in a remote ancestor. An example, all the plainer for being somewhat gross, is that human molar teeth have evolved more slowly and are therefore much more primitive than horse molar teeth, while human brains have evolved more rapidly and are less primitive than horse brains. But no molars just like human molars ever occurred in the horse's ancestry, and no brains just like horse brains in human ancestry. Moreover, in each case, teeth and brains are unlike any markedly earlier stage in their own ancestry. These problems by no means vitiate the method.

The comparative method requires the uniformitarian assumption that processes themselves have been the same in past as in present even if their rates have been different. Historical inference depends less on projection into the past of the immanent, construed in a static sense, than on projection of processes, which of course do depend on immanent characteristics. For the most part, these processes are recognized and characterized as they occur in the present. The record of the past and comparative results at present can then be interpreted as involving the same processes, on the uniformitarian principle that the processes have indeed been the same. That method also involves a confrontation of the record with knowledge of present processes. If known processes are evidently insufficient to explain the record, a minimal inference is that knowledge of present processes is incomplete. A maximal inference, which up to now, at least, has not proved to be necessary, would be that the uniformitarian principle is here incorrect, that there have been past processes not now operative. If the record should prove to be inconsistent with some supposed present processes, that, too, could be interpreted as indicating nonuniformity in the sense of being contrary to actualism. However, that conclusion is not necessary or acceptable in any instances known to me. In all such cases it turns out that one has a choice between alternative hypotheses or theories about processes. The uniformitarian viewpoint is then to reject the reality of supposed present processes that are inconsistent with the historical record.

In the geological field, this confrontation of record and present processes was initiated, in the main, by Hutton and fully developed by Lyell. That was their really essential accomplishment, and it made possible an effective, coherent science of geohistory. In the field of biohistory the confrontation was primarily performed by Darwin. Knowledge of present processes has considerably increased since Darwin, and later students have continued the confrontation that he began. One of my own earlier studies (Simpson,

1944) was an attempt to demonstrate that knowledge of the biohistorical record, vastly increased since Darwin, is consistent with one body of evolutionary theory (the synthetic theory), deriving in considerable part from Darwin but also with greatly increased knowledge of present processes. It was also maintained that the record is inconsistent with different, rival theories, such as those of Goldschmidt (1940) or of Schindewolf (1936).

In the total study of evolution, or indeed of any history, there are three phases: (1) obtaining and studying the historical data, for biohistory especially but not exclusively the fossil record; (2) determination of present processes, for biohistory especially but not exclusively those of population genetics; and (3) confrontation of (1) and (2) with a view to ordering, filling in, and explaining the history. Bock and von Wahlert (1963) have made almost the same distinction in somewhat different words. They emphasized the fact that (2) their (*b*) ("The study of the mechanisms of evolutionary modifications in organisms"), is definitely nonhistorical. It is a study of *present* ("actual") processes. That, more than incidentally, shows that the philosopher Marjorie Grene (1958) was quite wrong when she insisted that the study of evolutionay *theory* (processes) must be by historical methods. The main point in my present context is that (3) the explanation of evolution by confrontation of historical description and actual processes, is uniformitarian. It involves the principle that actual (present) processes are applicable and sufficient for interpretation and explanation of biohistory.

The element of explanation brings in another series of problems, both procedural and philosophical, lately much discussed by biologists and philosophers of science. Anything approaching adequate discussion of them cannot be included here and eventually would lead away from the topic of uniformitarianism. It is nevertheless necessary to indicate the existence of an issue and to refer to some treatments of it. Claims that the hypotheticodeductive method (e.g., Popper, 1959) is the only one allowable in science (e.g., Medawar, 1967) are almost absurdly extreme, but it obviously is *an* allowable method. Deduction from a hypothesis yields predictions, and the fulfillment of those predictions increases confidence in the hypothesis. If all evident predictions are fulfilled, the hypothesis becomes an accepted theory. That is a usual, perhaps for science as a whole the most usual, method of theory formation. In a paper now famous, Hempel and Oppenheim (1953) maintained that there is no difference in principle between prediction from (in layman's language) cause to effect, and explanation of the effect by retrodiction (or postdiction) from it to its cause.

Application of that supposed equality or temporal symmetry to nonhistorical procedures, for example in physical and chemical experimentation, may not involve serious difficulties. It breaks down badly, however, in historical application. By retrodiction, and especially by the method of uniformitarian confrontation specified above, it is possible to explain past events or present configurations that cannot be predicted. There is thus an evident asymmetry between prediction and *historical* retrodiction. That is certainly, indeed obviously, true in practice. Some philosophers (e.g., Scriven, 1959) and biologists (e.g., Mayr, 1961) maintain that it is also true in principle. On the other hand, some philosophers maintain that "Hempel-explanation" is applicable to biohistory, either in spite of temporal asymmetry (Grünbaum, 1963a, 1963b; see also rebuttal by Scriven, 1963) or by applying additional or "hedging" implications to the straight Hempel model (Beckner, 1967). "Hempel-explanation" seems to me quite inacceptable in the practice of historical science, and it seems highly improbable that prediction and retrodiction are really equivalent even in principle. However, neither point of view necessarily involves contradiction of any aspect of uniformitarianism. It could be maintained that the view I consider *less* probable is *more* uniformitarian, but the point need not be pursued further here. (There is some further discussion of it in Simpson, 1964, Chapter 7.)

Throughout the present essay emphasis has been on the historical aspect of almost all the acceptable principles that have at various times gone under the loose designation of uniformitarianism. In closing, brief attention may therefore be given to the opinion that there is no such thing as a historical science or, at least, that there is no difference in principle between so-called historical and nonhistorical aspects of science.

An extreme example is provided by R. A. Watson (1966). He first argues that *only* particular events occur, with the implication that there is no difference in principle between formation of a particular mountain range (usually considered historical) and a particular performance of a chemical experiment (the *reaction,* not the performance of it, usually considered nonhistorical). However, particular events are (Watson says) trivial. The real stuff of science is said to be "types" of events, "abstractions" or "inventions," like the laws of geological tectonics or of chemical valence. But, having said that all real events are particular, that is, unique and nonrecurrent, Watson goes on to claim that there is no such thing as a "one-of-a-kind" event that cannot "recur." Using the Grand Canyon and *Homo sapiens* as a geological and a biological example, his reasons for claiming that they are not one-of-a-kind and can recur are that for all we know canyons just like the Grand Canyon and men just like us may be present elsewhere in the universe, and even if they do not exist Watson can *imagine* them in his mind! His general conclusion is that history is not science.

It is somewhat surprising that one thoroughly competent geologist, Siever (1968), has agreed with Watson to the extent of saying ". . . We do not really care ... how the Grand Canyon of the Colorado River was formed. We only care how the generic class of Grand Canyons forms and has formed in the past, assuming that canyon-cutting was not a unique event." Here Siever, one might say "inadvertently," has put his finger right on a major distinction between historical and nonhistorical science. The historical scientist *does* care how the Grand Canyon of the Colorado River was formed. He takes that particular canyon's present configuration as a historical document; he also takes the nonhistorical knowledge of erosional processes, in which he and Siever are both interested; and then by confronting the two on uniformitarian principles he reaches a historical interpretation and explanation of the unique phenomenon. ("Unique," indeed, because no science is properly concerned with what is not known to exist, and only psychology is concerned with what may be imagined in Watson's mind.)

Along similar lines, extreme reductionists have denied the validity of historical explanation. Schaffner (1967), for instance, has maintained that only nonhistorical physicochemical laws are explanatory even in reference to unique phenomena such as constitute historical sequences. One of his examples is that "The Empire State Building is unique, nevertheless one would not expect that the laws of stresses and strains would not apply because of this uniqueness, and that the structure of the building would not be explicable on the basis of the principles of mechanics." Of course the "laws of stresses and strains" apply to this building, as they apply to every material object in relevant circumstances. Nevertheless it is surely obvious that those laws do not explain the actual structure of the building, the fact that it exists at all, and its uniqueness. Those are the interests of the historical scientist, and the needed explanaion is historical. Incidentally, in this example the explanation is not mechanical, as Schaffner assumes, but strictly biological. Historical causation of the Empire State Building is by actions of a species of animals.

In a footnote to the same study Schaffner (1967) tries to sweep the historical issue under the rug by saying "In neither the mechanical nor the biological case are we now concerned with the way in which the structure was formed; we only want to know why it is functioning as it does (the Empire State Building stands, the organism lives)." But historical principles and historical explanations are no less valid, necessary, and scientific just because Watson, Siever, and Schaffner are not interested in them.

Summary and Conclusion

Although Hutton is now usually considered the founder of uniformitarianism, he was not consistently uniformitarian and that was not the main issue in his theoretical discussions. Nevertheless he did banish preternaturalism from scientific inference, and he did show that the surface and crust of the earth can be treated as historical documents and interpreted on the principle that past geological processes were like those now observable. Those aspects of his work made the development of a valid natural science of geology possible. That development was carried out in great part by Lyell, who also made the first really full and explicit statement of uniformitarianism. In the most general terms, Lyell's proposition was that "the former changes of the earth's surface" can be explained "by reference to causes now in operation." That generalization has a number of facets and indeed involves several quite different principles still not always clearly evaluated and distinguished.

Lyell maintained not only that geological processes were the same in the past as at present but also that their average intensity has tended to remain constant. His was essentially a steady-state model of the earth, and his concepts of both geohistory and biohistory were at first anomalously nonhistorical. His opponents, among whom Conybeare is here taken as an example, proposed a historical model involving unidirectional change with continual decrease in the intensity of geological process. It was largely the latter point, which was in fact wrong, that Lyell was seeking to correct. However, in other respects the historical model of the catastrophists and some other nonuniformitarians was more nearly correct, and it was the background for the eventual acceptance of organic evolution. Although Darwin as a geologist was deeply indebted to Lyell, his evolutionary view of biohistory was a flat contradiction and not, as often stated, an outcome or application of Lyellian uniformitarianism.

As Hooykaas, Visotskii, Gould, and others have recently pointed out, much of the original and continuing confusion about uniformitarianism has been caused by failure to distinguish its two main aspects or divisions. One has to do with the immanent characteristics of the universe and the other with the configurations of matter and energy through time. Most important in the first category is actualism, the designation here accepted for the principle or postulate that the immanent characteristics of the universe are constant throughout the four dimensions of space and time. There is no apparent way to prove this proposition in a literal, absolute sense, but there is a great deal of observational evidence relevant to it, and all such evidence agrees with it. It is not correct to consider actualism as equally involved in all sciences, and it is not reducible to the method of induction and the principle of simplicity. Both the evidence for it and the applications of it are almost confined to the related and intergrading sciences of astrohistory, geohistory, and biohistory. The related principle or postulate that is relevant to all sciences is simply that at any one time (or specifically at the present time) the characteristics of the universe constitute a single, consistent system; that, in popular terms, the universe makes sense. Perhaps even more pervasive is the principle that scientific explanation must be naturalistic.

As regards configurations, the historical model that is essentially non-uniformitarian in origin is acceptable in one of its major points. The earth, moon, and sun constitute an essentially closed thermodynamic system, other sources or stores of energy being negligible as far as the earth is concerned. Configurational change follows a one-way course with increase of entropy from an energy-rich beginning toward an eventual end without available geological energy. Nevertheless Lyellian uniformitarianism was right in that unidirectional secular change in the intensity of geological processes is not evident in the course of recorded geohistory.

In biohistory a full steady-state model is completely untenable, as Lyell himself came to realize. Evolution is irreversible not only in the structural sense usually (but equivocally) ascribed to Dollo but also and especially in an overall historical sense. From the origin of life to the present day there has been great increase in the numbers and kinds of organisms and in various rates of evolution. Nevertheless these changes have not been even approximately constant or invariably in the same direction. Evolutionary acceleration continuing up to now is not shown by the record. Biomass and multiplicity of organisms in any one environment have probably tended to remain approximately constant once that environment was fully occupied. The directions of evolution have always been extremely multiple, without true central tendency toward unidirectionalism.

The old conflict between gradualism and catastrophism has taken a different form. Preternatural, worldwide catastrophes on the diluvial pattern have not occurred, but local catastrophes with actualistic and naturalistic causes have. Intensities and rates of processes have also varied greatly from time to time and place to place in both geohistory and biohistory. In geohistory a naturalistic doctrine of revolutionism involves belief that intensified, worldwide (but not ubiquitous in local detail), and relatively short episodes of tectonic activity have occurred. There now is, however, reason to think that supposed orogenic revolutions have been more long-continued, more multiple, and more local than previously supposed. In this respect, some reversion toward a Lyellian concept of an approximate statistical steady state may be justified.

In biohistory, each sufficiently recorded and studied major taxon has had times of accelerated evolution and times (not necessarily the same as the former) of greater diversity. Such times may be multiple within one taxon, and they are generally different for different taxa. Occasionally they do coincide for a number of large taxa and then there is a major episode of extinction, turnover, and proliferation, in that order. Notable well-documented examples are found in the Permian-Triassic and Cretaceous-Paleocene transitions. Such episodes may indeed be coincidental, but they probably represent intensification and broadening of some factors, especially those related to extinction, operative less strongly and less widely at other times. They probably do not represent factors operative only at these times. These episodes do not tend to be correlated with, and so are not caused by, the orogenic maxima of supposed geohistorical revolutions.

Darwinian gradualism in organic evolution is now well established as a general rule. There are minor exceptions among low toxonomic categories. Gradualism in evolution at the macrophenomenal level of populations is consistent with discontinuity in microphenomena within single organisms, that is, with the discrete nature of genes and mutations.

Actualism is an essential basis for historical inference, and that is the main reason for its interest and its acceptance. It is not, however, an adequate statement that uniformitarian method is simply the application of actualism to history. Procedural uniformitarianism requires that application and is not in any respect independent of it, but it does have other aspects. In its detectival aspect it is partly inductive but more extrapolative, as it reasons from the small and brief to the large and long. It depends in considerable part on documentation, the observation of present configurations that can be put into a temporal framework. Sequential documents, like tree rings or river terraces, are present configurations retaining features that originated sequentially in the past. The most important sequential bases for both geohistory and biohistory are, first, the principle of superposition of sediments, and then the biotic sequence worked out on those grounds. Other important documents are those that, instead of retaining a past configuration, change configuration at more or less constant rates. The elements used in radiometric dating are most important in this respect.

The comparative method of historical inference depends on comparison of related present configurations that have been more or less affected by uniform processes. That may be because the processes began to operate on the different configurations at different times in the past, or acted at different rates, or both. That is an essential method in itself and also useful as a validation of previously mentioned extrapolation.

Interpretation of the past involves confrontation of its record and of comparisons of its present results with knowledge of relevant processes. The general procedure of historical research has three phases: (1) obtaining and ordering historical data; (2) determining present processes; (3) confronting (1) and (2). The result is largely retrodictive. It involves one kind of explanation of past and also of present configurations. This is not the same as hypotheticodeductive explanation or "Hempel-explanation," and it is not symmetrical with prediction. Retrodictive interpretation and explanation are almost unique to the historical sciences, astrohistory, geohistory, and biohistory, and are not characteristic of science in general.

They require application of uniformitarian principles in a broad, collective sense. Some of those principles, notably actualism, must be postulated as absolute. Others, such as gradualism, are relative and applicable only as warrented by the given data. The results often remain uncertain and subject to correction, sometimes but not always more so than in the so-called exact sciences, which also do remain subject to change even in their "exact" results. The difference in depth between a rill and a canyon can be objectively measured, and an n–dimensional coefficient of likeness between the teeth of men and horses has no uniformitarian presuppositions. But such nontheoretical quantifications of the obvious are trivial, indeed really meaningless, unless put into a historical framework on grounds largely uniformitarian.

References

ABEL, O. 1911. Grundzüge der Palaeobiologie der Wirbeltiere. Stuttgart, E. Schweizerbart'sche Verlagsbuchhandlung (Erwin Nägele).
ADAMS, F. D. 1938. The Birth and Development of the Geological Sciences. Baltimore, The Williams & Wilkins Co.
ALBRITTION, C. C., JR., ed. 1963. The Fabric of Geology. Reading, Addison-Wesley Publishing Co., Inc. [A useful annotated and indexed bibliography by the editor includes most of the basic references on uniformitarianism, and the chapters by Kitts, McIntyre, and Simpson, here cited under author's names, bear on the subject.]
——— 1967. Uniformity and Simplicity. Geology Society of America, Special Paper No. 89. [Chapters by Hubbert, Wilson, Newell, and Goodman are here separately cited.]
ARNOLD, C. A. 1947. An Introduction to Paleobotany. New York, McGraw-Hill Book Company.
BADASH, L. 1968. Rutherford, Boltwood, and the age of the earth: the origin of radioactive dating techniques. Proc. Amer. Philos. Soc., 112: 157-169.
BAILEY, E. E. 1967. James Hutton—the Founder of Modern Geology. New York, Elsevier. [A convenient summary of Hutton's "Theory of the Earth," with useful but not wholly satisfactory commentary.]
BEADLE, G., and M. BEADLE. 1966. The language of life. Garden City, New York, Doubleday & Company, Inc.
BECK, L. W. 1953. Constructions and inferred entities. *In* Feigl and Brodbeck, 1953, pp. 368-381. [Originally published in 1950; consulted by me in the 1953 reprint.]
BECKNER, M. 1967. Aspects of explanation in biological theory. *In* Morgenbesser, 1967, pp. 148-159.
BLACK, M. 1967. The justification of induction. *In* Morgenbesser, 1967, pp. 190-200.
BOCK, W. J., and G. VON WAHLERT. 1963. Two evolutionary theories—a discussion. Brit. J. Philos. Sci., 14: 140-46. [A refutation of Grene, 1958.]
BRADLEY, W. H. 1963. Geologic laws. *In* Albritton, 1963, pp. 12-23.
BROUWER, A. 1959. Algemene Palaeontologie. Zeist, de Haan. [In Dutch; unrevised English translation by R. H. Kaye, London, Oliver and Boyd, 1966, and Chicago, University of Chicago Press, 1967.]
CANNON, W. F. 1960. The uniformitarian-catastrophist debate. Isis, 51: 38-55.
CLOUD, P. E., JR. 1968. Atmospheric and hydrospheric evolution on the primitive earth. Science, 160: 729-736. [Also discusses Precambrian biohistory.]

CONYBEARE, W. D. 1841. [See Rudwick, 1967.]
CUVIER, L. C. F. D. G. 1812. Recherches sur les Ossemens Fossiles de Quadrupèdes. Tome 1. Discours Préliminaire. Discours sur les Révolutions de la Surface du Globe. Paris, Deterville.
DARLINGTON, P. J., JR. 1957. Zoogeography: the Geographical Distribution of Animals. New York, John Wiley & Sons, Inc.
────── 1965. Biogeography of the Southern End of the World. Cambridge, Mass., Harvard University Press.
DARWIN, C. 1958. The Autobiography of Charles Darwin 1809-1882 with Original Omissions Restored. Edited with Appendix and Notes by his Grand-Daughter Nora Barlow. London, Collins. [The only complete edition of the autobiography written in 1876 and first published in extensively expurgated form in 1887.]
────── 1959. The Origin of Species by Charles Darwin. A Variorum Text Edited by Morse Peckham. Philadelphia, University of Pennsylvania Press. [Collates the original editions published from 1859 to 1878; Peckham supplies a system of reference to passages and editions that I have used here.]
DELEVORYAS, TH. 1962. Morphology and Evolution of Fossil Plants. New York, Holt, Rinehart & Winston, Inc.
DOBZHANSKY, TH. 1951. Genetics and the Origin of Species. 3d ed. New York, Columbia University Press.
────── 1962. Mankind Evolving. New Haven, Yale University Press.
DOLLO, L. 1893. Les lois de l'évolution. Bull. Soc. Belge Géol., 7: 164-166.
DUNBAR, C. O., and J. RODGERS. 1957. Principles of Stratigraphy. New York, John Wiley & Sons, Inc.
EARDLEY, A. J. 1951. Structural Geology of North America. New York, Harper & Row, Publishers.
EMBERGER, L. 1944. Les Plantes Fossiles dans leurs rapports avec les Végétaux Vivants. Paris, Masson.
FARRAND, W. R. 1961. Frozen mammoths and modern geology. Science, 133: 729-735.
────── 1962. Frozen mammoths. Science, 137: 450-452. [A reply to Lippman, 1962.]
FAUL, H. 1966. Ages of Rocks, Planets, and Stars. New York, McGraw-Hill Book Company.
FEIGL, H., and M. BRODBECK, eds. 1953. Readings in the Philosophy of Science. New York, Appleton-Century-Crofts. [Authors cited from this collection are listed separately in this bibliography.]
GEIKIE, A. 1897. The Founders of Geology. London, Macmillan & Co. Ltd. [With a somewhat unsympathetic view of Lyell's uniformitarianism.]
GILLISPIE, C. C. 1951. Genesis and Geology. Cambridge, Mass., Harvard University Press. [A historical account of the period 1790-1850; see especially Chapter V, "The uniformity of nature."]
GILLULY, J. 1949. The distribution of mountain-building in geologic time. Bull. Geol. Soc. Amer., 60: 561-590. [This now classic presidential address is the anti-revolutionist manifesto.]
────── 1967. Chronology of tectonic movements in the western United States. Amer. J. Sci., 265: 306-331. [A richly documented detailing of data for a single area relevant to the general thesis of Gilluly, 1949]
GLAESSNER, M. F. 1962. Pre-Cambrian fossils. Biol. Rev. 37: 467-494.
GOLDSCHMIDT, R. 1940. The Material Basis of Evolution. New Haven, Yale University Press.
GOODMAN, N. 1967a. Uniformity and simplicity. In Albritton, 1967, pp. 93-99.
────── 1967b. Science and simplicity. In Morgenbesser, 1967, pp. 68-78.
GOULD, S. J. 1965. Is uniformitarianism necessary? Amer. J. Sci., 263: 223-228.
GRANT, V. 1963. The Origin of Adaptations. New York, Columbia University Press.
GRENE, M. 1958. Two evolutionary theories. Brit. J. Philos. Sci., 9: 110-127, 185-193. [The theories are Schindewolf's and the synthetic theory as expounded by Simpson.]
GRÜNBAUM, A. 1963a. Temporally asymmetric principles, parity between explanation

and prediction, and mechanism versus teleology. *In* Induction: Some Current Issues, Middletown (New York), Wesleyan University Press, Chap. VI, pp. 114-149.
———— 1963b. Philosophical Problems of Space and Time. New York, Alfred A. Knopf, Inc.
HAPGOOD, C. H., and J. H. Campbell. 1958. Earth's Shifting Crust. New York, Pantheon Books, Inc. [Mr. Hapgood's unusual catastrophic hypothesis was also published in the Saturday Evening Post for 10 January 1959 and was endorsed by Ivan Sanderson in the same journal on 19 January 1960.]
HEMPEL, C. G., and P. OPPENHEIM. 1953. *In* Feigl and Brodbeck, 1953, pp. 319-352. [Originally published in 1948; consulted by me in the 1953 reprint.]
HENBEST, L. G., ed. 1952a. Distribution of evolutionary explosions in geologic time J. Paleont., 26: 298-394. [A symposium of seven papers with extended discussion; contributions by Henbest, Newell, and Simpson are here cited separately.]
———— 1952b. Significance of evolutionary explosions for diastrophic division of earth history. *In* Henbest, 1952a, pp. 299-318.
HERSCHEL, J. F. W. 1841. Whewell on inductive sciences. Quart. Rev. (London), 68: 177-238. [Ridicules the concept of configurational uniformitarianism.]
HOLMES, A. 1927. The Age of the Earth. Benn's Sixpenny Library, No. 102. London, Benn.
HOOYKAAS, R. 1956. The principle of uniformity in geology, biology, and theology. J. Trans. Victoria Inst., 88: 101-116.
———— 1959. Natural Law and Divine Miracle. Leiden, Brill. [Essentially an expansion of Hooykaas, 1956.]
———— 1963. The Principle of Uniformity. Leiden, Brill [Reissue and renaming of Hooykaas, 1959.]
HUBBERT, M. K. 1967. Critique of the principle of uniformity. *In* Albritton, 1967, pp. 3-33.
HUTTON, J. 1788. Theory of the earth. Trans. Roy. Soc. Edinburgh, 1: 209-304. [The first generally available form of Hutton's famous work, although it is now known that an abstract had been printed in 1785; the 1788 paper became the first part of the following edition.]
———— 1795. Theory of the Earth. 2 vols. Edinburgh, William Creech. [A facsimile was published in 1959 by Hafner Publishing Co., Inc., New York. A third volume was written but was discovered only in incomplete manuscript form and much later; it was published in 1899 by the Geological Society of London; it is summarized in Bailey, 1967.]
JASTROW, R. 1967. Red Giants and White Dwarfs. New York, Harper & Row, Publishers.
KELVIN, LORD [William Thomson]. 1899. The age of the earth as an abode fitted for life. Science, n.s., 9: 665-674, 704-711. [See also Thomson, 1862.]
KITTS, D. B. 1963a. The theory of geology. *In* Albritton, 1963, pp. 49-68. [Uniformitarianism is discussed on pp. 62-67.]
———— 1963b. Historical explanation in geology. J. Geol., 71: 297-313.
KULP, J. L. 1961. Geologic time scale. Science, 133: 1105-1114.
LADD, H. S., ed. 1957. Treatise on marine ecology and paleoecology. Vol 2. Paleoecology. Geology Society of America, Memoir 67, 2: i-x, 1-1077.
LAMARCK, J. B. M. de. 1809. Philosophie Zoologique. Paris, Dentu.
LIPPMAN, H. E. 1962. Frozen mammoths. Science, 137: 449-450. [An attack on Farrand, 1961.]
LOVEJOY, A. O. 1936. The Great Chain of Being. Cambridge, Mass., Harvard University Press.
LYELL, C. 1830-1833. Principles of Geology. 3 vol's. London, Murray.
MÄGDEFRAU, K. 1967. Die Geschichte der Pflanzen. *In* Heberer, G., ed. Die Evolution der Organismen. 3rd ed. Stuttgart, Gustav Fischer. Vol. 1, pp. 551-588.
MAYR, E. 1961. Cause and effect in biology. Science, 134: 1501-1506.
———— 1963. Animal species and evolution. Cambridge, Belknap Press.

McCrae, W. H. 1968. Cosmology after half a century. Science, 160: 1295-1299.
McIntyre, D. B. 1963. James Hutton and the philosophy of geology. In Albritton, 1963, pp. 1-11.
Medawar, P. B. 1967. The Art of the Soluble. London, Methuen. [See especially the (unnumbered) chapter on "Hypothesis and Imagination," pp. 131-155, for a violently negative view on induction as a scientific method.]
Merrill, G. P. 1906. Contributions to the history of American geology. Report U.S. Nat. Mus. for 1904 [published 1906]: 189-734 [From the point of view of this paper most interesting for its evidence of how little early American geologists were influenced by Lyell and uniformitarianism.]
Meyer, F. 1954. Problématique de l'Évolution. Paris, Presses Universitaires de France. [Includes an extensive but fallacious account of supposed evolutionary acceleration.]
Moore, R. C., ed., 1953. Treatise on invertebrate paleontology. Lawrence, Geology Society of America and University of Kansas Press. [Distribution of recognized genera of fossil invertebrates, publication continuing and approaching completion.]
Morgenbesser, S. 1967. Philosophy of Science Today. New York, Basic Books, Inc. Publishers. [An excellent collection of essays, marred by absence of references; essays most relevant to the present study are here cited separately by authors.]
Newell, N. D. 1952. Periodicity in invertebrate evolution. In Henbest, 1952a, pp. 371-385.
────── 1962. Paleontological gaps and geochronology. J. Paleont., 36: 592-610.
────── 1966. Problems of geochronology. Proc. Acad. Natural Sci., 188: 63-89.
────── 1967. Revolutions in the history of life. In Albritton, 1967, pp. 63-91.
Pap, A. 1953. Does science have metaphysical presuppositions? In Feigl and Brodbeck, 1953, pp. 21-33. [Originally published in 1949; consulted by me in the 1953 reprint.]
Piveteau, J. 1952. Traité de Paléontologie. Paris, Masson. [Distributional data, nearly complete for families of animals. Publication continuing since 1952.]
Playfair, J. 1802. Illustrations of the Huttonian Theory of the Earth. Edinburgh, Cadell, Davies and Creech. [Facsimile edition, 1956: Urbana, Univ. Illinois Press.]
Popper, K. 1935. Logik der Forschung. Vienna.
────── 1959. The Logic of Scientific Discovery. New York, Basic Books, Inc. Publishers.
Romer, A. S. 1966. Vertebrate Paleontology. Chicago, University of Chicago Press. [Includes distribution of all recognized genera of fossil vertebrates.]
Rudwick, M. J. S. 1967. A critique of uniformitarian geology: a letter from W. D. Conybeare to Charles Lyell, 1841. Proc. Amer. Philos. Soc., 111: 272-287. [Conybeare's letter is not only reproduced but also usefully discussed at length with its historical background.]
Russell, B. 1953. On the notion of cause, with applications to the free-will problem. In Feigl and Brodbeck, 1953, pp. 387-407. [Originally published in 1928; consulted by me in the 1953 reprint.]
Russell, L. S. 1951. Age of the Front-Range deformation in the North American cordillera. Trans. Roy. Soc. Canada, 45: 47-69.
Rutherford, E. 1904. The radiation and emanation of radium. Technics, for 1904: 11-16, 171-175. [Contains the epochal suggestion that the energy budget for the earth on a long time scale could be balanced by contributions from radioactivity.]
Schaffner, K. F. 1967. Antireductionism and molecular biology. Science, 157: 644-647.
Schindewolf, O. H. 1936. Paläontologie, Entwicklungslehre und Genetik. Berlin, Borntraeger.
────── 1950a. Grundfragen der Paläontologie. Stuttgart, Schweizerbart.
────── 1950b. Der Zeitfaktor in Geologie und Paläontologie. Rev. ed. Stuttgart, Schweizerbart.
────── 1954. Über die möglichen Ursachen der grossen erdgeschichtlichen Faunenschnitte. Neues Jahrb. Geol. Pal., 10: 457-465.
────── 1963. Neokatastrophimus? Z. deutsch. geol. Ges., 114: 430-445. [In "Jahrgang

1962," and usually cited as of that date, but issued in 1963; an example of confusion of different meanings of "catastrophism."]

SCHOFF, J. W., and E. S. BARGHOORN. 1967. Alga-like fossils from the early Precambrian of South Africa. Science, 156: 508-512. [Descriptions of some of the oldest known fossils and citations of the literature on others.]

SCHUCHERT, C. 1931. Geochronology, or the age of the earth on the basis of sediments and life. Bull. Nat. Res. Council, No. 80: 10-64. [Espouses a theory of acceleration of geologic processes.]

SCOTT, W. B. 1891. On the osteology of *Mesohippus* and *Leptomeryx*, with observations on the modes and factors of evolution in the Mammalia. J. Morph., 5: 301-406.

SCRIVEN, M. 1959. Explanation and prediction in evolutionary theory. Science, 130: 477-482.

—— 1963. Comments on Grünbaum paper. In Induction: Some Current Issues, Middletown (New York), Wesleyan University Press, pp. 147-149. [A rebuttal of Grünbaum, 1963a.]

SHKLOVSKII, I. S., and C. SAGAN. 1966. Intelligent Life in the Universe. New York, Dell Publishing Co., Inc. [Also paperback from the same publisher, 1968.]

SIEVER, R. 1968. Science: observational, experimental, historical. Amer. Sci., 56: 70-77.

SIMPSON, G. G. 1928. A Catalogue of the Mesozoic Mammalia in the Geological Department of the British Museum. London, British Museum (Natural History).

—— 1944. Tempo and Mode in Evolution. New York, Columbia University Press. [A facsimile edition was published in 1965 by Hafner Publishing Co., Inc.]

—— 1952. Periodicity in vertebrate evolution. In Henbest, 1952a, pp. 359-370.

—— 1953a. Life of the Past. New Haven, Yale University Press. [Also paperback, Yale, 1961, and Bantam Books, Inc. 1968.]

—— 1953b. The Major Features of Evolution. New York, Columbia University Press. [Also paperback: New York, Simon & Schuster, Inc. 1967.]

—— 1961. Lamarck, Darwin, and Butler. Amer. Scholar, 30: 238-249. [Reprinted as Chapter 3 in Simpson, 1964.]

—— 1963. Historical science. In Albritton, 1963, pp. 24-48. [Uniformitarianism is discussed on pp. 31-33.]

—— 1964. This View of Life. New York, Harcourt, Brace & World, Inc. [Chapter 3 is a reprint of Simpson, 1961, and Chapter 7 is a revised version of Simpson, 1963.]

—— 1965. The Geography of Evolution. Philadelphia, Chilton Book Company.

—— 1967. The Meaning of Evolution. Rev. ed. New Haven, Yale University Press. [On rates and directionalism see especially chapters 8 and 11.]

—— 1968. Evolutionary effects of cosmic radiation. Science [in press].

STEPANOV, D. L. 1959. [Neocatastrophsm in paleontology of these days.] Paleont. Zhurn. Akad. Nauk S.S.S.R., No. 4: 11-16. [In Russian; not read; cited from Schindewolf, 1963, and Newell, 1967.]

TEILHARD, DE CHARDIN, P. 1959. The Phenomenon of Man. New York, Harper & Brothers, Publishers. [First published in France as Le Phénomène Humain, Paris Editions du Seuil, 1955.]

THOMSON, W. [Lord Kelvin]. 1862. On the secular cooling of the earth. Trans. Roy. Soc. Edinburgh, 23: 157-169. [See also Kelvin, 1899.]

TOULMIN, G. H. 1783. The antiquity of the world. 2nd ed. London, Cadell. [A rarely noticed forerunner of Hutton; the first edition (not seen) was published in 1780.]

UMBGROVE, J. H. F. 1942. The Pulse of the Earth. The Hague, Nijhoff.

VELIKOVSKY, I. 1955. Earth in Upheaval. New York, Doubleday & Company, Inc. [One of the more radical modern versions of catastrophism.]

VYSOTSKII, B. P. 1961. Problema aktualizma i uniformizma i sistema metodov v geologii. Vop. Filos. Akad. Nauk S.S.S.R., No. 3: 134-145. [There is a large literature in Russian on uniformitarianism; this one paper is cited because it makes the same distinction as do Hooykaas and Gould.]

WATSON, J. D. 1965. Molecular Biology of the Gene. New York, W. A. Benjamin, Inc.

WATSON, R. A. 1966. Is geology different: a critical discussion of "The fabric of geology." Philosophy of Science, 33: 172-185. [A blast against Albritton, 1963, and especially Simpson's contribution to that work.]
[WHEWELL, W.] 1832. [Review of Lyell, 1830-1833, vol. 2.] Quart. Rev. 47: 103-132.
WHEWELL, W. 1837. History of the Inductive Sciences. 3 Vols. London, Parker.
WILMARTH, M. G. 1925. The geologic time classification of the United States Geological Survey compared with other classifications accompanied by the original definitions of era, period and epoch terms. Bull. U. S. Geol. Surv. 769.
WILSON, L. G. 1967. The origins of Charles Lyell's uniformitarianism. *In* Albritton, 1967, pp. 35-62.
ZEUNER, F. E. 1950. Dating the Past. 2nd ed. London, Methuen.

3

Evolution of Matter and Consciousness and Its Relation to Panpsychistic Identism[1]

BERNHARD RENSCH

Zoologisches Institut, Wilhelms—Universität, Münster West Germany

Introduction 97
Natural Laws and Their Epigenetic Manifestation 98
Causal Analysis of Life Processes 101
Factors and Laws of Organismic Evolution 104
Psychogenesis and the Epistemological Analysis of Matter 109
Physical Analysis of Matter 112
The Psychophysical Substrate and the Problem of the Psychophysical Connection 113
The Panpsychistic, Polynomistic, and Realistic Identism 117
References 118

Introduction

Our comprehension of physical, chemical, and biological processes has been deepened in an extraordinary manner during the last decades. By microphysical investigations we have gained a wholly new insight into the nature of matter. The causal analysis of the processes of life has

---

[1] Dedicated with cordial wishes to Theodosius Dobzhansky on the occasion of his 70th birthday.

progressed so rapidly that many peculiarities of organisms have become comprehensible. This is especially the case with regard to most physiological processes and to the elucidation of the processes of heredity and evolution. We are even beginning to understand the origin of life from nonliving matter. Hence, at present it is much more promising to get a satisfactory picture of the world by uniting all this new knowledge than it was in the beginning of this century.

As all our sensations, however, only mediate certain indications concerning the peculiarities of the transsubjective world, and as all our knowledge depends upon the abilities of our human thinking, it will be indispensable to include an epistemological basis, in a philosophical picture of the world. We must therefore unite the results of physics, chemistry, and biology, which are normally gained without consideration of philosophical problems, with the statements of epistemology. We have to consider especially the basic statement that only our psychic processes, our phenomena, can be regarded as absolutely indubitable facts, whereas the peculiarities of matter can be revealed by means of our phenomena only by abstractions. When I try to outline briefly such a philosophical picture, this will certainly not be a definite solution, but a solution that corresponds to our present knowledge and is relatively free of contradiction.

Natural Laws and Their Epigenetic Manifestation

Our phenomena of consciousness are not totally chaotic, but we often reexperience identical or similar phenomena. By comprising such identities in our thinking we are able to form concepts that we normally connect and characterize by words. This forming of concepts and the development of the special concept of our own self leads to the idea that "things" and processes also exist outside our ego. By daily experience we are informed that most processes show a certain regularity in the sense of a steady following of cause and effect. Since Democritos we have regarded this causal principle as a universal law.

The law of causality was deduced from steady experience, and this means that it is based on induction. Strictly speaking, it is not an "absolute" law, but a theory of highest probability. However, in all natural sciences this law has been stated. In biology sometimes the general validity is doubted. Some authors have tried to trace back several processes that could not yet be analyzed (as, for instance, purposeful embryological development or regeneration) to noncausal "finalizing" vital factors. But progressive analysis always reveals causal relations (compare the following section).

The general validity of the law of causality has also been doubted by many microphysicists like Heisenberg, Born, P. Jordan, and others. In the microphysical realm our knowledge is restricted especially by the uncertainty principle of Heisenberg. Therefore many microphysical events are not predictable. In such cases only statistical laws can be established, laws in which rules of probability, that is to say universal laws of logic, are involved. As, however, all macrophysical processes proceed in a strictly causal manner, it is difficult to imagine that this principle should not implicitly be based on microphysical processes, the integration of which produces macrophysical events. Only because of the restriction of our means of analysis in the microphysical realm, is a prediction of the processes not possible. But predictability is not a necessary condition of causality. It is only its consequence, as Ziehen (1939, p. 305), Hartmann (1937) and other authors have emphasized. Also Planck (1933) and Einstein (compare perhaps Born, 1955) were convinced of the unrestricted validity of the law of causality. If the assumption would be right that the microphysical realm would not be governed by causality, but only by laws of probability, then in the macrophysical realm causality would originate from the integration of only probable events. As the laws of probability belong to the universal laws of logic, concerning only formal relations, this seems to be rather unlikely.

We can define the principle of causality as a gapless lawful occurrence, valid in the whole universe. One can not agree with Kant, who characterized causality as "pure knowledge a priori," which means knowledge before any experience. But his "a priori" in itself indicates that he probably had in mind a transsubjective existence of an absolute causality. The fact that the law of causality is bound to the isotropic course of time also means that the universe passes through a directed evolution.

The processes of the universe also follow other laws that can not be traced back to causal ones. There are first *the constants of the universe* delimiting, so to speak, the causal processes: (1) The speed of light ($c = 2.99793 \times 10^{10}$cm/sec) is the maximum speed for the motion of energy, which can not be surpassed. It is the maximum speed for signals. (2) No effect is possible that would be less than one quantum ($h = 6.7 \times 10^{-27}$erg sec). (3) Electromagnetic effects are based on a constant elementary charge ($e = 4.77 \times 10^{-10}$electrostatic units). (4) The effects of gravitation are based on a constant of gravitation (g).

Not dependent on the causal laws are also the *principles of symmetry,* summarized by the PCT–theorem. The P–invariance means, that there exists a parity of right and left and that a physical world must be possible in which all processes occur in the sense of a mirror-image. The C–invariance means that for all particles antiparticles also exist. As is well known, many of the antiparticles have been demonstrated. The T–invariance postulates a possible symmetry also with regard to the direction of time. It is possible to calculate such a reversal, but its existence is rather improbable epistemologically. Besides, it would abolish the law of causality.

Independent of the causal laws are also the *conservation laws,* that is to say, the laws of the conservation of energy, of impulse, of spin, and of motion of the centre of gravity. The law of the conservation of energy, also valid for nuclear reactions, is especially important for biological processes.

The *laws of logic,* characterizing the most general formal conditions of all relations, are universal laws. They are independent of the principle of causality. The laws of logic are often regarded only as laws of thinking. However, as stated especially by Husserl (1922) and Ziehen (1934, § 22), they are timeless objective laws of the universe. The results of natural science postulate that the theorem "If two objects equal a third object, then they are also equal to each other" must also have been valid with regard to three atoms in the universe, before thinking beings existed. The realm of logical laws also comprises the *statements of probability,* which characterize the exceptional position of the statistical laws in the microphysical realm. A basic principle of the laws of probability is demonstrated by the fact that on the average, after throwing a coin, one side will as often come to rest face upwards as the other side, and that series in which one side will lie upward five times successively will be much more frequent than that series in which this will be the case ten times successively.

It can be doubted whether the *psychical laws* or the *laws of coordination (or parallelism) of psychic phenomena to physiological processes* in the central nervous system have also to be regarded as special laws. My own identistic conception makes it possible to identify physiological brain processes and corresponding phenomena (compare p. 115).

Hence, we see that a whole series of basic laws exist that can not be traced back to other laws. *The basis of all events of the world is polynomistic.* Correspondingly, matter also shows ultimate nonreducible characters like mass or energy, respectively, charge, spin, speed, and spatial and temporal characters.

All special causal laws became realized in an epigenetic manner parallel to the evolution of the "material" world. As long as a celestial body consists only of gaseous matter, the laws of refraction in lenses, of communicating tubes, or the special laws of falling bodies can not be effective. This is possible only after fluid and solid matter have been formed. These special laws are, however, implicitly involved in the general law of causality. After the evolution of still more complicated "material" systems many *systemic laws* also become manifest. This is already the case when molecules arise. For instance, by combination of the explosive light metal sodium with the gas chlorine, common salt arises, which shows absolutely new characters in consequence of systemic laws. The same holds good when amino acids are combined to macromolecular proteins, and so on.

Causal Analysis of Life Processes

When we restrict our considerations to "real" living beings, the most primitive of which would be smallest mycoplasmas, bacteria, and cyanophycea, we can enumerate the following characteristic properties. Organisms are individualized open systems of relative constancy, being composed of different, mainly organic, compounds, and showing a developmental cycle that is repeated in rhythmical sequence. Proteins and chains of polynucleotides are always necessary compounds, because they are the basis for the hereditary information and for many syntheses involved in life processes. "Real" organisms have a cellular structure, and the cells have certain organelles, at least chromosomes or their prestages, ribosomes, effecting the synthesis of proteins, and normally also mitochondria or their prestages, necessary for energy change. They show metabolism and hence a typical activity. They are able to reproduce and transmit their characters to their offspring. They are subjected to evolution by mutations, alterations in the combination of genes and processes of selection and isolation. All structures and functions are prevailingly useful with regard to the maintenance of the individual and the species.

However, the delimitation of living beings and nonliving matter is not quite clear. Viruses, normally regarded as nonliving systems, already show a series of specific life characters. They are composed of proteins and polynucleotides, they propagate and transmit their characters to the offspring, they mutate and are therefore capable of evolution. The biochemical complication of certain larger organisms, the Mycoplasmatales -Microtatobiotes or Cysticetes, is already similar to the complication of the simplest bacteria. On the other hand, during anabiosis, which theoretically can go on for infinite time, main life characters of multicellular animals and plants are lacking. There is no metabolism, no activity, no individual development, propagation, and evolution. (Seeds of *Lupinus arcticus* for instance, which had lain in ever-frozen ground sheets near the upper Yukon for about 15,000 years, proved to be capable of germinating.) (Porsild et al., 1967.)

As organisms are mainly composed of macromolecules and form highly complex, individualized systems, *in most processes of life systemic laws are involved*. Hence, nearly every mutation has a pleiotropic effect, and nearly every structure and every organ is caused by a series of genes. Correspondingly alterations in size, structure, and function affect other structures and functions in consequence of innumerable mutual correlations. It is understandable that these complex relations render the causal analysis of life processes much more difficult than the analysis of nonliving matter. Special difficulties arise by the fact that organisms are open systems showing steady activity. Moreover, these systems seem to contradict the law of entropy, since they show an increase in order as well in ontogeny as in phylogeny.

It is comprehensible that formerly the highly complicated and prevailingly useful processes of life were traced back to unknown vital forces. Especially the facts of "purposeful" individual development, the processes of regeneration and regulation, and the progress during phylogeny have led to the assumption of noncausal autonomous principles and factors producing the wholeness of the body. Driesch (1905, 1928) called such vital forces "entelechy," a term by which unknown relations were circumscribed, but not in the least explained.

However, in biology the application of the principle of causality proved its value without exception, and during the last decades it was possible to show that nearly all "marvels" of life are causal events. Most processes of metabolism became understandable after the function of vitamins, catalyticly effective enzymes, and regulating hormones could be elucidated. Photosynthesis, being the decisive basis for the phylogeny of organisms, could be analyzed to a far-reaching degree. The "marvel" of energy production by respiration in the cells could be explained by a chain of biochemical processes associated mainly with the mitochondria. Light-microscopic and electron-microscopic, as well as electrophysiological investigations, innumerable operations, and further experiments revealed the functioning of sense cells and nerve cells, the conduction of excitations, and to some degree even the integration of excitations in central nervous systems. Instincts, too, could be analyzed by many experiments with dummies and by application of hormones revealing the function of releasing mechanisms.

After the discovery of the gamones the processes of fertilization could partially be traced back to macromolecular reactions. The origin and the tenacious maintaining of sexuality in the long course of evolution of plants and animals became conceivable as necessary means to make sufficient gene combinations available, so that the species could survive the many examinations imposed by natural selection. Mendel's rules of heredity, incomprehensible for a long time, are now explained by the chromosome theory of genetics. The mutation theory provided an exact basis for the understanding of evolution. When molecular structure of nucleic acids was elucidated, the secret of the transmission of genetic information was able to be analyzed to a far-reaching degree. The problem of identical replication as well as the effect of genes during ontogeny has become more and more comprehensible.

We also begin to get a causal understanding of how the purposeful development and differentiation of embryonic structures and functions could originate, although these structures and functions—such as the limbs and sense organs of mammals—are used only after birth. Sometimes this "finalizing" of ontogenetic processes was brought forward against the causal explanation of the processes of life. We must, however, bear in mind that all changes of developmental stages caused by mutation could be maintained only when such alterations and the adult stages which they produce are able to survive. The processes of selection always hit the total individual and hence the whole course of development. The origin of the "strategy of genes" (Waddington, 1957) can therefore be regarded as a casual process. Only when we look at a developmental process after it has been evolved by selection and proved to be tolerable, does the ontogeny seem to be a final process. But *after* their course all causal processes can be regarded as final ones, including the origin of our earth as well as the origin of new organs or new species.

All these brief remarks about biological investigations should only show *that causal analysis proved to be successful without exception.* Of course, in all realms of biology innumerable questions have still to be solved. But there is no doubt that all progress can only be reached by well-planned tracing back to the causal chains step by step. Although we are rather far away from a definite understanding of all processes of life, due to the extraordinary molecular, structural, and functional complication of organisms, the rapid scientific progress shows *that life proceeds by a complicated integration of causal processes.* Only in the microphysical realm are causal investigations limited, especially by the principle of uncertainty formulated by Heisenberg. But it is not necessary to assume that these processes do not follow the law of causality.

With the integration of many highly complicated biochemical and biophysical causal processes in the wholeness of an individual, many *specific laws of life* arose. They are mainly systemic laws. As these laws, however, often cross one another, exceptions normally exist, and it is therefore proper to speak only of *rules,* such as Mendel's rules, Rubner's rule, and so on.

There has been much discussion about the fact that organisms are subjected only in a restricted manner to the *law of entropy.* The second theorem of thermodynamics states that in a closed physical system every natural irreversible process leads to conditions of increasing probability that is to say, of increasing disorder until a stage of balance is reached. This theorem is also valid for many processes of living beings. In many physiological processes heat and heat emission—that is, increase in entropy— arises. The order of molecules decreases when solid or crystallized substances are dissolved in fluids of the body or when substances change into a gaseous stage, as for instance in carbonic acid. However, when many different types of tissue develop from a relatively simple germ cell, then higher levels of functional order successively arise. This apparent contradiction to the law of entropy is sometimes claimed to prove vitalistic conceptions.

We must, however, consider that the law of entropy is valid only in closed systems and that organisms are open systems with steady metabolism (von Bertalanffy, 1949, p. 110). But metabolism is just the means to maintain a relatively stationary level of entropy. Living beings get rid of surplus entropy by passing heat, which arises by metabolism, to the environment (Schrödinger, 1944). Moreover, many biochemical reactions proceed in such a manner that *order arises from order*. This is especially valid with regard to ontogeny. The DNA molecules, which mediate the information for hereditary characters, are capable of identical replication, and that means that they get their order from the order of their forerunners. The sequence of bases of the polynucleotides determines the order of the messenger-RNA, which for its part determines the synthesis of specific proteins by fitting their codons into the anticodons of the different transfer-RNA, when they are attached to the ribosomes. The proteins are partly effective as enzymes and direct further specific syntheses. In such a manner the ontogenetic process of differentiation represents a *stream of order*. And this stream of order continues further on, when gene activity becomes released in due time, when substances are produced that induce specific structures in neighbouring tissues, and when mutual correlations develop between different structures and organs. So far as this stream of order can already be analyzed, all stages have proven to be chains of causal processes.

Factors and Laws of Organismic Evolution

The rapid progress of biological knowledge and the steady confirmation of the principle of causality in all biological analyses lead to the assumption that *all* "material" processes of life are chains of cause and effect inserted in the gapless chains of causality in the realm of nonliving matter. We have, however, to ask how it was possible that such chains of biological reactions could originate, reactions guaranteeing the constancy of processes from generation to generation and leading well-ordered to purposefully developing and usefully organized living systems. We have further to ask whether and how far we can explain in a causal manner the phylogenetic splitting up of innumerable types of anatomical construction, the phylogenetic progress of many lines of descent, and the origin and the uniqueness of man. Finally, we have also to ask how the first living beings could have originated.

As all these processes went on in past times the solution of the above-mentioned problems requires greater use of deductive methods, in spite of the rich paleontological material and the possibility of proving the origin of races and species by genetic and selection experiments. It is therefore understandable that several questions are still disputed. However, for most of the mentioned phenomena, well-founded causal explanations are already possible.

The development of new hereditary races and species is rather well established on the basis of innumerable observations on living and fossil material (compare Huxley, 1942; Huxley, Hardy, and Ford, 1954; Mayr, 1942, 1963; Simpson, 1944, 1949, 1953; Rensch, 1947, 1960), and many genetic and selection experiments (compare especially Dobzhansky, 1951, and the many investigations on selection published by him and his school). The effective factors are mutation, gene combination, gene flow, fluctuations of the frequency of genes in a population (genetic drift), selection, isolation, and hybridization. These factors operate in different combinations and, therefore, different ways of speciation exist. Most of these changes are already seen as causal processes. This is especially the case when we analyze gene flow, the changing gene frequency in a gene pool, and selection or isolation. Concerning gene combinations, we usually say that they happen "at random," because we do not know why a chromosome goes to one or the other side when the chromosomes are paired for reduction. However, this means only that we do not know the special causal conditions, because it is still too difficult to analyze the complicated process in a dividing cell. We encounter a similar difficulty when we are playing at dice; the result will happen "at random," but we are convinced that the movements of the dice are causal events, guided by the die's shape and the wall of the dice box, and by the direction and impulse of shaking the dice box.

With regard to gene mutation our knowledge of the releasing causes is still rather insufficient. It may be possible that "spontaneous" gene mutations are microphysical processes, the primary causes of which we can not analyze. Jordan (1945) even held the opinion that microphysical noncausality (and even "liberty") became effective and is transmitted into the macrophysical processes of heredity. However, as already pointed out above (p. 103), noncausality of microphysical processes is not really proved and is very improbable. Moreover, in some cases a successful causal analysis of gene mutations has already proved to be possible. Alterations in the sequence of the bases in the DNA-molecules can be caused by tautonomous restoring in consequence of ionizations. Some genes can also be blocked by certain substances. Surely, many investigations will be needed to elucidate the release of mutations, but it is already probable that speciation can generally be regarded as a chain of causal processes.

It is the great merit of Charles Darwin (1859) that he made understandable in a causal manner how usefulness of the morphological and functional organization of plants and animals came about. In most cases natural selection weeds out all unfavorable organizations in the course of time and it promotes more favorable organizations. This selection is extraordinarily intense, a fact best illustrated by the necessity of an enormous overproduction of offspring. In most multicellular plants and animals 95 to 99.9 percent and more of the offspring perish before becoming capable of reproduction. In the course of life every individual has to pass through a series of "examinations" by different selective factors. All kinds of changes of the environment, especially uncommonly cold or hot temperatures, extreme drought, enemies, parasites, and competitors of the same species, have the effect that all structures and functions remain adaptive. The fact that in spite of these processes of selection some disadvantageous structures always exist (such as vestigial organs, excessive "luxurious" structures, historically caused imperfect constructions or imperfect modes of individual development) understandable, these disadvantages are biologically tolerable and are often correlated with more important favorable characters. The development of excessive antlers of the Pleistocene giant stags (*Megaceros*), for instance, was coupled with prolonged or more intense growth leading to larger body size, diminishing the loss of heat. In all cases selection does not only promote, diminish, or weed out the structures of the adult organism but also the whole chain of developmental processes—and hence also the purposiveness of development.

Although the mutations are undirected and the conditions of selection often change, evolution as a whole is not totally undirected. On the contrary, we can state many *evolutionary laws*. But as these laws often cross one another, they always show a larger or lesser percentage of exceptions. Hence we normally only speak of *rules*. However, the factors determining these rules are causal ones.

A number of such rules are valid for all organisms, others only for animals or plants, only for marine or land animals, or only for special taxonomic groups. In my "Biophilosophie" (Rensch, 1968, p. 109-114) I mentioned 100 evolutionary rules, and I chose this "round" number in order to indicate that many other rules exist. It may be sufficient to mention only some rules as examples here, and first of all those which are more or less valid in all groups of organisms. Concerning their causality, which is evident in most cases, references will be found in corresponding books on evolution, genetics, or ecology.

1. All organisms show undirected "spontaneous" mutations that are mainly disadvantageous because they disturb the harmony of development acquired in a long line of descent and because they often diminish vitality or fertility.

2. In sexually reproducing organisms, characters are transmitted to the offspring corresponding to Mendel's rules, the third rule being restricted to genes, which are localized on different chromosomes.

3. All organisms produce a surplus of offspring, of which in most cases 99 percent or more die before becoming reproductive. This process is connected with natural selection.

4. As mutations, new gene combinations, and often changing processes of natural selection, always occur, all species change into other species, although with very different speed.

5. The speed of speciation is greater in isolated, smaller populations than in larger populations.

6. Because of the relatively lesser changes in selective factors, and because of their normally larger populations, marine animals show a slower speciation than related land animals.

7. The steadiness of natural selection normally causes an increasing adaptation to the conditions of the habitat.

8. After new favorable types of construction have been evolved, in most cases a great radiation of species, genera, or families takes place, which is successively reduced in the same measure as the available habitats are occupied.

9. Normally, early ontogenetic stages are less affected by evolutionary alterations than are later stages. Ontogeny, therefore, provides indications of what course phylogeny has taken (biogenetical rule).

10. In warm and moist zones the abundance of species is greater than in cooler regions, which, however, show a relatively greater abundance of individuals in most species.

11. Lines of descent leading successively to evolutionary progress normally begin with relatively unspecialized types (Cope's law of the unspecialized).

12. Varieties or species that have more rational structures and functions are normally more successful in competition. This leads to evolutionary progress in many lines of descent.

More special rules, valid only for single categories of organisms, are still more numerous. Here, too, only a few examples may be mentioned.

13. Among warm-blooded animals the geographical races inhabiting cooler zones, especially those with low winter minima, are mostly larger than races of the same species inhabiting warmer zones (Bergmann's rule).

14. Geographical races of birds inhabiting cooler zones produce more eggs per clutch than races of the same species inhabiting warmer zones.

15. Geographical races of insects that remained in Pleistocene refuges show a greater variation of inherited characters than races from regions colonized in postglacial times.

16. Birds breeding in holes normally develop a light, mostly white color of eggs, while birds breeding in the open develop camouflaged egg colors.

17. Species showing special care of eggs or of the offspring have less progeny than related species in which such instincts are lacking.

18. Large warm-blooded animals have a longer embryonic period and become fertile later than related smaller species.

19. Larger vertebrates usually have relatively smaller hearts, livers, kidneys, brains, and eyes than related smaller species living in the same climate.

20. Large animals more often have excessive structures than related small animals. In mammals they more often show excessive processes of the skulls; in insects, processes of the head or thorax or prolonged legs or feelers.

Such evolutionary rules can be supplemented by numerous cytological, histological, anatomical, physiological, and developmental rules, which also influence evolution as far as they set limits for hereditary alterations. The walls of lung cells, for instance, can not become so thick that diffusion of sufficient oxygen would be hindered. An eye lens can not be changed during phylogeny in such a manner that it becomes dull or that it is no longer capable of focusing rays. During ontogeny the relative speed of development of an organ can not be slowed down in such a manner that it can not function when full grown. In the lines of descent of birds a certain body size can not be passed over as the ability to fly would be lost because the body grows by the third power while the efficiency of the wings expands only by the second power.

These few examples may show that all evolutionary change is much more determined by a great many rules than is normally supposed. Because most probably all changes in species are causal processes, *it is possible to assume that the whole of evolution is a forced process,* a process annexed to the causal history of the earth.

In the same manner, as this is the case in physical and chemical laws, *the special phylogenetical rules became successively efficient in the sense of an epigenesis.* Mutation and selection could become effective, as soon as prestages of organisms containing polynucleotides originated. Mendel's rules could become effective only after sexual reproduction had been developed. Bergmann's rule could become manifest only after warm-blooded animals had been evolved. However, in the last analysis all these rules are potential effects of the universal law of causality, and they already existed, so to say, implicitly, before they could become efficient in more complex stages of the development of matter. Such a conception leads to far-reaching philosophical consequences; it is possible to assume that the whole phylogeny of plants, animals, and man proceed as *predestined development.*

During the evolutionary progress of animals and plants successively higher levels of structural and functional order originated. This fact seems to contradict the law of entropy. However, as already mentioned, in the chain of generations order derived from order. *When during phylogeny new sequences of order arose, the increase of disorder, of entropy, was prevented by natural selection,* which more or less weeded out disorders in structure and function and supported favorable new sequences of order—which, for instance, came about by centralization of functions or by insertion of new mechanisms of regulation.

The *possible* assumption that the whole evolution of organisms is a determined process is of special interest with regard to the *origin of man*. Our phylogenetic development would also be a forced process. Our apelike ancestors were predestined to become man by their long phase of youth; by the structure of hands with opposable thumb and consequent ability to perform manifold manipulations; by erect poise when sitting; by a relatively large forebrain divided into many fields for different functions; and by their social instincts. The impulse to become hominids was probably caused by the transition to biped gait in the steppe, which led to an increase of head and brain, because the head could be carried above the center of gravity of the body. The relative small front teeth (incisors and canines) possibly forced the early hominids to use and improve tools. The consequence was a selection pressure, favoring variants and hordes with better brain. Such improvement was possible by an increase of the number of neurones in some parts of the forebrain and by differentiation of new fields of association in the frontal cortex and a motor center for speech.

Psychogenesis and the Epistemological Analysis of Matter

The evolution of organisms offers still a further problem, which is difficult to solve: the phylogeny of the mind. The main difficulty is the fact that everybody can only pretend for himself that he experiences processes of consciousness, that is to say phenomena like sensations, mental images, feelings, and processes of thinking. We are, however, sure that our fellow men have corresponding phenomena, because we are informed about this by our language. Many accomplishments of the brain of higher animals are so similar to human accomplishments that we may also assume with great certainty that these species see, hear, taste, and smell, that they have memory, are able to form averbal abstract concepts, and that at least apes and monkeys can act according to projected plan. We must, however, realize that the assumption of such phenomena in animals is based only on conclusions by analogy. Higher mammals, especially apes and monkeys, have brains and sense organs of a similar structure and function as brains and sense organs of man. Furthermore, several expressions of the face are similar. The brain achievements of lower vertebrates like fish are markedly less, but fish still show memory and an ability for generalization, and we do not doubt that they can see, taste, smell, and partly hear.

But some invertebrates like cephalopods, bees, and bumblebees also have brains composed of a great many neurones (a honeybee worker has about 850,000; Witthöft, 1967), and such animals can be trained to learn optical tasks, and can even master three tasks simultaneously. Hence, we can conclude with some certainty that they also have sensations and memory. Central nervous systems and sense organs, the structure and function of which are similar although more primitive, are also found in lower invertebrates. And these animals are also capable of retaining simple tasks. I previously discussed these questions in more detail in the first volume of *Evolutionary Biology* (Rensch, 1967), and it will therefore not be necessary to repeat the statements.

Now we have to put the decisive question: On which phylogenetic level did psychic phenomena, at least sensations, first appear? Even with regard to protozoans we use to speak of "sense reactions." So far as we can judge, phylogenetic progress in animals occurred little by little. Should we suppose that psychic phenomena, something fundamentally new, differing in principle from all physiological functions and all characters of matter, suddenly arose? And from which roots would these new phenomena come, phenomena that very probably do not occur on other celestial bodies of our solar system and which did not exist before living beings originated?

It is therefore possible and by no means improbable to assume that consciousness in a general sense (of course not in the sense of reflecting self-consciousness) is a faculty of all living beings. In animals without a central nervous system, in unicellular organisms, and plants, we could eventually suppose isolated sensations or prestages of sensations as *protophenomena,* which are not yet connected with one another. As it is possible to assume with increasing probability that life had gradually been developed from nonliving matter, it is even possible to attribute such protophenomenal characters or a protophenomenal nature to all matter.

Such a conclusion might appear rather speculative, although many well-known philosophers have held such an opinion—for instance, Fechner, Paulsen, Erdmann, Wundt, Ziehen, Haeckel, Francé, Hartshorne, and others. However, this conception is supported by quite different statements. At first we have to consider that we have to deal with a corresponding problem in *psychontogenesis.* In a human fetus some weeks before birth certain reactions appear, which point to arising sensations, but which seem not yet to be connected in a general stream of consciousness. Probably first sensations arise parallel to the differentiation of nerve and sense cells.

However, the embryological development begins with a fertilized egg cell. Hence, a gradual appearance of psychic phenomena would be best conceivable when we would admit that even an egg cell and the first embryonic tissues are of protophenomenal nature.

Furthermore, the fact that many psychic characters of man—such as musical, mathematical, and other endowments—are inherited is important. But these psychic characters are transmitted from generation to generation by material germ cells, especially by DNA molecules. Consequently these molecules must have characters that are able to produce psychic phenomena during the following development. It is reasonable to suppose protophenomenal characters or a protophenomenal nature of these molecules.

Finally the epistemological analysis of "matter," which we suppose to be the "objective" basis of all living and nonliving "things," seems to be decisive for our problem. When we want to recognize the nature of matter, we must start from something that is indisputable reality: our perceptions and the corresponding mental images and connections of our thinking. An orange, for instance, produces visual sensations of color and shape, tactile and smelling sensations, and, when we let the fruit fall, also auditive sensations. Many experiences with such fruits have led us to the assumption that an orange is a "material thing" that releases our corresponding sensations. To this objectized "thing," the naive realist attributes the characters of his sensations. He is sure that the fruit "is" orange and hard, that it "has" a certain smell, and so on. The reflecting philosopher, however, concludes that visual, tactile, and other sensations only correspond to processes in the realm of our brain or perhaps also our sense organs. The orange only reflects electromagnetic waves of a certain length, and only their reaction with substances in the sense cells of the retina releases excitations to which color- and shape-sensation correspond. The same holds for other qualities of sensation. These "secondary qualities" do not belong to matter itself, as Democritos and Plato (in his "Theaitetos") recognized in principle and as Locke (1690, chapt. VIII) proved in more detail.

Berkeley (1710) and Kant (1787) also doubted that the primary qualities, the spatial and temporal characters, could be attributed to the transsubjective "things" (the *Ding an sich*). However, it seems to be much more probable that these characters belong to matter itself, because of their comformity in the visual, tactile, and auditive realm, the homogeneity of the spatial and temporal characters, and the possibility of melding together all these different experiences in the conception of an "absolute space" and an "absolute time." But we have to admit that possibly the experienced spatial and temporal characters of our sensations might be a little different from the corresponding characters of matter, because all sensations are based on the statement of relations, the mutual coordination of which may be influenced by the brain processes.

As these interpretations show, we develop an adequate concept of matter by *reduction*. Matter has no color, hardness, or smell, and no characters that correspond to our positive or negative feelings. Matter is a spatial and temporal structure without such qualities. But in this process of reduction we do not abstract from conciousness as such (compare Ziehen, 1913). By epistemological reduction matter still keeps consciousness in a most general sense. Matter is of *protophenomenal nature*. This conception allows us to interpret psychogenesis as a development of protophenomena to phenomena—that is to say, to sensations, mental images, and thinking processes parallel to the development of organic matter forming nerves, sense organs, and central nervous systems. To a certain degree such a panpsychistic conception is also supported by the physical analysis of matter.

Physical Analysis of Matter

Chemical and physical investigations have shown that the transsubjective "things" are a matter composed of molecules and atoms, being integrations of elementary particles. Hence, the qualitative multiplicity of the material world is brought about by systemic relations in the combination of basic elements differing mainly *quantitatively*.

The elementary particles are different only in few basic characters: mass, charge, spin, speed, and spatial and temporal characters. Einstein was able to show that mass is equivalent to energy. Resting mass is potential energy and can therefore be transformed into active energy ($E = mc^2$). The earlier conception of matter as something "solid" and "substantial" had to be given up. Elementary particles can be transformed into one another to a certain degree. Mass can become radiation and radiation can become matter. Following Schrödinger's (1954) works, material "particles" can be regarded as more or less transitory structures of the wave field.

It is therefore misleading to define matter philosophically as a "substance." This term arose because it was a useful practice of thinking to refer all characters to a "carrier." This was sufficient for practical life. But what we really know is only the statement that *matter is a complex of relations*. Such relations can rather well be expressed by mathematical formulas. It is therefore possible to try to formulate the spatial, temporal, and energetic relations of matter in a "world formula," as was attempted in Einstein's general theory of relativity and the formulas of Heisenberg.

It is important that also the spatial and temporal characters can be grasped only in relation to a fixed system of relations. The spatial and temporal characters are dependent upon the motion of the basic system to which they are related. "Real is only the combination, the unity of space, time and things, each of these three conceptions separately is an abstraction" (Schlick, 1920, p. 37). Without "things" there would be no space and time. This corresponds to the fact that we experience only spatial and temporal *characters* of sensations and mental images. The conceptions of an absolute space and an absolute time are based only on conclusions, because spatial and temporal characters are the same in visual, tactile, and auditive sensations and mental images.

Furthermore, matter has to be considered as a complex of relations, which no longer includes "solid material" in the former sense, but only energy ("power" in the former sense). This replacing of "substance" by fields of energy makes it easier to insert the results of modern physics into a panpsychistic conception (compare corresponding suggestions of Eddington, 1939, and Jeans, 1943). Even the former assumption of "psychic energy" in the sense of Ostwald (1908) can become conceivable, when we identify the physiological brain processes with psychic phenomena (see p. 115ff).

The complex of relations that we call matter is goverened by several universal laws, briefly discussed above (p. 112): the law of causality; the laws of conservation of energy, impulse, spin, and of motion of the centre of gravity; and the principles of symmetry. Matter is also based on several characters, which are to be regarded as irreducible at present: mass or energy respectively, charge, spin, speed, spatial, and temporal characters. Hence, the principle of monism, which proved to be so successful in reducing qualitative differences of complicated matter and of material processes to quantitative basic elements and laws at last meets with a limit. *The basis of matter and of all processes of the universe is polynomistic.*

The Psychophysical Substrate and the Problem of the Psychophysical Connection

We can distinguish two types of matter in the human body: parts of our brain are connected with psychical phenomena, the rest of the brain and body does not show such connection. The first type of matter, which I call the psychophysical substrate, can be delimited only in an insufficient manner. We do not know whether consciousness is connected only with the cortex of the forebrain, or also with subcortical regions—which is more probable—or even with other parts of the nervous system. But we know that no psychic phenomena are connected with motor excitations of the brain and nerves. There is no accordance of opinion about the question as to whether sensations and their corresponding mental images are connected with functions of the same or of different parts of the forebrain. We can not even pretend with absolute certainty that sensations could not be connected also with processes in the sense organs (compare Cibis and Nothdurft, 1948; Rensch, 1952). When I experience pain in a finger or in a toe, I do not feel it in the brain, but in spatial relations that correspond to the physical relations of finger or toe. This is indisputable phenomenological reality.

However, it is also possible to explain this localization "in" the limbs in a relatively comprehensible manner by a theory that supposes corresponding inherited characters of brain neurones (*nativistische Lokalzeichentheorie;* compare Ziehen, 1934, p. 152-178; Rensch, 1968, p. 135-137). According to this theory an inherited pattern of neurones exists to which spatial sensations and corresponding mental images are coordinated, representing the experienced sensations of the limbs without themselves having congruent spatial relations in the brain. Electrophysiological investigations of Hubel and Wiesel (1965) proved that in the brain of cats series of neurones exist that are excited only by horizontal or vertical shifting or turning of streaks of light. It is probable that such inherited arrangements of neurones, which are correlated with specific sensations, have been adapted to the physical reality of the world by natural selection in the course of evolution.

When we suppose that consciousness is correlated only with the psychophysical substrate, we must, however, consider that the atomic basis of this substance is the same as that of the remaining parts of the nervous system and of the body. During the development of the brain these atoms are derived from the food, and in the full-grown stage the molecules are exchanged by metabolism from time to time. Hence, the origin of psychic phenomena must be caused by a certain systemic integration of the elements and the processes in the psychophysical substrate.

The only specific peculiarity of brain processes is the existence of a complicated connection of excitations running parallel to the experienced "stream of consciousness." Would it be possible to comprehend the complex of excitations as a basis of consciousness? At first several facts seem to contradict this idea. Complexes of excitations also exist in the cerebellum, in the spinal cord, and in other nerve nets—that is to say, in regions in which we believe consciousness does not exist. However, it is possible to assume also that these excitations are correlated with "basic" consciousness. But such consciousness can not be experienced, because it is not connected with the stream of consciousness of the brain. We must also consider that the stream of consciousness is not only represented by the excitations along the fibers of neurones but also by *the electrical field filling the whole forebrain,* which we can record as electroencephalograms by putting electrodes on the skin of the head.

To many excitations no psychic phenomena are coordinated, because they have been stopped before reaching the central field of consciousness by the reticular formation, or by inhibitory mechanisms in the forebrain. We are therefore not aware of many biologically unimportant excitations such as tactile excitations of the clothes, excitations of the periphery of the retina, and so on. Important afferent excitations of the sense organs enter into the central field of consciousness, whereas efferent motor excitations, losing the connection with the central system of consciousness, remain unconscious. Only after such motor excitations have caused a reaction of muscles or glands do we become informed by corresponding secondary sense reactions (reafference). Of course all these explanations are still speculative, as long as the neurophysiological processes are not fully elucidated.

However, our hypothetical interpretation fits well into an *identistical conception*. As already stated above (p. 112), we can comprehend matter as a protopsychical complex of relations. The physiological analysis of matter, replacing the former "solidity" by fields of energy, does not contradict this conception. As all matter is supposed to be of protopsychic nature, there is no difference in principle between the psychophysical substrate and the matter of the remaining body. But *the protopsychic elements of the psychophysical substate flow together and form a stream of consciousness*. The protopsychic elements of the remaining body, on the other hand, do not enter into this stream of consciousness.

It becomes understandable, by such an identistical conception, that the active neurones of the psychophysical substrate in the brain are built up by molecules of the food and that they can be replaced in the course of cell metabolism. It also becomes understandable how a field of consciousness can originate that corresponds to the successive development of central nervous systems in the course of evolution. The same holds true for the successive development of self-consciousness during ontogeny.

When we do *not* accept an identistical conception and regard matter and psychic phenomena as something fundamentally different, then we are forced to assume mutual psychophysical effects or a special law effecting the parallelism of specific psychic processes to corresponding physiological processes in the brain. The concept of mutual effects between soul and body, assumed by many psychologists, philosophers, and physicians (who use a psychosomatic treatment) meets with great theoretical difficulties with regard to the law of the conservation of energy. When psychic processes would affect the physiological processes, then energy would be added, and, correspondingly, energy would be lost when physiological processes would have an "effect" on the soul. All this would contradict the conception of a gapless course of causality.

Such difficulties cease to exist when we assume a psychophysical parallelism. But if we suppose that matter and mind are something totally different, it remains incomprehensible how this "prestabilized harmony" (Leibniz) of the parallel physiological and psychical processes, which had to be regarded as needless "epiphenomena," could arise in the course of evolution. For selection would not promote unnecessary characters.

The panpsychistic conception of Ziehen (1913, 1934, 1939) approaches my own conception a little closer. His detailed analysis of the characters of the psychic phenomena leads to the assumption that two kinds of constituents exist: (1) those which are determined by causality and are found by reduction of the secondary characters and possible modifications of the primary characters of the sensations (called *Redukte*) and (2) the constituents corresponding to the qualities of the phenomena (like red, sweet, warm, etc.). Both constituents have to be regarded as conscious in the most general sense of the word. Ziehen assumes that the qualities that he calls *Parallelkomponenten* are coordinated with the corresponding physiological processes of the brain by a "law of parallelism," which we have to acknowledge as irreducible datum.

In spite of the panpsychistic character of matter, Ziehen's conception is still dualistic—although only by assuming two laws. In my opinion it is a little unsatisfactory that the law of parallelism, which has to be regarded as a universal law of the world, appears only "pointlike" in space and time in our solar system, and that it became effective on our planet only in the last phase of its development and only in the central nervous systems of living beings.

It seems to me more conceivable to assume *that the qualities of sensations and mental images do not run parallel to complexes of causal relations in the brain, but that they are identical with them.* The assumption of a "given" law of parallelism cannot be regarded as less hypothetical than the assumption of identity. Hence, in my opinion the qualities of colors, sounds, smells, and so on *are* the causal reality of the brain processes (of course not of the transsubjective "things").

A first consequence of such a *realistic, panpsychistic identism* will be the opinion that we have to regard the psychological laws also as physiological laws, and that each causal physiological explanation of relevant brain processes would also explain the psychological relations. In such a way a part of psychology would become pure natural science. Such realistic conception would also mean that the spatial and temporal characters of psychic phenomena would represent the causal relations of the corresponding brain processes. However, the relations of the transsubjective world could eventually differ to some extent from the relations of the brain processes.

As already mentioned, the qualitative differences of matter (of the constituents obtained by reduction of secondary qualities) can be traced back, to a large degree, to quantitative differences. The psychical peculiarities (or the parallel components of Ziehen), on the contrary, are characterized by the nearly infinite multitude of irreducible qualities. This seems to contradict our assumption that we could regard these psychic qualities as being identical with the causal components of matter.

But this apparent contradiction can be solved. We can analyze matter on all levels of integration—as structures of organisms, as organic compounds, as molecules, and as atoms. But we can analyze the qualities of our sensations and mental images only in their latest stage of phylogenetic development: as characters of human psychic phenomena. Only in higher animals can we judge some psychic phenomena to a certain degree by analogy. However, it is absolutely possible and it is even probable that qualities of phenomena have had a phylogenetic development by successive differentiation. Organisms have descended to more and more complicated levels of integration, have developed sense organs, neurones, and central nervous systems, and therefore new "material" characters. We can suppose that in a corresponding manner the protopsychic nature reached new stages of integration by systematic laws. However, we cannot analyze these different levels of integration, because we cannot sufficiently judge psychic phenomena of lower animals. We can sufficiently analyze only our human psychic phenomena. But we are the product of a long phylogeny, and we experience, therefore, a large number of very different qualities that we cannot trace back to more undifferentiated qualities in lowest organisms.

The Panpsychistic, Polynomistic, and Realistic Identism

The philosophical conception outlined in the preceding sections is based on the indubitable reality of the psychic phenomena, as analyzed with the help of logical laws of thinking, which were developed phylogenetically by adaptation to the universal logical laws of the world. At the same time, the results of science, which are normally obtained without considering the epistemological basis, were utilized. This combination led to the establishment of a panpsychistic and realistic identism. No contrast between mind and matter could be accepted, and psychical processes were identified with physiological processes of the brain. In such a manner epistemological panpsychism is combined with the realism of natural science, and a bridge is built to functional materialism. Among biologists the assumption of panpsychism seems to be rarely accepted. At present, especially the well-known geneticist Sewall Wright (1964) supports this opinion (although not in my realistic version).

The qualitative multitude of psychic phenomena, as well as the multitude of protopsychic matter, are finally based on the restricted multitude of irreducible characters of the complex relations that we call matter (mass or energy, respectively, charge, spin, speed, spatial, and temporal characters) and on the restricted multitude of irreducible laws (law of causality, universal logical laws, laws of conservation, principles of symmetry). The astronomical and geological formation of the innumerable types of matter as well as the phylogenetic development of mind are regarded as the result of systematic laws on different levels of integration of protopsychic matter. Also the human mind has to be regarded as being determined by the universal laws that finally produced the psychophysical complex of the human brain. The difference between transsubjective being and self-consciousness derives only from this high level of integration of brain neurones, which produces the stream of consciousness of an ego. *Human thinking, successively developed phylogenetically, is a part of the reality of the entity.*

References

BERKELEY, G. 1710. Principles of Human Knowledge.
BERTALANFFY, L. VON. 1949. Das biologische Weltbild. Vol. 1. Bern, Francke.
BORN, M. 1955. Albert Einstein und das Lichtquantum. Naturwissenschaften, 42: 425-431.
CIBIS, C., and H. NOTHDURFT. 1948. Experimentelle Trennung eines zentralen und eines peripheren Anteils von unbunten Nachbildern. Pflügers Arch. ges. Physiol., 250: 501-540.
DARWIN, C. 1859. On the Origin of Species by Means of Natural Selection. London, Murray.
DOBZHANSKY, TH. 1951. Genetics and the Origin of Species. 3rd ed. New York, Columbia University Press. (1st ed., 1937.)
DRIESCH, H. 1905. Der Vitalismus als Geschichte und Lehre. Leipzig, J. A. Barth, Publishers.
———— 1928. Philosophie des Organischen. 4th ed. Leipzig, Quelle u. Meyer. (1st ed. 1909).
EDDINGTON, A. 1939. The Philosophy of Physical Science. Cambridge, Mass., Cambridge University Press.
HARTMANN, M. 1937. Philosophie der Naturwissenschaften. Berlin, Springer-Verlag.
HUBEL, D. H., and T. N. WIESEL. 1965. Receptive fields and functional architecture in two non-striate visual areas (18 and 19) of the rat. J. Neurophysiol., 28: 229-289.
HUSSERL, E. 1922. Logische Untersuchungen. 3rd ed. 3 vols. Halle. (1st ed., 1900).
HUXLEY, J. 1942. Evolution, The Modern Synthesis. London, Allen and Unwin.
———— A. C. HARDY, and E. B. FORD. 1954. Evolution as a process. London, Allen and Unwin.
JEANS, J. 1943. Physics and Philosophy. New York, The Macmillan Company.
JORDAN, P. 1945. Die Physik und das Geheimnis des organischen Lebens. 4th ed. Braunschweig, Vieweg (1st ed., 1941).
KANT, I. 1787. Kritik der reinen Vernunft. 2nd ed., Riga (*In* Cassirer, E., ed. 1923. Vol. 3, Berlin, Werke.)

LOCKE, J. 1690. An Essay Concerning Human Understanding. *In* St. John, J. A., ed. Philosophical Works. Vol. I, London, 1877.
MAYR, E. 1942. Systematics and the Origin of Species. New York, Columbia University Press.
——— 1963. Animal Species and Evolution. Cambridge, Mass., Harvard University Press.
OSTWALD, W. 1908. Grundriss der Naturphilosophie. Leipzig, Reclam.
PLANCK, M. 1933. Die Kausalität in der Natur. *In* Wege zur physikalischen Erkenntnis, 223-259, Leipzig, Hirzel.
PORSILD, A. E., C. R. HARINGTON, and G. A. MULLIGAN. 1967. *Lupinus arcticus* Wats. grown from seeds of Pleistocene age. Science, 158: 113-114.
RENSCH, B. 1947. Neuere Probleme der Abstammungslehre. Die transspezifische Evolution. Stuttgart, Enke. (2nd ed., 1954).
——— 1952. Psychische Komponenten der Sinnesorgane. Stuttgart, Thieme.
——— 1960. Evolution above the Species Level. New York, Columbia University Press.
——— 1967. The evolution of brain achievements. *In* Dobzhansky, Th., Hecht, M. K. and Steere, W. C., eds. Evolutionary Biology, Vol. 1. New York, Appleton-Century-Crofts. pp. 26-68.
——— 1968. Biophilosophie auf erkenntnistheoretischer Grundlage. (Panpsychistischer Identismus). Stuttgart, G. Fischer.
SCHLICK, M. 1920. Raum und Zeit in der gegenwärtigen Physik. Berlin, Springer-Verlag.
SCHRÖDINGER, E. 1944. What is Life? Cambridge, Mass., Cambridge University Press.
SIMPSON, G. G. 1944. Tempo and Mode in Evolution. New York, Columbia University Press.
——— 1949. The Meaning of Evolution. New Haven, Yale University Press.
——— 1953. The Major Features of Evolution. New York, Columbia University Press.
WADDINGTON, C. H. 1957. The Strategy of the Genes. London, Allen and Unwin.
WITTHÖFT, W. 1967. Zahl und Verteilung der Zellen im Hirn der Honigbiene. Z. Morphol. Ökol. Tiere, 61: 160-184.
WRIGHT, S. 1964. Biology and the philosophy of science. *In* Reese, R., and E. Freeman, eds. Process and Divinity. LaSalle, Ill., Open Court Publishing Co. pp. 101-125.
ZIEHEN, TH. 1913. Erkenntnistheorie auf psychophysiologischer und erkenntnistheoretischer Grundlage. Jena, G. Fischer.
——— 1934, 1939. Erkenntnistheorie. 2nd ed. 2 vols. Jena, G. Fisher.

4

Competition, Coexistence, and Evolution[1]

FRANCISCO J. AYALA

The Rockefeller University, New York, N.Y.

Introduction	121
The Ecological Niche	122
The Concept of Competition	124
The Principle of Competitive Exclusion	128
The Logistic Theory of Population Growth	130
Volterra's Equations for Competitive Exclusion	132
Coexistence of Related Species in Nature	136
Coexistence in the Laboratory	138
Conditions for Competitive Coexistence	146
Selection and Life Cycles	146
Frequency Dependent Selection	148
Niche Heterogeneity and Temporal Variations	150
The Role of Natural Selection	151
Summary	154
References	155

Introduction

One of the basic questions in ecology is the relationship between species living at the same trophic level, and in particular the relationship between species that may compete for the available resources of food and places to live. One generalization relevant to this problem is known as the com-

---

[1] Supported by PHS career development award K3 GM37265 from the National Institute of General Medical Sciences, and by contract AT-(30-1)-3096, U. S. Atomic Energy Commission.

petitive exclusion principle, or Gause's principle. According to this principle, two species competing for the same limited resources cannot coexist in the same locality; that is, one or the other species will sooner or later be eliminated. Considerable disagreement exists among ecologists concerning the validity and the importance of this principle. It is my purpose to show that the principle of competitive exclusion cannot be defended in its generality: two species may compete for limited resources and still coexist. Some of the conditions that allow competitive coexistence are discussed in the latter part of this paper. An investigation of the concepts of ecological niche and competition, which are essential for a precise formulation of the principle of competitive exclusion, precedes the consideration of the principle itself, and of the Volterra equations for species competition, which constitute the mathematical foundation of the principle.

The Ecological Niche

The term *niche* is used in the ecological literature with a variety of meanings, not always well defined. Grinnell (1904, 1917, 1924) was apparently the first naturalist to use the term. For Grinnell, the niche of a species is a subdivision of the habitat; it comprehends all the essential components of the environment in which the organisms live. Elton (1927; also Elton and Miller, 1954) contributed greatly to the popularity of the term, but modified its meaning. For Elton, the niche of an animal is its characteristic role in the ecosystem, especially its energy relations, e.g., the animal's relationships to the food it eats and the significance of this for other organisms in the ecosystem. It has become a commonplace to say that for Grinnell the niche is the "address" of the species, while for Elton the niche is the "function" or role of the organism.

Hutchinson (1957, 1965) gave a formalized definition of niche in terms of set theory. The niche of a species is conceived as a set of points in an abstract hypervolume defined by coordinates each of which represents an environmental factor such as temperature, humidity, kind of food, and so forth. Take two environmental variables, say temperature and relative humidity, measured along rectangular coordinates. The limiting values, within which the species can survive and reproduce, define an area each point of which corresponds to an environmental state permitting the species to exist indefinitely. If the two variables are independent in their action on the species, the area so defined will be a rectangle, but even without such independence the area will exist whatever the shape of its sides. A third variable may now be introduced to define a volume, and then further variables until the ecological factors relevant to the life of the species have all been considered. The n-dimensional hypervolume so defined, which contains all the points corresponding to environmental states that permit the species to exist, is called the *fundamental niche* of the species. "The fundamental niche of any species will completely define its ecological properties. The fundamental niche defined in this way is merely an abstract formalisation of what is usually meant by an ecological niche" (Hutchinson, 1957, p. 416).

The concept of ecological niche as defined by Hutchinson includes all the ecological relations of the species to the total physical and biotic environment. The limitations of this concept of niche come from the impossibility of being certain that all relevant variables of the environment have been included, and from the difficulty of measuring certain variables (e.g., food quality, food size, and many others) along continuous coordinates. Also, this definition of niche ignores the fact that the probability of survival and reproduction is not the same for all individuals at all points of the hypervolume, nor is it the same for all individuals at any point. It is not at all clear how a distribution of reproductive fitnesses could be incorporated for each point of every variable. Moreover, the set of points defined by the fundamental niche of the species is likely to change with time as the population evolves owing to natural selection. Although of considerable theoretical interest, Hutchinson's concept of ecological niche has, at the present level of knowledge, little operational value.

I suggest using the concept of niche in its maximum generality, to include all the relationships of an organism to the physical and biotic environment. The niche so defined includes the physical conditions of temperature, light, humidity, and so forth, required for the survival of the organism, as well as its requirements of food, place to live, and its relationships to other organisms of the same and other species. It is in this general sense that the term is most frequently used in the biological literature. Heuristically, then, it is possible to speak of the niche of animals in general terms; operationally, however, quantitative or qualitative statements about the niche of an individual or a species can be made only by referring to one (or more) specified component(s) of the ecological niche. For instance, it is possible to measure the survival and the optimal temperature range of organisms of a certain genetic constitution, or of a species; or to compare the food requirements of two species and to conclude that one species has a broader *food* niche than another, if the diet of the former includes all the foods acceptable to the latter, and in addition some others. Statements such as given in these two examples have a precise meaning subject to empirical verification. Attempts can be made to consider all the components of the niche that are relevant in a specific instance, since obviously not all aspects of the niche are equally significant. General statements can, then, be formulated concerning the niche, such as that the niche of a species S_1 is broader than the niche of another species S_2. But without explicit or implicit reference to one or several specific components of the niche, general statements concerning the niche of species have no empirical validity.

The concept of niche as defined above is analogous to the accepted usage of the concept of phenotype. Both concepts include an indefinitely large number of components. They can be used heuristically in general statements and they can also be employed in precise empirical statements by referring to specific components of the generalized concept. The phenotype is the physical makeup of the individual, resulting from the interaction of its genotype with the environment. The niche comprehends all the relationships of the organism to its environment. The niche and the phenotype are both subject to change in the time dimension. For instance, the temperature tolerance and the food requirements of a certain organism may change through its lifetime, as well as its shape and size.

Like the phenotype, the niche refers primarily to the individual organism. Also like the phenotype, the concept of niche can be extended to a population or a species. The niche of the species is described by the significant moments of the frequency distribution of the individuals that make up the species. In a loose sense it is also possible to speak of the niche of a taxon larger than the species, again by specification of the significant moments of the frequency distribution of all the organisms comprised in the taxon. Thus, it is possible to speak of the temperature tolerance and geographic distribution of the species *Drosophila melanogaster,* of the genus *Drosophila,* and of the order *Diptera.*

The Concept of Competition

The term competition is applied in biological literature to a variety of phenomena, among them direct struggle among organisms, interactions between organisms trying to secure the same resources, relationships between predator and prey, and others. Birch (1957, p. 6) argued convincingly that "competition" in biology should have the following restricted meaning: "Competition occurs when a number of animals (of the same or of different species) utilize common resources the supply of which is short; or if the resources are not in short supply, competition occurs when the animals seeking that resource nevertheless harm one or the other in the process."

The "short supply" in the definition may be an *absolute* shortage of a resource, i.e., if there is not enough food for all the animals present, or it may be a *relative* shortage, when the resource is not immediately available to the organisms or they fight over the same morsel, even if the resource is plentiful. Del Solar (1968) and Del Solar and Palomino (1966) have shown that females of *Drosophila pseudoobscura* and *D. melanogaster* have a "gregarious" tendency in oviposition. In the presence of many sites suitable to lay eggs, female *Drosophila* select preferentially the sites in which eggs have been laid by other *Drosophila* females of the same or of a different species. In experimental population cages, the females laid eggs only in about half of the food cups available, and one cup had 50 percent or more of all the eggs laid. There were considerably more eggs in the cup with the largest number than the number of individuals that could develop from the amount of food in the cup. Competition for food among *Drosophila* larvae occurs in the cup with the largest number of eggs even when plenty of food is available in the surrounding cups. Similarly, competition for food occurs when two birds choose to fight for the same carcass even though there may be enough other carcasses around for both of them.

Two aspects of the process of competition can be distinguished, which have been called by Park (1954) the *exploitation* and the *interference* components of competition. The exploitation component of competition refers to the use of a limited resource by the competing individuals. Interference occurs if the resource is exploited less efficiently when two species or genotypes compete, than when only one is present. Interference between *D. serrata* and *D. pseudoobscura* occurred in the experiments reported below (see p. 125). When only *D. pseudoobscura* AR flies were present in the population, 269 flies emerged per food unit; where only *D. serrata* were present, 535 flies emerged per food unit. When both species competed for food only 133 *serrata* and 105 *pseudoobscura* flies, or a total of 238 flies, emerged per food unit. Weisbrot (1966) found that diffusable metabolites produced by larvae of *D. melanogaster* decreased the survival of *D. pseudoobscura*. Dawood and Strickberger (1969) observed a similar phenomenon in intraspecific competition between *D. melanogaster* larvae: the metabolic wastes of larvae of either one of two genotypes decreased the viability of larvae of any one of three other gentoypes.

The exploitation component of competition is illustrated by the careful experiments of Bakker (1961) with *D. melanogaster*. About 0.60 mg of yeast are required by each larva to reach the minimum weight necessary to metamorphose into an adult fly. With 0.30 mg of yeast per larva, no adult flies emerge; with 0.45 mg, 60 percent, and with 0.60 mg, 90 percent, of the larvae become adults. Increasing the amount of food from 0.60 to 2.40 mg per larva does not alter the percentage of larvae becoming adults, but the dry weight of the individual adults increases proportionately to the amount of food (from 0.08 to 0.30 mg per fly). Competition between larvae of *D. melanogaster* and *D. simulans* occurs by exploitation without interference (Miller, 1964). At densities ranging from 10 to 120 larvae per culture, the same number of adults emerge when both species are present in equal numbers as when only *D. melanogaster* or only *D. simulans* is present. Larvae of two species of blowflies, *Lucilia sericata* and *Chrysomya chloropyga*, reared on a limited quantity of meat, gave no evidence of interference, although both competed for the exploitation of the same food (Ullyett, 1950). When larvae of *Lucilia sericata* competed with larvae of *Chrysomyia albiceps* the death rate of *Lucilia* larvae became higher with increasing larval density. Ullyett observed that the larvae of *C. albiceps* prey on *Lucilia* larvae. This, then, is an example of predation rather than of interference in competition. Predation and parasitism are, doubtless, forms of interference in the nontechnical sense of this word, but predation and parasitism have well-defined meanings in the ecological literature and should not be confused with competition if the latter term is to have a precise and useful meaning in biological discourse.

Facilitation is the antonym of interference in reference to competition. Weisbrot (1966) observed that the presence in the food medium of metabolites produced by *D. pseudoobscura* increased the viability of larvae of *D. melanogaster,* although, as pointed out above, metabolic products of *D. melanogaster* interfered with the development of *D. pseudoobscura.* Facilitation among certain genotypes of *D. melanogaster* was observed by Dawood and Strickberger (1969).

Symbiosis represents the extreme of facilitation. The notion of facilitation, however, should be restricted to situations in which two (or more) species or individuals compete for the same resource. There is facilitation when the effective utilization of the shared resource, i.e., its transformation into biomass, is greater when two species are present than when the same amount of resource is distributed among the two species and exploited separately.

In some cases, facilitation occurs at intermediate densities. The proportion of surviving larvae of *D. busckii* is generally greater when the density of larvae per vial is 32 or 128 than when the density is 2 or 8 larvae per vial (Lewontin and Matsuo, 1963). Facilitation at intermediate densities occurs also in *D. melanogaster* (Lewontin, 1955). In these cases facilitation exists although presumably there is no competition proper since at the lower densities there is no shortage of food.

Nicholson (1954) has distinguished between two types of competition: scramble and contest, which have different effects upon the outcome of the competition. *Scramble* is the kind of competition in which success is usually incomplete; the resource is shared in various proportions by the competing animals. *Contest* occurs when each object of the competition is completely appropriated by one of the competitors (whether the competitor is one individual, as in most cases, or a group of individuals, as in some cases of competition for a territory) so that the competitors are either fully successful or unsuccessful.

In the scramble type of competition some, and at times all, of the requisite secured by the competing animals may not contribute at all to sustaining the population, if it is dissipated by animals that do not obtain sufficient amount of the resource for survival. In the experiments of Bakker (1961), when the initial number of larvae of *D. melanogaster* in a fixed amount of food (120 mg of yeast) was increased from 200 to 400, the proportion of survivors decreased from 90 percent to nearly zero. At the higher density nearly all food was wasted although there was enough for the development of 180 individuals. On the other hand, in the contest type of competition the object of the competition contributes fully to the maintenance of the population; the successful competitors use it for their own survival while the losers obtain no part of the resource.

Competition for food occurs frequently as scramble, for instance in the competition among *Drosophila* larvae or among blowfly larvae. Competition for a home territory or for a nesting place occurs usually as contest. Competition for food sometimes takes the form of contest, as in some parasitic wasps; only one wasp larva develops in each prey larva. A wasp approaching a potential prey will not lay any egg on it if the larva already contains a developing wasp.

Competition and natural selection are different concepts, although natural selection occurs in many instances of competition. Natural selection occurs when there is differential reproduction. Natural selection is the result of at least three component processes—differential survival, differential mating success, and differential fecundity. Differential reproduction, and therefore natural selection, can take place without shortage of resources, i.e., without competition in the sense defined above. Genes frequently affect the relative reproductive success of their carriers even when there is no competition. Dobzhansky and Spassky (1953) found that 21 percent of the chromosomes of *D. pseudoobscura* collected in the field produced complete lethality of the individuals that carried them in homozygous condition; among the nonlethal chromosomes 21 percent decreased the probability of survival of the homozygous carriers, although there was no or little competition.

The reproductive fitness of the carriers of a gene or a gentoype may be affected by the intensity of the competition. Birch (1955) found that in *D. pseudoobscura,* the "Chiricahua" genotype had higher fitness than the "Standard" genotype at low larval densities, while at high larval densities the "Standard" phenotype was superior. Chitty (1965) and Pimentel (1961) have suggested that selection favoring certain genotypes at low densities and others at high densities may be a mechanism contributing to the regulation of animal numbers. The reproductive fitness of a genotype may also be affected by competition with another species. Dawson (1968) found a mutant gene in *Tribolium castaneum* whose frequency, relative to its wild-type allele, increased gradually when the population competed with *T. confusum.* As the frequency of the mutant gene increased, the absolute numbers of *castaneum* in the population decreased gradually until eventually they were completely eliminated by *confusum.* The most likely explanation of this observation is that the mutant gene had higher intraspecific fitness than its allele but was worse in interspecific competition than the latter.

Genotypes with higher fitness in intraspecific but lower in interspecific competition also occur in *D. willistoni* (Levin, 1967).

The Principle of Competitive Exclusion

The notion that two species with similar ecological requirements cannot coexist indefinitely was formulated in modern terms for the first time by Grinnell. In 1904 (p. 375-376) Grinnell wrote, "Two species of approximately the same food habits are not likely to remain long evenly balanced in numbers in the same region." Grinnell drew the conclusion, which follows immediately from that statement: "It is, of course, axiomatic that no two species regularly established in a single fauna have precisely the same niche relationships" (Grinnell, 1917, p. 443). Volterra (1926), Lotka (1932), and also Haldane (1924) concluded from their mathematical investigations of the problem that two species cannot coexist at an equilibrium if they utilize a common resource available in limited amount. This concept was elaborated by Gause (1934; Gause and Witt 1935). In his excellent monograph, *The Struggle for Existence,* Gause reviewed the mathematical investigations and provided experimental and field observations supporting the conclusions derived from them. The notion that two species competing for common resources cannot coexist has come to be referred to as Gause's Principle (Hardin, 1960; Miller, 1967, p. 3) or Gause's Axiom (Slobodkin, 1961, p. 68), and also as the Principle of Competitive Exclusion (Hutchinson, 1965, p. 27; Hardin, 1960), and the Principle of Competitive Displacement (De Bach, 1966).

The British Ecological Society held in 1944 a symposium to examine Gause's contention "that two species with similar ecology cannot live together in the same place, and the bearing of this, if true, on the origin and persistence of species" (Anonymous, 1944). A considerable number of publications dealing with the experimental studies of interspecies competition with various organisms appeared, mostly in British and American journals, during the immediately following years (for instance, Crombie, 1945, 1946, 1947; Elton, 1946; Williams, 1947; Park, 1948, 1954; Bagenal, 1951). Interest on the theoretical and experimental aspects of the principle of competitive exclusion has continued actively to the present day. Examples of recent general treatments are Hutchinson (1965), De Bach (1966) and Miller (1967).

Hardin (1960) stated that ecology stands "at the threshold of a renaissance of understanding, a renaissance made possible by the explicit acceptance of the competitive exclusion principle." Hutchinson (1965), De Bach (1966), and many others share Hardin's conviction that the principle of competitive exclusion is one of the central and most fertile concepts in ecological and evolutionary theory. To the contrary, Andrewartha and Birch (1954), and Cole (1960) think that the principle is either invalid or trivial, depending on how it is formulated. According to some other authors (e.g., Patten, 1961; Miller, 1967) competitive exclusion is a useful concept which must, however, be placed within a more general theoretical framework that includes the cooperative as well as the competitive aspects of interspecific relationships.

One source of confusion in the discussions of the competitive exclusion principle is that it means different things to different people. It is perhaps not unfair to include the various formulations in one of two categories, depending on whether the two species are required to be ecologically identical for the principle to apply, or whether competition for at least one limiting resource is taken as sufficient condition for coexistence to be impossible.

The most inclusive formulation states that two species with identical niches cannot coexist. Or, conversely, that two coexisting species must occupy different niches. For instance, Hutchinson (1965, p. 27) states that "in equilibrium communities no two species occupy the same niche," and according to Slobodkin (1961, p. 122), "if two species persist in a particular region it can be taken as axiomatic that some ecological distinction must exist between them." Thus formulated, the principle of competitive exclusion is true, but trivial. If niche is understood, in Hutchinson's broad sense, as including all the ecological relationships of the organism, it is true that no two species occupy the same niche; but two individuals do not occupy the same niche either. The theory of evolution implies unambiguously that two species, whether they coexist or not, will not have exactly the same ecological requirements. Slobodkin's (1961, p. 123) axiom that some ecological distinction must exist between two species which persist indefinitely at (or close to) a steady state in the same region, can be extended to any two species, whether they coexist or not. The requirement of identical ecological requirements is sometimes relaxed by saying that two species cannot coexist if they have "similar" ecology. Unless it is specified what is meant by "similar" ecology, this formulation is uninformative. Completely circular reasoning occurs when the degree of similarity required is said to be determined empirically: if two species cannot coexist they are considered similar ecologically; if they can coexist they are said not to be sufficiently similar.

A less inclusive formulation of the principle, and therefore more stringent is given by Elton (1927), Lack (1944), Hardin (1960), De Bach (1966), Van Valen (1960) and many others. Although the actual formulations vary, these authors claim that two species cannot coexist if they share one or more resources essential for the survival of the species. Frequently, though not always explicitly, the shared resource is required to exist in limited supply. That is, according to these authors, two species cannot coexist if they compete for the same essential resource, usually food or a place to live. Competition is here understood in the sense defined above: it occurs when a number of animals share a resource that exists in short supply, or when the animals sharing a resource that is not in short supply harm each other in the process of sharing. Competition for a limited resource, and not necessarily complete ecological identity, is also all that is required in the mathematical investigations of Volterra, Lotka, and Haldane. This is also Gause's concept: "It is admitted that as a result of competition two similar species scarcely ever occupy similar niches, but displace each other in such a manner that each takes possession of certain peculiar kinds of food and modes of life in which it has an advantage over its competitor" (Gause, 1934, p. 19).

The Logistic Theory of Population Growth

The logistic theory of population growth was developed independently by Verhulst (1839) and by Pearl and Reed (1920). Pearl (1926), Volterra (1931), Lotka (1932), Gause (1934) and many others have used the logistic theory to describe the growth of populations of various organisms, like man, *Drosophila, Paramecium,* and yeast. The logistic theory expresses quantitatively the idea that the growth of a population at any time is determined by the interaction between the potential rate of increase of the organisms and some limiting environmental factor. The rate of growth of a population at any time can be described by the differential equation:

$$\frac{dN}{dt} = rN\frac{K-N}{K} = rN\left(1 - \frac{N}{K}\right) \tag{1}$$

where r is the capacity of the organisms to increase in numbers under the specified environmental conditions of temperature, humidity, quality of food, and so on, when there are no limiting resources, N is the density of the population at any time, and K is the maximum density of the population when the limiting resource is saturated.

The logistic theory assumes that a population has a geometric capacity for increase proportional to the number of individuals present, rN, and that every individual added to the population decreases the capacity to increase per individual by a constant $c = \frac{1}{K}$. Biologically, this assumption means that every individual added to the population decreases the amount available of the limited resource by a certain quantity; when the resource is saturated the population ceases to grow.

The population is at equilibrium with the environment, that is, it ceases to grow when

$$\frac{dN}{dt} = 0 \qquad (2)$$

There are two trivial solutions of (2): when $r=0$ (the organisms have no capacity to reproduce) or $N=0$ (there are no organisms). The nontrivial solution occurs when $K-N=0$, or $K=N$. K is then the saturation value, the maximum value that N can reach given the specified amount of the limiting resource.

The most attractive feature of the logistic theory is its extreme simplicity and, therefore, its mathematical tractability. Simplicity is a desirable quality of any scientific theory, since it helps to make the world comprehensible. Simplicity, however, is no proof of validity. "Our strategy must be to try out simple hypotheses and theories first, but be ready for more complex ones if evidence so indicates" (Dobzhansky, unpublished). Examination of the simplifying assumptions of the logistic theory indicates a number of potential sources of invalidation of the theory. Some of the assumptions are:

1. That the innate rate of increase of the organisms remains constant while the population grows. This assumption implies genetic uniformity of the organisms, or that natural selection is not operative. Yet the relative fitness of genotypes may change as the density of a population changes (Chitty, 1960, 1965; Pimentel, 1961, 1968; Ayala, 1968a, c).

2. That all the organisms in the population have at any time identical demographic properties, that is, that the probability of dying, giving birth, and so on, is the same for all the individuals; or, else, that the age-distribution of the population remains constant.

3. That there is a linear relationship between density and rate of growth, that is, that the rate of growth decreases by a constant amount for each individual added to the population independently of its density. Experiments with a number of insects, particularly *Drosophila*, have shown that the rate of growth of a population is not linearly related to density (Pearl, 1926; Pearl, Miner, and Parker, 1927; Sang, 1950; and others).

4. That there is no time lag in the response of the organisms to the conditions of the population, that is, that the population responds instantaneously to the addition of each individual. A lag in the response of the organisms leads to oscillations of the population around its saturation point, K, which have been investigated by Utida (1957) and others.

These assumptions are likely to restrict or invalidate the application of the logistic equation to many, if not all, kinds of organisms. For a critical examination of the assumptions of the logistic theory see Andrewartha and

Birch (1954, pp. 362-386). How restrictive the assumptions are depends, of course, on the kinds of organisms being considered. Assumptions 2, and 3, for instance, are more likely to be invalid in organisms with complex life cycles, like insects or man, than in organisms with simple life cycles like yeast or bacteria. Thus, it is not surprising that attempts to fit population growth to the logistic equation have been more nearly successful with unicellular organisms (Loka, 1925; Gause, 1934) than with higher organisms (Andrewartha and Birch, 1954). Yet, even for populations of microorganisms, Feller (1940) has pointed out that other sigmoid equations fit the experimental data as well as, or better than, the logistic curve.

Volterra's Equations for Competitive Exclusion

The logistic equation can readily be extended to two or more species living in a limited universe, that is, two species competing for a limited resource. Competition between two species in the same microcosm was considered theoretically by Volterra (1926) and, independently, by Lotka (1932). Given two species, S_1 and S_2, living together in a limited microcosm, the rate of growth in numbers of one species, say S_1, will be proportional to its innate rate of increase, r_1, under the given environmental conditions, and to the number of individuals present, N_1. It is assumed, as in the logistic theory, that each individual of S_1 added to the population decreases the rate of growth per individual by a constant amount $\frac{1}{K_1}$, where K_1 is the saturation density when only S_1 is present in the microcosm. It is also assumed that each individual of the competing species, S_2, decreases the rate of growth of S_1 by a constant amount $\frac{\alpha}{K_1}$. Similarly, each individual of S_1 decreases the rate of growth of S_2 by a constant amount $\frac{\beta}{K_2}$. The parameters α and β, which may be called the coefficients of competition, measure the inhibitory effect of an individual of one species on the rate of growth of the other species.

The instantaneous rate of growth of two species competing for a common limiting resource is expressed by the differential equations

$$\frac{dN_1}{dt} = r_1 N_1 \frac{K_1 - N_1 - \alpha N_2}{K_1};$$

$$\frac{dN_2}{dt} = r_2 N_2 \frac{K_2 - N_2 - \beta N_1}{K_2} \quad (3)$$

where the symbols represent the parameters just described, and the subscripts 1 and 2 apply respectively to S_1 and S_2.

The application of equations (3) to actual populations is restricted by the same assumptions as the logistic equation for the growth of a single species, plus the additional assumption that each individual of one species inhibits the growth of the other species by a constant quantity which is independent of the density of either species. Whether these assumptions invalidate or not the application to natural populations of the equations (3) can best be decided by examining whether, as it is sometimes claimed (Slobodkin, 1961, p. 62), the equations and the conclusions derived from them are in good agreement with experimental and field observations. It must be emphasized that the strongest case for the competitive exclusion principle rests on the alleged general validity of certain conclusions derived from the equations (3).

The system described by the equations (3) will be at equilibrium when

$$\frac{dN_1}{dt} = 0, \text{ and } \frac{dN_2}{dt} = 0 \qquad (4)$$

Trivial equilibria exist when r_1, r_2, N_1, or $N_2 = 0$
A nontrivial solution exists when

$$\begin{aligned} K_1 - N_1 - \alpha N_2 &= 0, \\ K_2 - N_2 - \beta N_1 &= 0 \end{aligned} \text{ and} \qquad (5)$$

Equations (5) can be expressed as

$$\begin{aligned} N_1 &= K_1 - \alpha N_2, \\ N_2 &= K_2 - \beta N_1 \end{aligned} \text{ and} \qquad (6)$$

which describe the equilibrium density of each species as a linear function of its saturation density, K, and the equilibrium density of the competing species. Since equations (6) are algebraic representations of straight lines, the possible outcomes of competition can be best investigated graphically in a plane limited by the rectangular coordinates N_1 and N_2. By making first $N_1 = 0$, and then $N_2 = 0$, two values that intercept the coordinates are obtained for the first equation (6). The straight line joining these two intercepts contains all the points at which $\frac{dN_1}{dt} = 0$. Inside the area defined by this zero isocline and the two coordinates, $\frac{dN_1}{dt}$ is positive, and either N_1 or both N_1 and N_2 will increase. Outside that area, $\frac{dN_1}{dt}$ is negative and

N_1, N_2, or both will decrease. Similarly, by making first $N_1 = 0$, and then $N_2 = 0$ in the second equation (6), the two intercepts with the coordinates of the zero isocline defined by this equation are obtained. All the points at which $\dfrac{dN_2}{dt} = 0$ are contained in the straight line joining the two intercepts. The simultaneous condition $\dfrac{dN_1}{dt} = \dfrac{dN_2}{dt} = 0$ will occur at the point or points, if they exist, where the two isoclines intercept each other.

Leaving aside the trivial case where $K_1 = K_2$, and $\alpha = \beta$, the outcome of the competition may be investigated by expressing the two coefficients of competition, α and β, as inequality functions of K_1 and K_2. There are four possible outcomes, which are represented graphically in Figure 1:

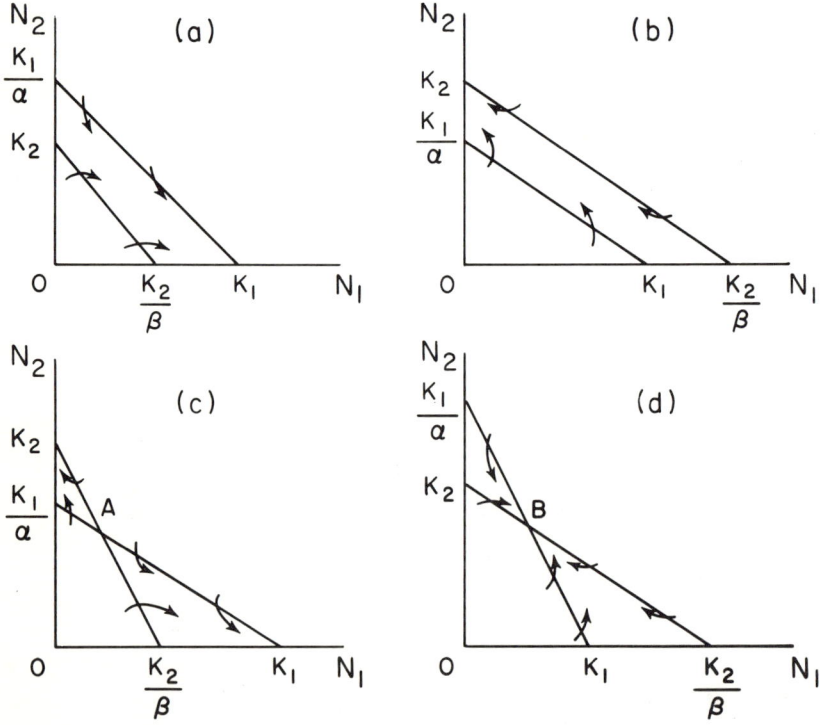

Fig. 1. Competition between two species. The coordinates N_1 and N_2 indicate the numbers of the two competing species S_1 and S_2 respectively. The lines K_1, K_1/α and K_2, K_2/β contain the saturation points for species S_1 and S_2 respectively. The arrows indicate the change in number of the species: below the saturation line each species will increase; above the saturation line each species will decrease. (a) S_1 wins; at equilibrium $N_1 = K_1$, $N_2 = 0$. (b) S_2 wins; at equilibrium $N_1 = 0$, $N_2 = K_2$. (c) There is one point of unstable equilibrium at A. (d) Stable equilibrium occurs at B.

(a). If $\alpha < \frac{K_1}{K_2}$ and $\beta > \frac{K_2}{K_1}$, the system will reach equilibrium only when $N_1 = K_1$, and $N_2 = 0$. S_1 is the only survivor.

(b). If $\alpha > \frac{K_1}{K_2}$ and $\beta < \frac{K_2}{K_1}$, equilibrium is reached only when $N_2 = K_2$, and $N_1 = 0$. S_2 is the only survivor.

(c). If $\alpha > \frac{K_1}{K_2}$ and $\beta > \frac{K_2}{K_1}$, the zero isoclines intercept at one point (A in the Figure). The outcome of the competition depends on the initial conditions. If N_1/N_2 is at any time greater than a certain value, equilibrium will be reached when $N_1 = K_1$, and $N_2 = 0$. If N_1/N_2 is at any time smaller than that value, equilibrium occurs when $N_2 = K_2$ and $N_1 = 0$. Thus either S_1 or S_2 is the only survivor, depending on the initial conditions. The intercept of the isoclines gives an unstable equilibrium with both species coexisting. Any departure, however small, from the point of equilibrium will lead to fixation with either N_1 or N_2 equal to zero.

(d). If $\alpha < \frac{K_1}{K_2}$ and $\beta < \frac{K_2}{K_1}$, the two isoclines intercept at one point (B in the Figure), which gives the outcome of the competition if N_1 and N_2 are initially greater than zero. (The model is deterministic; stocastic events are ignored.) When

$$N_1 = \frac{K_1 - \alpha K_2}{1 - \alpha\beta} > 0, \text{ and } N_2 = \frac{K_2 - \beta K_1}{1 - \alpha\beta} > 0 \qquad (7)$$

the system reaches equilibrium and S_1 and S_2 coexist. The equilibrium given by (7) is stable if $\alpha < \frac{K_1}{K_2}$ and $\beta < \frac{K_2}{K_1}$. Any departure from equilibrium will be counteracted by the competition process until the equilibrium is restored. The inequalities $\alpha < \frac{K_1}{K_2}$ and $\beta < \frac{K_2}{K_1}$ can be expressed as $\frac{\alpha}{K_1} < \frac{1}{K_2}$ and $\frac{\beta}{K_2} < \frac{1}{K_1}$. According to the definitions given above, $\frac{\alpha}{K_1}$ is the inhibitory effect of one individual of S_2 on the growth of S_1, while $\frac{1}{K_2}$ is the inhibitory effect of one S_2 individual on the growth of S_2. Similarly, $\frac{\beta}{K_2}$ and $\frac{1}{K_1}$ measure the inhibitory effect of one S_1 individual on the growth of S_2 and S_1, respectively. Thus, the conditions $\alpha < \frac{K_1}{K_2}$ and $\beta < \frac{K_2}{K_1}$ mean that any one organism inhibits the growth of its own species more than it inhibits the growth of the other species. This is taken to mean that the two species occupy different niches or, more precisely, that the two species are not limited by competition for the same shared resource. This argument constitutes the basis of the competitive exclusion principle: two species competing for the same limiting resource cannot coexist indefinitely; and the reverse formulation: two species coexisting at a stable equilibrium are controlled by different limiting resources. The principle can, of course, be extended by the same argument to more than two species.

Coexistence of Related Species in Nature

Species closely related phylogenetically and ecologically have frequently been observed to coexist in the same habitat, apparently exploiting the same resources. Some instances have been recently reviewed by Hutchinson (1965), and Miller (1967). I shall limit myself to three examples.

Ross (1957) studied a monophyletic group of six species of the leafhopper *Erythroneura* living on sycamores in Illinois. Individuals of up to five species may occupy the same tree. There are slight differences in tolerances to dryness among the species. *E. lawsoni* is the only consistent inhabitant of trees in open, dry, wind-swept situations, while the tolerance to dryness decreases from *E. arta*, through *E. usitata*, and *E. torella* to *E. morgani* and *E. bella*. The last five species rarely exist outside humid valleys or well-protected trees. However, even on trees in sheltered, humid valleys *E. lawsoni* is also frequently the dominant species. There is little evidence of competition and Ross suggests that rainfall may be the most important factor regulating these populations. All the species survive in at least some favorable foci from which they invade neighboring habitats when the conditions become favorable.

Fourteen species of the genus *Drosophila* live in the Sierra Nevada Mountains in the Yosemite region of California (Cooper and Dobzhansky, 1956). Some species, like *D. pseudoobscura, D. persimilis, D. occidentalis,* and *D. pinicola* occur at all elevations at which collections have been made ranging from 850 to 11,000 feet (250 to 3,300m). Some other species are restricted to part of the range. Every species is relatively more abundant in some altitudinal zones than in others. Within a locality, the relative abundance of different species varies from season to season and from year to year. There is evidence of competition among *D. pseudoobscura, D. persimilis,* and *D. miranda,* and perhaps also *D. azteca,* whose larvae feed on the same slime fluxes, mostly of oak trees (Dobzhansky, personal communication). The range of *D. pseudoobscura* includes the range of the other three species, although it is not the dominant species in all localities.

There is also evidence of competition between *D. occidentalis* and *D. pinicola*, which feed on fungi.

Dobzhansky and his colleagues have conducted since 1955 extensive studies of the four monophyletic species, *D. willistoni, D. equinoxialis, D. tropicalis,* and *D. paulistorum,* which live in the tropics of the New World.

TABLE I

Number of Females of Four Sibling Species of *Drosophila* Collected in Various Localities of Central and South America

Locality	Date	D. willistoni	D. tropicalis	D. equinoxialis	D. paulistorum
Costa Rica					
Turrialba	August 1956	14	10	22	18
La Lola	August 1956	2	42	89	5
Panama					
Locality not recorded	November 1955	27	17	418	47
Barro Colorado	August 1956	8	32	22	12
Cerro Campana	August 1956	3	3	2	6
Locality not recorded	February 1958	24	5	21	1
Locality not recorded	August 1959	5	10	12	3
Cerro Campana	September 1961	12	1	11	40
Barro Colorado	September 1961	4	31	12	4
Cerro Azul	June 1962	8	2	2	8
Darien	December 1962	4	1	72	2
Colombia					
Bucaramanga	September 1956	97	2	22	3
Llanos, Locality 1	March 1958	53	7	6	2
Llanos, Locality 2	March 1958	61	9	16	2
Turbo	February 1967	1	275	31	4
Leticia	February 1968	29	8	97	102
Valparaiso	February 1968	228	33	29	89
La Macarena	February 1968	72	37	177	2
Mitu	February 1968	116	6	29	121
Teresita	February 1969	94	240	388	15
Venezuela					
Cumanacoa	November 1956	2	1	8	1
Tuy	January 1963	12	9	88	3
Sarare	November 1967	12	28	37	2
Trinidad					
Locality not recorded	April 1958	59	2	2	19
British Guyana					
Locality not recorded	August 1957	1	2	2	2
Brazil					
Marco	February 1968	16	62	58	14

The species are morphologically indistinguishable, although Spassky (1957) discovered some consistent differences in the genitalia of the males. The species are also ecologically similar: they feed on a variety of tropical fruits, the same fruit frequently containing individuals of two or more species. Nevertheless, the four species coexist over a wide geographic area, from Central America and Trinidad, through Columbia, Venezuela, and Guyana, down to the Matto Grosso in Brazil. Table I has been compiled from the records kept by Mr. Spassky of collections made by several investigators from 1955 to 1968.

Those who propose the general validity of the principle of competitive exclusion argue that there is considerable evidence that species drawing upon similar resources exclude one another from the same region. Elton (1946) analyzed published ecological surveys of 55 animal communities and 27 plant communities from a wide range of habitats. Eighty-four to 86 percent of the genera were represented by only one species in any particular community. Surveys for larger regions indicated that the average number of species of the same genus in any community should be greater than the observed number if the species associated at random. Elton concluded that similar species tend to exclude one another from the same community by competition. This analysis by Elton has been convincingly invalidated (Williams, 1947; Bagenal, 1951), but it can still be argued that when two or more species coexist they either do not compete for the same resources, or if they do compete, each species must draw from additional resources so that the coexisting species are not limited by the resources that they share.

It is obvious that no amount of empirical evidence indicating coexistence of competing species obtained from field studies will be a conclusive argument against this sort of argument. The evidence can always be said to be incomplete, with the comment that finer analysis would have indicated significant ecological differences among the coexisting species. "If they seem identical the study is incomplete" (Slobodkin, 1961, p. 123).

Coexistence in the Laboratory

Park (1948) studied the dynamics of mixed populations of the flour beetles, *Tribolium confusum* and *T. castaneum* at 29°C and 65 to 70 percent relative humidity. In some populations *T. confusum,* and in others *castaneum,* was eliminated. Yet both species coexisted for a considerable length of time, dependent on the amount of food provided. With 8 g of flour, the mean number of days to the extinction of one of the species was 548 days; with 80 g of flour, the mean was 1155 days. That is, the two species coexisted for a mean number of 16 generations with 8 g of flour, and for a mean of 33 generations with 80 g.

The outcome of the competition between *T. confusum* and *T. castaneum* depends on the conditions of temperature and relative humidity (Park, 1954). High temperature and high relative humidity favor *T. castaneum*, while low temperature and humidity favor *T. confusum*. Thus at 34°C and 70 percent relative humidity, *confusum* is always eliminated first, and at 24°C and 30 percent relative humidity, *confusum* always wins. At intermediate temperature and/or relative humidity the outcome is not determined for every population, although the probability of one or the other species winning depends on the specific conditions. At 29°C, *T. castaneum* was the winner species in 86 percent of the populations when the relative humidity was 70 percent; when the relative humidity was 30 percent, *T. castaneum* won in only 13 percent of the populations. It seems that at intermediate temperature and humidity, the two species are nearly equal food competitors, and can coexist for many generations, the final outcome being largely decided by stochastic events. (The indeterminacy of the outcome has been shown by Lerner and Ho, 1961, to be dependent on the genetic constitution of the populations.)

Drosophila funebris and *D. melanogaster* can coexist at 20°C for long periods of time. Merrell (1951) found that *D. melanogaster* is favored by the addition of fresh food, while the proportion of *D. funebris* increases with the age of the food. When food is added at regular intervals, a fairly stable equilibrium occurs, with *D. melanogaster* predominating. The two species coexisted until the experiment was terminated after nearly two years or about 30 generations. Although the two species differ ecologically, there is no doubt that they competed for the same basic source of energy.

I have studied competition in the laboratory between *Drosophila serrata* and *D. nebulosa*. At 25°C *D. serrata* is eliminated in a few generations, but at 19°C both species coexist with *nebulosa* predominating (Ayala, 1966a). In another experiment, two populations were started at 19°C with 200 flies of each species. The populations were maintained by the serial transfer technique (Ayala, 1965a). The adult flies were introduced into one-half pint (0.47 liter) bottles with food. Every seven days they were etherized, counted, and transferred to a fresh bottle. When emergence of adult flies began in the containers where the flies had deposited eggs, the newly emerged flies were etherized, counted, and added to the container with the adult population. The adult ovipositing flies were thus in a single container while five other containers for each population had eggs, larvae, pupae, and young adults. This technique allows easy measurement of two parameters, namely, productivity, or number of flies emerging per container, and survival, or average longevity.

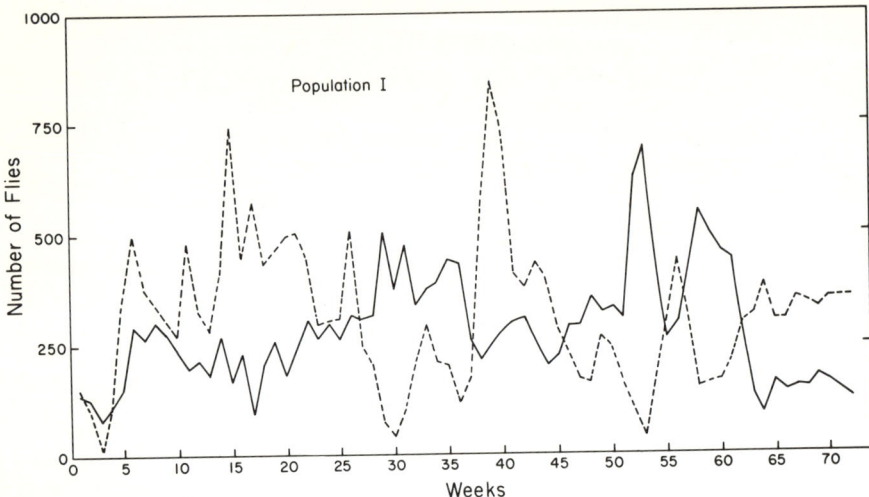

Fig. 2. Numbers of Drosophila serrata (solid line) and of D. nebulosa (broken line) in experimental population I.

The process of competition between the two species can be followed in Figures 2 and 3. The means for the number of flies emerging per week ("newborn" flies), the number of flies surviving from the previous week ("old" flies), and the total of both are given in Table II. *D. serrata* and *D.*

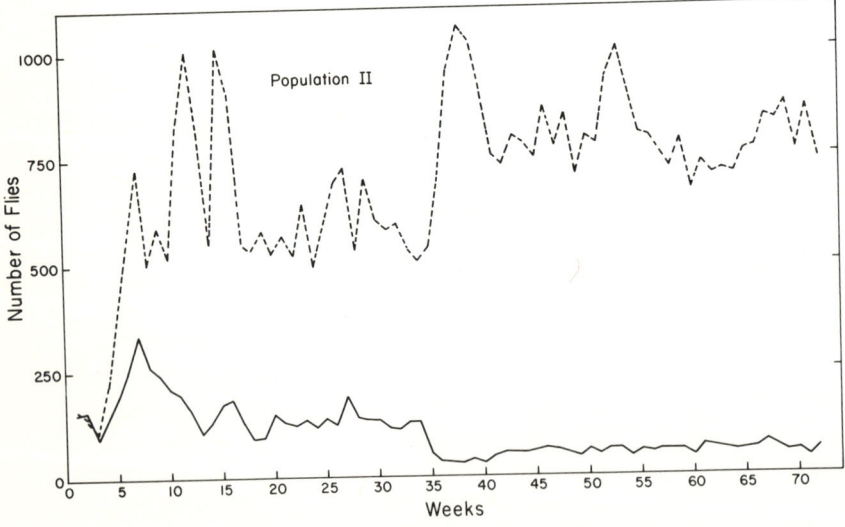

Fig. 3. Numbers of Drosophila serrata (solid line) and of D. nebulosa (broken line) in experimental population II.

TABLE II

Mean Number of Flies of *Drosophila serrata* and *D. nebulosa* in Each of Two Experimental Populations

Population	Species	Newborn	Old	Total
I	*D. serrata*	126 ± 9	164 ± 9	290 ± 15
	D. nebulosa	146 ± 13	162 ± 13	308 ± 22
II	*D. serrata*	26 ± 3	49 ± 7	75 ± 10
	D. nebulosa	267 ± 13	412 ± 17	679 ± 23

nebulosa coexisted in both populations for 72 weeks, or about 25 generations.

In population I there were wide oscillations in the numbers of the two species. These oscillations can be interpreted as indications that the equilibrium between the two species is unstable (see below), as due to uncontrolled environmental fluctuations or as the result of changes in the competitive ability of the species by natural selection. The first two hypotheses are unlikely, because populations I and II were treated similarly and simultaneously, and yet, the numbers of the two species remained reasonably constant in population II throughout the experiment. By genetic tests, it was demonstrated that the increase in *D. serrata* in population I, which started around week 22, was due to natural selection of genotypes with improved competitive ability (Ayala, 1969a).

The two species compete in these populations for the available resources of food and space. Clear evidence of competition comes from the inverse relationship between the numbers of the two species. The average number of *D. serrata* is considerably less in population II than in population I, while the reverse is true of *D. nebulosa*. Definite evidence of competition can be obtained from Figure 2; the increase in numbers of *D. serrata* from week 22 to 30 is accompanied by a decrease in *D. nebulosa*. In general, it can be observed from Figures 2 and 3 that when one species increases, the other species living in the same microcosm decreases, indicating that the two species compete for limiting resources.

The competition experiments of *Drosophila* just reported did not include measurement of the saturation levels of each species when growing alone. That is, no attempt was made to measure the parameters K_1 and K_2 in equations (5), and therefore there is no way to estimate α and β. To obtain this information, I have studied the competition between *D. serrata* and *D. pseudoobscura* in experiments which will be reported in full elsewhere.

At 19°C *D. pseudoobscura* eliminates *D. serrata* in a few generations and at 25°C *D. pseudoobscura* is rapidly eliminated (Ayala, 1969b). At 23.5°C the two species coexist at relative frequencies that depend on the genetic composition of the particular strains of the two species. Two strains of *D. pseudoobscura* were used, one homokaryotypic for the third chromosome inversion "Chiricahua" (CH), the other homokaryotypic for the third-chromosome inversion "Arrowhead" (AR) (Dobzhansky, 1944). The strain of *D. serrata* was collected in Popondetta, New Guinea (Ayala,

TABLE III

Mean Number of Flies of *D. serrata* and *D. pseudoobscura* in the Experimental Populations

Population	Species	Newborn	Old	Total
44	*D. serrata*	125 ± 14	137 ± 14	262 ± 25
	D. pseudoobscura AR	105 ± 10	149 ± 16	255 ± 24
45	*D. serrata*	101 ± 8	87 ± 7	187 ± 11
	D. pseudoobscura AR	119 ± 8	169 ± 11	288 ± 17
46	*D. serrata*	162 ± 15	168 ± 16	331 ± 27
	D. pseudoobscura AR	92 ± 9	115 ± 14	207 ± 20
47	*D. serrata*	131 ± 8	161 ± 10	292 ± 14
	D. pseudoobscura AR	108 ± 8	171 ± 12	279 ± 18
48	*D. serrata*	129 ± 9	160 ± 13	289 ± 18
	D. pseudoobscura AR	107 ± 8	154 ± 11	261 ± 17
49	*D. serrata*	149 ± 9	155 ± 11	304 ± 13
	D. pseudoobscura AR	97 ± 8	127 ± 10	223 ± 16
Average	*D. serrata*	133.0 ± 7.5	144.5 ± 9.1	277.6 ± 12.8
	D. pseudoobscura AR	104.6 ± 7.1	147.4 ± 11.0	252.0 ± 16.8
64	*D. serrata*	268 ± 15	278 ± 17	547 ± 27
	D. pseudoobscura CH	61 ± 8	69 ± 8	129 ± 15
66	*D. serrata*	245 ± 18	295 ± 23	539 ± 33
	D. pseudoobscura CH	78 ± 10	66 ± 9	145 ± 17
67	*D. serrata*	269 ± 20	319 ± 21	588 ± 33
	D. pseudoobscura CH	74 ± 7	75 ± 8	149 ± 14
68	*D. serrata*	248 ± 19	314 ± 25	562 ± 38
	D. pseudoobscura CH	55 ± 7	70 ± 8	125 ± 13
69	*D. serrata*	254 ± 19	254 ± 21	508 ± 36
	D. pseudoobscura CH	49 ± 12	61 ± 9	110 ± 20
Average	*D. serrata*	257.0 ± 13.5	292.0 ± 16.0	549.1 ± 25.6
	D. pseudoobscura CH	63.3 ± 6.2	68.3 ± 7.4	131.7 ± 13.1

1965b). Six replicate populations were started where *D. serrata* competed with *D. pseudoobscura* AR, and five replicates in which it competed with *D. pseudoobscura* CH. Every population was started with 300 flies of each species. The populations were maintained by the serial transfer technique described above.

In six weeks, the two species reached equilibria at levels dependent on the genetic composition of *D. pseudoobscura*. The populations were studied for 40 weeks, or about 15 generations. Table III gives the means and standard errors of each species for the following parameters: number of flies emerging per week ("newborn"); number of flies surviving from the previous week ("old"); and the total of both. The total count of each species in two populations, one of each type, chosen at random are represented in Figures 4 and 5. At equilibrium, *D. pseudoobscura* flies were 18 percent of the total number in the CH populations and 45 percent in the AR populations. Homozygous AR/AR flies are better competitors with *D. serrata* than flies CH/CH. The increase of *D. pseudoobscura* in the AR compared to the CH populations occurs at the expense of a decrease in the number of *D. serrata*.

After the populations reached equilibria, flies of each species were sampled from every population. With the F_1 progenies of these samples, the following "single-species" populations were established: three *D. pseudo-*

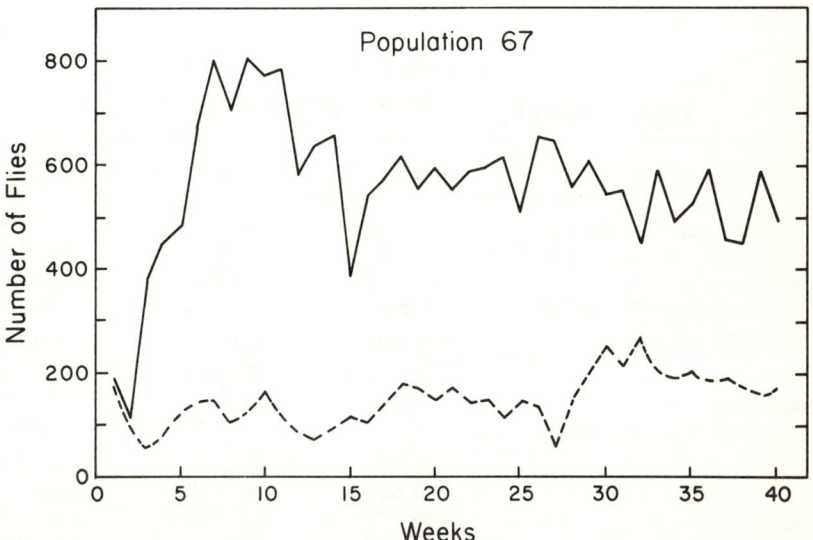

Fig. 4. Numbers of Drosophila serrata *(solid line)* and of D. pseudoobscura CH *(broken line)* in experimental population 67.

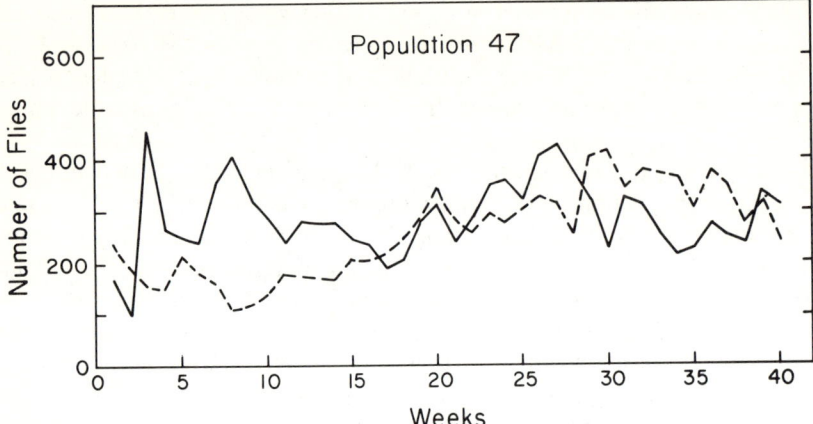

Fig. 5. Numbers of Drosophila serrata (solid line) and of D. pseudoobscura AR (broken line) in experimental population 47.

obscura CH and three AR populations; six *D. serrata* populations. Every population was started with 600 adult flies and maintained at 23.5°C by serial transfer. The mean numbers of flies after equilibrium and the standard errors are given in Table IV. The number of flies of one species is more than double in the single-species than in the two-species populations.

Equations (3) can now be solved for α and β:

TABLE IV

Mean Number of Flies in the Single-Species Populations

Population	Newborn	Old	Total
D. pseudoobscura AR, 40	272 ± 13	418 ± 14	690 ± 18
D. pseudoobscura AR, 41	262 ± 25	378 ± 29	640 ± 27
D. pseudoobscura AR, 42	274 ± 17	388 ± 42	662 ± 43
D. pseudoobscura AR, average	269.3	394.9	664.1
D. pseudoobscura CH, 60	208 ± 32	271 ± 35	480 ± 35
D. pseudoobscura CH, 61	281 ± 32	373 ± 50	654 ± 60
D. pseudoobscura CH, 62	300 ± 29	374 ± 55	674 ± 69
D. pseudoobscura CH, average	263.2	339.4	602.7
D. serrata, 40	508 ± 46	682 ± 59	1190 ± 63
D. serrata, 41	566 ± 51	669 ± 87	1235 ± 74
D. serrata, 42	530 ± 37	692 ± 100	1222 ± 93
D. serrata, 60	524 ± 40	689 ± 70	1213 ± 49
D. serrata, 61	530 ± 76	821 ± 97	1351 ± 104
D. serrata, 62	563 ± 42	730 ± 108	1294 ± 101
D. serrata, average	536.8	713.9	1250.7

$$\alpha = \frac{K_1 - N_1 - \frac{dN_1}{dt}\left(\frac{K_1}{r_1 N_1}\right)}{N_2}$$

$$\beta = \frac{K_2 - N_2 - \frac{dN_2}{dt}\left(\frac{K_2}{r_2 N_2}\right)}{N_1}$$

(8)

At equilibrium, $\frac{dN_1}{dt} = \frac{dN_2}{dt} = 0$, and therefore

$$\alpha = \frac{K_1 - N_1}{N_2} \qquad \beta = \frac{K_2 - N_2}{N_1}$$

Designating *D. pseudoobscura* as S_1 and *D. serrata* as S_2 we obtain for the CH populations: $K_1 = 602.7$, $K_2 = 1250.7$, $N_1 = 131.7$, $N_2 = 549.1$. Therefore:

$$\alpha_{\text{CH}} = \frac{602.7 - 131.7}{549.1} = 0.858$$

$$\beta_{\text{CH}} = \frac{1250.7 - 549.1}{131.7} = 5.327$$

$$\frac{K_1}{K_2} = \frac{602.7}{1250.7} = 0.482$$

$$\frac{K_2}{K_1} = \frac{1250.7}{602.7} = 2.075$$

For the AR populations: $K_1 = 664.1$, $K_2 = 1250.7$, $N_1 = 252.0$, $N_2 = 277.6$. Therefore

$$\alpha_{\text{AR}} = \frac{664.1 - 252.0}{277.6} = 1.485$$

$$\beta_{\text{AR}} = \frac{1250.7 - 277.6}{252.0} = 3.861$$

$$\frac{K_1}{K_2} = \frac{664.1}{1250.7} = 0.531$$

$$\frac{K_2}{K_1} = \frac{1250.7}{664.1} = 1.883$$

The α's and β's are, in both populations, greater than K_1/K_2 and K_2/K_1, respectively, giving evidence of competition. However, the conditions for coexistence according to the Volterra equations are $\alpha < \dfrac{K_1}{K_2}$ and $\beta < \dfrac{K_2}{K_1}$. The Volterra equations, then, are not necessarily valid for all instances of competition between species. Two species competing for limiting resources may coexist.

Conditions for Competitive Coexistence

The widespread acceptance of the principle of competitive exclusion has led to considerable neglect in the exploration of the conditions that may allow coexistence of competing species. I shall now consider briefly some of the mechanisms that may lead to competitive coexistence.

Selection and Life Cycles

Among insects and other organisms with complex life cycles survival and reproduction may depend on different resources at different stages of development. The larvae of certain mosquitoes are aquatic and have a different diet from that of the terrestial adults. Larval survival depends on the availability of certain plants and microorganisms; the fecundity of the adult females depends instead on their ability to obtain a protein meal from vertebrate blood. (The possibility that wild mosquitoes substitute hemolymph from larvae of other insects for vertebrate blood has been suggested recently by Harris, Riordan, and Cooke, 1969.) Butterfly larvae feed on leaves and other soft parts of their food plants, while the imagoes feed on flower sap. Food shortages may limit the survival of larvae; adults, however, are frequently controlled by predators (Kettlewell, 1955; Ford, 1964).

In laboratory experiments with the sheep-blowfly *Lucilia cuprina,* Nicholson (1954, 1957) has shown that two limiting factors, one operating at the larval and the other at the adult stage, may interact and regulate jointly the density of the population. Slobodkin (1954, 1957) showed that two limiting factors, food and predation, may jointly determine population size in *Daphnia.* In serial transfer populations of *Drosophila serrata* Ayala (1966b) found that the number of flies emerging per food unit was determined by the amount of food. In these very crowded populations, the adults competed for living space. Population size was controlled by the interaction between the competition for food among the larvae and the competition for living space among the adults. When space was kept constant an increase of 50 percent in the amount of food resulted in an increase of 39 percent in the number of adults emerging per time unit, yet the average adult population increased only by 17 percent. Similar results were obtained in experiments with *D. pseudoobscura, D. melanogaster,* and *D. birchii* (Ayala, 1967, 1968b). The interaction between food and space in the regulation of the size of the *Drosophila* populations was further investigated in an experiment in which the amount of food and the amount of space were both varied (Ayala, unpublished). When space was kept constant an increase of 50 percent in the amount of food resulted in an increase of 36 percent in the average population size; when food was constant an increase of 100 percent in the amount of space produced an increase of 20 percent in the average population numbers (Ayala, 1968c).

If the survival of larvae and adults is limited by different factors, it is possible that one species may be a better competitor than the other species at one stage of the life cycle but worse at some other stage. This seems to be the case in the competition between *D. serrata* and *D. nebulosa,* and between *D. serrata* and *D. pseudoobscura.* For instance, in Table II it can be observed that in population I the average number of adults emerging per week is greater for *D. nebulosa* than it is for *D. serrata.* Yet the average number of "old" flies is about the same for the two species, indicating that the average longevity of adult flies is greater for *D. serrata* than it is for *D. nebulosa.* Similarly (Table III) more *D. serrata* flies emerge per unit food than *D. pseudoobscura* AR, yet this difference is neutralized by the greater average longevity of the *D. pseudoobscura* flies.

A simple theoretical model can be constructed that allows the coexistence of two competing species if the relative fitness of one species is lower at one stage of development but higher at another than the relative fitness of the competing species. Assume that the fitness of one species, S_1, relative to another species, S_2, is W_1 for survival from zygote to adult; and that the fitness of S_2 relative to S_1 is W_2 for the adults (fecundity included). Let $P = \dfrac{N_1}{N_2}$ be the proportion of S_1 to S_2 zygotes at a certain time, t_0. We have:

	S_1	S_2	
Proportion of zygotes at t_0	P	1	
Relative fitness from zygote to adult	W_1	1	
Proportion of adults after larval selection	PW_1	1	
Relative fitness of the adults	1	W_2	
Proportion of zygotes after one generation	PW_1	W_2	(9)

After one generation of competition the proportion of S_1 to S_2 will be PW_1 to W_2 or N_1W_1/W_2 to N_2. If $W_1 > W_2$ the proportion of S_1 zygotes will increase after each generation of competition; if $W_1 < W_2$ the proportion of S_1 will decrease; and if $W_1 = W_2$ the proportion of S_1 to S_2 will remain constant from one generation to the next.

Frequency-Dependent Selection

The condition $W_1 = W_2$ leads to an unstable equilibrium. The equilibrium between the two species remains at the proportion determined by the initial conditions. If the ratio of N_1 to N_2 changes for whatever reason, the original equilibrium will not be restored but rather a new equilibrium will be established at the ratio of N_1 to N_2 determined by the change. Stochastic events will ultimately lead to the elimination of one or the other species. An unstable equilibrium probably occurred in the experiments by Park (1948, 1954) with *Tribolium*. One or the other species was eliminated in every experiment, although the probability of elimination of each species was determined by the conditions of temperature and moisture. Indication that competition at the larval stage between *D. melanogaster* and *D. simulans* may lead to an unstable equilibrium has been obtained by Miller (1964, 1967).

The condition $W_1 = W_2$ appears also to be highly restrictive. It seems unlikely that the relative advantage of one species over the other at one stage of development will be compensated exactly by its disadvantage at some other stage. However, $W_1 = W_2$ will be less unlikely if the relative

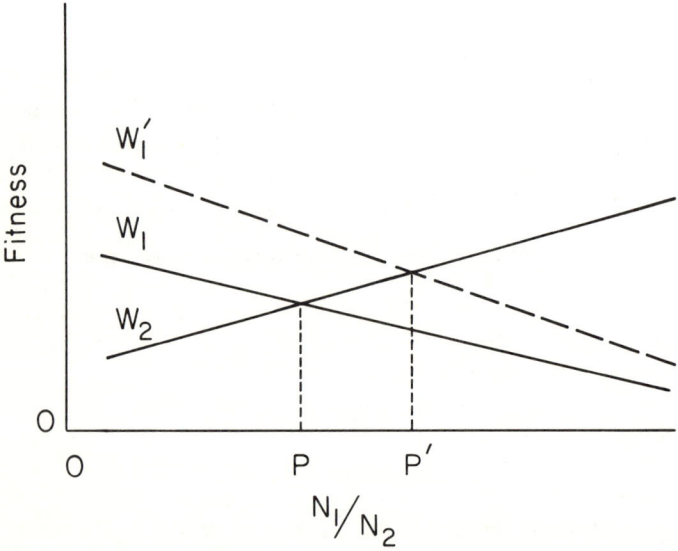

Fig. 6. Relative fitness, W_1 and W_2, of species S_1 and S_2 respectively. The fitnesses are linear functions of the relative numbers of the two species. W_1' is the fitness of S_1 after selection. P gives the ratio of N_1 to N_2 at which a stable equilibrium is reached. P' is the equilibrium point after selection.

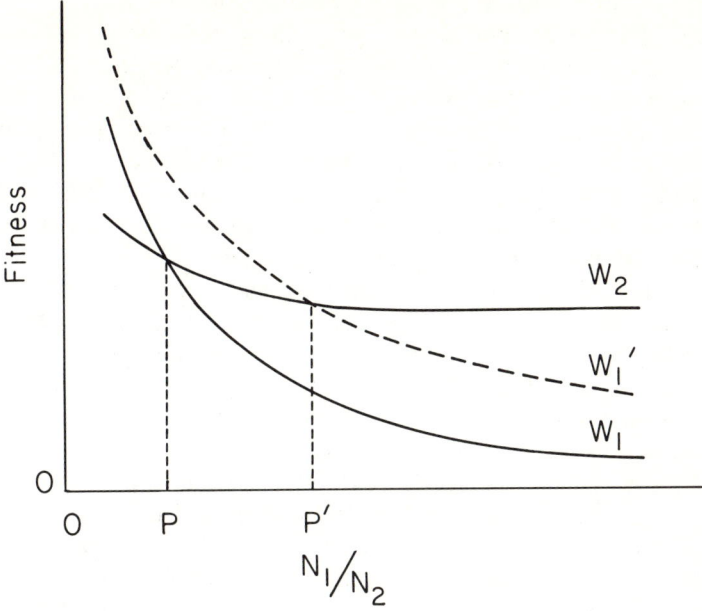

Fig. 7. The relative fitnesses are nonlinear functions of the relative numbers of the two species. See Fig. 6 for meaning of the symbols.

fitnesses of the two species are in some fashion inversely related to their relative numbers. This requirement of frequency-dependent relative fitness has the additional interest of adding to the equilibrium in (9) the property of stability. Let W_1 in (9) be inversely related, and W_2 be directly related to N_1/N_2. The two species will reach a stable equilibrium at $P = N_1/N_2$ if there is a P at which $W_1 = W_2$. This relationship is illustrated in Figure 6 and 7. In Figure 6, the W's are linear functions of N_1/N_2; in Figure 7 the W's are related to N_1/N_2 by some unspecified nonlinear function. If the W function is of the type $W = a + bN$, where a and b are constants, the fitness curve will be linear; a nonlinear fitness curve will be given, for instance, by the function $W = a + \dfrac{b}{N}$.

The only necessary condition for an equilibrium to exist in this case is that the two fitness curves intersect at least at one point. If the relative fitnesses of the two species are inversely related to their relative numbers the equilibrium determined by the intersection of the two fitness curves will be stable. This inverse relationship between relative numbers and fitness implies that $W_1 > W_2$ to the left of P, and $W_1 < W_2$ to the right of P. Thus, whenever $N_1/N_2 > P$, W_2 will be greater than W_1 and N_1/N_2 will decrease until $N_1/N_2 = P$. Similarly, if $N_1/N_2 < P$, then $W_1 > W_2$ and

N_1/N_2 will increase until $N_1/N_2 = P$. Mathematical models of frequency-dependent fitness for intraspecific competition have been discussed by Lewontin and White (1960), Clarke (1964), and Anderson (1969).

Variations in the relative fitness of alternative genotypes as a function of the frequency of the genotypes have been observed, particularly in *Drosophila* studies (L'Heritier and Teissier, 1937; Levene, Pavlovsky, and Dobzhansky, 1954; Petit, 1966; Kojima and Yarbrough, 1967). For a recent review see Petit and Ehrman (1969). In the competition between *Drosophila melanogaster* and *D. simulans,* Narise (1965) observed that the competitive ability of a species increased as the relative number of its parents increased. Pimentel et al. (1965) studied competition between the housefly and the blowfly and observed that natural selection increased the competitive fitness of the rare species. Ayala (1966a, 1969a) observed the same phenomenon in the competition between *Drosophila* species.

Frequency dependent selection has been scarcely investigated in the field. Of great interest is Clarke's (1962) observation that bird predators learn to search for prey of a certain appearance, and therefore kill a disproportionally high percentage of the common varieties of certain snails and a low percentage of the rare ones. As the frequency of a certain snail type increases, its relative fitness decreases. This mechanism may lead to a stable equilibrium between two competing species. The selective advantage of one species over another in the exploitation of food or other resources may be upset by its being preyed on disproportionally more than the other species. How common the phenomenon of frequency dependent selection is in nature remains to be investigated.

Niche Heterogeneity and Temporal Variations

The W_1 and W_2 curves in Figures 6 and 7 may represent the mean overall fitnesses of the two competing species when all the stages of the life cycle are considered. Frequency-dependent selection may then lead to an equilibrium between competing species in other cases besides the situation where the disadvantage of one species at some stage of development is compensated by an advantage at some other stage. Moreover, the model developed by (9) can easily be modified to apply to a variety of situations. Assume, for instance, that the two competing species draw from two (or more) resources that exist in different amounts and are exploited by the species with different efficiency. For instance, S_1 may exploit food A more efficiently than S_2, but S_2 may be more efficient than S_1 in the exploitation of food B. W_1 and W_2 in (9) may be taken now to represent the relative efficiency of S_1 and S_2 in the exploitation of A and B. If the two resources exist in different amounts, W_1 and W_2 are each multiplied by a factor determined by the relative abundance of the two resources. An equilibrium may be reached at a frequency of S_1 and S_2 determined by the magnitude of W_1 and W_2 and of the two resources. The equilibrium proportion of the two species may be stable if the amount of the two resources remains constant, but a new equilibrium may be obtained if the proportion of A and B changes. Mathematical models leading to stable equilibrium among competing genotypes when several resources exist have been developed by Levene (1953) and Prout (1968).

Oscillations in the competitive ability of the species from one generation to another may also lead to competitive coexistence. The diurnal cycle may modify from one generation to the next the competitive fitness of species with a generation time smaller or somewhat larger than one day. Seasonal changes will affect the fitness of species whose generations last several weeks. Cycles of several years of amplitude in the weather pattern or food availability are also known to occur. Haldane and Jayakar (1963) have given mathematical models leading to stable equilibria among genotypes when their relative fitnesses oscillate in time. Of considerable interest is their observation, which can readily be extended to two competing species, that two genotypes can coexist when the arithmetic mean of the fitnesses is greater (even much greater) for one of the genotypes, if their geometric mean is smaller than that of the other genotype.

Migration between populations, and still other mechanisms, may also lead to competitive coexistence. It is not my purpose to explore these mechanisms in detail, but rather to emphasize that under a variety of circumstances two species may compete for limited resources and yet coexist.

The Role of Natural Selection

In the foregoing discussion I have largely ignored the genetic composition of the competing species. Populations have been treated as if they were genetically homogeneous. Yet natural populations of sexual outbreeding organisms are known to be highly polymorphic genetically.

Moore (1952a), studying competition between *Drosophila melanogaster* and *D. simulans,* has shown that populations with improved competitive ability can be developed by selection, and that this can be done in relatively few generations. Pimentel (1961) and Pimentel et al. (1965) have investigated theoretically and experimentally the operation of natural selection on populations of competing species, and have concluded that natural selection may contribute to the stability of the system and lead to competitive coexistence. When two species compete with each other for a number of generations, the dominant species is at an evolutionary disadvantage in the sense that intraspecific competition is the main selective force acting on it, while the sparse species is being selected for interspecific competitive ability. Individuals of the rare species compete mostly with individuals of the abundant species, while the individuals of the abundant species compete primarily among themselves. In the rare species selection will favor genotypes with high interspecific competitive fitness, while in the dominant species selection favors genotypes with high intraspecific fitness. As the interspecific competitive fitness of the sparse species evolves, this species may increase in relative numbers and become the dominant one. Selection will now favor the other species. Oscillations in the dominance of the species may occur; given enough time a genetic adjustment between the competing species should result.

Pimentel et al. (1965) observed reversal of dominance in an experimental study of competition between the housefly, *Musca domestica,* and the blowfly, *Phaenicia sericata.* In a 16-cell population system, the housefly was dominant during the first 50 weeks of the experiment. From week 50 on, the blowfly increased sharply in numbers, and it became clearly dominant from week 57 until week 65 when the housefly became extinct. It was shown that selection had enhanced the genetic competitive ability of the blowfly.

Ayala (1966a) studied competition of *Drosophila serrata* with *D. pseudoobscura, D. nebulosa,* and *D. melanogaster.* In each case, *D. serrata* was initially at a disadvantage but eventually it became dominant in each of the three combinations. In another experiment, represented in Figure 2, the competition between *D. serrata* and *D. nebulosa* reached an equilibrium early in the experiment, with a frequency of *D. serrata* around 30 percent. On week 22 *D. serrata* started a gradual increase in absolute and relative numbers, reaching a frequency of 90 percent by week 31. No reversal of dominance occurred in the parallel experiment (population II) represented in Figure 3, which was carried out simultaneously. Between weeks 31 and 33 flies of both species were sampled from populations I and II. With the F_1 progenies of these samples five types of test populations were established (Ayala, 1969a): (1) *D. serrata* from population I and *D. nebulosa* from the stocks; (2) *serrata* from population II with *nebulosa* from the stocks; (3) *serrata* from the stocks with *nebulosa* from the stocks; (4) *serrata* from the stocks with *nebulosa* from population I; (5) *serrata* from the stocks with *nebulosa* from population II. Four replicate populations of each type were made; all started with 300 flies of each species. The process of competition in these "test" populations was studied for 12 weeks or three to four generations. Comparison among the first three types of populations showed that the competitive ability of *D. serrata* flies from population I was greater than that of *D. serrata* from population II or from the stocks. Comparison among the last three types of populations indicated that the competitive ability of *D. nebulosa* had increased through selection in both populations I and II.

The selection in population I of genotypes of *D. serrata* with greatly enhanced competitive ability did not result in the extinction of *D. nebulosa*. This is consistent with the hypothesis advanced earlier that competitive fitness may be frequency dependent. The effect of natural selection may then be interpreted as a displacement of the fitness curve in Figures 6 and 7. The fitness curve after selection (represented by the broken line in the figures) intercepts now the fitness curve of the competing species at a point to the right of the previous intercept. A new equilibrium is established at $P' > P$, with the relative numbers of N_1 greater than at the previous equilibrium.

The reversals of dominance observed in the competition between *Drosophila* species appear to be in agreement with Pimentel's (1961, 1968) postulate that natural selection favors the sparse species. Pimentel's explanation, however, cannot be accepted without modification. If two species coexist in a limited environment both may share some of the available resources of food and space, while each species may also exploit resources that the other species is not able to use, or does not exploit efficiently. In the *Drosophila* populations both species may share some food resources, like carbohydrates and certain yeast species, and they also share the available space. But it is likely that larvae of one species eat some yeast or mold species that are not exploited by the other species, and vice versa (Da Cunha et al., 1951; Merrell, 1951). Similarly, adult flies of the two species may utilize differentially certain components of the available space, and show preference for certain oviposition sites (Moore, 1952b; Del Solar, 1968). It is also possible for one species to utilize catabolites produced by, and useless to, the other species. In the extreme situation, each species may live on the metabolites of the other species, as in the phenomenon of symbiotic mutualism.

Selection for interspecific competitive ability may, then, occur in two different ways. First, selection may improve the ability of one species to exploit the resources also exploited by the second species. This may be called selection for "positive competitive ability" or, simply, selection for competitive ability proper. Second, genotypes may be selectively favored that allow the population to exploit resources not utilized by the competing species. This second process may be called selection for "avoidance of competition" since it tends toward decreasing the intensity of the competition. Obviously, intermediate situations can exist, such as the case of improvement of the ability of one species to exploit one resource that the other species exploits inefficiently or to a limited extent.

If two coexisting species exploit mostly the same limited resources, it is likely that one or the other species will eventually become extinct. However, selection for avoidance of competition increases the probability of coexistence by leading the two species towards ecological differentiation.

Therefore, selection for avoidance of competition operates as a positive feedback mechanism. The longer it proceeds the more likely are the two species to coexist, and therefore the greater the probability that the selection will continue. From the evolutionary point of view only selection for avoidance of competition is likely to continue for a large number of generations. If this reasoning is correct, one can make the prediction that populations of two species in localities where the two species coexist will be ecologically more divergent than populations of the two species in localities where they do not coexist. The accentuation of differences between species wherever they are sympatric has been named "character displacement" (Brown and Wilson, 1956).

Selection for competitive ability, i.e., selection for exploitation of the resources utilized by both species, will affect the two species independent of their relative frequencies. If the same yeast species is equally eaten by larvae of two *Drosophila* species, selection for intraspecific competition will be equivalent to selection for interspecific competition. The intensity of the selection will depend on the absolute numbers of both species together and not on their relative frequencies. On the contrary, selection towards avoidance of competition will preferentially occur in the sparse species. If one species is more efficient in the exploitation of the resources shared by both, those genotypes of the other species will be favored that allow the flies to exploit resources not utilized by the dominant species. Reversals of dominance are likely to be the result of selection for avoidance of competition.

Levin (1967) has investigated theoretically the role of natural selection in interspecies competition. The addition of genetic variability and selection to the Volterra equations makes possible the coexistence of two competing species. Oscillations in species dominance occur when at least at one of the loci controlling competitive performance one of the homozygotes has maximum interspecific competitive ability, but minimum intraspecific fitness, or vice versa.

Summary

The Volterra equations for competition between species and the principle of competitive exclusion are based on oversimplified assumptions about the nature of the competition process between species. Experiments with *Drosophila* species show that two species can compete for limited resources and yet coexist. Some of the biological mechanisms leading to competitive coexistence are frequency-dependent fitness, fitness interactions between stages of the life cycle, and temporal oscillations of relative fitness. Natural selection leads towards ecological differentiation of competing species and, therefore, promotes the stability of the ecosystems.

References

ANDERSON, W. W. 1969. Polymorphism resulting from the mating advantage of rare male genotypes (in press).
ANDREWARTHA, H. G., and L. C. Birch. 1954. The Distribution and Abundance of Animals. Chicago, University of Chicago Press.
ANONYMOUS. 1944. Symposium on "The ecology of closely allied species." J. Anim. Ecol., 13:176-177.
AYALA, F. J. 1965a. Relative fitness of populations of Drosophila serrata and Drosophila birchii. Genetics, 51:527-544.
——— 1965b. Sibling species of the Drosophila serrata group. Evolution, 19:538-545.
——— 1966a. Reversal of dominance in competing species of Drosophila. Amer. Natural., 100:81-83.
——— 1966b. Dynamics of populations. I. Factors controlling population growth and population and population size in Drosophila serrata. Amer. Natural. 100:333-344.
——— 1967. Dynamics of populations. II. Factors controlling population growth and population size in Drosophila pseudoobscura and Drosophila melanogaster. Ecology, 48:67-75.
——— 1968a. Evolution of fitness. II. Correlated effects of natural selection on the productivity and size of experimental populations of Drosophila serrata. Evolution, 22:55-65.
——— 1968b. Environmental factors limiting the productivity and size of experimental populations of Drosophila serrata and D. birchii. Ecology, 49:562-565.
——— 1968c. Genotype, environment, and population numbers. Science, 162:1453-1459.
——— 1969a. Evolution of fitness. IV. Genetic evolution of interspecific competitive ability in Drosophila. Genetics (in press).
——— 1969b. Genetic polymorphism and interspecific competitive ability in Drosophila. Genet. Res. (in press).
BAGENAL, T. B. 1951. A note on the papers of Elton and Williams on the generic relations of species in small ecological communities. J. Anim. Ecol., 20:242-245.
BAKKER, K. 1961. An analysis of factors which determine success in competition for food among larvae of Drosophila melanogaster. Arch. Neerl. Zool., 14:200-281.
BIRCH, L. C. 1955. Selection in Drosophila pseudoobscura in relation to crowding. Evolution, 9:389-399.
——— 1957. The meanings of competition. Amer. Natural., 91:5-18.
BROWN, W. L., and E. O. WILSON. 1956. Character displacement. System. Zool., 5:49-64.
CHITTY, D. 1960. Population processes in the vole and their relevance to general theory. Canad. J. Zool., 38:99-113.
——— 1965. Predicting qualitative changes in insect populations. Proc. XIII Int. Congr. Entom., London (1964). pp. 384-386.
CLARKE, B. 1962. Balanced polymorphism and the diversity of sympatric species. Systematics Assoc. Publ., 4:47-70.
——— 1964. Frequency-dependent selection for the dominance of rare polymorphic genes. Evolution, 18:364-369.
COLE, L. C. 1960. Competitive exclusion. Science, 132:348-349.
COOPER, D. M., and Th. DOBZHANSKY. 1956. Studies on the ecology of Drosophila in the Yosemite region of California. I. The occurrence of species of Drosophila in different life zones and at different seasons. Ecology, 37:526-533.
CROMBIE, A. C. 1945. On competition between different species of graminivorous insects. Proc. Roy. Soc. [Biol.], 132:362-395.

——— 1946. Further experiments on insect competition. Proc. Roy. Soc. [Biol.], 133: 76-109.
——— 1947. Interspecific competition. J. Anim. Ecol., 16:44-73.
DA CUNHA, A. B., TH. DOBZHANSKY, and A. SOKOLOFF. 1951. On food preferences of sympatric species of *Drosophila*. Evolution, 5:97-101.
DAWOOD, M. M., and M. W. STRICKBERGER. 1969. The effect of larval interaction on viability in *Drosophila melanogaster*. III. Effects of biotic residues (in press).
DAWSON, P. 1968. Mutation as a source of indeterminism in *Tribolium* "competition" experiments. Genetics, 60:172.
DEBACH, P. 1966. The competitive displacement and coexistence principles. Ann. Rev. Entom., 11:183-212.
DEL SOLAR, E. 1968. Selection for and against gregariousness in the choice of oviposition sites by *Drosophila pseudoobscura*. Genetics, 58:275-282.
——— and H. PALOMINO. 1966. Choice of oviposition in *Drosophila melanogaster*. Amer. Natural., 100:127-133.
DOBZHANSKY, TH. 1944. Chromosomal races in *Drosophila pseudoobscura* and *Drosophila persimilis*. Carnegie Instit. Washington Publ. 554, pp. 47-144.
——— and B. SPASSKY. 1953. Genetics of natural poulations. XXI. Concealed variability in two sympatric species of *Drosophila*. Genetics, 38:471-484.
ELTON, C. S. 1927. Animal Ecology. London, Sidgwick and Jackson.
——— 1946. Competition and the structure of animal communities. J. Anim. Ecol., 15:54-68.
——— and R. S. MILLER. 1954. The ecological survey of animal communities: with a practical system of classifying habitats by structural characters. J. Ecol., 42:460-496.
FELLER, W. 1940. On the logistic law of growth and its empirical verifications in Biology. Acta Biotheor. [A] (Leiden), 5:51-66.
FORD, E. B. 1964. Ecological Genetics. London, Methuen.
GAUSE, G. F. 1934. The Struggle for Existence. Baltimore, The Williams & Wilkins Co. (Repr., 1964. New York, Hafner Publishing Co., Inc.)
——— and A. A. WITT. 1935. Behavior of mixed populations and the problem of natural selection. Amer. Natural., 69:596-609.
GRINNELL, J. 1904. The origin and distribution of the chestnut-backed chickadee. Auk, 21:364-382.
——— 1917. The niche-relationships of the California thrasher. Auk, 34:427-433.
——— 1924. Geography and evolution. Ecology 5:225-229.
HALDANE, J. B. S. 1924. A mathematical theory of natural and artificial selection. Trans. Cambridge Philos. Soc., 23:19-41.
——— and S. D. JAYAKAR. 1963. Polymorphism due to selection of varying direction. J. Genetics, 58:237-242.
HARDIN, G. 1960. The competitive exclusion principle. Science, 131:1292-1298.
HARRIS, P., D. F. RIORDAN, and D. COOKE. 1969. Mosquitoes feeding on insect larvae. Science, 164:184-185.
HUTCHINSON, G. E. 1957. Concluding remarks. Sympos. Quant. Biol., 22:415-427.
——— 1965. The Ecological Theater and the Evolutionary Play. New Haven, Conn., Yale University Press.
KETTLEWELL, H. B. D. 1955. Selection experiments on industrial melanism in the Lepidoptera. Heredity (London), 9:323-342.
KOJIMA, K., and K. YARBROUGH. 1967. Frequency dependent selection at the Esterase 6 locus in *Drosophila melanogaster*. Proc. Nat. Acad. Sci. (U.S.A.), 57:645-649.
LACK, D. 1944. Ecological aspects of species formation in passerine birds. Ibis, 86:260-286.
LERNER, I. M., and F. K. HO. 1961. Genotype and competitive ability in *Tribolium* species. Amer. Natural., 95:329-343.
LEVENE, H. 1953. Genetic equilibrium when more than one ecological niche is available. Amer. Natural., 87:331-333.
——— O. PAVLOVSKY, and TH. DOBZHANSKY. 1954. Interaction of the adaptive values

in polymorphic experimental populations of *Drosophila pseudoobscura*. Evolution, 8:335-349.
LEVIN, B. R. 1967. Genetic variability and selection in a system of species competition. Ph.D. thesis. Ann Arbor, University of Michigan.
LEWONTIN, R. C. 1955. The effects of population density and composition on viability in *Drosophila melanogaster*. Evolution, 9:27-41.
―――― and Y. MATSUO. 1963. Interaction of genotypes determining viability in *Drosophila busckii*. Proc. Nat. Acad. Sci. (U.S.A.), 49:270-278.
―――― and M. J. D. WHITE. 1960. Interaction between inversion polymorphisms of two chromosomal pairs in the grasshopper, *Moraba scurra*. Evolution, 14:116-129.
L'HERITIER, P., and G. TEISSIER. 1937. Elimination des forms mutants dans les populations des Drosophiles. Cas des Drosophiles *Bar*. C. R. Soc. Biol. (Paris), 124:880-882.
LOTKA, A. J. 1925. Elements of Physical Biology. Baltimore, The Williams & Wilkins Co.
―――― 1932. The growth of mixed populations: Two species competing for a common food supply. J. Washington Acad. Sci., 22:461-469.
MERRELL, D. J. 1951. Interspecific competition between *Drosophila funebris* and *Drosophila melanogaster*. Amer. Natural., 85:159-169.
MILLER, R. S. 1964. Larval competition in *Drosophila melanogaster* and *D. simulans*. Ecology, 45:132-148.
―――― 1967. Pattern and process in competition. Advances Ecol. Res., 4:1-74.
MOORE, J. A. 1952a. Competition between *Drosophila melanogaster* and *Drosophila simulans*. II. The improvement of competitive ability through selection. Proc. Nat. Acad. Sci. (U.S.A.), 38:813-817.
―――― 1952b. Competition between *Drosophila melanogaster* and *Drosophila simulans*. I. Population cage experiments. Evolution, 6:407-420.
NICHOLSON, A. J. 1954. An outline of the dynamics of animal populations. Aust. J. Zool., 2:9-65.
―――― 1957. The self-adjustment of populations to change. Sympos. Quant. Biol., 22:153-172.
NARISE, T. 1965. The effect of relative frequency of species in competition. Evolution, 19:350-354.
PARK, T. 1948. Experimental studies of interspecies competition. I. Competition between populations of the flour beetles, *Tribolium confusum* Duval and *Tribolium castaneum* Herbst. Ecol. Monographs, 18:265-307.
―――― 1954. Experimental studies of interspecies competition. II. Temperature, humidity, and competition in two species of *Tribolium*. Physiol. Zool., 27:177-238.
PATTEN, B. C. 1961. Competitive exclusion. Science, 134:1599-1601.
PEARL, R. 1926. The Biology of Population Growth. New York, Alfred A. Knopf, Inc.
―――― J. R. MINER, and S. L. PARKER. 1927. Experimental studies on the duration of life. XI. Density of population and life duration in *Drosophila*. Amer. Natural., 61:289-318.
―――― and L. J. REED. 1920. On the rate of growth of the population of the United States since 1790 and its mathematical representation. Proc. Nat. Acad. Sci. (U.S.A.), 6:275-288.
PETIT, C. 1966. La concurrence larvaire et le maintien du polymorphism. C. R. Acad. Sci. (Paris), 263:1262-1265.
―――― and L. EHRMAN. 1969. Sexual selection in *Drosophila*. *In* Dobzhansky, Th., Hecht, M. K., and Steere, W. C., eds. Evolutionary Biology, Vol. 3. New York, Appleton-Century-Crofts (in press).
PIMENTEL, D. 1961. Animal population regulation by the genetic feed-back mechanism. Amer. Natural., 95:65-79.
―――― 1968. Population regulation and genetic feedback. Science, 159:1432-1437.

―――― E. H. FEINBERG, P. W. WOOD, and J. T. HAYES. 1965. Selection, spatial distribution, and the coexistence of competing fly species. Amer. Natural., 99:97-109.

PROUT, T. 1968. Sufficient conditions for multiple niche polymorphism. Amer. Natural., 102:493-496.

ROSS, H. 1957. Principles of natural coexistence indicated by leafhopper populations. Evolution, 11:113-129.

SANG, J. H. 1950. Population growth in Drosophila cultures. Biol. Rev., 25:188-219.

SLOBODKIN, L. B. 1954. Population dynamics in *Daphnia obtusa* Kurz. Ecological Monographs, 24:69-88.

―――― 1957. A laboratory study of the effect of removal of newborn animals from a population. Proc. Nat. Acad. Sci. (U.S.A.), 43:780-782.

―――― 1961. Growth and Regulation of Animal Populations. New York, Holt, Rinehart & Winston, Inc.

SPASSKY, B. 1957. Morphological differences between sibling species of *Drosophila*. Univ. Tex. Publ. No. 5721:48-61.

ULLYETT, G. C. 1950. Competition for food and allied phenomena in sheep-blowfly populations. Philos. Trans. Roy. Soc. London [B]., 234:77-174.

UTIDA, S. 1957. Population fluctuation, an experimental and theoretical approach. Sympos. Quant. Biol., 22:139-150.

VAN VALEN, L. 1960. Further competitive exclusion. Science, 132:1674-1675.

VERHULST, P. F. 1839. Notice sur la loi que la population suit dans son accroissement. Corr. Math. Phys., 10:113-121.

VOLTERRA, V. 1926. Variazioni e fluttuazioni del numero d'individui in specie animali conviventi. Mem. Accad. Lincei, ser. 6, 2:31-113. (English transl., Variations and fluctuations of the number of individuals in animal species living together. *In* Chapman, R. N., Animal Ecology, 1931. New York, McGraw-Hill Book Company. pp. 409-448.

―――― 1931. Leçons sur la theorie mathematique de la lutte pour la vie. Paris, Gauthiers-Vallars.

WEISBROT, D. R. 1966. Genotypic interactions among competing strains and species of *Drosophila*. Genetics, 53:427-435.

WILLIAMS, C. B. 1947. The generic relations of species in small ecological communities. J. Anim. Ecol., 16:11-18.

5

Adapted Molecules

MARCEL FLORKIN and E. SCHOFFENIELS

Department of Biochemistry, University of Liège, Liège, Belgium

Adaptation at the Molecular Level 159
Molecular Adaptation and Phylogeny 163
Molecular Adaptation at the Population Level........................ 166
Conclusions 170
References 171

Adaptation at the Molecular Level

The increasing interest shown in the molecular approach to biology induces us to consider a species as consisting of groups of individuals having closely related combinations of purine and pyrimidine bases in their macromolecules of deoxyribonucleic acid (DNA), and in which the systems of operators and repressors bring about the biosynthesis of similar sequences of amino acids. Their integration in the cell leads to similar structural and functional characteristics, adapted to the ecological niche in which the species succeeds.

Evolution may be defined as the changes that affect the relative proportions, the associations, and the nature of genetic factors within a population. Speciation is the means through which new aspects of evolution are introduced; it is worked upon by the nature of the environments, and what is generally called adaptation results from this influence. The environment is one of the agents of natural selection that produces its effects by affecting the rate of reproduction. A chemical modification of genetic factors is the most complete achievement of the molecular aspect of evolution. Ontogeny and phylogeny thus result from the reading of the code transmitted by the genotype, while evolution is a consequence of the existence of new codes or of new decipherment of the code.

Since the time of Lamarck, we have learned to consider the diversity of species as being the result of adaptations to different environmental conditions. Organisms have responded to the vicissitudes of the surroundings either by the diversification of the genotypes or by an adjustment of the translations of the code to the new conditions.

Adaptation, as understood here so far, is an organismic or evolutionary concept. It explains why a species survives in, or is able to colonize, a given environment. In a more teleological sense one could also say that adaptation functions for the survival of the species (Williams, 1966).

Within the complex picture of all the adaptive features it is possible to individualize biochemical traits: one thus defines the biochemical adaptations: e.g., the molecule(s) or molecular mechanism(s) that explain if not all at least part of the evolutionary adaptation.

If, for example, we consider the oxygen absorption curves of the few species for which we know the respiratory cycle (Fig. 1), we see that in the different cases, even if the external medium is the same, very different values of the gradient of oxygen pressures obtain between the environ-

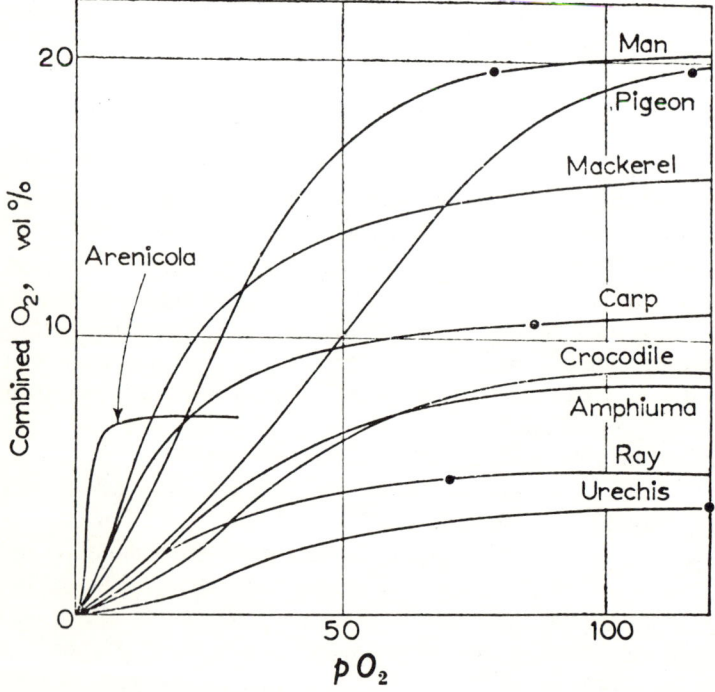

Fig. 1. Oxygen absorption curves for various bloods and coelomic fluids containing hemoglobin under "arterial" conditions reproduced in vitro. The black circle indicates the value of the arterial $p^0{}_2$ (Florkin, 1960).

ment and the *milieu intérieur*. The partial pressure of oxygen corresponding to "arterial" conditions is obviously related to the rate of exchange through the respiratory epithelium, gills, or lungs, as the case may be, as well as to that of the circulation of the oxygen carrier. In spite of this, the degree of saturation of the arterial blood corresponds in each case to the higher part of the dissociation curve. As a matter of consequence, any lowering of pO_2 in the medium results in the delivery of oxygen from the carrier (Florkin, 1948). Such a biochemical adaptation is evidently only one among many features that explains the relationship a given species establishes with its surroundings, and it is thus of prime importance to define the hierarchy of adaptations not only at the organismic or evolutionary level but also at the molecular scale.

A proper approach to the study of adaptation must start from the consideration of the relation organism-environment at the level of the community or of the organism, and proceed progressively from this organismic starting point to the underlying molecular aspects. Such a study, to be fully convincing, must be very detailed. It has been accomplished in the case of the adaptation of euryhaline invertebrates (Florkin and Schoffeniels, 1965, 1969; Schoffeniels, 1967). In certain of these euryhaline forms, the adaptation results from the combination of the effects of an anisosmotic regulation of the body fluids kept at a higher concentration than the external medium, and of an *isosmotic intracellular regulation* bringing the cells into osmotic equilibrium with the body fluids.

Isosmotic intracellular regulation results from several biochemical adaptations, one of which, the regulation of the total concentration of free amino acids and taurine, is always playing an important role in the adaptation. The analysis of this regulation has led to its location at the level of a differential action of inorganic cell constituents on several enzymes related to amino-acid metabolism and to the production of reducing equivalents. In this case, the evolutionary adaptation is partly explained by the peculiar properties of some protein molecules. The biochemical adaptation thus lies in certain features of the enzyme that place its catalytic properties under the control of the inorganic constituents of the cell.

In other words, this adaptation is at the level of protein structure. Consequently one is tempted to establish a direct relationship with the genotype.

The significant conclusion to be drawn from the study of a metabolic pathway throughout the animal kingdom or, when considering the same species, in various organs, is that an important aspect of adaptation rests not only on the existence of peculiar catalytic steps leading to specific metabolic products, but also on the development of specific means of control of a catalytic step common otherwise to any type of cell. This control, generally allosteric in nature, explains how the activity of a metabolic route may have various functional significances according to the differentiation considered. This conclusion, stemming from the careful investigations of research workers involved in the comparative study of enzyme systems, has an important consequence. It indeed focuses attention on the problem of the structure of the enzyme catalyzing the same reaction in a large diversity of cells. If according to the origin of the cell the enzyme is variously affected by controlling factors, this suggests that the differentiation or the adaptation, as the case may be, deals primarily with the development of control sites separated from the catalytic centers. Available data suggest that for a number of enzymes the genetic information responsible for the amino-acid sequence determines automatically the secondary and tertiary protein structure. The assumption that a folding template carries information differing from that contained in the amino-acid code has not yet been supported by experimental findings. We may thus expect that the emergence of allosteric sites is related to the primary structure of the protein. The possibility that cytoplasmic environmental influences may, in a given cell, affect the protein configuration that is responsible for the appearance of allosteric sites has still to be investigated. Another factor that may modify profoundly the biological function of an enzyme is its intracellular location. It all remains to be elucidated, as far as the factors that may induce this distribution are concerned.

Rather paradoxically, the concept of "molecular biology" is really an organismic concept. The authority of its inventors forces us to accept this view. If we read the Instructions to Authors of the *Journal of Molecular Biology,* we learn that molecular biology is the domain of studies related to the nature, production, and replication of *biological structures considered at the molecular level* and to the relations of these structures with the *functions of organisms.* It appears clear, at least in current theories, that heredity (a function of the organism) is linked to the nature of certain sequences of purine and pyrimidine bases within the macromolecules of DNA of the genes (e.g., to a biological structure considered at the molecular level).

If animals are generally endowed with motility (function of organism), this is related to the fact that certain cells biosynthesize macromolecules of contractile proteins (molecular structures related to function considered at the level of the organism). In fact, molecular biology, in spite of the concentration of its studies at the molecular level, brings us back to the organismic viewpoint too often neglected in biochemical studies.

This attractive and fruitful tendency should, nevertheless, not be considered as being universally applicable. Some macromolecules are active not at the level of the organism, but at another level of organization. We hope to be able, at some time, to climb the staircase of causal relations through the gradation: molecule, cellular organelle, cell, tissue, organ, individual, population, species, community, ecosystem—but this time has not yet arrived. The macromolecules studied at the structural level by orthodox molecular biologists are those whose functions are kept through all levels of organization, and remain unchanged at the level of the organism as a whole. There are exceptions and there are many cases of different functions assumed at the level of the organism by the same macromolecules (physiological radiations). This does not tend to contradict the primary interest of biological structures considered at the molecular level and of their relations with whatever level of organization they contribute to.

On the other hand, the authors do not intend to question the deep interest and importance of population dynamics. It is certainly true, as Grant (1963) has pointed out, that the presence of coyotes is a factor in the determination of the number of rabbits living in a given territory. As Grant rightly states, an increase in the population of coyotes will reduce the rabbit population. The expansion of the population of coyotes will eventually stop when the number of rabbits they eat decreases to zero. No molecular approach will reveal these causal relations. Everybody will agree to this. But whatever patience and ingenuity may be devoted to counting trouts and insects living in a stream, it is only by a molecular approach and through the knowledge of the properties of trout hemoglobin that we shall understand why trouts live in streams and not in marshes, where insects also exist.

It is at the level of the organism that the impact of natural selection takes place. In the case of an evolutionary adaptation allowing a species to live in the brackish water of estuaries as well as in seawater, the advantage conferred by the adaptation at the molecular scale of dimension is obvious, as it allows the colonization of new biotopes. It is this aspect and its nutritional consequences which give rise to an increase in the number of individuals reaching the age of reproduction and in the number of their offspring. This is the level where the phenomenological aspect of the evolution obtains, while the causal factors are located at the molecular level of protein structures.

All these considerations show that in the study of adaptations it would certainly be ill-advised to dissociate the molecular approach from its organismic counterpart.

Molecular Adaptation and Phylogeny

Molecular evolution takes on a number of forms in the history of living organisms. In descent, it can take the form of a change in structure of the molecule or macromolecule. The known aspects of molecular changes along the lines of phylogeny of multicellular organisms have recently been reviewed (Florkin, 1966). We are concerned with phylogeny of a protein when we establish comparisons between homologous forms of this protein at different levels of the same phyletic series of the organisms synthesizing these forms. The elegant studies of Ingram and his colleagues on the evolution of hemoglobin provides us with a good example. It has been clearly emphasized that this has a great ecological impact. Data of this kind have also been collected for the phylogeny of fibrinopeptides, of the peptide liberated by the activation of trypsinogen, of the neurohypophyseal hormones, and so forth.

Other forms of molecular phylogeny can be detected among molecules that result from the extensions of a biosynthetic pathway. In that case the various steps leading to the final product repeat its phylogeny. Molecular changes in descent can often be related, in these cases, to adaptive aspects. More and more examples will certainly be found as our knowledge of the biochemical continuum increases.

But are the molecular changes that are detectable at the level of primary structure along the paths of phylogeny[1] functional and adaptive? This certainly does not appear to be the case when one considers the phylogeny of the proteins so far studied. Nevertheless as stated by Simpson (1964), "it is certainly not true as a generalization that molecular differences among species are commonly nonfunctional or inadaptive." In the field of indirect homology,[2] molecular differences are clearly adaptive in many cases. Considering the biochemical continuum we meet problems of the biosynthesis of new molecules that are clearly of an adaptive nature, the integrated action of which is linked to a specific action on the receptor organism.

In direct homology, our ignorance of the possible adaptive nature of the changes appearing in the primary structure may be the only reason leading us to conclude that the change is not adaptive. Therefore, instead of reasoning from the molecular up to the organismic level, the problem must, at least tentatively, be tackled the other way around: adaptations clearly defined at the organismic level should be analyzed at the molecular level of dimension.

A change of primary structure in a protein chain always depends on a change of sequence in a nucleic acid. While the change of structure in homologous nucleic acids, and consequently of proteins, is the leading

---

[1] Bacteria are left out of the picture as their phylogeny is derived from molecular aspects.

[2] A high degree of isology (similarity in primary structure) is used to detect homology of proteins at least in the limits of a short section of a phylogenetic tree. In the case of molecules other than proteins or nucleic acids, and resulting from a biosynthetic enzymological pathway, the molecules will be called homologous (indirect homology) if the enzymes controlling their biosynthetic pathways are themselves homologous.

thread of the phylogeny of macromolecules, this form of change is not the only method of molecular evolution. Even if it has kept, more or less modified, some characteristics of an ancient prototype in parts of its aminoacid sequences, a protein may acquire new kinds of activities and properties by changes in its secondary or tertiary structures. Therefore, the primary structures, at least in the cases of marked isology, will give us information concerning molecular phylogeny while the truly adaptive property will reside in the configurational change. The exact contribution of the aminoacid sequence and of possible cytoplasmic factors in determining the tertiary structure of the protein has still to be defined.

The guide of molecular phylogeny remains, at least for the time being, the phylogeny of organisms as built by generations of naturalists. Evolution, with change of structure, has probably been most active at the level of the precellular eobionts, when the architecture we now find as common to all cells had been acquired by natural selection in the first reproducing organisms. But since then, this kind of molecular evolution has continued to take part in the ecobiochemical development of relationships between organisms and the medium during the expansion of the biosphere. Change of structure is eventually accompanied by a change of action.

Other forms of molecular evolution have played an ecological role. Systems of macromolecules can be modified by the change of structure of one or several of their components, by modifications in the relative concentrations of these components, by the loss of one or several of these, by the introduction of an existing macromolecule into another system. One of the aspects of molecular evolution in vertebrates is the acquisition, at the level of the mesodermal cells, of new steroid biosynthesis. One of the consequences of this is ionic regulation by the action of corticosteroids at the level of the nephridia. This action is due to the introduction of the corticosteroids in the systems of transport at the level of the urinary tubules. A new form of molecular evolution, taking place in amphibians, consists of the introduction of pituitary hormones into the system, with the consequence that water is reabsorbed—an important aspect of the factors allowing for terrestrial life in toads, for instance.

On the other hand, a biochemical system, whether it remains the same or undergoes a change of structure and eventually of *action,* may undergo a change of *function* when considered at a level higher than molecular. This results from a different insertion in the tridimensional system of the cells, and constitutes another form of adaptive radiation.

The system of the biosynthesis of thyroid hormones appears in Tunicata, and its course, at least at the start, remains the same in vertebrates, although no particular organ is differentiated for this function in tunicates, and although the role played by the thyroid hormones in these animals has not yet been identified. In cephalochordata, such as *Branchiostoma (Amphi-*

oxus lanceolatum), iodine concentrates in a special organ, the endostyle (which may be homologous to the thyroid gland). Here, the secretion takes on the holocrine mode, the endostyle cells in which the hormones are synthesized going toward the digestive tube where they are hydrolyzed and where the thyroid hormones are recovered by intestinal absorption. In larval cyclostomes, the same mode prevails, but when they reach the adult stage, the thyroid hormones are liberated by the merocrine secretion oriented toward the circulating blood. This physiological radiation brings the thyroxine and the triiodotyrosine to the fulfillment of the important function they play in the higher vertebrates (Roche, 1963).

The change of function of an enzyme system such as the chitinolytic one, formed by chitinase and chitobiase, shows a number of ecological aspects, its effect having the role of entering into digestion, or molting, or the penetration of parasites into their host (Jeuniaux, 1963).

From what has been stated above, it appears that different aspects of molecular evolution have been used in taking advantage of environmental fitness. They can all provide ecobiochemical viewpoints, whether or not they are of taxonomic or of phylogenic importance. The examples of analogous substances, as well as evolution with a change of molecular structure, are ways of ensuring the diversity of properties of biochemical components playing similar roles. This diversity is one of the conditions of the integration and of the regulation of the organism and of its integration into the ecosystem. In the history of comparative physiology, homologous and analogous components have not been distinguished. In our present consideration, this fundamental distinction must be present in the first rank, for the benefit of a sounder appreciation of the adaptive nature of molecular evolution and of the site of impact of natural selection.

Molecular Adaptation at the Population Level

The concept of the biosphere defined as representing the portion of the earth and its atmosphere that is capable of supporting life is useful in the field of biogeochemistry. From the ecological viewpoint it is also useful to consider a biochemical continuum of the total of the living mass and its metabolic extensions. The oxygen of the atmosphere has been produced by living plants, and the CO_2 it contains has a clear relationship with the metabolism of organisms. The sediments at the deepest bottoms of the oceans, and the organic matters coating sand or mud particles there, are linked to the surface of the seas by a cloud of molecules—this cloud being of variable density. The same notion applies to soils, to fresh waters, and even to sedimentary rocks in which a part of the organic matters of sediments persists in the form of kerogen. The whole of the biochemical continuum forms a population of molecules densely associated in parts and more separated in other regions. Some of the dense parts are architectured in the form of living cells. At this level of organization, the population of molecules is kept away from the state of equilibrium, and a mechanism has been developed for reproduction and information transfer. Monocellular and pluricellular organisms associate in communities whose populations maintain themselves and form—in association with different components of the environment—ecosystems, inside which and between which currents of matter and energy take place.

The biochemical continuum is the result of a slow evolution starting from an organic abiogenic continuum. The transition from the abiogenic organic continuum to the biochemical continuum is characterized by an important step: the use of organic molecules by organized aggregates (precell or protocell). This is the origin of nutrition. After the invention of photosynthesis, an organized aggregate becomes a producer of organic molecules serving as "food" to the forms that were lacking the systems of photoautotrophy. Progressively, the reciprocal relations between the various systems became more and more elaborate, which led to the exchanges of molecules that were not only of nutritional importance but also acquired regulatory and control properties on metabolism.

In an ecosystem, besides the contribution of the trophic chains in supplying molecules endowed with nutritive or regulatory functions, one may describe molecules active in the constitution and the maintenance of the biotic community. It is convenient to record them under the general denomination of *ecomones*. An ecomone may act in different respects. For instance, the concentration of dissolved carbohydrates in seawater greatly varies from place to place and may have an influence on the nature and growth of the phytoplankton. The ecological importance of these carbohydrates has also been detected with respect to the pumping rate of oysters. Collier (1953) has shown that each oyster has a threshold limit to carbohydrate, above which it will pump, and this threshold is raised with increasing temperature.

Other ecomones are recognized as being specifically active in the process of the coaction of the organisms upon each other. Such specific substances or *coactones* are determinant in the relationship of the *coactor* (active or directing organism) to the *coactee* (passive or receiving organism). This implies the existence of chemosensory functions that are, however, far from being fully understood. Behavioral studies indicate that the antennae in arthropods, the mantle cavity in molluscs (Kohn, 1961), the skin of batrachians, and so forth, are the sites of the chemoreception. The electrophysiological data providing the clearest demonstration of the role of these organs are still lacking.

A number of coactones are liberated by the coactor in the medium and reach the coactee. These we may call *exocoactones*. Among these are, for instance, the molecular factors of the orientation of animals through perception of the odor of the animals taken as food.

Secondary plant substances may act on insects without being liberated in the medium: the plant growth substance gibberellic acid and the insect growth substance ecdysone-λ have similar effects on both plants and locusts (Carlisle et al., 1963). The feeding behavior when the animal is on the plant onto which it may have been attracted by other factors may also be unlocked specifically by plant products. Such *endocoactones* are largely responsible for the food specificity of insects.

In the ecological relationship of plants and insects, it is important to consider, as suggested by Fraenkel (1959), that the primary components of plants, resulting from the biosynthetic pathways also existing in bacteria and animals, are the same in every plant species. The basic requirements of phytophagous insects are covered by these constituents. But the plants contain a vast array of "secondary substances" and Fraenkel suggests that these substances (e.g., glucosides, saponins, tannins, alkaloids, and essential oils) have been developed as repellents, unpalatable compounds, or poisons for such insects. Had these aspects of adaptive evolution been fully successful, insects would no longer be dangerous for plants. But according to Fraenkel, insects have, on their part, responded to this chemical control. When a given insect species, by genetic selection, overcame the repellent effect it gained a new source of food, further selection producing new species attracted by the former repellent. This points to the importance of the reciprocal selective responses in the originization of organic diversity (Ehrlich and Raven, 1964).

A plant may be attractive and poisonous, but if the insect develops an adaptation to particular secondary components of a particular species of plant, these components may increase the food specialization by their function as endocoactones, such as biting factors, chewing factor, and so forth, and all these will lead to a consumption of the plant food sufficient to ensure the welfare of the insect.

According to these views, secondary substances of plants are repellents, attractants, or we may add, endocoactones, either making the plant unpalatable or even poisonous, or making the plant bitable, chewable, palatable, and therefore consumed in amounts covering the quantitative needs of the insects.

It applies also to the relationship existing between an animal and some of its parasites. The peculiar stimulation of mosquitoes which ensures gorging and subsequent dispatch of the food to the midgut is chemical in nature and comes from the blood cells (Hosoi, 1959). It is mainly adenosinephosphates which act as stimulants, although the presence of sodium ions is also required (Galun et al., 1963).

According to this conception, food specialization is the result of the existence of coactones ensuring a sufficient consumption of a given food and not of special nutritive characteristics of the food—all living creatures are equivalent in this respect, as a consequence of their remarkable unity of composition with regard to their primary constituents. This can certainly not be generalized as representing the nutritional interrelation in all living species. Some primary constituents of certain organisms may be required by some other organisms to cover their specialized chemical needs. Among Crustacea, euphausiids are exceptional with respect to their vitamin-A content. In its nutrient value in the food of whales, the chemical need of vitamin A is covered in these animals by the substance itself, while in other mammals it is mainly covered by the use of provitamin A in the form of carotenes.

A special category of excoactones is represented by those substances produced by a coactor and active on a coactee, both coactor and coactee belonging to the same species. These exocoactones are called pheromones. According to Karlson and Lüscher (1959), "Pheromones are defined as substances which are secreted to the outside of an individual and received by a second individual *of the same species,* in which they release a specific reaction, for example a definite behavior or a developmental process."

As suggested by Karlson (1960), it is useful to distinguish in animals between the pheromones acting olfactorily and those acting orally. Sex attractants, marking scents, and alarm substances belong to the first category.

Alarm or alerting substances are induced by threatening stimuli and communicate the presence of danger to members of the same species (review in Pfeiffer, 1963; Eisner and Meinwald, 1966). The response is very often flight, for example in the tadpoles of *Bufo,* or in fish (Eibl-Eibesfeldt, 1949; Schutz, 1956; Pfeiffer, 1966a, 1967). The specificity is rather high since the pheromone produced by *Bufo bufo* or *B. calamita* is uneffective on anurans not belonging to the Bufonidae (Pfeiffer, 1966b).

Ants generally migrate on trails indicated by a marking scent laid down by the workers as part of the colony organization (see, for instance, Lindauer, 1963; Blum and Ross, 1965; Cavill and Robertston, 1965). Many mammals release scent to mark out territories and home ranges. Indol and skatol, present in excrement, are used in this function by a number of carnivorous mammals. Other odorants are secreted by exocrine glands located in different regions of the body, according to the species. The wolf rubs his back on the trees limiting his territory and marks them with a specific scent; some prosimians have special humeral glands that are rubbed against branches of trees.

Sex pheromones releasing sexual attraction or sexual behavior, or both, are known in many animal taxa (literature in Wilson and Bossert, 1963)—

insects (see also Butler, 1967; Jacobson, 1965), crustaceans, fish, salamanders, snakes, and mammals, as well as in the lower phyla.

Other pheromones act orally rather than olfactorily. The queen substance of the honey bee is in this category. The chemical structure has been recognized as that of an unsaturated keto-acid (Butler et al., 1959; Callow and Johnston, 1960; Barbier et al., 1960).

The information carried to the colony by this substance is the presence of the queen. Ingestion of this substance by the worker bee licking the body of the queen inhibits the development of ovaries in workers and influences their behavior by preventing queen-cell construction (see also Rembold, 1964).

As to the nature of the reception mechanism, all has still to be done. The extremely small amount of substance necessary to produce a response from the coactee renders the task very arduous. In the case of sex attractants produced by the female of arthropods, 10^{-17} g of the compound, or even less, is still effective on the male (Wharton et al., 1962).

This problem is essentially the same as that dealt with in permeability studies when considering the chemical nature of the active sites in the membranes (Schoffeniels, 1967). It remains largely unexplained, although it represents one of the main factors in ecophysiological relations at the molecular level. The most remarkable feature of the studies concerned with the chemical regulation of courtship and other social activities in invertebrates, and more specifically in insects, is that they outdate the philosophical arguments of psychological thesis that lures the lazy mind with a dangerous word: instinct. They definitively show that a highly sophisticated social organization such as that exhibited by a colony of bees finds its explanation at the molecular scale of dimensions.

Conclusions

To summarize briefly, when one deals with the problem of adaptation, it is certainly ill-advised to dissociate the organismic and the molecular aspects. It is true that some function of an organism may be traced down to specific properties of a given molecule. But few adaptations can be related to the properties of one single molecule, thus indicating the frequent polygenic nature of the phenomenon. It remains certain, however, that we owe to what is generally called molecular biology the knowledge that enables us to define the concept of homology at the molecular level. This gives biologists the possibility of seizing the concepts of adaptation, no longer in their usual phenomenological context, but at the physicochemical scale of dimensions, thus permitting the unraveling of the molecular evolution that holds the key to morphological and physiological adaptations.

References

BARBIER, M., E. LEDERER, T. REICHSTEIN, and O SCHINDLER. 1960. Auftrennung der sauren Anteile von Extrakten aus Bienenköniginnen (*Apis mellifica* L.); Isolierung des als Königinnen-Substanz bezeichnetes Pheromones. Helv. Chim. Acta, 43:1682.

BLUM, M. S., and G. N. ROSS. 1965. Chemical Releasers of Social Behaviour— V. Source, Specificity, and Properties of the Odour Trail Pheromone of *Tetramorium guineense* (F.) (Formicidae : Myrmicinae). J. Insect Physiol., 11:857.

BUTLER, C. G. 1967. Insect Pheromones. Biol. Rev., 42:42.

────── R. K. CALLOW, and N. C. JOHNSTON 1959. Extraction and Purification of "Queen Substance" from Queen Bees. (London), 184:1871.

CALLOW, R. K. and N. C. JOHNSTON. 1960. The Chemical Constitution and Synthesis of Queen Substance of Honeybees (*Apis mellifera*). Bee World, 41:152.

CARLISLE, D. B., D. J. OSBORNE, P. E. ELLIS, and J. E. MOOREHOUSE. 1963. Reciprocal Effects of Insect and Plant-growth Substances. Nature (London), 200:1230.

CAVILL, G. W. K., and P. L. ROBERTSON. 1965. Ant Venoms, Attractants, and Repellents. Science, 149:1337.

COLLIER, A. 1953. The Significance of Organic Compounds in Sea Water. Trans. 18th Amer. Wildlife Conf., p. 463.

EIBL-EIBESFELDT, I. 1949. Über das Vorkommen von Schreckstoffen bei Erdkrötenquappen. Experientia, 5:236.

EISNER, T., and J. MEINWALD. 1966. Defensive Secretions of Arthropods. Science, 153:1341.

EHRLICH, P. R., and P. H. RAVEN. 1964 Butterflies and Plants: A Study in Coevolution. Evolution, 18:586.

FLORKIN, M. 1948. La biologie des hématinoprotéines oxygénables. Experientia, 4:176.

────── 1960 Unity and Diversity in Biochemistry. New York, Pergamon Press, Inc.

────── 1966. A Molecular Approach to Phylogeny. New York, Elsevier Publishing Co.

────── and E. SCHOFFENIELS. 1965. Euryhalinity and the Concept of Physiological Radiation. *In* Munday, K. A., ed. Studies in Comparative Biochemistry. New York, Pergamon Press, Inc., pp. 6-40.

────── and E. SCHOFFENIELS. 1969. Molecular Approaches to Ecology. New York, Academic Press, Inc.

FRAENKEL, G. 1959. The Chemistry of Host Specificity of Phytophagous Insects. *In* Levenbook, L., ed. Biochemistry of Insects. New York, Pergamon Press, Inc. pp. 1-14.

GALUN, K., Y. AVI-DOR, and N. BAR-GEEV. 1963. Feeding Response in *Aedes aegypti:* Stimulation by Adenosine Triphosphate. Science, 142:1674.

GRANT, V. 1963. The Origin of Adaptations. New York, Columbia University Press.

HOSOI, T. 1959. Identification of Blood Components which induce Gorging of the Mosquito. J. Insect Physiol., 3:191.

INGRAM, V. M. 1963. The Hemoglobins in Genetics and Evolution. New York, Columbia University Press.

JACOBSON, M. 1965. Insect Sex Attractants. New York, John Wiley & Sons, Inc.

JEUNIAUX, CH. 1963. Chitine et chitinolyse, un chapitre de la Biologie moléculaire. Paris, Masson.

KARLSON, P. 1960. Pheromones. Ergebn. Biol., 22:212.

────── and M. LÜSCHER. 1959. "Pheromones": a New Term for a Class of Biologically Active Substances. Nature (London), 183:55.

KOHN, A. J. 1961. Chemoreception in Gastropod Molluscs. Amer. Zool., 1:291.
LINDAUER, M. 1963. Allgemeine Sinnesphysiologie. Orientierung im Raum. Fortschr. Zool., 16:58.
PFEIFFER, W. 1963. Alarm Substances. Experientia, 19:113.
——— 1966a. Die Schreckreaktion der Fische und Kaulquappen. Naturwissenschaften, 53:565.
——— 1966b. Die Verbreitung der Schreckreaktion bei Kaulquappen und die Herkunft der Schreckstoffes. Z. Vergleich. Physiol., 52:79.
——— 1967. Schreckreaktion und Schreckstoffzellen bei Kneriidae und Phractolaemidae (*Isospondyli, Pisces*). Naturwissenschaften, 54: 177.
REMBOLD, H. 1964. Die Kastenenstehung bei der Honighiene, *Apis mellifica* L., Naturwissenschaften, 51:49.
ROCHE, J. 1963. Biochimie des hormones thyroidiennes et évolution. *In* Oparin, A. I., ed. Evolutionary Biochemistry, Proc. of the 5th International Congress of Biochemistry, Moscow 1961. New York, Pergamon Press, Inc., pp. 313-326.
SCHOFFENIELS, E. 1961. Cellular Aspects of Membrane Permeability. New York, Pergamon Press, Inc.
SCHUTZ, F. 1956. Vergleichende Untersuchungen über die Schreckreaktion bei Fischen und deren Verbreitung. Z. Vergleich. Physiol., 38:84.
SIMPSON, G. G. 1964. Organisms and Molecules in Evolution. Science, 146:1535.
WHARTON, D. R. A., E. D. BLACK, C. MERRIT, M. L. WHARTON, M. BAZINET, and J. T. WALSCH. 1962. Isolation of the Sex Attractant of the American Cockroach. Science, 137:1062.
WILLIAMS, G. C. 1966. Adaptation and Natural Selection. Princeton, N. J., Princeton University Press.
WILSON, E. O., and W. H. BOSSERT. 1963. Chemical Communication among Animals. Recent Progr. Hormone Res., 19:673.

6

Variation and Evolution in Plants: Progress During the Past Twenty Years

G. LEDYARD STEBBINS

Department of Genetics, University of California, Davis, California

Introduction	174
Variation Patterns	174
Exploring and Charting Patterns of Variation	174
Cytological, Serological, and Distributional Characters	175
The Concept of the Population	177
Examples of Variation Patterns	177
The Ecotype Concept	177
Patterns of Variation on the Level of the Species and Genus	180
The Basis of Individual Variation	180
Environmental Modification and its Effects	180
Types of Mutations and their Significance	181
Genetic Effects of Mutation	182
Rates of Mutation	182
Natural Selection and Variation in Populations	182
Experimental Evidence for Natural Selection	182
The Adaptive Value of Diagnostic and Distinguishing Characteristics	184
The Dynamics of Selection and Random Variation	184
Genetic Systems as Factors in Evolution	185
Isolation and the Origin of Species	187
Isolating Mechanisms	188
Origin of Isolating Mechanisms	191
Hybridization and its Effects	192
Polyploidy and Apomixis	193
Structural Hybridity and Chromosomal Evolution	197

Evolution above the Species Level 200
 Some Common Evolutionary Trends 200
 Recapitulation and Embryonic Similarity 201
 The Principle of Irreversibility 201
 Orthogenesis, Specialization, and the Differentiation of Plant
   Families 201
 The Nature and Value of Paleobotanical Evidence 202
 Patterns of Distribution, Localities of Origin, and
   Rates of Evolution 202
References 203

Introduction

Professor Dobzhansky's "Genetics and the Origin of Species" was a milestone of progress in our understanding of evolution for two reasons. On the one hand, it was the first book to synthesize into a coherent whole the basic facts about traditional and population genetics, chromosomal variation, and natural selection. It therefore represents the birth of the modern synthetic theory of evolution. In addition, it attracted the attention of many biologists trained in disciplines quite different from his own, who then extended the synthetic theory in a variety of different directions. The present author was stimulated to apply the theory to plants. Even more than by the book, I was inspired by many exciting discussions of evolutionary theory with Professor Dobzhansky himself. My own book (Stebbins, 1950) is now old enough to be almost obsolete, and various other commitments make a complete revision unlikely during the next few years. I should like, therefore, to dedicate to Professor Dobzhansky on his 70th birthday a discussion of some of the most important topics of my book, as they appear to me now, based upon my interpretations of research results and theoretical discussion which many evolutionists have published during the past twenty years. In order that readers may correlate the discussions easily with the original text, each topic is discussed under the same chapter and subject headings that are used in the book, and page references are given.

Variation Patterns

Exploring and Charting Patterns of Variation

The most important advance in this field has been the development of computer methods in taxonomy. These methods have been used chiefly by zoologists, and in the United States are identified to a great extent with Dr. R. R. Sokal, whose book on the subject has been widely read and discussed (Sokal and Sneath, 1963). Taxonomists have followed the progress of these methods with varied and mixed feelings. The subject is truly controversial. Three attempts to treat plant groups in this fashion can be mentioned: those of Rogers and Fleming (1964), which deal with the algal genus *Halimeda;* of El-Gazzar et al. (1968) on *Salvia;* and the treatment by Heiser, Soria, and Burton (1965) of *Solanum* subg. *Morella*. The first two articles have extensive discussions of technique, while Heiser and his associates were content to apply the technique developed by Sokal and Sneath directly to their material. Heiser, Soria, and Burton give a critical evaluation of the method in relation to the older and more familiar methods of taxonomy. Since I find myself in almost complete agreement with their point of view, I take the liberty of quoting the last paragraph of their discussion.

> It seems evident to us that numerical taxonomy does not live up to the extravagant claims made for it by Sokal and Sneath (1963); on the other hand, it is probably not as bad as some of the critics have claimed. We feel that it might be a valuable adjunct in taxonomic studies and it may be of particular application in groups composed exclusively of cultigens, apomictic complexes, and the like, in which the type of classification given by numerical taxonomy is all that can be hoped for. It has previously been suggested that taxonomists should use more characters in their work, and that computers are a great aid in analyzing the large bodies of data which result. Perhaps the greatest good that has come out of numerical taxonomy to date is that it has forced many taxonomists to give more thought to their method and objectives. Our hope in the present paper is that, by presenting actual data rather than discussion alone, certain issues might be clarified.

A further discussion of computer techniques has been made by Rogers, Fleming, and Estabrook (1967). Here they give us a critical review of a recent attempt to use computer techniques for deducing phylogenetic pathways of evolutionary ancestry (Camin and Sokal, 1965). After reading this review, I have reached the conclusion that computer methods are no worse but also no better for constructing phylogenies than are the traditional methods of taxonomists. In both instances, the accuracy of the "tree" depends upon the correctness of various assumptions which must be made about the degree of primitiveness or advancement of certain morphological characteristics. Without fossils to guide us, such assumptions are always hazardous guesses. Whether or not computers are used, phylogenies are at best unprovable hypotheses, and should not be attempted without very thorough knowledge of the group of organisms that is being studied.

Cytological, Serological, and Distributional Characters

The advances in our knowledge of biochemistry are of great importance for both taxonomy and evolution. Modern refined methods of electrophoresis are making possible an analysis of specific differences between proteins at a much more precise level than has previously been possible. Using total seed proteins, Ove Hall and B. Lennart Johnson have applied this method successfully to species relationships in the tribe Triticinae, which contains the wheat plant (Hall, 1959; Johnson and Hall, 1965). An even greater refinement has been the study of individual enzyme "families," most if not all of which are now known to consist of series of isozymes. This approach was first used by Drew Schwartz for studying the genetics of maize (Schwartz, 1964; Schwartz, Fuchsman, and McGrath, 1965), and such studies have been continued by other workers (Beckman, Scandalios, and Brewbaker, 1964; Scandalios, 1964, 1965). In *Drosophila,* Hubby and Lewontin (1966) and Lewontin and Hubby (1966) have found an extraordinary amount of polymorphism for variant isozymes in natural populations, and there is every reason to believe that the same situation exists in many wild plant species.

In addition, the discovery of the deoxyribonucleic acid (DNA) structure of genes and the genetic code has opened the way to direct comparisons between genes of related organisms via artificial "hybridization" between the DNA extracted from different species. This technique has been developed chiefly by Hoyer, MacCarthy, and Bolton (1964). It is difficult to use, and some investigators have expressed to me orally their dissatisfaction with it. Nevertheless, it deserves much attention, in hopes that it can be further refined. At present, we can say that closely related species have DNA's so similar that significant differences in hybridization cannot be found, while members of different phyla, and probably of different classes, cannot be analyzed because there is too little hybridization between their DNA's. Within an intermediate range of affinities, however, it is proving useful. The method may prove to be of greater value for determining relationships between groups of bacteria and blue-green algae than in higher organisms. This is because in the latter the DNA of chromosomes is usually compacted and folded in such a way that much of it cannot easily be prepared for free hybridization.

Comparing methods employing protein differences with those which concentrate on nucleic acids, we can say that for determining relationships between races and species proteins are much the better of the two. Consequently, it is through studying differences in proteins, particularly isozymes that play important roles in plant development, whereby we can obtain the most significant information about processes of evolution at the molecular level. Comparisons between DNA's may, however, shed light upon the relationships of families and orders of plants of which the affinities are now obscure, such as the amentiferous families of woody plants.

Another way in which biochemistry has amplified our understanding of taxonomy and evolution is through the use of secondary organic compounds, such as phenolic compounds, alkaloids, terpenes, volatile oils, and the like. These are differentiated by the methods of chromatography, of which paper chromatography is the simplest and most used. The rather extensive literature in this field, including descriptions of methods used, is reviewed by Alston and Turner (1963) and by Alston (1967). The methods have been particularly useful in the recognition of certain races within species, in verifying the suspected hybrid nature of natural populations, and in detecting the probable ancestry of hybrid polyploids.

The Concept of the Population

Although a number of review papers, chapters of books, and entire volumes have been written during the past 20 years on population genetics, none of them has defined the concept of population in terms sufficiently precise so that the word has maximal meaning for discussions of evolution. The concept of interbreeding between members of the same population is fundamental to the definition of the term, as I stated in the book. If, however, one considers a very large, continuously distributed group, such as an extensive forest consisting of a single species of tree, its delimitation as a population becomes difficult. In any one generation, interbreeding will be almost entirely among individuals located only a few score or a few hundred meters away from each other. Nevertheless, an adaptively beneficial gene or gene combination can be transported throughout the forest by successive crossings between neighboring trees, given a sufficient number of generations. Consequently the time factor can be very important in an evolutionarily significant definition of a population. Over a small number of generations, isolation by distance can be very important even in different parts of a continuously distributed population. If, however, a very large number of generations is considered, this factor will be important only if there is a wide gap in which no potentially interbreeding indivduals are found.

Examples of Variation Patterns

The Ecotype Concept

In discussing this concept, I asked two questions (1950, p. 43): "First, to what extent are the biotypes of a species grouped into . . . distinct ecotypes, and to what extent do they form a continuous series? Second, what is the relation between the ecotype concept in plants and the concept of *polytypic species* or *Rassenkreise*, as it has been developed by modern zoological systematists . . . ?"

The large amount of experimental work that has been done in this field during the past 20 years has been well reviewed by Heslop-Harrison (1964). He has concluded that both clinal variation and mosaics of distinctive ecotypes can be recognized, depending upon the geographical and ecological distribution pattern of the species concerned. Species of forest trees, prairie grasses, and others that are found over large areas in which environmental conditions change gradually, tend to have clinal variation. On the other hand, species that occupy territories that are broken up into mosaics of sharply distinct ecological conditions tend to have mosaic-like patterns of recognizable ecotypes.

This latter condition is most conspicuous in plants that occupy radically different kinds of soils. The recognition and study of such edaphic ecotypes is one of the most important advances in genecological study during the past 20 years. Examples are the ecotypes adapted to serpentine soils, which were revealed by Kruckeberg (1951) in a number of Californian species, and the studies of Bradshaw and his coworkers on species of grasses adapted to differing levels of calcium, as well as to lead salts, which are toxic to most species of plants (Snaydon and Bradshaw, 1961; Bradshaw, 1962; Jowett, 1964; Bradshaw, McNeilly, and Gregory, 1965).

We can, therefore, answer the first question posed above in essentially the same way as it was answered in my book. Genetic variation within most species of plants is extensive, and the great bulk of this variation is adaptive to the various ecological conditions which the species encounters over its geographical range. An answer to the second question can be obtained by comparing my discussion and that of Heslop-Harrison with the discussion of variation within animal species, recently presented by Mayr (1963) in his book, "Animal Species and Evolution." Mayr has shown that in animals as in plants, extensive adaptive variation exists in most if not all species, and geographic races represent adaptations to particular sets of ecological conditions. There is, however, a difference between the variation pattern found in plants and that in many kinds of animals, such as birds and other vertebrates, upon which Mayr's conclusions are largely based. In these mobile animals gene flow between neighboring populations can be extensive, and the insulation of the young from the immediate effects of their environment tends to minimize the selective effects of this environment upon early stages of development. Consequently, populations can maintain distinctive constellations of genes only if they are well isolated spatially or by other means from neighboring populations. In plants, however, their sedentary mode of life tends to restrict gene flow. More important, however, is the fact that selection acts very strongly upon young seedlings, eliminating all that are not well-adapted to their immediate environment (Harper, 1965). Under these conditions, populations can maintain distinctive constellations of adaptive genes in spite of extensive gene flow from differently adapted populations.

Two examples of direct observation support this generalization. The first is the eight-year study by myself and various students of a hybrid swarm of *Helianthus* (Stebbins and Daly, 1961). These plants are annuals that because of self-incompatibility are obligate outcrossers. They are normally pollinated by strong-flying bees. Nevertheless, two parts of the same hybrid swarm, which had a common origin but later became separated from each other by a distance of only 120 meters, maintained over a period of eight years, and hence as many generations, distinctive constellations of genes in association with different environmental conditions which prevailed in the two areas. Far from converging in response to gene flow, they not only maintained their distinctive characteristics but actually became increasingly different during this period. Unfortunately, the population has now been obliterated by further habitat disturbance, so that we shall never know what would have been its eventual fate. In another study, Lyman Benson and his associates (Benson et al., 1967) have shown that a hybrid swarm of *Quercus* in southern California contains different variants on a south-facing slope from those existing on the north-facing slope of the same mountain.

We can, therefore, answer my second question, concerning the connection between the concept of ecotypes and that of polytypic animal species, as follows. In general, the concepts of infraspecific categories are very similar in animals and plants. Nevertheless, differences exist because of the very different modes of life found in plants as compared to animals. Differences between related populations of plants are much more closely correlated with differences in the inanimate, nonbiological environment than are comparable differences between animal populations. The strong selective pressures exerted by this environment upon early developmental stages enable plant populations, even when normally outcrossed, to resist gene flow and so to evolve adaptive constellations of genes to quite different habitats. Such variation forms the basis of the ecotype concept. Operationally, it is revealed by transplant techniques, which are much more difficult to perform in most animals than they are in plants.

The most important addition to our knowledge of the genetic nature of ecotypes has been the series of analysis of F_2 progenies of hybrids between ecotypes of *Potentilla glandulosa,* made by Clausen and Hiesey (1958). From this analysis, they have reached two conclusions of fundamental importance. First, most of the differences between ecotypes segregate in such a way that each character difference must be determined by many different genes. This is particularly true of such adaptively important differences as size, earliness, and extent of winter dormancy, all of which are determined by multiple genetic factors. Second, the segregation patterns of these F_2 progenies show extensive correlations between different characters. Clausen and Hiesey believe that these correlations are all due to genetic linkage. Their data, however, do not exclude the possibility that some correlations are developmental, and represent pleiotropic effects of certain genes upon different morphological characteristics. The relative importance of genetic linkage and pleiotropic gene action in producing correlations between morphological characteristics is a question that is not yet decided, and deserves more attention.

Patterns of Variation on the Level of the Species and Genus

The most noteworthy of the studies on patterns of variation in the last 20 years (cf. Stebbins, 1950, pp. 52-71) are the work of T. H. Goodspeed and his associates on *Nicotiana* (Goodspeed, 1954), of Blakeslee and his group on *Datura* (Avery, Satina, and Rietsema, 1959), of Grant on the phlox family and particularly the genus *Gilia* (Grant, 1959, 1963, 1964), of Lewis and his group on the genus *Clarkia* (Lewis and Lewis, 1955), and of Ehrendorfer (1964, 1965) on the Dipsacaceae. These examples verify and amplify the conclusions that I stated on pp. 70-71 (1950). The study of the comparative evolution of plants is progressing well, but we still do not have enough information so that we can make generalizations about the patterns of variation that we find, nor can we yet understand why different genera have evolved in different ways.

The Basis of Individual Variation

Environmental Modification and Its Effects

The phenomenon of phenotypic plasticity, or the ability of certain genotypes to respond to varying environments by producing radically different phenotypes, has recently been well reviewed by Bradshaw (1965). He has shown that plasticity is, in general, an adaptive characteristic. If, as in certain aquatic forms, there is an adaptive advantage to the production by the same genotype of strikingly different phenotypes under the influence of certain environmental stimuli, mechanisms for bringing this about have been evolved. On the other hand, many characteristics, such as seed size and the form of flowers, remain constant over a variety of environmental conditions. For them, morphological constancy is an adaptive trait maintained by natural selection.

Natural selection for phenotypic plasticity exerted by habitats that are subject to drastic and unpredictable alterations is well illustrated by the experiments of Cook (1968) on *Ranunculus flammula*. This species has long been known for its heterophylly, which includes the ability of a single plant to produce very different kinds of leaves when exposed to relatively dry conditions, as compared to those produced in water. Cook has shown that populations of this species that occupy immature and unpredictable environments are relatively homogeneous genetically, but contain gene combinations that promote phenotypic plasticity. On the other hand, populations of "terrestrial specialists" have less heterophylly but more intrapopulational variability of a genetic nature.

As a general principle, the inheritance of acquired characters is completely discredited, largely because attempts to produce evidence for it have either failed completely or have been proved incorrect by other workers. Furthermore, the demonstration of a self-replicating code of DNA has extended to all organisms, including bacteria, Weismann's principle of the separation of the germ line from the somatic or metabolic activities of the cell and body. The last fortress of neo-Lamarckism, the school of Lysenko, has been overwhelmed by the facts of modern genetics.

Nevertheless, there is one way in which environmental modification can contribute to evolution. This is by revealing certain genetic potentials, and making them available to the action of natural selection. This phenomenon has been termed by C. H. Waddington genetic assimilation, and he has conducted a number of experiments on *Drosophila* that demonstrate its effectiveness (Waddington, 1960). I have suggested a possible explanation of the phenomenon on the basis of our present knowledge of genes and enzyme action (Stebbins, 1966a, p. 79).

Types of Mutations and Their Significance

The discovery of the genetic code and of the fine structure of genes has given us a completely new insight into the nature and significance of the different kinds of mutations. We can now distinguish clearly between those mutations (in the broader sense of the word) which add to the genetic variability present in populations, and so contribute directly to the potentiality for evolutionary change, and those which affect principally the mechanisms for gene recombination and gene flow between populations. Chromosomal translocations and inversions clearly belong in the latter category. This conclusion, which I already reached in my book, has been amply supported by a great body of evidence that has accumulated since then.

The mutations that provide almost all of the variation found in natural populations are point mutations, which consist of the substitution of one base pair for another in the chemical composition of the DNA molecule. The consequence of such substitutions is the alteration of the triplet codons so that they code for a different amino-acid residue at a particular position on the polypeptide chain of the protein molecule. We can, therefore, determine the past occurrence of such mutations by comparing the sequence of amino-acid residues in homologous molecules found in related organisms. When such comparisons are made on molecules such as the hemoglobin of different vertebrates (see the summary made by Jukes, 1966), we find that differences that must have resulted from alterations in single base pairs are 10 times as common as any other differences.

The only other kinds of mutations that have been of primary importance for evolution have been duplications of entire sequences of base pairs, which code for individual polypeptide chains. Such duplications, followed by mutational divergence of the duplicated genes, have greatly enriched the battery of enzymes possessed by all higher organisms. The classical example of this duplication–differentiation sequence is the evolution of hemoglobin and related molecules, as worked out by Ingram (1963). Contemporary research is now revealing many similar examples.

Genetic Effects of Mutation

The conclusions reached in the book (1950, pp. 85-96) have been borne out by more recent investigations. In particular, studies of segregation in progeny of hybrids between races and species, such as that of Clausen and Hiesey (1958), have shown that most of the differences between such populations are governed by large numbers of genes, each of which has individually a small effect on the phenotype. One can conclude from such data that most of the mutations that have contributed to the evolutionary divergence of populations from each other have been those with small effects on the phenotype. This conclusion is also supported by data such as those of Gaul (1965) on the mutagenic effects of radiation. He has shown that this agent produces mutations with small effects at a considerably higher frequency than the more familiar mutations with large effects on the phenotype.

Rates of Mutation

Contemporary research in population genetics has forced me to retract the statement (1950) made on p. 96, that ". . . the mutation rate may under some conditions be a limiting factor" in evolution. Studies of many natural populations of plants, even those which because of self-fertilization might be expected to be genetically homozygous and homogeneous (Allard, 1965), have revealed in them a vast, complex, and hitherto unsuspected store of hidden genetic variability. Under these conditions, the amount that single, newly occurring mutations contribute to this variability is always so little that their absence can never be a limiting factor. Evolution is guided by natural selection, acting upon variability which has resulted from the occurrence and establishment of mutations over many generations.

Natural Selection and Variation in Populations

Experimental Evidence for Natural Selection

During the past 20 years a large number of experiments have been performed dealing with natural selection and competition. I have summarized several of them elsewhere (Stebbins, 1965, 1966a, chap. 4). Some of them have followed changes in frequency of individual genes or chromosomes over many generations. Because of their more rapid reproduction, bacteria, molds, and flies (*Drosophila*) are more favorable subjects for such experimentation than are higher plants, and so have been by far the most used. In them, the universal presence of natural selection has been demonstrated, and the rate of selection under different conditions has been measured.

On the other hand, plants have proven to be particularly favorable for analyzing another feature of natural selection: the competition that brings it about. The research of Sakai (1965) and that of Harper and his associates (1965) are both particularly important in this respect. They have shown clearly the complex nature of competition, and the difficulty of associating competitive success with any particular morphological characteristic. From his studies of competition, Sakai (1965, p. 236) has concluded ". . . that selection in a colonizing species may involve (1) loss owing to chance, (2) selection for fitness in the given environment, (3) interspecific competition and interactions, (4) selection for density response, (5) selection for intraspecific competitive ability, and (6) selection for migratory activity." Most of these characteristics can vary independently of each other, and all of them are determined genetically by large numbers of interacting genes. The results of Harper's experiments fully support these conclusions of Sakai. Furthermore, they focus our attention upon the great hazards that accompany the establishment of seedlings under natural conditions, and the high intensities of selection that occur during seedling stages. Because of this fact, the characteristics of the seed and seedling are of great importance for the success of a species. These results confirm the opinions that I expressed in my book, in which I devoted an entire section to the adaptive nature of seed and ovary characteristics in *Camelina*, and another section to a review of Salisbury's studies on the adaptiveness of seed characteristics. The work of Harper's school has greatly refined the techniques for this kind of experimentation. As a result, they have shown that in genera such as *Rumex* and *Plantago* interspecific differences in the characters of their seeds are clearly adaptive to different microhabitats, which the species can colonize.

One result of these experiments on competition has been to show that natural selection can be effective through differential fecundity even though no mortality occurs at all between seed germination and plant maturity. An example is the series of experiments by Lee (1960) on competition between varieties of cultivated barley. Previous data had shown that when a mixture of four varieties, each of them constituting one-fourth of the mixture, were sown in the field, differences in the number of seeds that each plant produced caused the proportions of the four varieties in the seeds harvested to be very unequal. If a small sample of this harvest was planted in the same field the following year, and the same procedure was repeated for several generations, the less successful varieties were completely eliminated from the mixture. By studying the behavior of each plant in mixed plantings at known densities of spacing, Lee showed clearly that these effects are produced without the loss of a single plant or the failure of any seeds to germinate.

The Adaptive Value of Diagnostic and Distinguishing Characteristics

Evolutionists have come to realize that we cannot demonstrate the effectiveness of selection in natural populations by focusing our attention solely upon the adaptiveness of individual morphological traits. Dobzhansky (1956) has clearly expressed the situation in the following words: "A visible 'trait' is the outcome of the occurrence of certain developmental, physiological, and ultimately physico-chemical processes in the organism. . . . A trait is an aspect of the whole or of a certain portion of the developmental pattern of the organism. An adaptive trait is, then, an aspect of the developmental pattern which facilitates the survival and/or reproduction of its carrier in a certain succession of environments."

Nevertheless, experimental evidence has been obtained to show that certain morphological differences between genotypes of which the adaptive significance was previously either in doubt or unsuspected, clearly have an adaptive value. This evidence is derived chiefly from competition experiments between races differing by a single Mendelian factor, and raised under controlled conditions. Examples are the demonstration of Scheibe (1955) that the waxy cuticle normally present in the field pea (*Pisum arvense*) helps the plant to resist drought, and demonstrations of the insect-repellent properties of such characteristics as pubescence on the ovaries of cotton (Stephens, 1957), alkaloids in the leaves of potatoes (Schreiber, 1957), and silica in grass leaves (Sasamoto, 1958).

The Dynamics of Selection and Random Variation

The "supposition," stated in my book (1950, p. 146) that in insular populations "many nonadaptive differences have been established by random fixation . . ." seems now to have been incorrect. Recent studies of these island habitats have revealed their great diversity at all levels, so that natural selection must be acting both strongly and differentially in them. A reexamination by Bailey (1956) of widely cited data purporting to show the effect of drift on variation in land snails has forced him to conclude that differential selection, reinforced by isolating mechanisms, has been more important in producing these variation patterns than have chance events. Observations on continental populations of land snails have reinforced this conclusion, by showing that both polymorphism within populations and genetic differences between them are produced and maintained chiefly by the action of natural selection (Cain and Sheppard, 1954; Lamotte, 1959). The highly diverse and polymorphic plant genera of some oceanic islands have not yet been carefully studied, and deserve attention.

Similar conclusions have been reached by Carl Epling and his associates (Epling et al., 1960) from their continued investigation of the annual desert species, *Linanthus parryae,* the original investigation of which I summarized on p. 151. They have shown that the pattern of distribution of blue- and white-flowered forms has remained remarkably constant for 15 years, in spite of great fluctuations in population size. This suggests intense local selection. Genetic drift, if important at all, has been of only local consequence. The nature and action of the selective forces on this population are, however, completely unknown.

The random association of adaptive and selectively neutral characters which I emphasized on pp. 148-150 is similar in some important respects to the founder principle as discussed in various publications by Mayr (1963). In both phenomena, chance differences are not due to random depletion of variability in a population that remains in the same place, but to random effects that accompany the colonization of new habitats. These represent the most probable ways in which chance events affect the course of evolution, acting always in association with selection. In addition, as Mayr (1963, p. 213) has emphasized, chance may often determine which of several gene combinations having similar adaptive values will be established in a particular local population. This effect can produce evolutionary divergence when similar environmental changes are exerting their effects on different segments of a genetically homogeneous series of populations in regions that are partly or completely isolated from each other.

Genetic Systems as Factors in Evolution

The material presented in the first part of Chapter 5 (1950) has been amplified and modernized in a later publication (Stebbins, 1960), in which our recent knowledge about genetic recombination in microorganisms has been taken into account. In that paper I presented my reasons for believing that genetic recombination existed already in the earliest organisms, and that the predominantly asexual condition, which appears to prevail in such groups as the blue-green algae, some bacteria, and many groups of flagellates, has been secondarily derived through loss of function. In respect to the evolution of the alternation of generations, I have amplified my reasons for accepting the hypothesis that the haploid condition is primitive, isomorphic alternation of generations is intermediate, and both the sporophyte dominance of vascular plants and the gametophyte dominance of bryophytes are derived from isomorphy by divergent evolution. In addition, I revised my hypothesis concerning the selective pressures that established the diploid condition in many groups of organisms, and which have ensured the continuing success of most of these groups. The initial establishment of diploidy was probably due to its success in maintaining certain adaptive heterozygous gene combinations, which could not be maintained in the haploid state. Its continuing success depended upon the enrichment of the genotype by adding genes with complementary functions, particularly with respect to the development of the organism. Modern research in population genetics has repeatedly revealed great stores of genetic variability that are perpetuated in the heterozygous state. The more complex are the gene interactions in development, the more important is this variability for enabling populations to evolve further in response to continuing environmental change.

The material in the second half of Chapter 5 was amplified in two review articles (Stebbins, 1957a, 1958a). In the first of these I gave additional evidence for the hypothesis that predominant or habitual self-fertilization is a derived condition, and suggested three kinds of selective pressure which could have favored its development. The first of these is the necessity in most annual species for producing at least some seeds every year, in order to maintain the population. The second is the great value of self-fertilization in individual plants that have been transported by long distance dispersal to a new area, and are beginning to colonize that area. This condition has brought about the correlation first pointed out by H. G. Baker (1955), which I have called "Baker's law." It states that when a group of plant species occurs in two or more widely separated areas, as on Eurasian and the North American continents, those species found in the area where the group originated will be predominantly cross-fertilizing, while those found in the area into which they have migrated secondarily will be predominately self-fertilizing. The third kind of selective pressure for self-fertilization, already suggested in Chapter 5 of the book (p. 177), is its advantage for the actual process of colonization. It is based upon a principle that could be called the "infective principle," since it is particularly significant for races of pathogenic organisms that are infecting a new host. If a single successful invader or colonizer has entered an unoccupied niche that could be filled by its descendants, the invasion or infection will take place most efficiently if its descendants share all of the genetic qualities that have adapted it particularly well to that specific habitat. Such a resemblance between the initial colonizer and its descendants will be most likely to exist if genetic recombination is severely restricted, as in asexual reproduction or self-fertilization.

In the later paper (Stebbins, 1958a) I amplified this concept to explain the spread of apomictic reproduction in *Crepis, Taraxacum,* and other genera of angiosperms, and to account for the drastic alterations in karyotype that are found in many groups of annual species. The latter changes are explained on the basis of the adaptive advantage of certain clusters of linked genes, which are protected from recombination both by being placed on the same chromosome arm and by various devices that reduce the amount of crossing over in the chromosomal arms that bear them.

The evolutionary future of species groups with predominant self-fertilization or apomixis was discussed in both of these papers, and more recently by Allard (1965). The latter author has shown that self-fertilizing species contain nearly or quite as much genetic variation in their total populations as do cross-fertilizers, and differ from them chiefly in the method by which this variability is released over short periods of time. Based upon the progeny analyses made by Clausen and his group (Clausen, 1954) on the facultatively apomictic species of *Poa,* we can conclude that such species also contain great stores of genetic variability. Nevertheless, the fact remains that all or nearly all groups of self-fertilizing or apomictic angiosperms are specialized subgenera or sections of genera that also contain cross-fertilizing species. This condition suggests that both self-fertilization and apomixis are specialized conditions that have arisen many times in the evolution of angiosperms, but that groups consisting of predominant self-fertilizers of apomicts have never been able to continue their evolution so far that they have evolved into new genera or families. They have been highly successful in evolving large numbers of aggressive and successful species, but have been severely limited in respect to long time evolutionary progress.

Isolation and the Origin of Species

In reexamining Chapters 6 to 11 (1950), I find little reason for changing the overall form or the basic hypotheses, which were worked out for them 20 years ago. In respect to the concept of species (pp. 188-195), numerous discussions of the problem have not weakened the evolutionary argument of Muller (cf. p. 195) in favor of recognizing the degree of reproductive isolation as the principal criterion by which we judge whether two populations should be placed in the same or in different species. If two populations can exist sympatrically and still retain their identities and their own ways of adapting to the environment, they are likely to evolve further toward new divergent adaptations. If they lose their identity through hybridization and gene flow whenever they occur together, they will evolve, if at all, both in the same direction.

The principal argument that plant systematists still raise against reproductive isolation as the primary basis for recognizing species is the difficulty of applying it in individual examples. I would be more impressed with this argument than I am if I found that different systematists working on the same group by the use of morphological criteria usually reached the same opinion regarding the limits of species. Actually, in most critical or difficult genera, the differences in opinion between "lumpers" and "splitters" among morphological taxonomists seem to me to be as great today as they were 40 years ago when I first became acquainted with plant taxonomy. If, therefore, morphological criteria don't permit taxonomists to reach an agreement on the limits of species in difficult groups, why object to reproductive isolation because it, too, can't do this?

As Darwin and many modern evolutionists have pointed out, the difficulty of defining the limits of many species, which will always be with us, is what one would expect on the assumption that evolution has occurred. On the other hand, evidence from both fossils and geographic distribution tells us that we cannot regard all examples of imperfect or variable degrees of isolation between two groups of populations as "incipient speciation" (Dobzhansky and Pavlovsky, 1966), which will eventually progress toward full speciation, or complete reproductive isolation. An instructive example in this connection is that of certain California oaks, as worked out by Tucker (1952) and reinforced by my personal observations. In northern California, one of the commonest species is the blue oak (*Quercus Douglasii*) a tree with deciduous leaves, which may become as much as 12 to 15 meters tall. In the inner coast-range mountains, it grows together with the scrub oak (*Q. dumosa*), an evergreen shrub which is usually 3 to 4 meters high. North of San Francisco Bay, these species grow side by side in many places, and only occasionally hybridize, never forming hybrid swarms. In central and southern California, however, hybrid swarms involving *Q. Douglasii, Q. dumosa,* and the closely related species or subspecies, *Q. turbinella,* are very common. In many places, the distinction between these three entities becomes completely lost in the extensive hybrid swarms which are formed. These three entities are certainly on the borderline between species and subspecies, and one might argue that they are incipient species. Nevertheless, fossil evidence (Axelrod, 1939; Chaney, Condit, and Axelrod, 1944) indicates that the counterparts of both *Q. Douglasii* and shrubs of the *Q. dumosa-turbinella* alliance existed as separate entities already in the Miocene or the beginning of the Pliocene epoch, 10 million or more years ago! Here we have apparently a case of arrested rather than incipient speciation. Less complete evidence in many other groups indicates that this type of arrested speciation is common in many genera of woody plants.

Isolating Mechanisms

The classifications of isolating mechanisms given recently by Mayr (1963, p. 92), Grant (1963, pp. 353-354), and myself (Stebbins, 1966a, p. 97) are essentially the same as the one on p. 196 of my book (1950), except that more appropriate terms are used. Many more examples of the various kinds of mechanisms are now known, but they have not changed the overall picture very much. The phenomenon of pseudocopulation, discussed on pp. 211-212, has been thoroughly studied in an excellent monograph by Kullenberg (1961).

The inviability of interspecific hybrids is a phenomenon that deserves investigation on the basis of our new knowledge about the genetic code and the molecular basis of differential gene action. In the light of these facts, even the review that I made of the subject 10 years ago (Stebbins, 1958b) is completely out of date. The best avenue to an attack on this problem would be investigations of the two kinds of reciprocal differences in the success of interspecific hybrids. One of these, which is based upon cytoplasmic differences, and is well exemplified by certain species of *Epilobium* as well as the chlorotic hybrids between some species of *Oenothera,* is discussed in pp. 216-217 of my book. These phenomena should be reexamined at the biochemical level, based upon the now well-known fact that subsidiary genic codes of DNA exist in both plastids and mitochondria. A likely hypothesis is that these examples, like all others, are due to disharmony of gene action in development. The probable difference is that whereas in most examples of hybrid inviability the disharmony is between different nuclear genes, in *Epilobium, Oenothera,* and similar cases it is between certain nuclear genes and others that are located in the DNA of the cytoplasmic organelles.

The second type of hybrid inviability which deserves reinvestigation was extensively reviewed in Stebbins (1958b). This involves reciprocal differences in viability and developmental behavior in crosses between certain diploids and tetraploids, as well as between some diploid × diploid crosses in genera such as *Primula* (Valentine, 1953). A new insight into this problem has been provided by the investigations of Wangenheim (1961, 1962) on diploid × tetraploid crosses in the genera *Solanum* and *Oenothera.* He showed that although the diploid × tetraploid cross is never successful if the embryo sac and egg nuclei have the normal haploid number of chromosomes, it proceeds normally in those rare instances when, through failure of meiosis, diploid embryo sacs are produced. This and other evidence led Wangenheim to the conclusion that the chief disharmony in crosses of this type is between the endosperm and some extranuclear factor that exists in the embryo sac before fertilization. An hypothesis to explain this situation can now be advanced. In animals, particularly echinoderms and amphibians, messenger-RNA produced by gene action during oöcyte development is active in synthesizing proteins during the early stages of cleavage and blastula formation of the fertilized egg (Glisin et al., 1966). Although this long-lived messenger has not yet been detected in the early embryogeny of angiosperms, it has been found as an accompaniment of embryo development during seed germination (Dure and Waters, 1965; Waters and Dure, 1966). Consequently, as a working hypothesis, I should like to suggest that the reciprocal differences found in diploid × tetraploid crosses, as well as in such combinations as *Primula elatior* × *veris,* are due to disharmony between messenger-RNA produced by the maternal gametophytic nuclei and carried over into the embryo and endosperm; and new messenger RNA produced by genes in the hybrid nuclei of the developing tissues after fertilization. On the basis of this hypothesis, the "extra-nuclear factor" of Wangenheim is long-lived messenger RNA.

In respect to hybrid sterility, a new terminology is needed, and has been adopted by some authors, including myself (Stebbins, 1966a, p. 103). The terms "genic" and "chromosomal sterility," first used by Dobzhansky (1951), have always been difficult to define, and have now become meaningless because of our new understanding of the nature of the gene. Originally, chromosomal sterility was defined as the sterility of hybrids resulting from structural differences between the parental chromosomes, and is accompanied by irregular pairing and subsequent abnormal behavior of the chromosomes at meiosis. Operationally, it could be recognized by the fact that doubling the chromosome number eliminates it and produces a fertile allopolyploid. Genic sterility, on the other hand, was considered to be due to disharmonic interaction between parental genes, and not eliminated by polyploidy. When I found that in certain grasses there exist examples of hybrid sterility that are not accompanied by reduction in the amount of chromosomal pairing and other meiotic abnormalities, but which nevertheless can be partly eliminated by chromosome doubling, I postulated (Stebbins, 1950) that in these examples we were dealing with cryptic structural differences. These were believed to be just the same as the visible translocations and inversions which were the basis of Dobzhansky's concept of chromosomal sterility, but so small that they did not cause obvious abnormalities of pairing. Now, however, we recognize that the gene itself is a linear double helix containing hundreds of nucleotide pairs. Once this structure is recognized, we realize that inversions as well as translocations of groups of base pairs can occur within the confines of a single gene, and thus be classified as gene mutations. Moreover, crossing-over within genes is known in microorganisms, and may well take place in higher plants. In this way, two genes that are both functional, but contain mutations at different positions along their length, could produce, by crossing-over, an entirely new gene, which might well be nonfunctional. Consequently, the "cryptic structural differences" that I postulated could perfectly well exist within individual genes and so be called genic rather than chromosomal differences.

Consequently, the most clear-cut difference is between hybrid sterility that results from abnormal development of the reproductive structures, and that which results from abnormal segregation of chromosomes or genes at meiosis. The former can be called *developmental,* the latter *segregational,* sterility. The two kinds can be distinguished operationally by the fact that segregational sterility can be partly or completely overcome by doubling the chromosome number, while doubling has no effect on developmental sterility. Developmental sterility can be either diplontic or haplontic, since the development of microspores and male gametophytes can be blocked by abnormal development of the tapetum. Segregational sterility is always haplontic.

Origin of Isolating Mechanisms

The findings of the past 20 years have contributed but little to this subject, in my opinion. The hypothesis that species originate through the occurrence of single "macromutations" or "systemic mutations," which was championed by Richard Goldschmidt and was popular among some biologists during the 1940's, has now almost completely disappeared from the scientific literature owing to the lack of supporting evidence. The similar views of Herbert Lamprecht have been recently published as a book (Lamprecht, 1966), but in this form are no more convincing than they were when originally published as research papers. Doubt has been cast upon Lamprecht's example, *Phaseolus coccineus* and *P. vulgaris,* by the genetic studies of Wall and York (1957), whose evidence indicates that one of the principal "interspecific genes" supposedly separating these two species actually does not exist.

Another kind of instantaneous or "saltational" speciation has been postulated by Lewis (1962, 1966). In the genus *Clarkia* (which in Stebbins, 1950, was designated by its then current name *Godetia*), he has found a number of examples of sympatric species that are closely related to each other but differ with respect to a large number of chromosomal translocations. In each example, one species of a pair has a very restricted distributional range, lying entirely within that of the other. This evidence has led Lewis to conclude that the chromosomal changes separating the species all occurred simultaneously, perhaps triggered off by the action of an extreme environment.

While this hypothesis is plausible, an alternative one is not excluded by his evidence. This is that the translocations occurred and became established in rapid succession, perhaps over a period of a few hundred years, when the species, owing to different climatic conditions from those which now prevail, had different and more widely separated ranges of geographic distribution. On this hypothesis, the present situation is regarded as due to secondary sympatric occurrence, followed by extinction of the rarer species in those localities where they originated. The successive establishment of different translocations would then be explained, as in Stebbins (1950), by their advantage as devices for keeping together adaptive combinations of linked genes. The recent studies of Kyhos (1965 and unpublished) on the annual species of *Chaenactis,* which Lewis cites in support of the hypothesis, indicate that for this genus the second explanation is the most likely one.

Lewis recognizes that his hypothesis of sudden speciation through simultaneously occurring translocations cannot be applied to all genera of higher plants. The opinion that I expressed in my book (p. 235) that *Clarkia* ("Godetia") is one of a few genera in which gross chromosomal differences between species are exceptionally highly developed, still holds. Consequently, whatever may turn out to be the dominant method of speciation in this genus, it cannot be regarded as a basis for generalizations about speciation in most groups of angiosperms.

Among the various hypotheses about the origin of isloating mechanisms, which were presented (1950) on pp. 236-250, those in which the selective action of the environment features prominently have held up the best. The importance of adaptive differences in developmental pattern or rhythm as basis for hybrid inviability was pointed out in my review paper (Stebbins, 1958b). Good evidence favoring the adaptiveness of chromosomal differences within species, which through their accumulation may evolve into isolating mechanisms separating distinct species, has been found by M. Kurabayashi and his coworkers in the genus *Trillium* (Kurabayashi, 1958).

Hybridization and Its Effects

In Stebbins (1959a) I presented an evolutionary definition of hybridization, characterizing it as crossing between populations having different adaptive requirements, regardless of their taxonomic status as different subspecies, species, or even genera. This helps further to emphasize the fact that hybridization occurring between two well-established species in an old, undisturbed habitat will have little or no effect on evolution, but that when two populations with different adaptive requirements hybridize in a disturbed, rapidly changing habitat that offers new ecological niches to certain hybrid derivatives, the effects of hybridization may be profound and far-reaching. The extensive literature which has accumulated during the past two decades has served to emphasize repeatedly this all-important fact.

The hypothesis that wide hybridization may lead to new, fertile, stabilized adaptive populations (Stebbins, 1950, pp. 279-289)—and this without change in chromosome number—is now supported by three artificially produced examples, in *Elymus* (Stebbins, 1957b), *Nicotiana* (Smith and Daly, 1959), and *Gilia* (Grant, 1966). In the first two examples the parental species had relatively similar chromosomes, so that a high degree of pairing was possible in meiosis of the F_1 hybrid. In *Gilia malior* × *modocensis,* on the other hand, chromosome pairing in the F_1 hybrid was very poor, and the stabilized derivative was very difficult to obtain. Because of this fact, Grant believes that this manner of speciation cannot have occurred often under natural conditions. I would interpret the three examples mentioned to mean that hybridization without chromosome doubling can serve as an evolutionary catalyst if the parental species are closely related to each other, but not if they have diverged to such an extent that their chromosomes have few homologous segments in common. In the former case, the initial segregation will produce barriers between the newly segregated population and its parents, which are in general weaker than those which seperated the parental species from each other. Successful speciation by this method, therefore, probably requires a later reinforcement of the barriers by divergent evolution. The hybridization and subsequent segregation must be regarded strictly as catalytic agents, and not as responsible by themselves for the entire process of speciation.

Polyploidy and Apomixis

Since 1950, relatively little, in my opinion, has been added concerning these phenomena, and I see no reason for modifying any of the major conclusions in the book (Stebbins, 1950, Chapters 8, 9, and 10, especially the latter). Swanson (1957) has presented a good, brief treatment of the polyploid complex, including a corrected version of my Figure 34 (1950, p. 315; on the bottom line of this figure, right center, the symbol under B_1B_2 should read 7-II's, not 14-II's).

In respect to the evolution of segmental allopolyploids (Stebbins, 1950, p. 325), the work of Riley (1966) on *Triticum* is particularly significant. He has shown that the polyploid wheats, which were first thought to be typical or genomic allopolyploids, are actually segmental allopolyploids, with considerable homology still present between the chromosomes of the different genomes, particularly A and B. Their failure to form multivalents is due to the presence of a special gene on Chromosome 5, which specifically inhibits multivalent association. Evolution toward stability and fertility in these polyploids has not proceeded by the accumulation of additional differences between genomes at the chromosomal level, as postulated on p. 325 of my book, but by occurrence and establishment of particular gene mutations that have specific effects on meiosis. In wheat, the presence of the gene affecting pairing could be detected because of the nullisomic strains which had previously been produced by E. R. Sears. It seems likely that if similar information could be obtained about species in other genera that are believed to be typical allopolyploids, some of them would also prove to have the same kind of evolutionary history as the wheat species.

In respect to the relative geographic distribution of diploids and polyploids, the discussion on pp. 342-350, and particularly the conclusions reached in the final paragraph, are fully supported by our more recent findings. In particular, the increasing percentages of polyploidy associated with higher north latitudes in western Europe and the islands to the northward, which formed the original basis of the hypothesis that polyploidy arises in response to increasingly cold climates, are now unquestionably interpreted best in another way, as I and several other authors have pointed out (Reese, 1958; Johnson, Packer, and Reese, 1965). Two additional sets of data have strengthened the alternative interpretation that increased percentages of polyploidy in these northern regions are the result of the great shifts in climatic and edaphic factors, which accompanied the repeated advances and retreats of the Pleistocene ice sheets, and of the greater ability of polyploids than related diploids to colonize the new habitats made available by these drastic changes.

In the first place, Favarger (1957, 1961) has shown that the percentage of polyploidy in the floras of the high Alps, at and above the snow line, is no higher than it is in the surrounding lowlands. Second, the data on frequency of polyploids in floras at increasing latitudes along the Pacific coast of North America clearly support the ease of colonization hypothesis. Along this coast, the factors of glaciation and of increasing severity of the climate are separated from each other. The Queen Charlotte Islands, at 53° north latitude, were extensively glaciated, although a few of their peaks and headlands may have protruded as nunataks above the ice sheet. Their present climate is, however, mild, with mean winter temperatures for the coldest month about 0°C. Having an abundant rainfall, they support a luxuriant forest, containing many species that range as far southward as northern California, at latitude 40°. On the other hand, northwestern Alaska at latitude 68° was largely unglaciated, but has a very severe climate, with winter temperatures reaching −40° and permafrost extending to depths of over 300 meters. It supports only a scanty tundra type vegetation. In Pacific North America, therefore, the greatest difference in the severity of the climate exists between latitudes 53° and 68°, while the difference between maximal and minimal effects of the Pleistocene glaciation exists between latitudes 40° and 53°. In respect to percentages of polyploidy, the figure for the Queen Charlotte Islands is 53 percent (Taylor and Mulligan, 1968), and that for northwestern Alaska is 56 percent to 59 percent (Johnson and Packer, 1965), a slight and hardly signficant increase. My own estimate of the frequency of polyploidy in the Coast Ranges of Northern California, at 38° to 42° north latitude, is 36 percent or less. This figure is based only upon herbaceous perennials, which have higher percentages of polyploidy than other life forms, and which form the great bulk of the flora at the higher latitudes. There is little doubt that along this coast, where the effects of past glaciation and of increasingly severe climate are largely separated from each other, the increase in polyploidy is associated far more with the effects of Pleistocene glaciation than with increasing severity of the present climate.

Two objections have been raised to the general hypothesis that the establishment and spread of polyploids is associated chiefly with the colonization of newly available habitats. One of these is that the percentage of polyploids among weeds is no higher than the overall percentage of polyploids in a given flora (Heiser and Whitaker, 1948; Heiser, 1950; Mulligan, 1960, 1965). The second is that high percentages of polyploidy are now known to exist in tropical floras, many of which certainly are very ancient. Each of these objections can be answered, albeit in quite different ways.

The fact that weeds do not include a signficantly high percentage of polyploids is due to the fact that most weeds are annuals, and in the floras of northern mesic regions they are almost the only annuals present. Annuals consistently have lower percentages of polyploids than herbaceous perennials. This fact, which I established many years ago (Stebbins, 1938), is fully supported by estimates that I have made on the basis of the much more extensive data now available. The question which we should ask, therefore, is not: Are polyploids in general more abundant than diploids among weeds?, but rather: If polyploidy has arisen in a group of annuals, are the polyploids of this group any more likely to become weeds than their diploid ancestors?

This second question can be answered positively. In a study of those genera native to California, of which some species have become weedy during the past hundred years, the weedy diploid annuals were found to have no non-weedy polyploid relatives. On the other hand, all of the weedy polyploid annuals have non-weedy annual relatives, which are either diploid or have lower degrees of polyploidy (Stebbins, 1965). Furthermore, the frequency of polyploids among weedy California annuals is 42 percent, as compared with 17 to 20 percent for native California annuals in general.

More recently, I have compared chromosome numbers in weedy species introduced into eastern North America from Europe with the numbers of their relatives that did not invade North America (Stebbins, unpublished). This comparison revealed the fact that the bulk of the 118 genera studied either contained only diploids (44 genera) or contained diploids and polyploids on both continents (36 genera). Nevertheless, in 28 genera polyploids invaded America from Europe while related diploids did not, while in only

10 genera the reverse was the case. Even among weeds, therefore, polyploidy does increase the chances that a species will become a successful colonizer.

The high percentages of polyploidy in ancient tropical floras must have a completely different explanation. The best data on such a flora are those of Mangenot and Mangenot (1962) on the rain forest of the Ivory Coast in western tropical Africa, but they are supported by the data of Manton (1953) and Mehra (1961) on tropical ferns, and of Morton (1961) on various angiosperms.

The relationships of most of these tropical polyploids to diploids belonging to the same or related genera are quite different from those found in most temperate groups. Polyploid series within the same genus or subgenus, and particularly polyploid "races" or "cytotypes" within the same species, are much less common than they are in temperate floras. Many of the numbers judged to be of polyploid origin characterize entire genera with pantropical distributions, or even groups of related genera. In the family Leguminosae, for instance, a high proportion of the species belonging to the pantropical genus *Erythrina* have been counted; all of them have $2n = 42$. Hence the entire genus is judged to be of polyploid origin. In the family Bombacaceae all of the species counted have more than 100 chromosomes, the modal number being $2n = 144$. The data from tropical floras are supporting with increasing emphasis the hypothesis that I suggested on p. 365 of my book (1950) that increasing polyploidy accompanied the establishment and spread of new groups of angiosperms during the early stages of their evolutionary history. This hypothesis is further supported by continued cytological investigations of the most primitive angiosperms, the woody Ranales, as summarized by Ehrendorfer et al. (1968). With the exception of the Annonaceae, all modern taxa of this group may be of ancient polyploid origin.

Three different cytologists (Raven and Thompson, 1964; DeWet 1965, 1968) have challenged the widely accepted belief that polyploidy is a unidirectional phylogeny from diploids to polyploids. Their reasons are that in laboratory cultures or garden plots reversions from polyploids to diploids have been recorded many times, and in some instances the reverted diploids have nearly or quite normal meiosis. From the evolutionary point of view, however, the important question is not whether such reversions occur, but whether they affect materially the overall phylogenetic trend. These two questions are not the same, since polyploid phylogeny involves not only doubling the chromosome number, but also various secondary changes, such as mutations, chromosomal rearrangements, and hybridizations, which are essential for the successful adaptation of polyploids to new habitats. The evidence obtained by DeWet on the grass genera *Dichanthium* and *Bothriochloa* indicates that successful reversions to diploidy occur only in autopolyploid populations that appear to be of relatively recent origin and are living sympatrically with their diploid progenitors. Such reversions usually do nothing more than add to the already large gene pool of the diploids, although they may retard the divergent evolution of the polyploids. The tetraploids tested by DeWet that had spread beyond the range of their diploid ancestors were either incapable of producing viable diploid offspring, or the revertants were weak, sterile, or both.

This result is expected. The secondary processes mentioned above tend to convert gene loci from the tetrasomic to the disomic condition. When this has happened, the diploids derived from modified tetraploids are monosomic at many gene loci. This condition inevitably gives rise to their weakness or sterility.

The geographic distribution of diploids and tetraploids in many polyploid complexes also supports the concept of the essential irreversibility of polyploid phylogenies. Many examples are now available of large genera that contain only polyploids in extensive parts of their distributional areas, where their widespread occurrence and diversification into many species and races indicate that they have existed for a long time, and have evolved extensively at polyploid levels. The New World species of such genera as *Festuca, Poa, Danthonia, Silene,* and *Senecio* can be mentioned here. Evidence from many directions indicates that although individual events of polyploid chromosome doubling can be reversed, this reversal is no more than a form of oscillatory "noise" in the general trend of evolutionary phylogeny. The reversion to the diploid state of successful, highly evolved polyploids is either impossible or so rare that it can be disregarded as a factor of evolution.

Structural Hybridity and Chromosomal Evolution

The general theme which was developed in Chapters 11 and 12 (Stebbins, 1950) has been greatly reinforced by research in the field of population genetics, and can now be stated positively as follows. Adaptation depends to a large extent upon epistatic interaction between genes at different loci, and is often greatest in genotypes heterozygous for several pairs of alleles. Consequently, cytological mechanisms that maintain adaptive gene combinations in the heterozygous state have in many species an exceptionally high adaptive value. Chromosomal inversions and translocations are the principal types of these mechanisms. Complete treatments of this theme, with review of the evidence in favor of the statements just made, can be found in papers and books such as those of Dobzhansky (1951, 1957), Carson (1961), Bodmer and Parsons (1962), and Lewontin (1964).

Research on the genus *Oenothera*, the principal example in my Chapter 11, has greatly declined in recent years, and little more has been learned about it. The interesting hypothesis that the balanced lethal system that has evolved in the genus is based upon the modified action of self incompatibility alleles, was developed by Steiner (1956, 1957).

I have discussed aneuploid changes in chromosome numbers in a recent review paper (Stebbins, 1966b). One facet of this problem that was not adequately discussed there is the influence of diffuse kinetochores or centromeres. Detailed studies of chromosome morphology in the genus *Luzula* (de Castro, 1952; Brown, 1954; Nordenskiöld, 1956) have shown that in this genus the chromosomes do not have a single fiber attachment point or kinetochore, but several. As a consequence, the products of transverse chromosomal fragmentation can still have functional kinetochores, which enable them to move normally to the poles of the spindle at meiotic anaphase, and so to be retained in the genotype. Nordenskiöld has concluded that many species of *Luzula* having chromosome numbers that are twice those of other species are not true polyploids, but "fragmentation polyploids," which have originated by chromosomal fragmentation rather than replication of entire chromosomes. Since diffuse kinetochores probably exist in a number of other genera, particularly *Carex* and other genera of Cyperaceae, chromosomal fragmentation may have played a large role in the evolution of the extensive aneuploid series of chromosome numbers found in these genera.

An hypothesis for the evolutionary trend toward increasing asymmetry of the karyotype (discussed in Stebbins, 1950, pp. 458-465), was presented in a later paper (Stebbins, 1958a). If changes in karyotype morphology are due to unequal inversions and translocations that become established because of the clusters of adaptively interacting genes that they assemble on the same arm of a chromosome, then we would expect that certain arms on which such adaptive gene clusters had begun to accumulate would tend to receive additional chromosomal material containing more adaptive genes, and thus would increase in length at the expense of arms that did not contain such complexes.

As stated in my recent review (Stebbins, 1966b), we are still far from explaining the great differences that exist between related species in respect to chromosome size, which in general reflects their content of nuclear DNA (Baetcke et al., 1967). It is completely illogical to suppose that eight times as many different genes are required to code for the development of a plant of rye (*Secale*) as of rice (*Oryza*) or panic grass (*Panicum*), yet the former species contains about eight times as much DNA in its nuclei as the latter two.

Three general hypotheses may be suggested to explain these differences. (1) Organisms containing large amounts of nuclear DNA contain a high proportion of DNA molecules that have "nonsense" sequences of nucleotides, and therefore are devoid of any genetic function. (2) Organisms containing large amounts of DNA contain chromosomes that are multistranded. They are, therefore, genetically polyploid, even though they have a diploid number of centromeres and of somatic metaphase chromosomes. (3) These organisms contain gene loci that are replicated many times, tandem fashion.

Each of these hypotheses is difficult to reconcile with certain available facts. If the "nonsense DNA" hypothesis were generally true, we should expect that, over a long period of time, species that had initially acquired a large amount of DNA would gradually lose it by random deletions. Such species should exhibit a great amount of variability between individuals with respect to DNA content, and even greater differences between species that have been isolated from each other over long periods of time. This, however, is not the case. The genus *Trillium,* for example, contains related species in temperate eastern North America and eastern Asia, as well as in western North America. These species are everywhere associated with the ancient Arcto-Tertiary deciduous, or mixed coniferous forest, which has existed almost unchanged since early in the Tertiary period. On this basis, we can hardly escape the conclusion that the genus *Trillium* is at least 40 to 50 million years old. Yet all of the diploid species of *Trillium* agree with each other in having five pairs of very large chromosomes, both chromosome size and gross morphology being very constant throughout the genus. It is hard to see how this constancy could have been maintained for such a long time unless it has some kind of adaptive significance.

The hypothesis of multistrandedness apparently fits the data obtained from the genus *Vicia,* in which *V. faba,* with very large chromosomes, appears to possess a multistranded condition, while *V. sativa,* with about half the nuclear-DNA content of *V. faba,* does not (Wolfe and Martin, 1968). On the other hand, chromosomal volumes in such genera as *Crepis* appear to form a graded series rather than a series of modes at values that are multiples of each other, which one would expect on the basis of the hypothesis of multistrandedness. In animals, cellular-DNA content has been obtained for more than 200 species of fishes. The resulting values, which range from 0.4 to 4.4 picograms—showing a tenfold difference between the lowest and highest values—are distributed in a monomodal fashion (Hinegardner, 1968). Among animals also, various salamanders have very high cellular-DNA contents. In them, careful studies of chromosomes at the "lamp brush" state of oöcyte development has failed to reveal a multistranded condition (Callan, 1967). Multistrandedness may, therefore, explain the occurrence of large chromosomes in some groups of plants, but it can hardly be postulated as a universal explanation for differences in chromosome size and DNA content among related groups of higher organisms.

The hypothesis of tandem duplications of individual gene loci has been championed by Callan (1967), who has devised an elaborate and ingenious scheme to explain how such duplicated loci could either remain genetically identical with each other or else mutate simultaneously. As yet, no evidence for or against this hypothesis has been obtained from plants. This problem has been discussed in other publications (Stebbins, 1966b and in press).

Evolution Above the Species Level

Some Common Evolutionary Trends

I have recently published an up-to-date version of my ideas on this subject (Stebbins, 1967). The principal difference between this version and the one in the book is my present emphasis on the probability that many trends, particularly those in respect to numbers of parts, can go in both directions, toward either increase or decrease. Hence, if there exists in a family a series ranging from few to many morphological parts, such as stamens per flowers or flowers per inflorescence, we must not assume that there are only two possible ways of reading the series in a phylogenetic sense, i.e., from few (primitive) to many (advanced) or from many (primitive) to few (advanced). As shown by the examples given in Stebbins (1967), in many series, such as the number of florets per capitulum in the Compositae, and the number of stamens per flower in the Rosaceae, the primitive number is most probably intermediate. High as well as low numbers are due to adaptive radiations upward and downward, in separate evolutionary lines, from this primitive intermediate number.

As pointed out also in that paper, there appears to be a fundamental difference in this respect between trends in numbers of parts and those trends grouped under the terms "fusion" or adnation." In respect to the former, different directions of evolution are not only possible but occur often. In respect to the latter, trends away from fusion or adnation toward separate parts are rare. Trends from radial to bilateral symmetry are also usually unidirectional. This difference can probably be explained on the basis of gene interaction during development. Changes in numbers of parts can probably take place through relatively simple adjustments in the timing of gene action. On the other hand, the alterations of developmental processes needed to produce fusions and adnations are so much more complex that an orderly reversal of these processes during phylogeny by the selection of appropriate mutations and gene recombinations is very difficult.

If these speculations are borne out by future research on gene action in development, we must recognize that morphological series that involve fusion or adnation have a much greater validity for determining phylogenetic positions than do those which involve changes in numbers of parts.

The significance of fusions and adnations for the phylogeny of the more primitive vascular plants has been thoroughly discussed in recent years by Zimmermann (1959, 1961, 1965), whose books deserve careful study by all botanists who are interested in phylogeny.

Recapitulation and Embryonic Similarity

I have more recently made a comparative study of seedling heterophylly, which was the principal topic treated in this section (Stebbins, 1959b; cf. 1950, pp. 488-492). This study showed that related species living in different climatic regions can differ greatly in the amount of difference between their seedling and adult leaves. On the basis of this study, I reached the conclusion that this type of "recapitulation" is best explained on the basis of natural selection in plants that are exposed to very different environmental conditions at the seedling stage, compared to those which they face when older.

The Principle of Irreversibility

A discussion of this principle has recently been given by Stebbins (1967), considerably amplified over that in Stebbins (1950, pp. 493-499). In particular, Ganong's principle of "metamorphosis along the lines of least resistance" has been restated as selection along the lines of least resistance, and several examples of it have been suggested. On the basis of these examples, as well as some of those given in the book, I now believe that the dictum "lost parts cannot be regained" should be modified to read "lost parts have a very low probability of being regained." The difference between these two statements becomes highly significant when we recognize that any population or species of plants faced with the problem of becoming adapted to a new set of conditions can solve this problem in a number of different ways. Recovering a lost part might be one way of doing this, but the mutations necessary for such recovery would probably be several, and each one of them would probably have a low chance of occurrence. Mutations or new gene combinations that would enable the plant to carry out the function of the lost part in some other way would be much more likely to occur.

Orthogenesis, Specialization, and the Differentiation of Plant Families

The data to which I referred in the book (1950) on these matters was published in Stebbins (1951). Further evidence that the combination of morphological characteristics found in angiosperm families respresent adaptive peaks related to the function of pollination and seed dispersal is presented in books and articles by the Grants (Grant and Grant, 1965) and Van der Pijl (1960, 1961, 1966).

The Nature and Value of Paleobotanical Evidence

The most important progress in this field has been the greatly increased use of microfossils, particularly pollen. Good reviews of this subject have been presented by Pokrovskaia (1950) and Waterbolk (1964). The strength of the palynological evidence lies in the fact that pollen can be identified with much greater certainty than leaves, and that pollen deposits are found in many strata that completely lack megafossils. Its weakness lies in the fact that only those species having pollen similar to modern forms can be identified. The extinct common ancestors of distantly related modern families will never be identified by their pollen.

By the use of megafossils, the evolutionary history of the Tertiary floras of western North America has been worked out in great detail by Axelrod (1958, 1959, 1968), and many valuable conclusions have been reached about it. On the other hand, the earlier evolution of angiosperms, and particularly their origin, remain as much of a mystery as ever. The conflict between those who believe that angiosperms originated during the Cretaceous period, when the first unquestionable remains of them are recorded (Scott, Barghoorn, and Leopold, 1960), and those who rely on questionable fossils plus a certain amount of logical deduction or speculation to suggest that they originated much earlier (Axelrod, 1961, 1967), is still unresolved.

Patterns of Distribution, Localities of Origin, and Rates of Evolution

The final sections of my book (1950, pp. 530-561) deal with subjects about which definite experimental evidence cannot be obtained. Hence, no viewpoint about them can ever be validated unquestionably, and room for differences of opinion will always exist. Botanists should not be surprised to find, therefore, that after 20 more years of securing additional information, arguing back and forth with their colleagues, and presenting new theories, they are still divided in respect to such questions as the importance of long-range dispersal in explaining patterns of distribution, the influence of continental drift, if any, and the likelihood of determining the place where a group originated by studying the distribution of its contemporary species. Since I still retain the point of view on these subjects that I expressed in my book, I shall not bother the reader with an account of the more recent literature in the field.

In respect to differential rates of evolution, I am more impressed than ever with the evidence indicating that there is no normal or modal evolutionary rate, and that related evolutionary lines can differ from each other by as much as a thousandfold in the rate at which they are evolving. Some of this evidence was presented in a paper dealing with the situation in two annual species of the genus *Plantago* (Stebbins and Day, 1967).

The existence of such divergent evolutionary rates should provide great encouragement to the experimental evolutionist. If natural rates can vary by as much as a thousandfold, we should be able artificially to increase the rate of evolution in some groups to such a degree that we can see speciation taking place under our eyes. This goal has almost been achieved by Dobzhansky and his group in *Drosophila* (Dobzhansky and Pavlovsky, 1966). It can also be done in plants, given the right species and enough knowledge about it.

References

ALLARD, R. W. 1965. Genetic systems associated with colonizing ability in predominantly self-pollinated species. *In* Baker, H. G., and Stebbins, G. L. eds. The Genetics of Colonizing Species, pp. 49-76.
ALSTON RALPH E. 1967. Biochemical systematics. *In* Dobzhansky, Th., Hecht, M. K., and Steere, W. C., eds. Evolutionary Biology. New York, Appleton-Century-Crofts. pp. 197-305.
——— and B. L. TURNER. 1963. Biochemical Systematics. Englewood Cliffs, N. J. Prentice-Hall, Inc. 404 pp.
ANDERSON, E., and G. L. STEBBINS. 1954. Hybridization as an evolutionary stimulus. Evolution, 8:378-388.
AVERY, A. G., S. SATINA, and J. RIETSEMA. 1959. Blakeslee: The Genus Datura. New York, The Ronald Press Company. 329 pp.
AXELROD, DANIEL I. 1939. A Miocene flora from the western border of the Mohave Desert. Carnegie Inst. Washington, Publ. N., 516:1-129.
——— 1958. Evolution of the Madro-Tertiary geoflora. Bot. Rev., 24:433-509.
——— 1959. Late Cenozoic evolution of the Sierran bigtree forest. Evolution, 13:9-23.
——— 1961. How old are the angiosperms? Amer. J. Sci., 259:447-459.
——— 1967. Drought, diastrophism, and quantum evolution. Evolution, 21:201-209.
——— 1968. Tertiary floras and topographic history of the Snake River Basin, Idaho. Bull. Geol. Soc. Amer., 79:713-733.
BAKER, H. G. 1955. Self-compatibility and establishment after "long-distance" dispersal. Evolution, 9:347-348.
BAETCKE, K. P., A. H. SPARROW, C. H. NAUMAN, and S. S. SCHWEMMER. 1967. The relationship of DNA content to nuclear and chromosome volumes and to radio-sensitivity (LD_{50}). Proc. Nat. Acad. Sci. (U. S. A.), 58:533-540.
BAILEY, D. W. 1956. Re-examination of the diversity in *Partula taeniata*. Evolution, 10:360-366.
BECKMAN, L., J. G. SCANDALIOS, and J. L. BREWBAKER. 1964. Genetics of leucine amino–peptidase isozymes in maize. Genetics, 50:899-904.
BENSON, L., E. A. PHILLIPS, P. A. WILDER, et al. 1967. Evolutionary sorting of characters in a hybrid swarm. I: Direction of slope. Amer. J. Bot., 54:1017-1026.
BODMER, W. F., and P. A. PARSONS. 1962. Linkage and recombination in evolution. Advances Genet., 11:1-87.
BRADSHAW, A. D. 1962. The taxonomic problems of local geographical variation in plant species. Systematics Assoc. Publ. No. 4, Taxonomy and Geography, pp. 7-16.
——— 1965. Evolutionary significance of phenotypic plasticity in plants. Advances Genet., 13:115-155.

——— T. S. McNeilly, and R. P. G. Gregory. 1965. Industrialization, evolution and the development of heavy metal tolerance in plants. Ecol. and the Indust. Soc., 5th Symp. of British Ecol. Soc. pp. 327-343.

Brown, S. W. 1954. Mitosis and meiosis in *Luzula campestris* DC. University of California Publ. in Botany, 27:231-278.

Cain, A. J., and P. M. Sheppard. 1954. Natural selection in *Cepaea*. Genetics, 39:89-116.

Callan, H. G. 1967. The organization of genetic units in chromosomes. J. Cell Sci., 2:1-7.

Camin, J. H., and R. R. Sokal. 1965. A method for deducing branching sequences in phylogeny. Evolution, 19:311-326.

Carson, H. L. 1961. Heterosis and fitness in experimental populations of *Drosophila melanogaster*. Evolution, 15:496-509.

de Castro, D. 1952. Nota sobre a perpetuaçao de fragmentos cromosomicos em *Luzula purpurea*. Agronomia Lusitana, 14:95-99.

Chaney, R. W., C. Condit, and D. I. Axelrod. 1944. Pliocene floras of California and Oregon Carnegie Inst. Washington Publ. No. 553:1-405.

Clausen, J. 1954. Partial apomixis as an equilibrium system in evolution. Caryologia, 469-478.

——— and W. M. Hiesey. 1958. Experimental studies on the nature of species. IV. Genetic structure of ecological races. Carnegie Inst. Washington Publ. No. 615:1-312.

Cook, S. A. 1968. Adaptation to heterogeneous environments. I. Variation in heterophylly in *Ranunculus flammula* L. Evolution, 22:496-516.

De Wet, J. M. J. 1965. Diploid races of tetraploid *Dichanthium* species. Amer. Natural., 99:167-171.

De Wet, J. M. J. 1968. Diploid-tetraploid-haploid cycles and the origin of variability in *Dichanthium* agamospecies. Evolution, 22:394-397.

Dobzhansky, T. 1951. Genetics and the Origin of Species, 3rd ed. New York, Columbia University Press. 364 pp.

——— 1953. Evolution in *Drosophila*. Evolution, 7:92-93.

——— 1956. What is an adaptive trait? Amer. Natural., 90:337-347.

——— 1957. Mendelian populations as genetic systems. Sympos. Quant. Biol., 22:385-393.

——— and O. Pavlovsky. 1966. Spontaneous origin of an incipient species in the *Drosophila paulistorum* complex. Proc. Nat. Acad. Sci. (U. S. A.), 55:727-733.

Dure, L., and L. Walters. 1965. Long-lived messenger RNA: evidence from cotton seed germination. Science, 147:410-412.

Ehrendorfer, F. 1959. Polyploidie, Hybridisierung und Evolution: Ergebnisse vergleichender phylogenetischer Studien an *Galium, Knautia* und *Achillea*. Proc. IX Int. Bot. Cong., 2:102.

——— 1964. Evolution and karyotype differentiation in a family of flowering plants: Dipsacaceae. Proc. XI Int. Cong. Genet., 399-407.

——— 1965. Dispersal mechanisms, genetic systems, and colonizing abilities in some flowering plant families. *In* Baker, H. G. and Stebbins, G. L. eds. The Genetics of Colonizing Species. New York, Academic Press, Inc. pp. 331-352.

——— F. Krendl, E. Habeler, and W. Sauer. 1968. Chromosome number and evolution in primitive Angiosperms. Taxon, 17:337-353.

El-Gazzar, A., L. Watson, W. T. Williams, and G. N. Lance. 1968. The taxonomy of *Salvia*: a test of two radically different numerical methods. London, J. Linnean Soc., 60:237-250.

Epling, C., H. Lewis, and F. M. Ball. 1960. The breeding group and seed storage: a study in population dynamics. Evolution, 14:238-255.

Favarger, C. 1957. Sur le pourcentage des polyploides dan la flore d l'étage nival des Alpes Suisses. Proc. VII Inter. Bot. Cong., pp. 51-58.

——— 1961. Sur l'emploi des nombres de chromosomes en géographie botanique historique. Ber. Geobot. Inst. Eidg. Techn. Hochschule Zürich, 32:119-146.

GAUL, H. 1965. The concept of macro- and micro-mutations and the results on induced micro-mutations in barley. *In* The Use of Induced Mutations in Plant Breeding. Rep. FAO and IAEA, pp. 407-428.
GLISIN, V. R., M. V. GLISIN, and P. DOTY. 1966. The nature of messenger RNA in the early stages of sea urchin development. Proc. Nat. Acad. Sci. (U. S. A.), 56:285 289.
GOODSPEED, T. H. 1954. The Genus *Nicotiana*. Massachusetts, The Chronica Botanica Company. 536 pp.
GRANT, V. 1959. Natural History of the Phlox Family. The Hague, Martinus Nijhoff. 280 pp.
———— 1963. The Origin of Adaptations. New York, Columbia University Press. 606 pp.
———— 1964. The biological composition of a taxonomic species in *Gilia*. Advances Genet., 12:281-328.
———— 1966. The origin of a new species of *Gilia* in a hybridization experiment. Genetics, 54:1189-1199.
———— and K. A. GRANT. 1965. Flower Pollination in the Phlox Family. New York, Columbia University Press. 180 pp.
HALL, O. 1959. Immuno-electrophoretic analyses of allopolyploid rye wheat and its parental species. Hereditas, 45:494-504.
HARPER, J. L. 1965. Establishment, aggression, and cohabitation in weedy species. *In* Baker, H. G., and Stebbins, G. L., eds. The Genetics of Colonizing Species. New York, Academic Press, Inc. pp. 243-265.
HEISER, C. B., Jr. 1950. A comparison of the flora as a whole and the weed flora of Indiana as to polyploidy and growth habits. Proc. Indiana Acad. Sci., 59:64-70.
———— and T. O. WHITAKER. 1948. Chromosome number and growth habit in California weeds. Amer. J. Bot., 35:179-186.
———— J. SORIA, and D. L. BURTON. 1965. A numerical taxonomic study of *Solanum* species and hybrids. Amer. Natural., 99:471-488.
HESLOP-HARRISON, J. 1964. Forty years of genecology. Advances Ecol. Res., 2:159-247.
HINEGARDNER, R. 1968. Evolution of cellular DNA content in teleost fishes. Amer. Natural., 102:517-523.
HOYER, B. H., B. J. MCCARTHY, and E. T. BOLTON. 1964. A molecular approach in the systematics of higher organisms. Science, 144:959-967.
HUBBY, J. L., and R. C. LEWONTIN. 1966. A molecular approach to the study of genic heterozygosity in natural populations. I. The number of alleles at different loci in *Drosophila pseudoobscura*. Genetics, 54:577-594.
INGRAM, V. I. 1963. The Hemoglobins in Genetics and Evolution. New York, Columbia University Press. 165 pp.
JOHNSON, A. W., and J. G. PACKER. 1965. Polyploidy and environment in arctic Alaska. Science, 148:237-239.
———— J. G. PACKER, and G. REESE. 1965. Polyploidy, distribution, and environment. Quaternary of the United States, 497-507.
JOHNSON, L., and O. HALL. 1965. Analysis of phylogenetic affinities in the Triticinae by protein electrophoresis. Amer. J. Bot., 52:506-513.
JOWETT, D. 1964. Population studies on lead-tolerant *Agrostis tenuis*. Evolution, 18:70-80.
JUKES, T. H. 1966. Molecules and Evolution. New York, Columbia University Press. 285 pp.
KRUCKEBERG, A. R. 1951. Intraspecific variability in the response of certain native plant species to serpentine soil. Amer. J. Bot., 38:408-419.
KULLENBERG, B. 1961. Studies in *Ophrys* pollination. Zool. Bidrag Uppsala, 34:1-340.
KURABAYASHI, M. 1958. Evolution and variation in Japanese species of *Trillium* Evolution, 12:286-310.

KYHOS, D. W. 1965. The independent aneuploid origin of two species of *Chaenactis* (Compositae) from a common ancestor. Evolution, 19:26-43.
LAMOTTE, M. 1959. Polymorphism of natural populations of *Cepaea Nemoralis*. Sympos. Quant. Biol., 24:65-86.
LAMPRECHT, H. 1966. Die Entstehung der Arten und höheren Kategorien. Vienna and New York, Springer Publishing Co., Inc. 452 pp.
LEE, J. A. 1960. A study of plant competition in relation to development. Evolution, 14:18-28.
LEWIS, H. 1962. Catastropic selection as a factor in speciation. Evolution, 16:257-271.
――― 1966. Speciation in flowering plants. Science, 152:167-172.
――― and M. E. LEWIS. 1955. The Genus *Clarkia*. University of California. Publ. in Bot., 20:241-392.
LEWONTIN, R. C. 1964. The role of linkage in natural selection. Proc. XI Inter. Cong. Genet., 518-525.
――― and J. L. HUBBY. 1966. A molecular approach to the study of genic heterozygosity in natural populations. II. Amount of variation and degree of heterozygosity in natural populations of *Drosophila pseudoobscura*. Genetics, 54: 595-609.
MANGENOT, S., and G. MANGENOT. 1962. Enquête sur les nombres chromosomiques dans une collection d'espèces tropicales. Rev. Cyt. Biol. Vég., 25:411-447.
MANTON, I. 1950. Problems of Cytology and Evolution in the Pteridophyta. New York, Cambridge University Press. 316 pp.
――― 1953. The cytological evolution of the fern flora of Ceylon. Symp. Soc. Exptl. Biol. No. 7, Evolution, 174-185.
MAYR, E. 1963. Animal Species and Evolution. Cambridge, Mass., Harvard University Press. 797 pp.
MEHRA, P. N. 1961. Cytological evolution of ferns with particular reference to Himalayan forms. Proc. 48th Indian Sci. Cong., II:1-24.
MORTON, J. K. 1961. The incidence of polyploidy in a tropical flora. Recent Adv. Bot., 900-903.
MULLIGAN, G. A. 1960. Polyploidy in Canadian weeds. Canad. J. Genet. Cytol., 2:150-161.
――― 1965. Recent Colonization by Herbaceous Plants in Canada. *In* Baker, H. G., and Stebbins, G. L. eds. The Genetics of Colonizing Species. New York, Academic Press, Inc. pp. 127-143.
NORDENSKIÖLD, H. 1956. Cyto-taxonomical studies in the genus *Luzula* II. Hybridization experiments in the *campestris-multiflora* complex. Hereditas, 42:7-73.
POKROVSKAIA, I. M. (red). 1950. Pyl'tsovoi analiz. Moskva (in Russian; French transl. 1958, Ann. Serv. Inform. Geol.) B.R.G.G.M. p. 435.
RAVEN, P. H., and H. J. THOMPSON. 1964. Haploidy and angiosperm evolution. Amer. Natural., 98:251-252.
REESE, G. 1958. Polyploidie und Verbreitung. Z. Botanik, 46:339-354.
RILEY, R. 1966. The genetic regulation of meiotic behaviour in wheat and its relatives. Proc. 2nd Int. Wheat Genetics Sympos., 2:395-408.
ROGERS, D. J., and H. FLEMING. 1964. A computer program for classifying plants. II. A numerical handling of non-numerical data. BioScience, 14:15-28.
――― H. S. FLEMING, and G. ESTABROOK. 1967. Use of computers in studies of taxonomy and evolution. *In* Dobzhansky, T., M. K. Hecht and W. C. Steere, eds., Evolutionary Biology, New York, Appleton-Century-Crofts, pp. 169-196.
SAKAI, KAN-ICHI. 1965. Contributions to the problem of species colonization from the viewpoint of competition and migration. *In* Baker, H. G. and G. L. Stebbins, eds. The Genetics of Colonizing Species, New York, Appleton-Century-Crofts. pp. 215-241.
SASAMOTO, K. 1958. Studies on the relation between the silica content in the rice plant and the insect pests. VI. On the injury of silicated rice plant caused by the rice stem borer and its feeding behavior. Japan. J. Appl. Entomol. Zool., 2:88-92.

SCANDALIOS, J. G. 1964. Tissue-specific isozyme variations in maize. J. Hered., 55:281-285.
——— 1965. Leucine aminopeptidase isozymes in maize development. J. Hered., 56:177-180.
SCHEIBE, A. 1955. Die Wirkung der natürlichen Auslese bei *Pisum arvense*-Formen mit und ohne Wachsschicht. Der Züchter, 25:97-103.
SCHREIBER, K. 1957. Natürliche pflanzliche Resistenzstoffe gegen den Kartoffelkäfer und ihr möglicher Wirkungsmechanismus. Der Züchter, 27:289-299.
SCHWARTZ, D. 1964. Genetic studies on mutant enzymes in maize. IV. Comparison of pH 7.5 esterases synthesized in seedling and endosperm. Genetics, 49: 373-377.
——— FUCHSMAN, L. and K. H. MCGRATH. 1965. Allelic isozymes of the pH 7.5 esterase in maize. Genetics, 52:1265-1268.
SCOTT, R. A., E. BARGHOORN and E. B. LEOPOLD. 1960. How old are the angiosperms? Amer. J. Sci., Bradley Vol., 258A:284-299.
SMITH, H. H. and K. DALY. 1959. Discrete populations derived by interspecific hybridization and selection in *Nicotiana*. Evolution, 13:476-487.
SNAYDON, R. W. and A. D. BRADSHAW. 1961. Differential response to calcium within the species *Festuca ovina* L. New Phytol., 60:219-234.
SOKAL, R. R. and P. H. A. SNEATH. 1963. Principles of Numerical Taxonomy. San Francisco, W. H. Freeman and Co., Publishers. pp. 359.
STEBBINS, G. L. 1938. Cytological characteristics associated with different growth habits in dicotyledons. Amer. J. Bot., 25:189-198.
——— 1950. Variation and Evolution in Plants. New York, Columbia University Press. 643 pp.
——— 1951. Natural selection and the differentiation of angiosperm families. Evolution, 5:299-324.
——— 1957a. Self-fertilization and population variability in the higher plants. Amer. Natural., 91:337-354.
——— 1957b. The hybrid origin of microspecies in the *Elymus glaucus* complex. Proc. Inter. Genet. Symp., Japan 1956, pp. 336-340.
——— 1958a. Longevity, habitat, and release of genetic variability in higher plants. Sympos. Quant. Biol., 23:365-378.
——— 1958b. The inviability, weakness, and sterility of interspecific hybrids. Advances Genet., 9:147-215.
——— 1959a. The role of hybridization in evolution. Proc. Amer. Philos. Soc., 103:231-251.
——— 1959b. Seedling heterophylly in California flora. Bull. Res. Counc. Israel, 7D:248-255.
——— 1960. The comparative evolution of genetic systems. *In* Tax, S., ed. Evolution After Darwin. Chicago, University of Chicago Press, Vol. 1. pp. 197-226.
——— 1965. The experimental approach to problems of evolution. Folia Biologica, (Prague) 11:1-10.
——— 1966a. Processes of Organic Evolution. Englewood Cliffs, N. J., Prentice-Hall, Inc. 191 pp.
——— 1966b. Chromosome variation and evolution. Science, 152:1463-1469.
——— 1967. Adaptive radiation and trends of evolution in higher plants. *In* Dobzhansky, T. Hecht, M. K., and Steere, W. C. eds. Evolutionary Biology. New York, Appleton-Century-Crofts. pp. 101-142.
——— and K. DALY. 1961. Changes in the variation pattern of a hybrid population of *Helianthus* over an 8-year period. Evolution, 15:60-71.
——— and A. DAY. 1967. Cytogenetic evidence for long continued stability in the genus *Plantago*. Evolution, 21:409-428.
STEINER, E. 1956. New aspects of the balanced lethal mechanism in *Oenothera*. Genetics, 41:486-500.
——— 1957. Further evidence of an incompatibility allele system in the complex-heterozygotes of *Oenothera*. Amer. J. Bot., 44:582-585.

STEPHENS, S. G. 1957. Sources of resistance of cotton strains to the boll weevil and their possible utilization. J. Econ. Entomol., 50:415-418.
SWANSON, C. P. 1957. Cytology and Cytogenetics. Englewood Cliffs, N. J., Prentice-Hall, Inc., 596 pp.
TAYLOR, R. L., and G. A. MULLIGAN. 1968. Flora of the Queen Charlotte Islands, Part 2. Cytological Aspects of the Vascular Plants. Research Branch Canad. Dept. Agric. Mon., 44, Pt. 2:1-148.
TUCKER, J. M. 1952. Evolution of the California oak *Quercus alvordiana*. Evolution, 6:162-180.
VALENTINE, D. H. 1953. Evolutionary aspects of species differences in *Primula*. Sympos. Soc. Exp. Biol., No. 7, Evolution, 146-158.
VAN DER PIJL, L. 1960. Ecological aspects of flower evolution. I. Phyletic evolution. Evolution, 14:403-416.
———— 1961. Ecological aspects of flower evolution. II. Zoophilous flower classes. Evolution, 15:44-59.
———— 1966. Ecological aspects of fruit evolution. I. A functional study of dispersal organs. Proc. Kon. Nederl. Akad. Wet. (C), 69:597-640.
WADDINGTON, C. H. 1960. Evolutionary adaptation. *In* Tax S., ed. Evolution After Darwin, I. The Evolution of Life. Chicago, University of Chicago Press. pp. 381-402.
WALL, J. R. and T. L. YORK. 1957. Inheritance of seedling cotyledon position in *Phaseolus* species. J. Hered., 48:71-74.
WANGENHEIM, K. H. VON. 1961. Zur Ursache der Abortion von Samenanlagen in diploid-polyploid-Kreuzungen. Zeit. Pflanzenzucht., 46:13-19.
———— 1962. Zur Ursache der Abortion von Samenanlagen in diploid-polyploid-Kreuzungen. II. Unterschiedliche Differenzierung von endospermen mit gleichem Genom. Z. Vererbungsl., 93:319-334.
WATERBOLK, H. T. 1964. Pre-quanternary pollen analysis. *In* Faegri, K. and Iversen, J. eds. Textbook of Pollen Analysis. New York, Hafner Publishing Co., Inc. pp. 140-153.
WATERS, L. C. and L. S. DURE. 1966. Ribonucleic acid in germinating cotton seed. J. Molec. Biol., 19:1-27.
WOLFE, S. L. and P. G. MARTIN. 1968. The ultrastructure and strandedness of chromosomes from two species of *Vicia*. Exp. Cell Res., 50:140-150.
ZIMMERMAN, W. 1959. Die Phylogenie der Pflanzen. Ein Überblick über Tatsachen und Probleme, Stuttgart, G. Fischer, 777 pp.
———— 1961. Phylogenetic shifting of organs, tissues, and phases in Pteridophytes. Canad. J. Bot., 39:1547-1553.
———— 1965. Die Blütenstände, ihr System und ihre Phylogenie. Ber. Deutschen Bot. Ges., 78:3-12.

7

Population Research in the Scandinavian Scots pine (*Pinus sylvestris* L.): Recent Experimentation

VILHELMS EICHE
Department of Forest Genetics,
Royal College of Forestry, Stockholm, Sweden

ÅKE GUSTAFSSON
Institute of Genetics, Lund University,
Lund (formerly Royal College of Forestry) Sweden

Introduction 209
Material and Methods 210
Climatic Conditions of the Experimental Plantations and the
  Natural Populations Involved 212
Cumulative Mortality and Survival 214
Cumulative Mortality in Selected Experimental Plantations 217
  EP 27, Kiruna 217
  EP 24, Kåbdalis 218
  EP 21, Långsjöby 219
  EP 20, Lycksele 220
  EP 14, Brämön 222
Single-Year Mortality 223
Height Growth 228
Conclusions and Summary 231
References 234

Introduction

Since the turn of the century population research, in the form of provenance experiments, has been in the foreground of forest genetics and breeding. The use and distribution of hardy and well-growing materials has formed an important object in advanced forestry practice. Norway spruce (*Picea Abies* (L.) Karst.) and Scots pine (*Pinus sylvestris* L.) are the two valuable conifer species in northern Europe. The background of work and the results have been discussed by numerous scientists. References are here made especially to papers by Langlet and by Eiche (see references).

Well-managed conifer forests in Sweden have a rotation time of up to 100 years or more. The stands ready for clear-cutting are the final results of continuous genotype-climate interactions combined with human interferences in the form of silvicultural treatments. Tree number per hectare is the outcome of *actual survival*, involving hardiness and parasite resistance, as well as *accidents*, and, finally, *silvicultural procedures* influencing the growing stock. The genotypic constitution of a population, its reaction to climate, and the type of management are responsible for the total productivity, calculated in cubic metres (and crowns) per hectare. The following study is based on experimental plantations laid out by the staff of the Department of Forest Genetics (Royal College of Forestry, Stockholm) during the years 1952 to 1955. The planning of field collections and methods of analysis go back to discussions between a series of experts in 1947 to 1949.

Population research is a fundamental branch of biological theory and its applications in practice. In a country like Sweden the analysis of conifer populations and their reaction to environmental conditions also constitute a basis for national economy and welfare. Accordingly, it is with considerable delight that we dedicate this article to the master of experimental population analysis, Dr. Theodosius Dobzhansky.

Material and Methods

In Figure 1 the Scandinavian populations used in the study are denoted by black dots. Seed material of *Pinus sylvestris* representing 100 natural stands and obtained after open pollination was collected in 1948 to 1950. Of the populations 90 were Scandinavian, 9 came from Germany, and 1 from the Netherlands. The extra-Scandinavian materials formed bulk collections. The Scandinavian materials, on the other hand, were represented by collections from individual trees, usually 20 to 30 trees per population (stand). Progenies from 2,200 individual trees were analyzed and partially utilized in the experimental plantations (EP). In addition, 8 Scandinavian and 10 extra-Scandinavian bulk collections were sown. Seedlings raised in two nurseries were distributed to 30 experimental plantations (EP) spread over Sweden in a special network manner. Dead plants were continually replaced by new ones as long as suitable young plants of equal age were available.

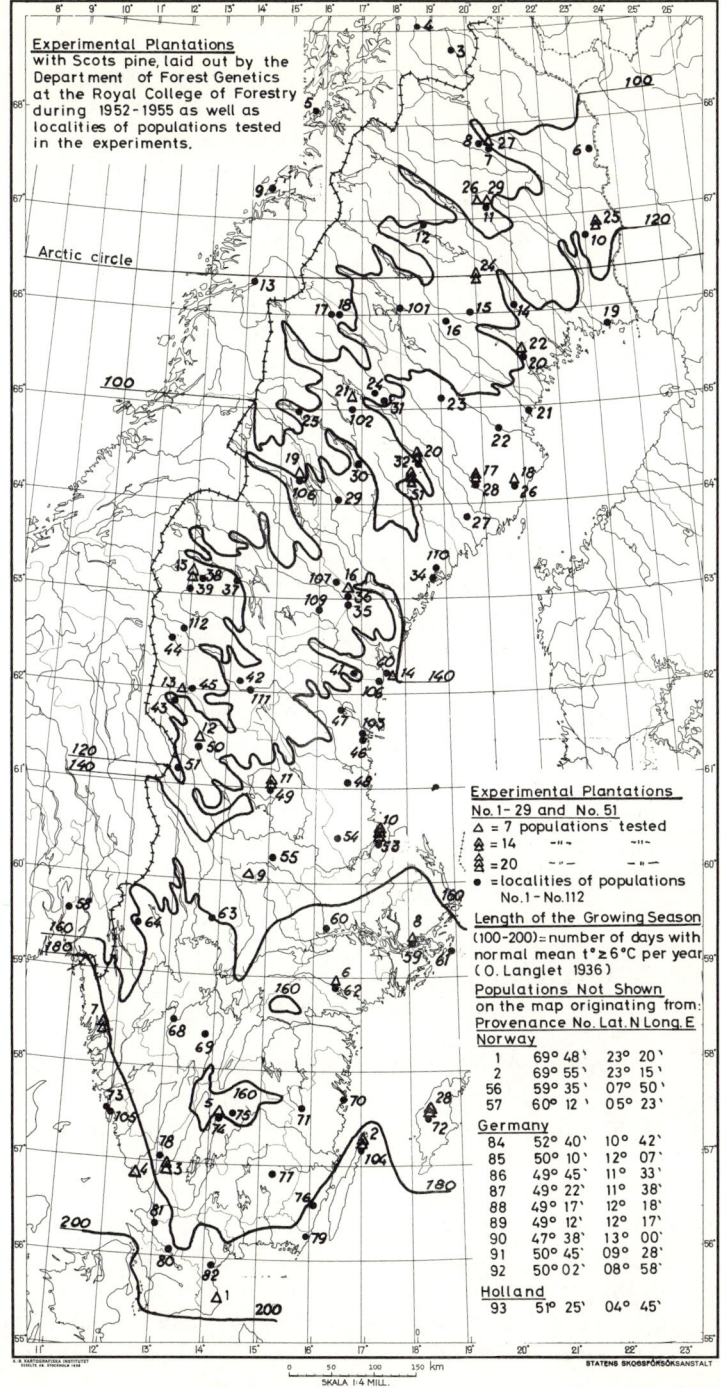

Fig. 1. Population analysis of Scots pine (Pinus sylvestris L.). Experimental plantations, denoted by triangles, were established in 1952-1955. Black dots indicate the populations studied in the land-wide Swedish experiments.

The plantations were managed in the most careful way possible, and many were fenced. In order to counteract accidental mass-killings by the snow-blight fungus (*Phacidium infestans* Karst.), the plantations were yearly sprayed with CaS-solutions according to methods devised by Björkman (1948) and further discussed by Eiche (1966) and Langlet (1968).

The EPs were "simple," containing 7 populations, "double" with 14 populations, or larger with a greater number of populations. The design is documented in articles by Eiche (1966) and Langlet (1968). It consists in each EP of four replications, two with wide spacing (2×2 m), two with dense spacing (1.25×1.25 m). The progenies were individualized except for the above-mentioned bulk populations (nos. 10, 23, 42, 74) used as standards. In each EP a population was represented by 20 trees and 13 offspring plants per tree; in all, 260 individuals.

In EPs with seven populations one was always local (or nearly local), two were standards represented on numerous EPs, and four (two + two) had their natural habitats north or south of the experimental locality. In these two-by-two instances one population originated from a higher, and one from a lower altitude compared to the locality of the EP. In EPs with 14 (or more) populations, the additional 7 (or more) populations were transferred from more remote habitats. This gave a possibility for the simultaneous testing in an EP of adjacent and remote populations.

It should be added here that this study, as well as the thesis by Eiche (1966), deal only with the 20 most northern EPs. This is for the reason that climatic conditions are rather severe in the forest and mountain regions north of latitude 61°. The northernmost EP is situated north of the Arctic circle at a latitude of almost 68°N.

Climatic Conditions of the Experimental Plantations and the Natural Populations Involved

In Figure 1 isolines for the length of the growing season (Y) are drawn indicating the number of days with an average temperature of at least 6°C (Langlet, 1936). In the EPs discussed here the growing season Y varies between 159 days, calculated in the Langlet manner, for EP 10; and 98 days for EPs 23 and 29. Latitude and altitude find their climatic expression in the length of the growth season. Local conditions, type of ground vegetation, microclimate, soil, winds, snow and ice also exert a great influence on success or failure in reforestation. Hardy materials are generally needed.

In fact, it is interesting to note that local populations are not always the most successful ones in reforestation, neither with regard to survival, nor with regard to productivity. An example of this was included in the paper by Gustafsson (1962), referring to data gathered by Eiche, and further discussed by the latter author in 1966.

Local climates may widely differ from generation to generation. The climatic conditions of a special area, for instance in the years 1950 to 1960, were certainly not entirely identical with those of the corresponding 10-year period 1850 to 1860. Especially the young plant, below an age of 15 to 20 years, is sensitive to harsh exterior conditions, including human, animal, and parasite interferences. Climate and fortuitous weather conditions vary from generation to generation, from century to century, and from decade to decade. Probably this is also true of the genotypic constitution of successive populations, due to different selection pressures responding to climatic differences and human interferences. Seed collection and natural sowing from a few seed trees of a stand, or from a few flowering trees, may in many instances and in special years lead to a certain amount of homozygotization, with various degrees of inbreeding depression and consequent decrease of viability. The present materials must be considered representative of the local populations, since at least 20 trees giving ample seed were used in the progeny plantations. However, it was not possible to define the degree of actual inbreeding in different stands and trees.

The distribution of the 19 EPs chiefly discussed in this study is as follows, with special regard to growing season:

```
98–100 days:  3 EPs,  65°43'–67°50',  2 simple,  1 double EP
105–110  "    3  "    64°33'–66°16',  1  "      2  "       "
113–120  "    4  "    61°54'–66°56',  2  "      2  "       "
121–128  "    4  "    61°27'–65°38',  3  "      1  "       "
136–145  "    3  "    60°04'–64°12',  2  "      1  "       "
150–159  "    2  "    60°32'–62°12',  1  "      1  "       "
```

In all, 71 different populations are tested in these 19 EPs. They show the following distribution with regard to the growing season of the local populations:

```
– 100  – 120  – 140  – 160  – 180  – 200  – 220 days
  3      21     20     10      9      4      4  =  71
```

Three EPs with a growing season of 98 to 100 days are situated in extremely severe climate,—one, in fact, above the timberline in the region of the birch forest (no. 23, Tärnaby, Y=98, 65°43', 630 m altitude). It is extremely difficult to obtain populations adapted to such extreme climates and localities. Also, populations from the extreme north of Norway and Sweden show considerable mortality under these conditions.

Seventeen populations south of the area of the 19 EPs examined have been included in order to analyze the possibility of an increase in growth properties still combined with a reasonable degree of survival. The 71 populations analyzed have to a varying degree been utilized in the individual experiments. Great attention has been paid to very hardy populations and these have a fair distribution over the entire experimental area. The distribution on EPs is as follows:

```
 2 populations on 16 experimental plantations
 1 "            "  14 "               "
 1 "            "   9 "               "
 2 "            "   6 "               "
 5 "            "   5 "               "
 5 "            "   4 "               "
 7 "            "   3 "               "
16 "            "   2 "               "
32 "            "   1 "               "
```

Cumulative Mortality and Survival

Before the discussion of some selected individual EPs a survey of the entire material will be presented, referring to the data by Eiche (1966).

Three EPs of extreme conditions (nos. 23, 26, 27) form a special group owing to the difficulty (or impossibility) of finding populations with a shorter growing season. Three southern EPs included in this study (9, 10 and 14) lie outside the risk zone. Even some long-distance materials show a fairly good survival cultivated there (e.g., a low cumulative mortality). Between these two climatic extremes there are 14 EPs with a more or less drastic increase in mortality of many alien populations; some of these, however, give an increased hardiness.

Dead plants were counted almost yearly. Analysis was in many cases carried on until 1968. A few examples will be given relating to the last census; otherwise the data refer to 1964. In addition to measurements of cumulative mortality (CM), also *single year mortalities* (SYM) have been analyzed, especially for the study of the causes of plant death and its connection with changes in the environment. In fact, there is, as pointed out by Eiche (1966), a more or less wavelike appearance of plant death, rising very high in sensitive materials after special catastrophic years.

The data are summarized in Table I. For a correct understanding of this table it should be emphasized that each EP is considered as one individual unit in the comparisons. Thus, several populations are included in the comparisons more than once. However, the population mortalities are always measured in terms of differences from the local population of each EP.

At a plant age of the 14 years (1964) the cumulative mortality of the local populations in the EPs with a growing season of 98 to 100 days

TABLE I

Average Cumulative Mortality (CM) in Relation to the Length of Growing Season (Y)

Growing season of EPs, days (Y)	No. of experimental plantations (EP)	CM of local pop. %	Cumulative mortality, percent													No. of comparisons (local-foreign)
			with an increase of $Y(Y_{EP}-Y_{pop})$ by					with a decrease of $Y(Y_{EP}-Y_{pop})$ by								
			50	40	30	20	10	0	10	20	30	40	50	60-70 days	
98-100	3	41.2						61.7 (1)^a	45.7 (5)	52.4 (6)	57.8 (6)				18
105-110	3	30.1					12.8 (1)	27.8 (5)	20.3 (8)	47.0 (10)	65.6 (4)	94.5 (2)	72.1 (2)		32
113-120	4	23.6				24.9 (4)	25.8 (9)	26.8 (5)	41.3 (12)	50.6 (5)	83.5 (1)	59.3 (2)	100.0 (1)		39
121-128	4	43.4			29.8 (3)	34.5 (7)	46.6 (9)	61.0 (5)	71.8 (5)	82.1 (3)	86.9 (1)	98.1 (2)	—	100.0 (1)	39 / 31
136-145	3	28.9			24.8 (6)	22.0 (2)	21.1 (5)	38.5 (1)	42.7 (2)	40.2 (5)	49.8 (2)	60.4 (1)	79.2 (1)		25
105-145	14	31.80			26.47 (9)	28.43 (14)	29.57 (24)	32.12 (18)	48.63 (29)	56.64 (17)	76.48 (6)	74.19 (7)	89.60 (2)	100.0 (1)	127
150-159(A)	2	14.0	27.7 (1)	14.4 (2)		14.2 (2)	27.8 (3)	11.8 (3)	14.2 (1)	20.5 (5)	—	—	31.9 (1)	48.9 (1)	19
159(B)^b	1	16.9							53.5 (1)		39.5 (3)	35.4 (1)	40.0 (1)		6

^{a} Brackets comprise number of comparisons.
^{b} This is an extra EP (10 B), situated close to EP 10 A (159 A), including one Dutch and five German bulk populations.

reached an average value of 41 percent. The average figure of the 18 comparisons was definitely above this value. In fact, it can be considered highly difficult to find alien populations with an increased hardiness. This is an important task then for further tree breeding. Selection should primarily be carried out within the most hardy populations by searching for individual very hardy genotypes useful in subsequent hybridization work. The necessity of such a procedure was pointed out by Gustafsson (1962).

With the 127 comparisons in 14 EPs of 105 to 145 days (Table I), it was found that mortality decreases if alien materials are introduced from very harsh localities. On the other hand, there is a steady decrease in survival related to the length of the growing season equal to or slightly below the mean value of the local populations and ending with -10 days. When alien materials are transferred from regions with a 10-day longer growing season or more, average mortality suddenly rises and ultimately reaches values of almost 100 percent. These comparisons indicate that in the total area, covered by the 14 EPs, populations from localities with a growing season increased by more than 10 days do not generally tolerate the new climate.

The diversified data of this group give similar results: In EPs of 105 to 110 and 113 to 120 days a striking increase of mortality sets in within populations having a native 20-day or more prolonged growing season. Mortalities, equal or smaller, are, however, found in the columns to the plus side and in the minus column closest to the zero point.

In EPs of 121 to 128 days there is a decrease in mortality of populations originating from very harsh regions and an equal mortality in the 0- to 10- day column to the plus side. Populations from milder regions show a high increase of mortality.

In EPs of 136 to 145 days mortality is decreased in all columns to the plus side but is considerably increased to the minus side.

In the third group of EPs, with 150 to 159 days of growing season, a long-range transfer of populations is possible. However, the introduction of German collections, with a native growing season extended by 30 or 40 days, almost always results in a very high mortality. A sister plot to EP 10 (159 B) with a control mortality of 17 percent showed a high increase in mortality of German populations having a native season 10 to 20 days longer, or more.

In his thesis of 1966, Eiche presented results of a regression analysis involving the two variables: X_2 = cumulative mortality, and X_1 = differences in length of growing season. The regression lines were extremely steep in hard regions, indicating the drastic killing influence of severe conditions even in the case of rather small population transfers. In areas with little average killing the regression lines were flatter (cf. Eiche, 1966, Figs. 61 and 59.)

Cumulative Mortality in Selected Experimental Plantations

Of the 19 EPs studied, 5 will be analyzed in some detail (EPs 27, 24, 21, 20, and 14). These comprise plantations with different degrees of mortality and populations with widely differing response to changes in climate.

EP 27, Kiruna, 67°50′, 360 m, Y=100 (Fig. 2)

The experiment comprises 7 populations with native growing seasons (Y) varying from 94 to 128 days. The local population has $Y = 100$.

The result of 1968 is shown in Figure 2. The local population (7), with a mortality of more than 50 percent, is nonetheless the hardiest one in the comparison. The sequence is:

7 with $Y = 100$, local population 67°50′
23 ″ $Y = 118$, Malå 65°08′
101 ″ $Y = 105$, Arjeplog 66°03′

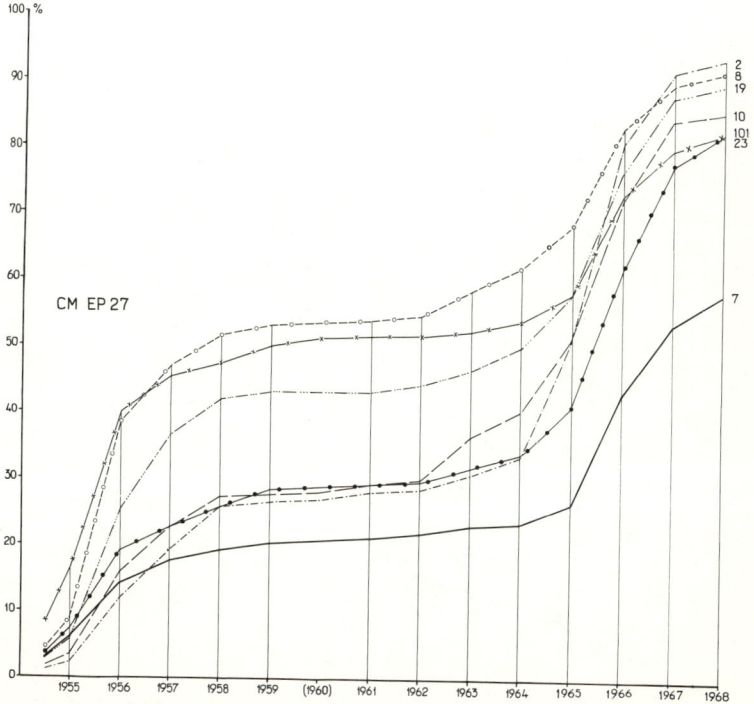

Fig. 2. Cumulative mortality (CM) in experimental plantation (EP) 27, Kiruna, 67°50′ N, altitude 360 m, growing season $Y = 100$ days. Local population no. 7.

10 "   Y = 144, Korpilombolo 66°50′
19 "   Y = 128, Kalix 65°50′
 8 "   Y = 94, Kiruna 67°50′
 2 "   Y = 104, Norway, 69°55′

It is interesting to note that population 8, from an even harsher region, has a mortality surpassing 90 percent and is definitely less hardy than the local population. Height growth is also less. This may indicate a weak original population or, possibly, some degree of homozygosity in the offspring as a result of partial inbreeding. (Populations from the extreme limit of distribution often gave rise to aberrant offspring. In fact, one population, no. 3 [Norway: 68°40′], showed a peculiar type of development, with plantlets weak and yellowish.)

Figure 2 also indicates that the local population was in a relatively stable state between 1956 and 1965. Then a sudden mortality set in.

EP 24, Kåbdalis, 66°16′, 440 m, Y = 105 (Fig. 3)

The experiment comprises 14 populations with a Y varying between 100 and 144 days. The local population (101) has Y = 105 and a mortality

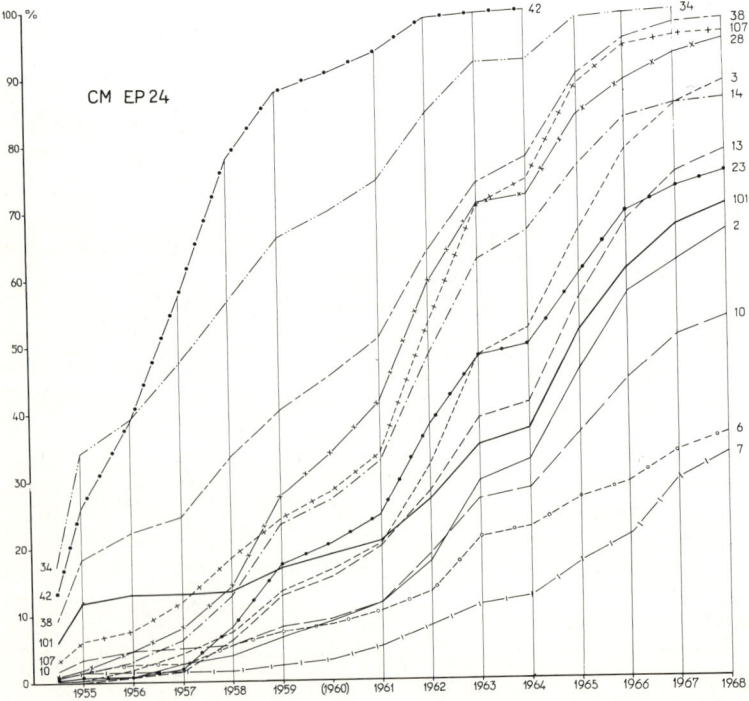

Fig. 3. Cumulative mortality (CM) in EP 24, Kåbdalis, 66°16′, 440 m, Y = 105. Local population no. 101.

of 71 percent with a steady rise since 1958. The wide spread of the other mortalities is obvious.

Four populations are superior in survival:

7 with Y=100, Kiruna 67°51'
6 "   Y=107, Kitkiöjoki 67°45'
10 "   Y=114, Korpilombolo 66°53'
2 "   Y=104, Norway 69°55'

Nine populations are more or less inferior:

23 with Y=118, Malå 65°08'
13 "   Y=123, Norway 66°18'
14 "   Y=121, Spikseleå 66°05'
3 "   Y=100, Norway 68°47'
28 "   Y=128, Vindeln 64°14'
107 "   Y=122, Bispfors 63°08'
38 "   Y=117, Vallbogården 63°10'
34 "   Y=144, completely killed 1967, Örnsköldsvik 63°15'
42 "   Y=132, "   "   1964, Sveg 62°03'

The differences between populations 7, 6, and 10, on the one hand, and 101, on the other, are statistically significant. Population 10, superior to 101, is significantly inferior to both 6 and 7.

The result implies a better adaptation in survival of the alien populations 7, 6, and 10 transferred to the harsh climate of EP 24.

Population 23, relatively hardy also in the previous EP 27, is not significantly different from the local population. The relatively low hardiness of the local population (101) may depend on some degree of spontaneous selfing, since the original stand consisted of widely spaced seed trees. Its height growth also is not very good.

EP 21, Långsjöby, 65°09', 465 m, Y=110 (Fig. 4)

The experiment comprises seven populations with a Y varying between 105 and 136 days; Y of the local population (102) is 111 days (one day longer than that of the EP itself).

The local population shows a high mortality amounting to more than 50 percent, with a conspicuous increase after 1964. This weakness possibly depends on a certain degree of inbreeding.

Interesting is the better survival of population 4 up to the year 1965. After that mortality rises steeply to the level of no. 102. No. 108, with Y=150, is completely killed.

The order of survival is:

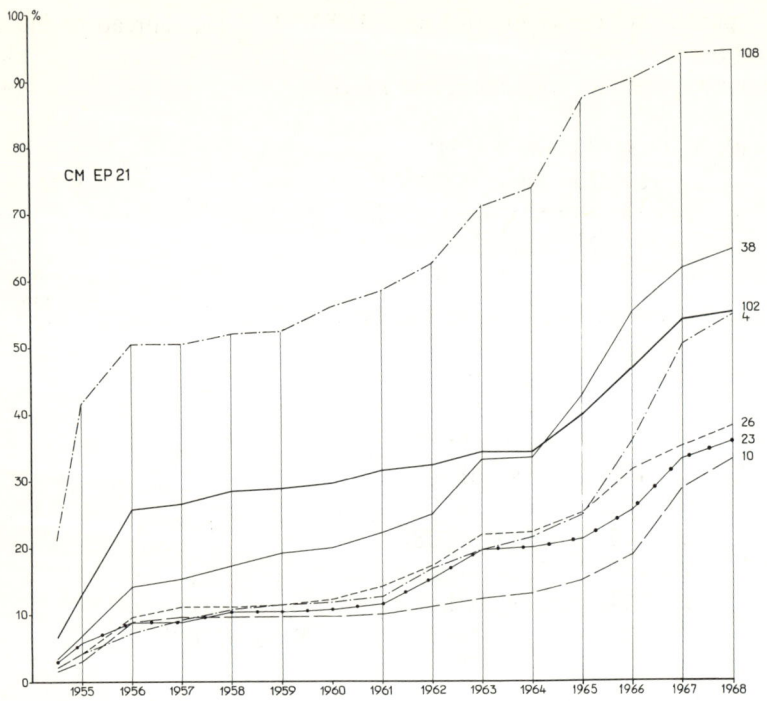

Fig. 4. Cumulative mortality (CM) in EP 21, Långsjöby, 65°09', 465 m, Y=110. Local population no. 102.

10 with Y = 114, Korpilombolo 66°50'
23 "   Y = 118, Malå 65°08'
26 "   Y = 136, Robertsfors 64°12'
 4 "   Y = 105, Norway 69°07'
102 "   Y = 111, local population, 65°09'
38 "   Y = 117, Vallbogården 63°10'
108 "   Y = 150, Galtström 62°10'

Apparently the northern latitude and short growing season of population 4 were not specially helpful in the case of this high altitude EP. Note the conspicuous superiority in hardiness of populations 10 and 23.

EP 20, Lycksele, 64°33', 555 m, Y = 109 (Fig. 5)

The plantation comprises 14 populations with a growing season varying between 98 and 150 days. The local population (32) has a Y of 110 days (one day prolonged compared to the site of the EP). Its mortality smoothly increases up to 35 percent.

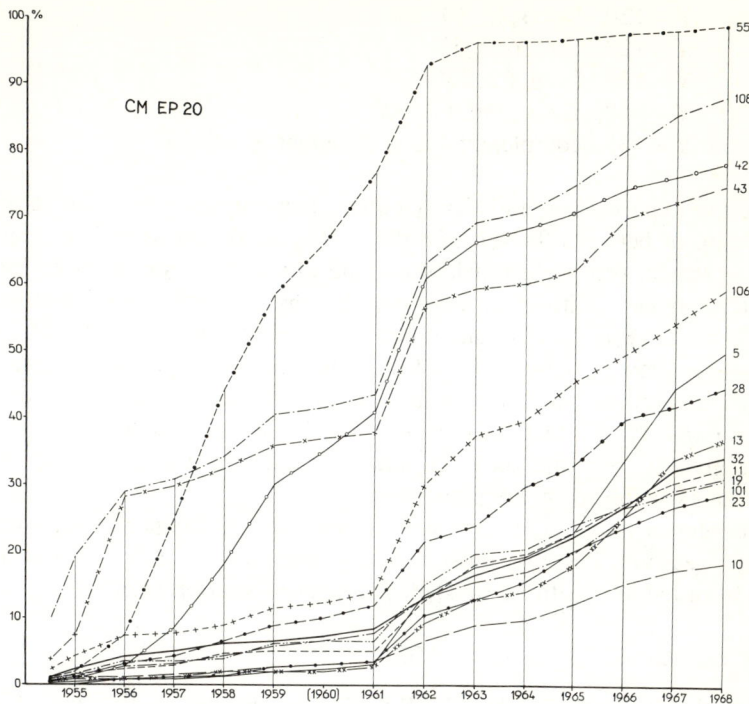

Fig. 5. Cumulative mortality (CM) in EP 20, Lycksele, 64°35', 555 m, Y=109. Local population no. 32.

One population (10, from Korpilombolo, cf. EPs 21 and 24) shows a definite increase in hardiness. Y was here 114 as compared to 109 of the EP. Also, population 23 (from Malå, with Y=118, transferred from the north) is increased in survival capacity.

Superior populations:

10 with Y = 114, Korpilombolo 66°53'
23 " Y = 118, Malå 65°08'
101 " Y = 105, Arjeplog 66°03'
19 " Y = 128, Kalix 65°50'
11 " Y = 98, Gällivare 67°14'

Inferior populations:

13 with Y = 123, Norway 66°18'
28 " Y = 128, Vindeln 64°14'
5 " Y = 115, Norway 68°11'

106 "   Y = 120, Harrsjön 64°18'
43 "   Y = 124, Idre 61°53'
42 "   Y = 132, Sveg 62°03'
108 "   Y = 150, Galtström 62°10'
55 "   Y = 145, completely killed Grangärde 60°16'

The five superior (or statistically equal) populations have a native growing season of between 98 and 128 days. The weakness as to length of the growing season (composed of latitude and altitude) is compensated by the definitely higher latitude in two populations. One superior population (Malå, no. 23) is remarkable, since its original habitat is characterized by a longer growing season and a lower altitude. As previously mentioned, this population has an outstanding hardiness and, in addition, a high growth capacity. In fact, in the fall of 1963 it was significantly better in height growth than the local population, and was the next best of all populations involved in the plantation.

The inferior populations invariably originated from regions with a longer growing season. The two most sensitive populations (nos. 55 and 108) came from habitats with 36 to 41 days longer vegetation period.

EP 14, Bråmön, 62°12', 5 m, Y = 150 (Fig. 6)

This plantation is situated on an island in the Gulf of Bothnia. Figure 6 indicates that after the first years, up to 1956, survival was stabilized on levels characteristic of the individual populations. Mortality of no population has surpassed 40 percent. The local population has a value of less than 20 percent mortality. It is clear from the figure that population no. 23 (from Malå) is superior to the others.

The order is:

23 with Y = 118, Malå 65°08'
40 "   Y = 150, local population, 62°12'

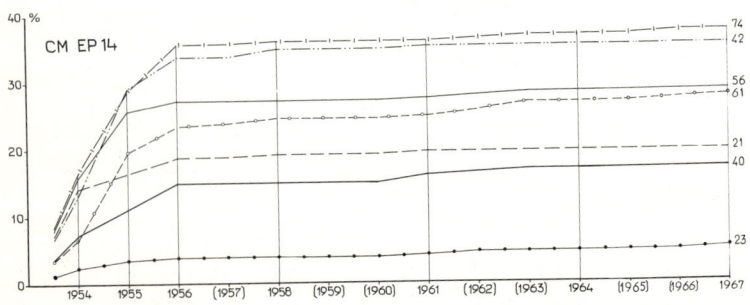

Fig. 6. Cumulative mortality (CM) in EP 14, Bråmön, 62°12', 5 m, Y = 150. Local population no. 40.

21 "   Y = 133, Byske 64°57'
61 "   Y = 167, Sandhamm 59°17'
56 "   Y = 137, Norway 59°35'
42 "   Y = 132, Sveg 62°03'
74 "   Y = 165, Eckersholm 57°36'

The local population in this favorable maritime locality is superior to nos. 21, 56, and 42, although these came from harsher climates (with shorter growing seasons). Again outstanding, as mentioned above, is the Malå population with less than 5 percent mortality.

Single-Year Mortality

Cumulative mortality gives the *end result* of killing at certain chosen periods. The exact *process of killing,* however, is better studied immediately after the death of a plant. The killing by snow blight (*Phacidium*), an often disastrous epidemic in certain parts of northern Sweden, was continually counteracted by spraying with CaS-solution up to the age when the fungus is no longer harmful. Plant death was accordingly caused by insufficient hardiness to extreme climatic conditions and can be referred to physiological, anatomical, and histological damages. Single-year mortality provides the possibility of showing more precisely the effects of individual years (or cycles of years) in a special site, EP, and population. The damage syndrome comprises primary and secondary features. Attacks by many fungi and insects can be referred to the secondary group and depend on the increased weakness of plants following the severe cold damage itself.

Three main types of damage to the above-ground part of the plants are distinguished: (1) injury of the main stem, consisting of different degrees of stem girdling and frost canker; (2) die-back of the top of plants and of the crown of the tree; and (3) distortion of the stem, branches, and crown by snow.

Damage type (1) is widely represented in all plantations showing a high amount of killing. Generally, it consists of the entire or partial death of the cambium, mostly at the base of the stem and close to the ground. Depending on the extent of the killed cambium all around the stem or in larger or smaller patches, the callus overgrowth and subsequent frost canker will vary in size and hazard to the plant. Some damaged plants may recover, if mild summers and winters follow. On the other hand, many plants perish even in the first period of growth after the cold damage. Some recover partially but are destroyed later on by another onset of severe winter cold or wavelike spring frosts during subsequent years. As amply illustrated in previous work (Eiche, 1962, 1966), stem girdling and frost-canker may develop and be effective in killing over an extended period

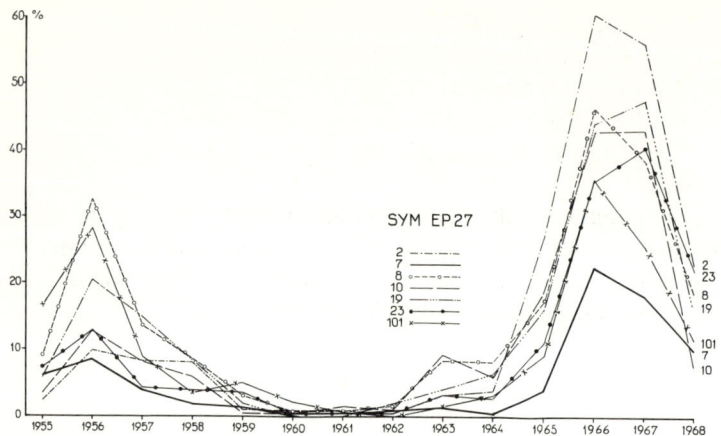

Fig. 7. Single-year mortality (SYM) in EP 27, Kiruna (cf. Fig. 2), 67°50', 360 m, Y=100. Local population no. 7.

of years following a single very severe attack. Sakai (1968) has tested and confirmed such a conclusion experimentally.

No doubt, however, the majority of injuries imply combinations of basal stem girdling with other damage patterns, also with types (2) and (3) involved, weakening plant vigour considerably. As far as possible each EP, studied in this set of experiments, and its partaking blocks, populations, and individual plants, have been examined as to damage pattern, with regard to both exterior and interior properties. A survey up to the year 1964 was published by Eiche (1966). Data from four EPs will be discussed here.

In *EP 27*, Kiruna, with its extreme conditions, there are three peaks of severe damage and killing (Fig. 7): one in 1956, a small one in 1963, and a very pronounced one in 1965 to 1967. The losses of plants in 1956 were due to the exceptional drought in 1955, as well as to severe spring frosts in 1956. In the period of 1958 to 1962 very little killing took place. The spring frosts in 1963 were especially harmful to populations 8 and 10, the former from a locality in the vicinity of the timberline. The local population (no. 7) was resistant throughout the period 1955 to 1964. Then, however, every population, although no. 7 the least, was extremely damaged by late winter cold and repeated spring frosts in 1965. Final killing of previously injured plants was extended over the years 1966 and 1967 as delayed result of the primary damage in 1965. It is noticeable how susceptible was population no. 2, from Norway, 69°55'.

The cause of the extreme killing in 1965 to 1968 was a combination of basal stem girdling and die-back of the top. The long-time killing process took place as aftereffects of the basal stem girdling in 1965. The winter

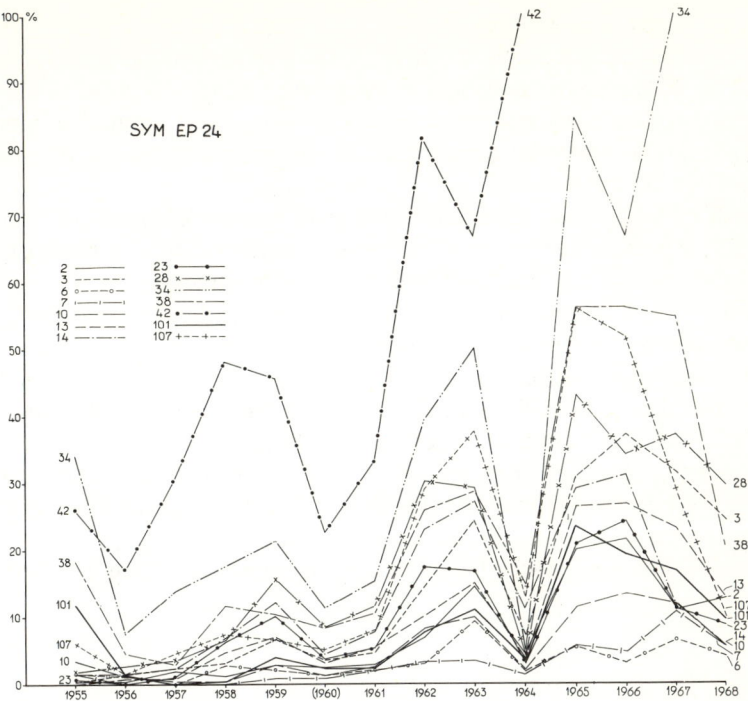

Fig. 8. Single-year mortality (SYM) in EP 24, Kåbdalis (cf. Fig. 3), 66°16′, 440 m, Y = 105. Local population no. 101.

of 1964-65 was very cold, ending with a long early spring in 1965, repeatedly changing from thawing to freezing and with great temperature changes between day and night.

The wavelike appearance of the single-year-mortality curves is especially clear in *EP 24,* Kåbdalis (Fig. 8). Here also remarkable differences between populations can be seen. The first peak of killing occurred in 1955 and was due to early frosts in August 1954 and to severe spring frosts in 1955. In 1958 and 1959 a second culmination after weak basal stem girdling took place. In 1962 and 1963 extreme damages occurred as aftereffects of spring frosts and basal stem girdling in 1961. In April 1962 sudden fluctuations of diurnal temperature due to strong solar radiation caused injuries of trees above the snow level. Except for population 42, which was completely killed, a striking decrease in killing was found in 1964.

Then, in late winter and early spring of 1965, new severe cold damages occurred, lasting over three to four years. The wide differences in aftereffects can be seen between a few populations (nos. 6, 7, and 10), which are very hardy, a couple of others (no. 2, the local population 101, and 23), which are medium sensitive, and populations (13, 14, 3, 28, 107, 38, 34), which show high to complete killing. In spite of the conspicuous variation in resistance between years, the cumulative mortality (Fig. 3) gives a parallel appearance of the mortality curves through the years 1954 to 1968.

Girdling and frost canker were the main causes of death of plants in this EP, accompanied by snow damage, expressed by die-back of tops, and other symptoms, and followed by parasite attacks.

In *EP 20,* Lycksele (Fig. 9), the material is divided into two parts, one comprising the sensitive populations, one the hardy ones. At the outset this EP was a great success. In October 1954 only 4 percent average losses were recorded. Excluding the 1955 and 1956 positions of populations 108 and 43, the first marked increase of killing appeared in 1958 to 1959 and became even more conspicuous in 1963 following a severe cold damage in 1961 given rise to abundant basal stem girdling. Death then primarily took place in 1962, both in the sensitive and the hardy populations. Also in this EP the year 1964 led to a certain recovery. Then in 1965 cold

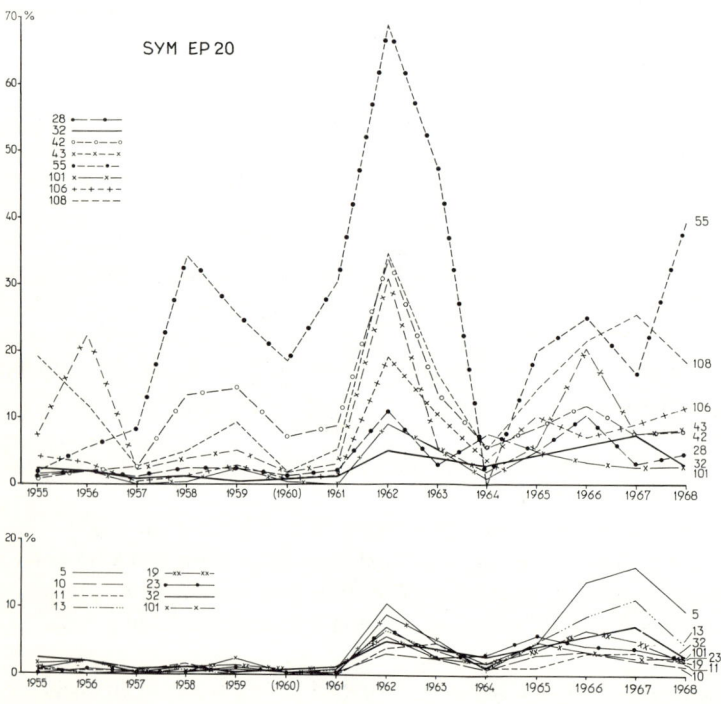

Fig. 9. Single-year mortality (SYM) in EP 20, Lycksele (cf. Fig. 5), 64°35′, 555 m, Y=109. Local population no. 32.

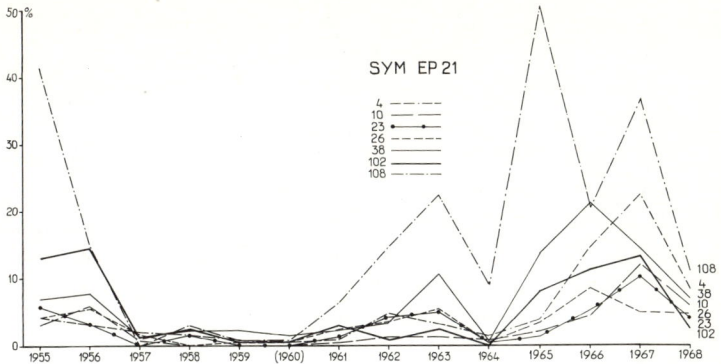

Fig. 10. Single-year mortality (SYM) in EP 21, Långsjöby (cf. Fig. 4), 65°09', 465 m, Y = 110. Local population no. 102.

damage, following late winter and early spring frosts and thawing, occurred with a killing protracted over the years 1965 to 1968. Populations 10, 11, 19, 23, and 101, however, are hardy and show an almost negligible fluctuation of the killing percentages, in agreement with the data on cumulative mortality.

In *EP 21,* Långsjöby (Fig. 10), there was already in 1955 a marked difference in killing. In the local population 102 this was due to weak transplants, possibly as a consequence of inbreeding depression. The stand itself consisted of rather isolated seed trees in an old spruce population. Later on, this population showed itself to be quite hardy, although less so than populations 23, 26, and 10. Number 108 showed a pronounced killing throughout the period of analysis with peaks in 1955, 1956, 1963, 1965, and 1967. It is a coastal population with much longer growth season.

The losses in 1956 were due to cold damage and summer drought in 1955, as well as spring frosts in 1956. During the entire period to 1964 the basal stem girdling occurring in 1961 was the chief cause of losses in plants. There is a distinct culmination in 1963. In a few populations there was another positive excess in 1963 probably intensifying the die-back of trees girdled and cankered in 1961. In the years 1965 to 1967 the profound killing effects were due to severe late winter and fluctuating spring temperatures in 1965. The basal stem girdling was in some cases so intense that even trees 5 m in height were damaged and died the following year.

The curve of the single-year mortality of this EP indicates that the cumulative mortality of population 102 was to a certain extent due to the high amount of killing in 1955 and 1956. After this early weakness 102 became definitely hardier than populations 4, 38, and 108 (the latter completely killed), but was still inferior to populations 10, 23, and 26.

This is also indicated by the single-year mortality (SYM) curves in 1965 and 1967.

Summing up these four examples it is obvious that the type of climate greatly influences the amount of killing, as well as its expression. In large parts of the northern Swedish inland it leads to mass killing of individuals and populations not suited to a type of climate characterized by repeated thawing and freezing of the snow cover due to a profound fluctuation of diurnal temperatures. Cambium damage is abundant, leading to basal stem girdling and canker. Hardy populations do exist, however, some even better adapted than the local populations.

Height Growth

Data accumulated by Eiche and published by Gustafsson (1962)[1] show that the local population is not always the one best adapted in survival capacity. Nor is it necessarily superior in growth properties. In EP 16 (Bispgården, 63°08′, altitude 470 m, Y = 121 days) the local population showed a survival of 51 percent in 1961. Four populations were definitely inferior in survival; two were superior, both of more northern origin (no. 6: 67°45′, Y = 107, and no. 23: 65°08′, Y = 118).

One of the less hardy populations (no. 42, Y = 132) was superior as to growth (**). Of the very hardy populations one (no. 23, Malå, Y = 118) was definitely superior in height (t = 3.4, ***). The most northern one (no. 6, Kitkiöjoki, 67°05′, Y = 107) was equal in height growth.

Later Eiche (1966) published data on height development in 7 EPs (9, 10, 12, 16, 20, 21, and 25).

In EP 9 (Gravendal, Y = 145) one population (42, Y = 132 was significantly superior (***), one (70, Y = 175) inferior (*) to the local material.

In EP 10 (Älvkarleby, A and B, Y = 159) a series of populations, both northern and southern, were significantly inferior, none was definitely superior.

EP 12 (Bunkris, Y = 125) showed five inferior populations out of six comparisons. The local population was, in addition to being very hardy, also growing well on this harsh site.

EP 16: see above.

EP 20 (Lycksele, Y = 109) has, as shown in Figure 5, a wide spreading

---

[1] $P > 0.05°$
$0.05 > P > 0.01*$
$0.01 > P > 0.001**$
$P < 0.001***$

of the involved populations with regard to hardiness. Two populations, more hardy than the local one and with lower Y (101: Y=105 and 11: Y=98), were statistically seen equal in height growth to the local material. Three more populations, also superior in hardiness, have growing seasons longer than the local population (10: Y=114, 23: Y=118, 19: Y=128). All three were superior in growth, two of them significantly (23, *** and 19, ***). Both are more northern in origin.

Of the eight remaining less hardy populations, two were significantly superior in growth (42: Y=132, ***, 28: Y=128, *), three significantly inferior (5: Y=115, ***, 43: Y=124, ***, 108: Y=150, ***). Three were, statistically equal in growth.

This EP thus indicates that it is possible by transfer of natural populations to combine increased hardiness and growth. The outstanding populations, one of them being 23: Malå, certainly have a slightly longer growing season than the local population. In this EP they originated from a more northern latitude.

Moreover, populations moved from the south, with longer growing season and decreased hardiness, may show a conspicuous superiority in growth. The population with the very best height growth in 1963 was no. 42. However, its hardiness was so decreased that in spite of its growth superiority it can scarcely be recommended in practice.

The data can be summarized in the following way:

Growing season increased by 10 – 20 days: S°
"   "   "   "   0 – 10 " : I°

Growing season decreased by 0 – 10 days: S***, S°, I***
"   "   "   "   10 – 20 " : S***, S°, I°, I°, I**
"   "   "   "   20 – 30 " : S***
"   "   "   "   30 – 40 " : I°
"   "   "   "   40 – 50 " : I***

(S = superior, I = inferior in growth.)

EP 21 (Långsjöby, Y=110) contains two populations superior in height to the local one. The best as to height growth was no. 26, Robertsfors (Y = 136, ***), and the next best one no. 23, Malå (Y=118, ***). The first is more southern in origin, the second is from a similar latitude. Both are also hardier than the local material. No. 10, which is the hardiest population (Fig. 4), has a Y=114, but originates from a more northern latitude. Height growth was in this case identical to that of the local material.

The remaining three populations were equal to the local one.

EP 25 (Korpilombolo, Y=113) originally consisted of 14 populations, five of which in 1968 were almost completely killed. Nos. 11, 4, 19 (Y = 105, 105, and 128 respectively) were superior in hardiness. Two were equal in height 1963. No. 19 was definitely superior (***). This originated from a more southern locality (Kalix).

The less hardy populations, excluding the ones almost completely killed, were in three cases significantly inferior in height growth, in two cases equal. The data can be summarized in the following way:

Growing season increased by 10 – 20 days: I***
" " " " 0 – 10 " : I°, I°, I*

Growing season decreased by 0 – 10 " : I*
" " " " 10 – 20 " : S***, H, I°
" " " " 20 – 30 " : I°

In 1967-68 further measurements of height growth were performed.

A final analysis is not yet ready. For some EPs an analysis of variance shows definite population differences in hardiness and growth, and an influence of spacing, as well as an interaction of spacing and populations.

Four individual EPs have so far been fully analyzed (29: Y = 98, 27: Y = 100, 24: Y = 105, 22: Y = 127).

In *EP 29,* Linalombolo, with almost complete extinction of most populations, a reasonable number of plants of three populations including the local one are still intact. No certain height differences were found, although population no. 10 (Y = 114) was slightly superior and no. 4 (Y = 105) slightly inferior.

EP 27, Kiruna, has a fair number of undamaged plants only in one population except the local one, namely, no. 101 (Y = 105). No significant height difference was found.

The 1967 survival and height measurements of *EP 24,* Kåbdalis (cf. Fig. 3) (66°16′, Y = 105) are specially informative and important.

Of the four hardiest populations (nos. 7, 6, 10, and 2, in the order given) two are significantly superior in height (6: 67°45′, Y = 107, ***; 10: 66°53′, Y = 114, **), two slightly superior (7: 67°51′, Y = 100, °; 2: 69°55′, Y = 104, °).

Populations 23, 13, 14, and 3 are less hardy than the local population in the order mentioned. Two are superior in growth (23: 65°08′, Y = 118, **; 14: 66°05′, Y = 121, *(*)), one slightly superior (13: 66°18′, Y = 123, °), and one slightly inferior (3: 68°47′, Y = 100, °).

Five populations are completely or almost completely killed.

Growing season increased by 0 – 10 days: S°, S°, I°
Growing season decreased by 0 – 10 days: S***, S**
Growing season decreased by 10 – 20 days: S**, S*(*), S°

Populations 7 and 6 originate from the same latitude, with slightly different growing seasons (100 and 107 days, respectively) and an almost identical cumulative mortality. Population no. 6 is superior to no. 7 (S*) in growth. In fact, it is the best population in this respect partaking in the plantation.

Populations 6 (from Kitkiöjoki), 10 (from Korpilombolo) and partially 23 (from Malå) imply highly valuable materials under harsh conditions like those of EP 24.

In *EP 22* (Älvsbyn, 65°38′, Y = 127) with good survival of all seven partaking populations the sequence of height increment in relation to growing season was as follows:

Growing season increased by 20 – 30 days: I*
Growing season increased by 10 – 20 days: I°, I**
Growing season increased by 0 – 10 days: S°
Growing season decreased by 10 – 20 days: I°, I***

Populations nos. 10, 23, 11, and 5 are superior in hardiness and two of them, nos. 10 and 23, equal the local population in height growth. No. 23 (Malå, 66°50′, Y = 118) is slightly superior to the local population (S°) and significantly better than no. 10 (Korpilombolo, 66°50′, Y = 114, **). Also here the advantageous qualities of the Malå population are evidenced. The inferior populations originated from Norway (no. 5, 68°11′, Y = 115, **), Gällivare (no. 11, 67°09′, Y = 105, *) and Färila (no. 47, 61°49′, Y = 145, ***).

Conclusions and Summary

In this chapter a series of natural populations of Scots pine have been analyzed with regard to their survival capacity and, to a minor degree, also their growth properties. A series of experimental plantations (EPs) were laid out in the years 1952 to 1955; they cover extremely harsh regions north of the Artic Circle, close to the tree limit (67°50′ N), down to the southernmost part of Sweden (55°37′ N). The growing season (Y) of the EPs varies between less than 100 and almost 200 days. Here some 20 EPs between 67°50′ and 60°04′ N lat. are discussed.

The wide climatic range of this region implies a complex network of genotype-milieu interactions. Conditions exist where a genetic adjustment to *maximal survival capacity* is essential. In some southern EPs, on the contrary, mortality due to climatic reasons is no serious problem; long-range transfer of material is indeed possible. Between those two extremes wide forest areas exist, where local populations are not necessarily adjusted to a maximal survival capacity. In fact, foreign populations hardier and tougher than the local ones can be found. In the highly mechanized modern silviculture of northern Sweden, with vast clear-cut areas, consequent wind erosion and deteriorating soil, a considerable strain is often exerted on the local populations, in the sense that progenies even of these are not hardy enough for a successful reforestation.

Such a strain on the local populations in evidenced by Figures 3 to 6, in which some populations, for instance nos. 6, 7, 10, 23, are superior in survival to the local ones. This is a result verified by the analysis of other plantations not included in this survey.

Table I indicates that transfers to moderately milder climates generally lead to better survival. If the growing season of introduced populations is increased in their new habitats up to 30 days, then average cumulative mortality of the introduced populations decreases considerably. In certain regions of the enormous central forest district survival capacity of short-seasoned populations gives even higher values, as in the EPs with a growing season of 121 to 128 and 136 to 145 days.

On the other hand, it is obvious that the transfer of long-season populations to harsh climates gives rise to a seriously increased mortality, in some cases complete killing. (Compare for instance the continuous rise of mortality in the compiled columns of the minus side in the EPs having a growing season of 105 to 145 days.) Transfer of populations with a native growing season 10 to 20 days longer is, on the whole, harmful. It becomes directly fatal with populations adapted to a longer season of 20 to 30 days and more.

However, with a detailed analysis it can be shown that these average results cover important exceptions: (1) local populations do exist that are extremely hardy and difficult to surpass in hardiness; (2) some short-seasoned populations, when transferred, may be less hardy than the local longer-season ones; and (3) some long-season populations show themselves very hardy also when transferred to localities with shorter growing seasons.

Case (1) is especially valid with regard to extreme latitude and altitude populations. In case (2), historical causes may be at work, such as chance of origin and migration, but also environmental conditions and fortuitous climatic effects, or purely genetical events following partial inbreeding. In case (3) historical factors, microclimate selection, and the like may lead to considerable hardiness. In all populations there exist pronounced population-milieu interactions.

Case (3) is specially interesting. In EP 24 (Fig. 3) populations 6 and 10 have longer growing seasons than the local population 101 but are more northern in origin. In EP 21 (Fig. 4) the same is true of populations 10, 23, and 26, having longer growing seasons than no. 102; one is more northern (10), one equal in latitude (23), and one more southern (26). In EP 20 (Fig. 5) 10 and 23 have longer growing seasons than the local population no. 32; both of them are more northern in origin.

In general, length of growing season affords a reasonable criterion for survival capacity.

In forest districts where maximal survival capacity is not required, i.e., in regions with a growing season of 105 to 145 days (or more), transfer of some alien populations may result not only in a better survival, as indicated above, but also in a better growth. Several populations in EPs 9, 16, 20, 21, 24, and 25 give such evidence. Certainly the material was rather young at the time of measurements (13 to 17 years: 1963 to 1967), but there seems to be no doubt that a combination of increased hardiness and height growth is possible to achieve. The silvicultural result of a transfer, also with regard to growth, is determined by EP locality, vigor of the local population, and origin and vigor of the transferred populations.

The transfer of some long-season populations has given a reasonable survival and, in addition, superior growth properties. On the whole, however, in spite of the exceptions found, a warning should be issued against fast-growing southern long-season populations for use in the northern Swedish forest inland with its severe winters, fluctuating spring temperatures and repeatedly melting and freezing spring snow. Such climatic conditions, more or less commonly, lead to a high amount of cambium damage, stem girdling, and frost canker, especially at the stem base.

Increased spacing has become an important feature in Swedish reforestation. High labor costs favor wide plant spacing, with subsequent few clearings and little thinning up to a relatively advanced stand age. A prerequisite for the planting success is the occurrence of hardy, disease-resistant and fast-growing materials. Until plant breeding can furnish all necessary improved seed and plant material, transfer of populations from one locality to another will be an important tool in reforestation.

Rules for such transfers have been worked out by many authors. The suitable Swedish ones were originally formulated by Langlet (1945, 1957; see also 1968). He has proposed the term "silvicultural product," e.g., the product of the percent of healthy surviving plants and their mean height at a certain age. Previously the terms "silvicultural result" (Eneroth, 1926) and "silvicultural quotient" (Stefansson and Sinko, 1967) had been used. Langlet applied the data published by Eiche (1966) in order to illustrate the use and significance of the term "silvicultural product." The discussion of these terms falls outside the scope of this chapter. But features like hardiness, disease resistance, productivity, and quality form the basic interests in practice and consequently also in plant breeding. For the extremely harsh conditions, near the tree limit, it may be almost impossible to find even local populations sufficiently hardy in certain year cycles. Here transfer of populations is largely excluded and breeding procedures have to rely on the biotype variability within the hardy populations themselves. The search for hardy populations and biotypes is an urgent future task.

An important problem in heterosis breeding is the degree of self-incompatibility of the selected plus trees. This problem is equally important in the collection of natural stand seed. Some cases of conspicuous mortality and poor growth properties, also of local populations, may depend on increased homozygosity for semilethal or viability-decreasing genes. Outcrossing certainly becomes more precarious when seed is collected from widely spaced seed trees that are remnants of former large populations.

Partial self-compatibility, and related crossing, may further lead to the formation of "isolates"—populations showing distinct differences in morphological and physiological characteristics. It is true, in our experience, that pine populations growing near the tree limit show a wide differentiation in stem and crown properties (cf. Holmgren, 1937, 1939), often with peculiarities or abnormalities in progeny tests. This may result from an effective isolation and homozygotization process, influenced by differences in time and type of flowering, pollen spreading, as well as degree of self-compatibility. The number of seed trees per hectare, their mutual spacing, seed formation, and seed development under harsh conditions (Gustafsson and Simak, 1956) influence the extent of self- versus cross-pollination. In our opinion, these important problems have neither been sufficiently considered in practice, nor in population and progeny analysis.

The conclusions of this study can be formulated in two opposite ways: *a positive one* implying that transfer of populations may increase hardiness and productivity, even in exposed areas (in fact, certain populations, for instance, nos. 6, 10, and 23 of this study) are surprisingly hardy over wide distances; and *a negative one* implying that without sufficient caution the transfer of populations may lead to complete failure. Forestry practice has to watch these alternatives continuously. Intensive studies of border populations, combined with analysis of selfing possibility and frequency in natural stands, will give valuable information on reforestation procedures in harsh regions. Population-milieu interactions ("genecology," Turesson, 1923) are of fundamental importance in forestry theory and practice.

References

BJÖRKMAN, E. 1948. Studier över snöskyttesvampens (*Phacidium infestans* Karst.) biologi samt metoder för snöskyttets bekämpande. (Studies on the biology of the *Phacidium*-blight (*Phacidium infestans* Karst.) and its prevention.) Medd. Statens Skogsforskn. Inst., 37:1-136.
——— 1963. Resistance to snow blight (*Phacidium infestans* Karst.) in different populations of *Pinus silvestris* L. Stud. For. Suec., 5:1-16.
EICHE, V. 1962. Nya aspekter å plantavgång och bristande vinterhärdighet i norrländska tallkulturer. (New aspects on plant mortality and inferior winter hardiness in North-Swedish pine plantations.) Skogen, 49:423-426, 432.

—— 1966. Cold damage and plant mortality in experimental provenance plantations with Scots pine in Northen Sweden. Stud. For. Suec., 36:1-219.
ENEROTH, O. 1926. Studier över risken vid användning av tallfrö av för orten främmande proveniens. (A study on the risks of using in a particular district pineseed from other sources.) Medd. Statens Skogsförsöksanstalt, 23:1-62.
GUSTAFSSON, Å. 1962. Genetik och växtförädling i skogsbrukets tjänst. (Genetics and plant breeding as applied to Swedish forestry.) Sv. Skogvårdsför. Tidskr., 60:111-150.
GUSTAFFSON, Å., and SIMAK, M. 1956. X-ray diagnostics and seed quality in forestry. Int. Union Forest Res. Org., 12th Congress, Oxford. Vol. 1: 398-413.
HOLMGREN, A. 1937. Något om krontypen hos ett tallbestånd i Fulufjäll. (Note on the crown type of a pine population in Fulufjäll.) Norrl. Skogsvårdsförb. Tidskr., 1937: 225-239.
—— 1939. Ett ärftlighetsförsök med tall. I-II. (A genetical experiment in Scots pine.) Norrl. Skogsvårdsförb. Tidskr., 1939: 95-151, 331-362.
LANGLET, A. 1936. Studier över tallens fysiologiska variabilitet och dess samband med klimatet. (Studien über die physiologische Variabilität der Kiefer und deren Zusammenhang mit dem Klima.) Medd. Statens Skogsförsöksanstalt, 29:219-470.
—— 1945. Om möjligheterna att skogsodla med gran- och tallfrö av ortsfrämmande proveniens. (On the possibilities of reforestation using spruce and pine seed of foreign descendance.) Sv. Skogsvårdsför. Tidskr., 43:68-78.
—— 1957. Vidgade gränser för förflyttning av tallprovenienser till skogsodlingsplatser i norra Sverige. (Wider limits for transfer of pine populations to reforestation localities in Northern Sweden.) Skogen, 44:319.
—— 1968. Om klimatiskt betingade gränser för förflyttning av tallprovenienser till skogsodlingsplatser i Norrland. (On climatic limits for transfer of Scots pine provenances for silviculture in Northern Sweden.) Sv. Skogsvårdsförb. Tidskr., 1968 (232.12): 503-532.
SAKAI, A. 1968. Frost damage on basal stems in young trees. Low Temp. Sci. Contrib., Ser. B, No. 15:1-14.
STEFANSSON, E., and SINKO, M. 1967. Försök med tallprovenienser med särskild hänsyn till norrländska höjdlägen. (Experiments with provenances of Scots pine with special regard to high-lying forests in Northern Sweden.) Stud. For. Suec., 47:1-108.
TURESSON, G. 1923. The scope and import of genecology. Hereditas, 4: 171-176.

8

Heterozygosity and Genetic Polymorphism in Parthenogenetic Animals[1]

M. J. D. WHITE

Department of Genetics, University of Melbourne, Australia

Introduction 237
Nature of Parthenogenesis 238
Parthenogenesis in Vertebrates 241
    Parthenogenesis in Poeciliid Fishes 241
    The Case of the Parthenogenetic Ambystomas 243
    Parthenogenesis in Lizards 244
Parthenogenesis in Insects 248
    Parthenogenesis in Beetles 248
    Parthenogenesis in Dipterous Flies 250
    The Case of *Moraba virgo* 253
    Parthenogenesis in an Embiid 255
Conclusions 256
References 259

Introduction

The name of Th. Dobzhansky will always be associated with the concepts of heterosis and adaptive genetic polymorphisms in natural populations

---

[1] Some of the investigations reported here have been supported by Public Health Service Grant No. GM-07212 from the Division of General Medical Sciences, U.S. National Institutes of Health, and by a grant from the Australian Research Grants Committee.

of sexually reproducing organisms. But the relationship between these concepts is still in part controversial and even after approximately 35 years of serious study of these phenomena in a variety of organisms, both in nature and in the laboratory, it is by no means clear how far the adaptive significance of most polymorphisms is to be ascribed to "pure heterosis" (i.e., superiority of the heterozygote in all environments) and how far to "annidation" (in the terminology of Ludwig, 1950)—i.e., an adaptive correspondence between the various genotypes present in the population and the alternative ecological niches present in the environment.

It has consequently seemed worthwhile, in a volume honoring Professor Dobzhansky, to examine the situation with regard to heterozygosity and genetic polymorphism in organisms whose reproduction is nonsexual (thelytokous or parthenogenetic). No attempt will be made to review all cases of parthenogenesis in animals, and the details of the cytological mechanisms will not be described. In general, the discussion will be limited to those instances in which detailed investigations (cytogenetic, biochemical, or studies on graft compatibility) have been carried out, which will enable us to draw some definite conclusions regarding the mode of adaptation of the system.

Nature of Parthenogenesis

Genetic systems in which reproduction is entirely parthenogenetic, i.e., in which populations are all-female or effectively so (rare nonfunctional males may sometimes be present) have traditionally been divided into *automictic* (or *meiotic*) and *apomictic* (or *ameiotic*). In the former type meiosis occurs, but a compensatory doubling of the chromosome number takes place at some other stage in the life cycle; in the apomictic type of system meiosis has been suppressed altogether, and there is no compensatory doubling process since there has been no reduction.

It will be evident that in neither type of parthenogenesis is there any recombination of genes present in different individuals. But in most types of automixis there is at least the theoretical possibility of genetic segregation, provided that the parent was heterozygous.

There has been a good deal of rather uninformed speculation as to the evolutionary potentialities of parthenogenetic systems. Opinions have ranged from the view that the genotypes of such organisms are virtually "frozen" and incapable of adaptive change, to the equally extreme standpoint according to which they are capable of a great deal of essentially chaotic genetic change. These divergent opinions find their reflection in the taxonomic treatment of parthenogenetic forms, some authors assigning species status to them, while others speak of "agamic complexes"—including a large number of biotypes differing to almost every conceivable degree and presenting a formidable and perhaps insoluble problem as far as nomenclature is concerned.

The distinction between automictic and apomictic systems, although based on a fundamental difference in the cytological mechanism, is not particularly meaningful in genetic terms. A more significant classification of parthenogenetic systems would be based on the distinction between mechanisms that favor heterozygosity and those which enforce homozygosity. "Heterozygosity," in this sense, includes both allelic differences at single loci and structural heterozygosity due to inversions, translocations and other types of chromosomal rearrangements. Mechanisms of parthenogenesis that favor heterozygosity generally (but not always) permit polyploidy, and a significant number of animal populations that reproduce parthenogenetically consist of triploids and a few are tetraploid or pentaploid.

The only mechanism of parthenogenesis that rigidly enforces complete homozygosity is the one in which we have meiosis leading to a haploid egg nucleus, with restoration of diploidy by a fusion of embryonic cleavage nuclei in pairs. This system seems to be a very rare one, almost all the known examples being in scale insects and white flies (order Homoptera), although the parthenogenesis of the mite *Cheyletus eruditus* has been said to fall in this category by Peacock and Weidmann (1961). Excellent examples of this type of mechanism have been described in the scale insect *Gueriniella serratulae* by Hughes-Schrader and Tremblay (1965) and in the soft scale *Pulvinaria hydrangeae* by Nur (1963). There are a considerable number of parthenogenetic forms in the scale insects (apparently all diploid) and Brown (1965) states that in the armored scales (family Diaspididae) the usual mechanism is apomixis, while among the Pseudococcidae most parthenogenetic systems are meiotic.

Obviously there can be no heterosis or adaptive polymorphism in those cases where the mechanism of egg maturation and subsequent development enforce homozygosity. The main significance of parthenogenesis in the cases referred to above is almost certainly that it enables the population to get along without the very fragile, short-lived and nonfeeding adult males. Some cases of parthenogenesis in moths (*Solenobia* spp.) which were formerly believed to depend on fusion of the cleavage nuclei in pairs are now interpreted differently.

Apart from the above instances of homozygosity-enforcing parthenogenesis, the great majority of cases of thelytoky, both in animals and plants, seem to involve considerable levels of heterozygosity and must be interpreted as genetic systems that exploit the advantages of heterosis and adaptive polymorphism. This conclusion does not seem to have been fully appreciated until now. However, heterozygosity in parthenogenetic organisms may have arisen in three different ways: it may reflect a hybrid origin by interspecific or interracial crossing, it may have been originally an adaptive polymorphism in a bisexual population, or it may have arisen by mutation since the adoption of unisexual reproduction.

Theoretically, triploidy may occur in association with parthenogenesis in a number of ways. One might imagine that in some instances it might arise simultaneously with the origin of parthenogenetic reproduction, through fertilization of an unreduced, diploid egg by a normal sperm. Such an event would only be likely to lead to a triploid parthenogenetic biotype if it took place in a species that already had a considerable potentiality for parthenogenetic development. Secondly, one could imagine autotriploidy arising in an automictic parthenogenetic biotype through fusion of three of the products of meiosis. Thirdly, there is the possibility of fertilization of a normally parthenogenetic female by a male of a related bisexual form. The first mechanism would lead to autotriploidy, the second either to autotriploidy or allotriploidy (if the diploid parthenogenetic form was of hybrid origin), and the third to allotriploidy (genotype AAB or ABC, combining haploid sets of two or three ancestral bisexual forms). It seems probable that the majority of triploid parthenogenetic taxa have arisen by the third mechanism.

Tetraploid parthenogenetic forms may presumably be either autotetraploid or allotetraploid. The apomictic tetraploid long-horned grasshopper *Saga pedo* (Matthey, 1941, 1946) from southern Europe may well be an autotetraploid; but its chromosome number ($4n = 68$) is too high for a detailed study of its karyotype. It is apparently smaller than related bisexual forms which have $2n \, \delta = 31$ and $2n \, \delta = 33$; its cells are stated to be the same size as those of its bisexual relatives. The few tetraploid parthenogenetic weevils that are known have numerous triploid relatives and may plausibly be interpreted as having arisen from apomictic triploids "accidentally" fertilized by males of a bisexual species.

The majority of organisms with meiotic thelytoky depend on a restoration of the original chromosome number by a fusion of two of the nuclei resulting from meiosis in pairs. There are endless variation of detail, which have been described by Narbel-Hofstetter (1964). In some instances anaphase separation may be more or less suppressed after either the first or the second meiotic division; alternatively four nuclei may be formed, and two of these subsequently come together and fuse. It will be convenient to consider the genetic consequences of two systems which we may designate as (1) fusion of second division sister nuclei and (2) fusion of second division nonsister nuclei. We may suppose that each chromosome arm regularly forms a single chiasma, as will usually be the case. In the first case heterozygosity may be maintained in respect of genetic loci distal to the chiasma. In the second case heterozygosity for proximal regions (i.e., those between the centromere and the most proximal location ever occupied by the chiasma) may be perpetuated from generation to generation.

Mechanisms of thelytoky that depend on a premeiotic doubling of the chromosome number in the oocyte, followed by a normal meiosis, may conserve heterozygosity provided that synapsis occurs only between sister chromosomes (i.e., ones that are exact molecular copies of one another). This is apparently what happens in the grasshopper *Moraba virgo* (see p. 253 ff.). To what extent this restriction on synapsis operates in other parthenogenetic forms with a premeiotic doubling mechanism is not known. Premeiotic doubling of the chromosome number does, however, permit coexistence of triploidy (or higher grades of polyploidy) with a regular meiosis in which only bivalents are formed. This is what happens in a number of species of earthworms (Omodeo, 1952, 1955) which are most probably of hybrid origin.

A relatively large number of cases of parthenogenesis are now known to involve gynogenesis (pseudogamy). In such systems the egg is stimulated to develop by the penetration of a sperm, which does not function in a genetic sense, i.e., it does not contribute any chromosomes to the developing egg. The sperm may come from a male of a related bisexual form (race or species)—in which case we have a situation of "reproductive parasitism," the thelytokous form being strictly dependent on the bisexual one as far as reproduction is concerned, and being unable to exist outside the geographic range of the latter. In certain groups such as oligochaetes, however, which are basically hermaphroditic, we may have thelytokous individuals that possess both ovaries and testes, and in such cases the sperm that stimulates the egg to development may come from the same individual or from one belonging to the same biotype.

Parthenogenesis in Vertebrates

Parthenogenesis in Poeciliid Fishes

Poecilia (formerly *Mollienisia*) *formosa* is an apparently diploid all-female species of fish that is widespread in southern Texas and northern Mexico (Hubbs and Hubbs, 1932). It is gynogenetic, the eggs being stimulated to develop by sperms derived from one of several closely related bisexual species. Electrophoretic serum protein analysis suggests that *P. formosa* arose by hybridization between *P. latipinna* and *P. mexicana*, since it has two serum albumin bands that correspond to the single bands present in those species (Abramoff et al., 1968); alternatively, it is certainly possible that it is of nonhybrid origin but has become a permanent heterozygote at this locus by mutation.

It has been supposed, but never unequivocally demonstrated, that *P.*

formosa is apomictic. Natural populations include several clones that differ genetically, as shown by their mutual histoincompatibility (Kallman, 1962a, b; Darnell, Lamb, and Abramoff, 1967). This is presumably due to divergent evolution of these clones as a result of mutation. If so, we must imagine them as inhabiting slightly different ecological niches. However, the mechanism of gynogenesis occasionally breaks down in *M. formosa,* i.e., the sperm nucleus is occasionally retained in the egg, leading to the production of true hybrids exhibiting some paternal characters (Haskins et al., 1960; Rasch et al., 1965). Apparently these hybrids are triploids, with two chromosome sets of maternal origin and one derived from the sperm. No triploid clones of *P. formosa* are known to occur in nature, but it is probable that occasional triploid hybrids are produced in nature, since they occur spontaneously, although infrequently, in the laboratory. It is thus conceivable that the various histocompatibility clones owe their origin to occasional transfers of paternal genetic material (perhaps single chromosomes or portions of chromosomes) to the karyotype of *P. formosa.*

It was claimed by Kallman (1964) that *P. formosa* must be homozygous for its incompatibility loci, since grafts from normal females into hybrids between *formosa* and *sphenops* are not rejected (thereby proving that all the tissue antigens of *formosa* are present in the hybrid). But this argument rests on the supposition that the hybrids have only one haploid set derived from *formosa,* and collapses if they have two. In view of the more recent work of Abramoff et al. (1968), cited above, it seems much more probable that, whatever the precise mechanism of parthenogenesis (apomictic or automictic), *P. formosa* is a stabilized hybrid and hence extensively heterozygous. The heterozygosity due to its hybrid ancestry may be assumed to be heterotic, while the mutations superimposed on this (some of which are responsible for the histocompatibility clones) have presumably proved adaptive in permitting a slight amount of annidation (if this were not so, only one clone would be bound to survive—at the expense of all the others—in each stream or river system). *P. formosa,* like other pseudogamous forms, remains a "reproductive parasite" dependent for its survival on geographic coexistence with related bisexual forms such as *P. latipinna* and members of the *sphenops* group, which includes *P. mexicana.*

Gynogenetic parthenogenesis seems to occur also in the triploid "*Cy*" biotype of *Poeciliopsis* in the rivers of Sinaloa, Mexico (Schultz, 1967). There are also two all-female strains designated "*Cx*" and "*Cz,*" but these are not parthenogenetic, since they depend on mating with males of the bisexual species *P. lucida* and the F_1 exhibit characters of both parents. However, these F_1 individuals apparently transmit only the maternal characters (Schultz, 1961, 1966). Laboratory crosses between *Cx* and the less closely related species *P. latidens* gave rise to progeny of both sexes; when females of *Cz* were mated with *latidens* they gave all-male progeny! Apparently, the "strength" of the sex chromosomes is different in these species. Although *Cx* and *Cz* cannot be regarded as parthenogenetic forms they may help us to understand how gynogenetic parthenogenesis has arisen in some instances. And the rejection of the male chromosomes in the oogenesis (if it has been correctly interpreted, and whatever the precise mechanism) may depend on a nuclear-cytoplasmic incompatibility of the same type as that which is responsible for the rejection of the male pronucleus in true gynogenesis such as exists in *Poecilia formosa*. Clearly, individuals of strains *Cx* and *Cz* may exhibit heterosis; there is presumably no crossing-over in their oogenesis, but this is not certain. An additional all-female strain of *Poeciliopsis* referred to by Schultz as *Fx* seems to have the same relation to an undescribed bisexual species *"F"* as *Cx* and *Cz* have to *P. lucida*. The unique mode of reproduction in strains *Cx, Cz,* and *Fx,* in which the maternal genome is transmitted unchanged from generation to generation, with the male genome simply "on loan" for each generation, may be regarded as a highly unique mechanism, analogous in certain respects to parthenogenesis, which ensures that all individuals have a highly heterozygous constitution.

The Case of the Parthenogenetic Ambystomas

Certain North American salamanders belonging to the *Ambystoma jeffersonianum* complex were shown by Uzzell (1963, 1964) to be parthenogenetic ($3n = 42$) triploids. Macgregor and Uzzell (1964) found that 42 lampbrush bivalents with chiasmata were present in the oocyte nuclei. There is consequently a premeiotic doubling of the chromosome number ($42 \rightarrow 84$), as in *Moraba virgo,* followed by synapsis. There are apparently about twice as many chiasmata per bivalent in the triploids as there are in diploid, bisexual members of the *jeffersonianum* group. Quadrivalents were seen occasionally in the lampbrush nuclei, proving that synapsis is at any rate not rigorously restricted to sister chromosomes derived from the doubling process, as it is in the grasshopper *Moraba virgo* (see p. 253 ff.).

Reproduction in these triploid parthenogenetic salamanders is gynogenetic, i.e., the eggs are stimulated to development by penetration of a sperm derived from a male of one of the bisexual species, *A. jeffersonianum* and *A. laterale*. But the role of the sperm is nongenetic, i.e., it does not contribute any chromosomes to the individual.

It has been shown by Uzzell and Goldblatt (1967) that there are two parthenogenetic "species" in this group, referred to as *A. tremblayi* and *A. platineum*. It was demonstrated by serum protein electrophoresis that both are triploids of hybrid origin. If a haploid set of *A. jeffersonianum* chromosomes is represented by *J* and one of *A. laterale* chromosomes by *L,* the constitution of *tremblayi* is *JLL* and that of *platineum* is *JJL.*

Uzzell and Goldblatt point out that four features of the genetic system of these triploids must have arisen in evolution: (1) hybrid origin, (2) triploidy, (3) the premeiotic doubling of the chromosome number, (4) rejection of the sperm nucleus. The case is considerably more complicated than that of *M. virgo,* in which only the premeiotic doubling mechanism needed to evolve. It seems likely that at some time in the past, diploid parthenogenetic biotypes of the constitution *JL* existed and that abnormal retention of a *laterale* sperm nucleus in a *JL* egg gave rise to *tremblayi,* while retention of a *jeffersonianum* sperm led to the origin of *platineum.* Since *platineum* has a disjunct distribution (New England and the Indiana-Ohio-Michigan region), there is a distinct possibility that it arose by two separate hybridizations. But no stabilized diploid hybrid biotypes are known to exist in this group, and the sequence of events postulated above must remain hypothetical until more evidence is available. In any event, however, the parthenogenetic biotypes of the *A. jeffersonianum* complex seem capable of competing with their diploid bisexual relatives, on which they are reproductively parasitic.

Parthenogenesis in Lizards

It has been known since the work of Minton (1958), Tinkle (1959), Maslin (1962), and McCoy and Maslin (1962), that a number of taxa belonging to the lizard genus *Cnemidophorus* reproduce parthenogenetically in Colorado, New Mexico, Arizona, and Mexico. Pennock (1965) discovered that one of these was a triploid. Lowe and Wright (1966b) list eight parthenogenetic "species"; one of these is a diploid, six are triploids, and one includes both diploid and triploid forms.

The bisexual species of the genus almost all have $n=23$. In spite of this relative uniformity of chromosome number (only one species is known to show the deviant number $n=26$), there is a considerable variety of karyotypes among the bisexual species and it is fairly clear that numerous chromosomal rearrangements have occurred. Lowe and Wright divide the chromosomes into three groups, a group I composed of large metacentrics, a group II of medium-sized acrocentrics or subacrocentrics, and a group III of small acrocentrics (microchromosomes). Certain species such as *C. tigris* are stated to have three chromosomes in group I, eight in group II, and twelve in group III, while others such as *C. inornatus* and *C. septemvittatus* have one in group I, twelve in group II, and ten in group III. The diploid parthenogenetic form *C. neomexicanus* appears to have a karyotype which includes one haploid set derived from *C. tigris* and one derived from

C. inornatus; it is interpreted as a stabilized hybrid between these two bisexual species. Occasional backcrossing of the diploid *neomexicanus* to the bisexual *inornatus* has led to the formation of a triploid parthenogen "*C. perplexus*" (Lowe and Wright, 1966a).

It is uncertain whether the method of parthenogenesis in these lizards is apomictic or automictic. The prevalence of triploid forms certainly suggests an apomictic mechanism, unless there is a premeiotic doubling of the chromosome number with restriction of synapsis to sister chromosomes.

Zweifel (1965) published a careful morphological study of the parthenogenetic "*Cnemidophorus tesselatus*." He distinguished six different pattern classes (*A* to *F*). In general these are allopatric: *E* occupies a large area of west Texas, New Mexico, and Chihuahua; *C* an area of the Texas panhandle, southern Colorado, and northern New Mexico; *A* and *B* are confined to small areas of southeastern Colorado; and *F* is known only from a restricted area of southwestern New Mexico. Form *D* is sympatric with *C* at one locality in southeastern Colorado and at another in northeastern New Mexico. The differences between these color forms seem to be distinct, and intermediates have not been reported. The variability of the "species" as a whole is comparable with that of a diploid species of *Cnemidophorus,* but the individual color forms are very uniform in phenotype.

Wright and Lowe (1967a, b) report that forms *C* to *F* are diploid ($2n = 46$). They are interpreted as having one haploid set derived from the bisexual species *C. tigris* and one derived from *C. septemvittatus.* Heterozygosity for additional rearrangements seems to occur in some populations of *C* and *D.* Forms *A* and *B* are triploids with an additional chromosome ($3n+1$); they are regarded as having arisen by hybridization between a parthenogenetic diploid hybrid and the bisexual *C. sexlineatus;* at what stage the additional chromosome became incorporated into the karyotype is not clear. As we might expect, skin grafts from the triploid form *A* are rejected by the diploid form *E;* but form *A* will accept grafts from form *E* (Maslin, 1967)—suggesting that its karyotype includes two haploid genomes essentially similar to those present in form *E.*

The geographic variation in *C. tesselatus* seems much more compatible with a multiple origin of forms *C, D, E, F* (by four hybridizations, each followed by a switch to parthenogenetic reproduction) than with a single hybridization and acquisition of parthenogenetic reproduction followed by divergent gene mutation. Similarly (if we accept Wright and Lowe's interpretation) the allotriploid forms *A* and *B* would have arisen on two separate occasions by hybridization between *C* or *D* (probably the former) and a male of *sexlineatus.* If gene mutation were responsible for the various diploid phenotypes, we would certainly expect a more chaotic type of variation basically similar to that observed in the bisexual species of the genus.

We must consequently imagine that, in the past, hybridization between *tigris* and *septemvittatus* was occurring from time to time and that on four different occasions a switch to parthenogenetic reproduction occurred in a hybrid. This "switch" may have been due to a newly arisen mutation or to the bringing together in the hybrid of genetic factors which, in combination, permitted parthenogenesis. In either case parthenogenesis fixed a hybrid genetic constitution that was presumably adaptive. The differences between forms C to F must be basically due to the fact that they originated from different individuals of the bisexual species; however, some phenotypic differences may be due to the occurrence and fixation of dominant mutations subsequent to the origin of parthenogenesis.

Parthenogenetic biotypes of *Cnemidophorus* occur in Yucatán (McCoy and Maslin, 1962; Maslin, 1968) as well as in the southwest of the U.S., but the karyotypes of the Yucatecan forms have not been described. Just what features of the genetic system of the genus *Cnemidophorus* have favored the repeated independent origin of parthenogenesis associated with hybridization is not clear. That the numerous parthenogenetic systems that have been developed in this genus exploit heterosis can hardly be doubted.

A considerable amount of sympatry exists in many areas between thelytokous "species" of *Cnemidophorus* and between these and some of their bisexual relatives. Milstead (1957) investigated the situation in the Sierra Vieja mountains of western Texas. Here the bisexual *C. inornatus* coexists with the parthenogenetic *C. sacki* (=*exsanguis*) and *C. tesselatus* (form E). In "plains" habitats the relative numbers of the three forms were as follows: *inornatus* 312 individuals, *sacki* 15, *tesselatus* 55. But in "roughland" habitats *inornatus* was not present and the ratio of *sacki* to *tesselatus* was 130:14. Zweifel (1965) states concerning one locality in New Mexico: "Of the four species of *Cnemidophorus* that occur in this immediate area, three are presumably parthenogenetic."

Maslin (1968) has expressed the view that "parthenogens from hybrid stock . . . will not be genetically adapted to any one habitat." He states that "They behave as opportunists, propagating rapidly in areas where competition is weak or absent. But where a species is well entrenched and an integral part of a community that has not been disrupted by man, the hybrid parthenogen is at a disadvantage and is excluded. Triploid parthenogens could conceivably have three different genomes thereby increasing their heterogeneous nature still further . . ." There does not seem to be any definite factual evidence for these somewhat dogmatic views. Obviously, parthenogens of hybrid origin that have been highly successful in spreading over large areas must be well adapted to one or more habitats; in fact they must usually be better adapted than either of the parental bisexual forms—otherwise they would not be able to spread in the first place. Because of their ability to respond to ecological change by alterations in the genetic composition of the population due to recombination, bisexual forms would seem far more likely to be able to colonize new, man-made or man-disturbed habitats.

Parthenogenetic reproduction also occurs in four taxa of lizards related to *Lacerta saxicola* in the Caucasus (Darevski and Kulikova, 1961, 1964). These are referred to as *armeniaca, dahli, rostombekovi,* and *unisexualis* by Darevski (1966). They are diploids, with $2n = 38$. This is a case of meiotic parthenogenesis, with restoration of diploidy by fusion of sister nuclei at the second meiotic division. As pointed out earlier, this mechanism should result in homozygosity for proximally located genes and heterozygosity for the distal loci in each chromosome. There are altogether 15 bisexual races of *L. saxicola* and it seems fairly certain that the four parthenogenetic forms have arisen in the first place as different hybrid combinations of some of these. They are broadly sympatric in a rather limited area of eastern Transcaucasia between Tbilisi and Erevan. Bisexual races of *L. saxicola* do not, in general, occur in the area occupied by the parthenogenetic forms, although zones of contact and narrow overlap do exist. About 5 to 7 percent of the eggs of the parthenogenetic females give rise to inviable monsters. Some of these were males. It is not clear whether the monsters are polyploids (i.e., result from the fusion of more than two of the products of meiosis), aneuploids, or mosaics. Possibly they, or some of them, may arise through the occurrence of cross-overs in a more distal location than usual, leading to homozygosity for a lethal allele.

In areas of sympatry, hybridization between females of the parthenogenetic forms and males of the bisexual races of *saxicola* occurs naturally from time to time, leading to the production of sterile triploid females with abnormal ovaries.

Clearly, if we consider the four parthenogenetic taxa as a whole, they have adapted to the environment which they occupy by some form of annidation, just as in the case of the fly *Lonchoptera dubia* (see p. 250). Each of them must be adapted to a different ecological niche, even though they all occur together (if it were not so, the best adapted would replace the other three in course of time). And, in general, they seem to occupy an area from which they have excluded the bisexual races. Their adaptive superiority (in a rather limited geographic area, it is true) over the bisexual forms may be due to true heterosis (depending on heterozygosity for distal loci). But forms of hybrid origin with a genetic system such as this will combine characters of both ancestral races, even if in the course of time they have become largely or completely homozygous. Thus the evolutionary success of these four parthenotaxa may be due to a combination of two genetic phenomena: (1) coexistence in one individual of some proximal chromosome regions derived from ancestral race *A* and some from race *B,* even though the proximal region of each chromosome pair is homozygous for *A* genes or for *B* genes; (2) heterozygosity for *A* and *B* alleles (and for mutations that have established themselves subsequent to the assumption of unisexual reproduction) in the distal chromosome regions. Unfortunately, we do not have any data on the frequency and distribution of chiasmata in the oocytes of these lizards.

There is some suggestion (Lantz and Cyrén, 1936; Darevsky, 1966) that in addition to the four strictly parthenogenetic taxa in this group there are some (*L. s. bithynica* and *L. s. mixta*) in which males are rare; they may be facultatively parthenogenetic. Although Darevsky has stated that he considers that the origin of parthenogenesis in this group of lizards was associated with hybridization, this suggestion is left vague and undocumented, and he apparently regards *L. unisexualis* as a direct descendant of *L. s. defilippii*. One thing seems clear—the four parthenogenetic taxa are quite distinct from one another and are not part of a chaotic assemblage of unisexual clones, such as was postulated by some earlier authors in other cases. Each clearly represents an independent origin of parthenogenesis, whatever the precise conditions under which this occurred (with or without hybridization). As in the case of the *Cnemidophorus* species studied by Zweifel (1965), the coefficients of variation for metrical characters are notably lower in the parthenogenetic forms than they are in the related bisexual races of *L. saxicola*.

There are at least two other all-female species of lizards—the teiid *Gymnophthalmus underwoodi* on the islands of Trinidad, Barbados, and St. Vincent (Thomas, 1965) and the gecko *Hemidactylus garneti* (Smith, 1935). Nothing is known of the cytogenetics of these, however.

Parthenogenesis in Insects

Parthenogenesis in Beetles

A fairly large number of parthenogenetic taxa are known in beetles, especially in the weevils (family Curculionidae). The extensive cytological studies of Suomalainen (1945, 1947, 1953, 1961, 1962) on European weevils, supplemented by those of Seiler (1947), and Mikulska (1953, 1960) and the work of Takenouchi (1966) on Japanese species provide a general picture of the situation. These insects are apparently all apomictic and virtually all of them are polyploids. They belong to the genera *Otiorrhynchus, Peritelus, Trachyphlorus, Polydrusus, Eusomus, Sciaphilus, Strophosomus, Barynotus, Tropidophorus, Liophloeus, Catapionus, Listroderes, Pseudocneothinus, Scepticus,* and *Tropiphorus* and have been assigned to species and "races" within those genera. The "races" are really grades of polyploidy, and there is no certainty, for example, that a triploid biotype in Finland has a common origin (either in the sense of bisexuality

→ thelytoky or diploidy → triploidy) with a similar but biometrically different form in Switzerland. The frequency of the various grades of polyploidy in these parthenogenetic "races" seems to be as follows:

$2n$ — 1
$3n$ — 23
$4n$ — 9
$5n$ — 4

37

There are examples of "species" including a diploid and a triploid biotype, a triploid and a tetraploid one, a triploid and a pentaploid one, and a tetraploid and a pentaploid one.

The basic chromosome number of all these forms is apparently $n = 11$. There is only a single mitotic maturation division in the egg. But this is apparently peculiar in that the spindle is compound, there being a number of contiguous metaphase plates, each carrying a haploid or a diploid set of univalent chromosomes. It is possible that this peculiar nuclear behavior is due to, and indicative of, a hybrid constitution. Polyploidy in these insects may be due to occasional fertilization of thelytokous females by males of related bisexual forms (adding a haploid genome to the previously existing karyotype). But it is also possible that the peculiar compound metaphase plates in the oocyte have permitted a form of autopolyploidy to develop.

Suomalainen (1961) has carried out some biometrical studies on the European forms *Otiorrhynchus scaber, O. chrysocomus, O. salicis, O. singularis*, and in *Peritelus hirticornis*. Since in each of these statistically significant differences in morphometric characters were found between populations collected at different localities, he concluded that "evolution does not come to a complete standstill in polyploid parthenogenetic populations" and that the observed biometric differences are due "to differences in the gene pools of the populations" (presumably resulting from accumulation of different mutations).

There are several criticisms that can be made of this work. In the first place, the analyses of *O. scaber* and *P. hirticornis* included both triploid and tetraploid biotypes and that of *O. salicis* included diploid and triploid forms. That of *O. singularis* is even more unsatisfactory in that in addition to six triploid parthenogenetic samples a sample of a related diploid bisexual biotype was included. Since all the material was collected in the wild, it is quite unknown what contribution environmental factors may have made to the observed biometrical differences. It also seems quite possible that many of the polyploid biotypes are polyphyletic, the polyploid constitution having originated several times, either from occasional hybridizations with bisexual species or by rare acts of autopolyploidy in developing eggs. The rarity of diploidy in these parthenogenetic weevils (only one diploid thelytokous form known) indicates fairly strongly that heterozygosity (whether resulting from actual hybridity or not) is the basis for their evolutionary success. But more critical studies of these weevils, involving biochemical investigations, DNA-value determinations, and detailed karyotype analysis, are necessary before the full meaning of their genetic system is made clear.

Thelytokous reproduction is known in a few beetles belonging to families other than the Curculionidae. An interesting case is that of the triploid spider-beetle *Ptinus mobilis* (Moore et al., 1956; Sanderson and Jacob, 1957; Sanderson, 1960). Here the oocyte nucleus undergoes a doubling of the chromosome number prior to meiosis ($3n = 27 \rightarrow 6n = 54$) and synapsis then follows, producing 27 bivalents (called "pseudobivalents" by Sanderson, for no apparent reason). There is only a single maturation division in the egg, but this is not completed until the egg is penetrated by a sperm from a male of the related diploid bisexual species *P. clavipes* ($2n = 18$, XY in the male). There is no fertilization in the genetic sense, however. It seems probable, although unproven, that the triploid state of *P. mobilis* is due to a hybrid origin.

Parthenogenesis in Dipterous Flies

Diptera reproducing exclusively by parthenogenesis include the diploids *Drosophila mangabeirai* (Carson et al., 1957; Carson, 1962, 1967) and *Lonchoptera dubia* (Stalker, 1956) and a number of triploid chironomids (Scholl, 1956) and simuliids. *D. mangabeirai* apparently consists of a single biotype, invariably heterozygous for the same three inversions. This is a case of automictic parthenogenesis in which restoration of the diploid number is by fusion of nonsister nuclei after the second meiotic division, thereby preserving the structural heterozygosity of the proximal segments.

Hatchability of the eggs in *D. mangabeirai* is only about 60 percent (Murdy and Carson, 1959); the remaining 40 percent that do not hatch may include a number that are homozygous for proximal regions as a result of crossing-over or as a result of fusion of second-division sister nuclei instead of nonsisters. Three males of *D. mangabeirai* were caught in the wild, but none have been obtained in the laboratory; these three may have been XO. In spite of its constant karyotype, *D. mangabeirai* occupies a considerable range of habitats in Central America, from sea level at least up to 4,000 feet. Carson (1962) suggests that the karyotype "confers homeostatic properties on its carriers such that they exploit these varied environments by means of a general adaptability."

The North American *Lonchoptera dubia* consists of four clones or biotypes, which differ in their karyotypes, as revealed in the polytene chromosomes of the ovarian nurse-cell nuclei. There are four chromosome arms in these nuclei, each consisting of two synapsed homologs. Arm IV was not analyzed by Stalker, but is apparently always heterozygous for complex rearrangements. There are altogether four known sequences (*A*, *B*, *G*, *K*) of arm I, four (*L*, *M*, *M'*, *O*) of arm II, and three (*P*, *P'*, *Q*) of arm III. The combinations of these that occur in the four known clones are as follows:

Clone	Arm I	Arm II	Arm III
1	*A/B*	*L/L*	*P/Q*
2	*B/K*	*M/O*	*P'/Q*
3	*B/K*	*M/M'*	*P'/Q*
4	*G/G*	*M/M*	*P'/Q*

The differences between the alternative sequences of a particular arm are generally complex, e.g., *A*, *B*, *K*, and *G* all differ from one another by multiple rearrangements. *M* and *M'* differ only by a single inversion, but the differences between these and sequences *L* and *O* are complex. *P* and *Q* differ by a subterminal inversion, and *P'* differs from *P* by a short median inversion.

The reproduction of *L. dubia* is apparently a form of meiotic (automictic) parthenogenesis; Stalker has suggested that diploidy is restored by fusion of nonsister nuclei after the second meiotic division, thereby preserving heterozygosity for all chromosome regions proximal to the chiasma (it being assumed that there is only one chiasma per arm). At least 25 percent of the eggs do not hatch; these may be homozygous for chromosome segments normally heterozygous, either as a result of a chiasma occurring more proximally than usual or because fusion has occurred between sister nuclei rather than between nonsisters.

There seems no reason to believe that these strains constituting the "species" *Lonchoptera dubia* have arisen by hybridization; it is surely more probable that they arose independently from a bisexual species that was highly polymorphic for chromosomal rearrangements and that the four karyotypes represent particularly well-adapted combinations of those rearrangements. *L. dubia* is associated with swards of *Trifolium repens* and *Plantago lanceolata*, which generally occur as lawns, roadsides, and pastures; its relationship to a European parthenogenetic form that has been named *L. furcata* is uncertain.

At many localities in the United States two of the biotypes of *L. dubia* coexist, and at Rochester, N.Y., all four occur. Since the numbers of individuals studied was relatively small, it is likely that sympatry is even more extensive than is revealed by the data. We may be certain, therefore, that there are physiological differences between the four biotypes, with differential adaptation to specific ecological niches in the general environment. Were it not so, we would expect that one of them would eventually displace the others, at each locality, as a result of the "mutual exclusion principle" (Gause, 1934; Hardin, 1960).

A much more complicated situation occurs in the triploid parthenogenetic simuliid *Cnephia mutata* in Canada (Basrur and Rothfels, 1959). This is apparently only one of a number of triploid all-female simuliids—others include *Prosimulium ursinum* in the North West Territory and Norway; *Gymnopais* sp. in the North West Territory; and *Prosimulium macropyga* in Norway. No details of the cytogenetics of these were published by Basrur and Rothfels. A further triploid species of *Prosimulium* from North Bay, Ontario, is stated to produce numbers of triploid males, whose role in the population is uncertain.

There are two "races" of *Cnephia mutata*—a diploid bisexual one and the triploid parthenogenetic biotype. The former is not polymorphic for any inversion sequences except that males are invariably heterozygous for the banding pattern of the centromere region of chromosome I, which must represent the XY sex chromosome pair.

The triploid *C. mutata* apparently never produces males and must be XXX. Individuals of this biotype exhibit extensive polymorphism for inversions. There are four different inversions in the long arm of chromosome I (the X chromosome) and two in the long arm of chromosome II. Obviously, in a triploid there are four possible genotypes in the case of each inversion: $St/St/St$, $St/St/In$, $St/In/In$ and $In/In/In$ (St is, of course, the sequence present in the bisexual form). As far as inversion IIL-1 is concerned all four of these were encountered in a sample of 105 individuals from Inglewood, Ontario, but in the case of most of the other inversions, the $In/In/In$ genotype was not found. There were altogether 15 different genotypes (i.e., combinations of inversions) detected in this sample, but there must actually have been more, since in the case of the inversions of Chromosome I the $St/In/In$ and $St/St/In$ genotypes were not distinguished. Rothfels (*in litt.*) states that as many as 700 genotypes were found in later, unpublished, studies. It will be evident that the population genetics of the triploid *Cnephia mutata* is fundamentally different from most other parthenogenetic organisms, with their great genetic uniformity.

Basrur and Rothfels have argued that this must be a case of automictic parthenogenesis, in which crossing-over and segregation lead to a multiplicity of different combinations of inversion sequences. They reject the alternative hypothesis of an apomictic system, since this would necessitate an independent evolutionary origin of each of the triploid parthenogenetic genotypes.

The particular meiotic mechanisms postulated by Basrur and Rothfels to explain this case (refusion of the products of the first meiotic division, or fusion of the central, i.e., nonsister, nuclei after two completed meiotic divisions) do not appear very probable. Both involve the assumption that the first meiotic division takes place with trivalents, or with bivalents and univalents, which does not seem very likely. It seems much more probable that this is a case where there is a premeiotic doubling of the chromosome number followed by a normal meiosis. Recombination of inversions would occur if synapsis was not restricted to sister chromosomes as it is in *Moraba virgo*; in fact synapsis between sisters could be the usual process, with pairing of nonsisters exceptional but sufficiently frequent to give rise to new combinations of inversions at a rate sufficient to account for the polymorphism observed in the natural populations. Basrur and Rothfels dismiss the possibility of a premeiotic doubling of the chromosome number because they did not observe any such process in the oogonial divisions they examined; it will be obvious that such a doubling could occur subsequent to the stages they studied but before the initiation of meiosis.

The triploid *Cnephia mutata* is unique among the parthenogenetic organisms that have been studied hitherto, in showing an extensive cytogenetic polymorphism comparable to that of such bisexual species as *Drosophila willistoni* or *D. subobscura*. The origin of parthenogenesis in this case is uncertain, but it seems likely that there was at one stage a diploid and highly heterozygous biotype that hybridized with a bisexual diploid, giving rise to the triploid. The adaptive significance of inversion polymorphism in this case (purely heterotic, heterotic-annidational, or mainly annidational) is obscure.

It will be evident that a system of parthenogenetic reproduction in which genetic recombination leads to the production of some homozygous offspring from heterozygotes will eventually lead to the population becoming entirely homozygous, unless there is a compensatory elimination of homozygotes, which would involve a considerable genetic load. This is one reason why we feel that the meiotic mechanisms postulated by Basrur and Rothfels for *C. mutata* are improbable. The same genetic load would have to be postulated for a meiotic mechanism with premeiotic doubling of the chromosome number if synapsis of homologs was random. But if synapsis of nonsister chromosomes was only a rare accident, the genetic load could be much less and *Cnephia mutata* could still be a species opportunistically exploiting the advantages of extensive inversion polymorphism with only a minimal genetic load associated.

The Case of Moraba Virgo

Moraba virgo is an Australian grasshopper, belonging to the family Eumastacidae, whose reproduction is exclusively parthenogenetic (White et al., 1963). No male has ever been seen and the mode of reproduction makes it extremely improbable that any could be produced. It is a stenoecous form, restricted to arid sandy areas in western New South Wales and northwestern Victoria, where it feeds on various species of *Acacia*, especially *A. wilhelmiana* and *A. loderi* (White, 1966). Its closest relatives occur approximately 1600 km away in Western Australia; these are bisexual species that are also *Acacia* feeders. A "putative ancestor" for *M. virgo* is the bisexual species "*P151*" (White and Webb, 1968), whose females show a very close similarity to those of *virgo*. It is not suggested that the existing species *P151* can be regarded as a direct ancestor to *virgo*, merely that the bisexual ancestor must have been a species with essentially the appearance and the karyotype shown by *P151*. It is postulated that during the climatic and ecological changes of the Pleistocene members of the *virgo* group became extinct over the 1600-km-wide territory between the Warburton Ranges in Western Australia and western New South Wales. We do not know, of course, whether the forms that died out included parthenogenetic biotypes—there is good evidence (see below) that at least one bisexual form became extinct.

M. virgo is a diploid that is heterozygous for a number of chromosomal rearrangements including a fusion between two of the small chromosomes and a pericentric inversion in the CD-chromosome. At 17 out of 19 localities at which *virgo* has been collected the "Standard" karyotype is present. At two other localities additional rearrangements have been superimposed on the Standard karyotype, thereby increasing the degree of structural heterozygosity. The chromosome number is $2n=15$ at all localities.

The mechanism of parthenogenesis in *virgo* consists of a doubling of the chromosome number (to 30) in the oocyte nucleus prior to meiosis, followed by a synapsis between sister chromosomes (which are molecular copies of one another). Fifteen chiasmate bivalents are formed and the general course of meiosis is entirely normal, except that the number of bivalents is approximately twice that in related bisexual forms; two polar bodies are given off. This mechanism ensures that all the progeny are genetically identical to one another and to their mother (except for newly arisen mutations) and preserves the heterozygosity intact from generation to generation; although formally an automictic system, it is genetically equivalent to apomixis.

The heterozygosity of *virgo* is not confined to the identifiable structural rearrangements of the karyotype. It has been shown by autoradiography of somatic mitoses that there are significant differences in DNA-replication pattern between the two members of the large *AB* pair of chromosomes and also between the members of the *CD* pair, which are also distinguishable because of the inversion previously mentioned. The late-labeling of one member of the AB pair and of the short arm of the "Standard" *CD* may be assumed to indicate a switching-off of some genetic loci in these segments of the karyotype. Thus *M. virgo,* although formally a diploid, may be functionally haploid for a number of its genetic loci, or even for whole segments.

The chromosomal rearrangements that are present in all individuals of *M. virgo* at the localities Monia Gap and Yatpool, but which do not exist in other colonies of the species, must clearly have arisen subsequent to the establishment of parthenogenetic reproduction. The inversion in the CD-chromosome and the fusion responsible for the little m_2-chromosome are heterozygous in every individual of *virgo*; they may consequently have arisen in "bisexual times." In fact it is quite possible that *M. virgo* arose from a species that was suffering from a particularly heavy genetic load (due to low selective values of homozygotes for the inversion polymorphism, the fusion polymorphism or one or more genic polymorphisms). If so, the "switch" to parthogenetic reproduction, by suddenly shedding this load, would have had a high adaptive value.

The heterozygosity for labeling pattern has presumably arisen by a long series of gene mutations, since the onset of parthenogenetic reproduction. It could hardly have been maintained in a bisexual species with free genetic crossing-over (*M. virgo* has chiasmata but since they are between chromosomes that are genetically indentical they do not lead to crossing-over).

Since *M. virgo* is highly heterozygous in several respects the question as to whether it could have arisen by hybridization (like the diploid form of *Cnemidophorus tesselatus*) must be asked. However, it seems that this question must be answered in the negative. Certainly *virgo* has two structurally identical X chromosomes which do not differ in labeling pattern (White and Webb, 1968). Both of them have a fusion (which we refer to as F10) between the original X and one of the small autosomes. This is our evidence for concluding that *virgo* arose, not directly from a form like *P151* with XO males, but from a species that possessed F10 and hence had neo-XY males. Thus if *virgo* arose by hybridization, this would necessarily have been between two different species with XY males, i.e, which both possessed F10. It is obviously much simpler to assume that no hybridization was involved in this case.

Parthenogenesis in an Embiid

The parthenogenetic "race" of the Mediterranean embiid *Haploembia solieri* is an apomictic diploid ($2n=22$), and possesses an extra pair of chromosomes by comparison with females of the bisexual form, which have $2n=20$. Stefani (1956), who has investigated this case, has claimed that the parthenogenetic females are tetrasomic for the X chromosome, but this interpretation does not seem very firmly established. A few triploid parthenogenetic females which laid triploid eggs ($3n = 33$) were found in Sardinia and on the peninsula of Argentario. It seems likely that they arise from time to time from the diploid parthenogenetic form and do not constitute a distinct and permanently reproducing biotype; they are unlikely to have arisen through hybridization with the bisexual form, which does not occur in Sardinia.

The unique interest of this case lies in the fact that Stefani (1956, 1959, 1964) has shown that populations of *Haploembia* are very subject to parasitization by a gregarine, *Diplocystis clerci*. This leads to abnormalities of the testis and male sterility and may consequently have strongly favored the "evolutionary leap" to parthenogenetic reproduction. In the Balearic Islands, in the south of France, in Liguria, and near Rome, the bisexual and parthenogenetic biotypes coexist; there is a strong positive correlation between the percentage of females at a locality that are parthenogenetic and the percentage of individuals of both sexes that are parasitized by *Diplocystis*.

Conclusions

Almost all recent work on the genetics of parthenogenetic organisms points to their high level of heterozygosity. The few exceptional examples of parthenogenetic reproduction leading to complete homozygosity are almost confined to scale insects (Coccidae) and related Homoptera of the family Aleurodidae. The heterozygosity of parthenogenetic organisms may, however, arise in several ways: (1) by direct inheritance from the genetic polymorphism of a pre-existing ancestral bisexual population, (2) as a result of hybridization between two bisexual species to give a diploid parthenogen, (3) by hybridization between a diploid parthenogen and a male of a related bisexual form to give a triploid parthenogen, (4) by mutation subsequent to the assumption of parthenogenetic reproduction. Many parthenogenetic forms will, of course, have a combination of several of these types of heterozygosity.

There have been very few biometrical or biochemical studies of populations of parthenogenetic organisms. However, the evidence seems to point to two main types of population structure. In the first, which is probably typified by *Moraba virgo* and *Drosophila mangabeirai,* there is a single karyotype and all the individuals in the population at any one locality are probably almost (but presumably not quite) identical, genetically. Heterozygosity, in this case, must be purely heterotic, and annidation to different ecological niches plays no part in the system.

In the second type of population structure, there are a number of sympatric genotypes, which may or may not possess visibly different karyotypes. Presumably, in all such cases the different genotypes are differentially adapted to specific ecological niches in the same general habitat. Typical of such systems are *Lonchoptera dubia* (four different karyotypes) and *Poecilia formosa* (genotypes distinguishable only by graft incompatibility). *Cnephia mutata* is unique in the large number of genotypes in the population, and in the fact (if indeed it is a fact) that they are being continually produced by meiotic segregation.

Apart from *Cnephia mutata,* parthenogenetic forms that include a number of distinguishable genotypes may have evolved in two ways. They may represent multiple origins of parthenogenesis from the same bisexual ancestral population or a complex of closely related bisexual populations. Alternatively, they may have arisen by disruptive selection acting on a population of parthenogenetic individuals faced with a number of alternative ecological niches or habitats. At least, these are the theoretical alternatives.

As far as animals are concerned, the idea that parthenogenetic forms exhibit essentially chaotic variation, and evolve into vast "agamic complexes," which are difficult or even impossible to treat taxonomically, does not seem to correspond to reality. Parthenogenetic taxa either consist of a single rather uniform biotype or a limited number of biotypes differing cytogenetically in quite clear fashion. In the latter case, we seem almost always to be dealing with multiple (polyphyletic) origin of parthenogenesis, rather than with the results of divergent evolution under conditions of uniparental reproduction.

Well-studied cases of parthenogenesis thus tend to support the view that adaptive evolution under conditions of parthenogenetic reproduction is limited. A parthenogenetic form may be adapted to a new ecological niche at the time of origin and may be very successful at conquering that niche. But it is unlikely to undergo much subsequent diversification adapting it to further environments. Where annidation seems to be a prominent feature of the adaptive system of a parthenogenetic taxon, we may reasonably suspect a polyphyletic origin of parthenogenesis.

The switch from bisexuality to parthenogenetic reproduction has, of course, demographic as well as genetic effects. Other things being equal, the innate rate of increase (Andrewartha and Birch, 1954; see also Dobzhansky, 1968) will be immediately doubled. But "other things" are not necessarily equal, and in certain cases parthenogenetic organisms seem to produce a rather large percentage of inviable progeny (at least 25 percent of the eggs do not hatch in *Lonchoptera dubia,* about 40 percent fail to hatch in *Drosophila mangabeirai,* and so forth). These rather large egg mortalities are presumably due to "imperfections" of the genetic system (e.g., cross-overs or incorrect fusion of postmeiotic nuclei leading to genotypes homozygous for lethal or sublethal alleles). They would presumably have been reduced if there had been natural selection for higher innate rates of increase.

TABLE I

Adaptive Advantages of Initiation of Parthenogenesis

1. Shedding a segregational genetic load in a previously genetically polymorphic species.	*Moraba virgo* *Drosophila mangabeirai* *Lonchoptera dubia*[a]
2. Taking advantage of heterosis shown in an interspecific or interracial hybrid.	
2a. Where the hybrid is diploid.	*Poecilia formosa*[a,b] Parthenogenetic forms derived from *Lacerta saxatilis*.[a] Diploid parthenogenetic forms of *Cnemidophorus*.
2b. Where the hybrid is triploid, as a result (probably) of hybridization between a diploid parthenogenetic form and a related bisexual one	"Cy" biotype of *Poeciliopsis* Numerous weevils *Ptinus mobilis*[b] Triploid parthenogenetic forms of *Cnemidophorus* *Cnephia mutata* *Ambystoma platineum and laterale*[b]
3. Getting rid of males that are biologically unsatisfactory	
3a. Because of lack of resistance to a parasite, with consequent frequent sterility	*Haploembia*
3b. Because of fragility and short life of adult males	Many scale insects

[a] In these cases, a number of sympatric parthenogenetic biotypes are apparently exploiting different ecological niches.
[b] Gynogenetic, i.e., 'reproductively parasitic' on a related bisexual form.

Information concerning the prolificity of parthenogenetic forms by comparison with their bisexual relatives is somewhat conflicting. The number of ovarioles in *Moraba virgo* (i.e., the number of eggs that can be laid in a batch) is very high for a morabine grasshopper, but has not been compared with that of its closest bisexual relative, species *P151*. Individuals of an arctic parthenogenetic simuliid, *Gymnopais* sp., lay only about 20 eggs compared with several hundreds in temperate bisexual forms (Downes, 1964).

A number of authors have claimed that parthenogenetic forms are especially characteristic of areas that were glaciated during the Quaternary, and more especially areas that became uncovered when the glaciers of the Würm age retreated. Evidence for this view has been put forward for moths of the genus *Solenobia* by Seiler (1946, 1961) for alpine weevils by Suomalainen (1953), and for lizards of the *Lacerta saxicola* group by Darevsky (1966). The implication is that the parthenogenetic biotypes were able to survive under especially cold or harsh conditions to which bisexual forms could not adapt. Darevsky specifically attributes the adaptability of parthenogenetic lizards in such habitats to their higher innate rate of increase.

It is unlikely that the true explanation of this apparently genuine correlation between the distributions of certain parthenogenetic forms and the areas glaciated during the Würm age is as simple as this. Harsh or cold conditions in themselves probably had nothing to do with the success of thelytokous genetic systems. But the fairly rapid retreat of the glaciers probably left relatively unoccupied and very uniform habitats. Under such circumstances, strongly heterotic biotypes with high general adaptation but little ability to undergo further adaptive genetic change may have got their chance. Innate rate of reproduction may have had no more than an incidental significance, or no significance at all in this connexion. Of course, a great many parthenogenetic forms live in areas that were never glaciated.

In Table I, we present a summary of the know or suspected advantages of the initiation of parthenogenesis. The demographic "advantage" has not been included here. Probably more significant than the increase in the innate rate of reproduction is the fact that every individual of a parthenogenetic form carried accidentally to a new locality, or left behind in an old locality after local extermination, can regenerate a new colony.

Those types of parthenogenesis which depend on fusion of meiotic products in the egg may have arisen gradually, through a steady increase in the frequency of such fusions. But those which depend on a cytogenetic *tour de force* such as a premeiotic doubling of the chromosome number are most unlikely to have originated in a series of stages; we must envisage the switch to parthenogenetic reproduction as having been quite sudden in these cases.

References

ABRAMOFF, P., R. M. DARNELL, and J. S. BALSANO. 1968. Electrophoretic demonstration of the hybrid origin of the gynogenetic teleost *Poecilia formosa*. Amer. Natural., 102:555-558.
ANDREWARTHA, H. G., and L. C. BIRCH. 1954. The Distribution and Abundance of Animals. Chicago. Chicago University Press. 782 pp.

BASRUR, V. R., and K. H. ROTHFELS. 1959. Triploidy in natural populations of the black fly *Cnephia mutata* (Malloch). Canad. J. Zool., 37:571-589.
BROWN, S. W. 1965. Chromosomal survey of the armored and palm scale insects (Coccoidea: Diaspididae and Phoenicococcidae). Hilgardia, 36:189-294.
CARSON, H. L. 1962. Fixed heterozygosity in a parthenogenetic species of Drosophila. University Texas Publ., 6205:55-62.
———— 1967. Permanent heterozygosity. Evol. Biol., 1:143-168.
———— M. R. WHEELER, and W. B. HEED. 1957. A parthenogenetic strain of *Drosophila mangabeirai* Malogolowkin. University Texas Publ., 5721:115-122.
DAREVSKY, I. S. 1966. Natural parthenogenesis in a polymorphic group of Caucasian rock lizards related to *Lacerta saxicola* Eversmann. J. Ohio Herpetol. Soc., 5:115-152.
———— and V. N. KULIKOVA. 1961. Natürliche Parthenogenese in der polymorphen Gruppe der kaukasischen Felseidechse (*Lacerta saxicola* Eversmann). Zool. Jahrb. Syst., 89:119-176.
———— and V. N. KULIKOVA. 1964. Natural triploidy in a polymorphic group of Caucasian rock lizards (*Lacerta saxicola* Eversmann) resulting from hybridization between bisexual and parthenogenetic forms of this species. Dokl. Akad. Nauk SSSR, 158:202-205.
DARNELL, R. M., E. LAMB, and P. ABRAMOFF. 1967. Matroclinous inheritance and clonal structure of a Mexican population of the gynogenetic fish, *Poecilia formosa*. Evolution, 21:168-173.
DOBZHANSKY, TH. 1968. On some fundamental concepts of Darwinian biology. *In* Dobzhansky, Th., Hecht, M. K., and Steere, W. C. eds. Vol. 2. New York, Appleton-Century-Crofts. pp. 1-34.
DOWNES, A. 1964. Arctic insects and their environment. Canad. Entom., 96:279-307.
GAUSE, G. F. 1934. The Struggle for Existence. Baltimore, The Williams & Wilkins Co.
HARDIN, G. 1960. The competitive exclusion principle. Science, 131:1291-1297.
HASKINS, C. P., E. F. HASKINS, and R. E. HEWITT. 1960. Pseudogamy as an evolutionary factor in the Poeciliid *Mollienisia formosa*. Evolution, 14:473-483.
HUBBS, C. L., and L. C. HUBBS. 1932. Apparent parthenogenesis in nature, in a form of fish of hybrid origin. Science, 76:628-630.
HUGHES-SCHRADER, S., and E. TREMBLAY. 1965. *Gueriniella* and the cytotaxonomy of iceryine coccids (Coccoidea: Margarodidae). Chromosoma, 19:1-13.
KALLMAN, K. D. 1962a. Gynogenesis in the teleost, *Mollienisia formosa* (Girard), with discussion of the detection of parthenogenesis in vertebrates by tissue transplantation. J. Genet., 58:7-21.
———— 1962b. Population genetics of the gynogenetic teleost, *Mollienisia formosa* (Girard). Evolution, 16:497-504.
———— 1964. Homozygosity in a gynogenetic fish. Genetics, 50:260-261.
LANTZ, L. A., and O. CYRÉN. 1936. Contribution à la connaissance de *Lacerta saxicola* Eversmann. Bull. Soc. Zool. France, 61:159-181.
LOWE, C. H., and J. W. WRIGHT. 1966a. Evolution of parthenogenetic species of *Cnemidophorus* (Whiptail lizards) in western North America. J. Arizona Acad. Sci., 4:81-87.
———— and J. W. WRIGHT. 1966b. Chromosomes and karyotypes of cnemidophorine teiid lizards. Mamm. Chrom. Newsl., 22:199-200.
LUDWIG, W. 1950. Zur Theorie der Konkurrenz. Die Annidation (Einnischung) als fünfter Evolutionsfaktor. Neue Ergebn. Probleme Zool. Kl. Festschr., pp. 516-537.
McCOY, C. J. JR., and MASLIN, T. P. 1962. A review of teiid lizard *Cnemidophorus cozumelus* and the recognition of a new race, *Cnemidophorus cozumelus rodecki*. Copeia, 1962:620-627.
MACGREGOR, H. C., and T. M. UZZELL JR. 1964. Gynogenesis in salamanders related to *Ambystoma jeffersonianum*. Science, 143:1043-1045.
MASLIN, T. P. 1962. All-female species of the lizard genus *Cnemidophorus*, Teiidae. Science, 135:212-213.

——— 1967. Skin grafting in the bisexual teiid lizard *Cnemidophorus sexlineatus* and in the unisexual *C. tesselatus*. J. Exp. Zool., 166:137-150.

——— 1968. Taxonomic problems in parthenogenetic vertebrates. System. Zool., 17:219-231.

MATTHEY, R. 1941. Étude biologique et cytologique de *Saga pedo* Pallas (Orthoptères: Tettigoniidae). Rev. Suisse Zool., 48:91-102.

——— 1946. Démonstration du caractère géographique de la parthénogénèse de *Saga pedo* Pallas et de sa polyploidïe, par comparaison avec les espèces bisexuées *S. ephippigera* Fisoh. et *S. gracilipes* Uvar. Experientia, 2:260-261.

MIKULSKA, I. 1953. The chromosomes of the parthenogenetic and thelytokian Weevil *Eusomus ovulum* Germ. (Curculionidae, Coleoptera). Bull. Acad. Polon. Sci. Classe II, 1951:293-307.

——— 1960. New data to the cytology of the parthenogenetic and thelytokian weevil *Eusomus ovulum* Germ. (Curculionidae, Coleoptera) from Poland. Cytologia, 25:322-333.

MILSTEAD, W. W. 1957. Some aspects of competition in natural populations of whiptail lizards (genus (*Cnemidophorus*). Texas J. Sci., 9:410-447.

MINTON, S. A. JR. 1958. Observations on amphibians and reptiles of the Big Bend region of Texas. Southwestern Natural., 3:28-54.

MOORE, B. P., G. E. WOODROFFE, and A. R. SANDERSON. 1956. Polymorphism and parthenogenesis in a ptinid beetle. Nature (London), 177:847-848.

MURDY, W. H., and H. L. CARSON, 1959. Parthenogenesis in *Drosophila* mangabeirai Malog. Amer. Natural, 93:355-363.

NARBEL-HOFSTETTER, M. 1964. Les altérations de la méïose chez les animaux parthénogénétiques. Protoplasmatologia, Band VI, F2. Vienna, Springer-Verlag. 163 pp.

NUR, U. 1963. Meiotic parthenogenesis and heterochromatinization in a soft scale, *Pulvinaria hydrangeae* (Coccoidea: Homoptera). Chromosoma, 14:123-139.

OMODEO, P. 1952. Cariologia dei Lumbricidae. Caryologia, 4:173-275.

——— 1955. Cariologia dei Lumbricidae. II. Contributo. Caryologia, 8:135-178.

PEACOCK, A. D., and U. WEIDMANN. 1961. Recent work on the cytology of animal parthenogenesis. [Polish, with English summary.] Przeglad Zool., 5:5-27, and 101-122.

PENNOCK, L. A. 1965. Triploidy in parthenogenetic species of the Teiid lizard genus *Cnemidophorus*. Science, 149:539-540.

RASCH, E., R. M. DARNELL, K. D. KALLMAN, and P. ABRAMOFF. 1965. Cytophotometric evidence for triploidy in hybrids of the gynogenetic fish, *Poecilia formosa*. J. Exp. Zool., 160:155-170.

SANDERSON, A. R. 1960. The cytology of a diploid bisexual spider beetle *Ptinus clavipes* Panzer and its triploid gynogenetic form *mobilis* Moore. Proc. Roy. Soc. Edinburgh [B] 67:333-350.

——— and J. JACOB. 1957. Artificial activation of the egg in a gynogenetic spider beetle. Nature (London), 179:1300.

SCHOLL, H. 1956. Die Chromosomen parthenogenetischer Mücken. Naturwissenschaften, 43:91-92.

SCHULTZ, R. J. 1961. Reproductive mechanisms of unisexual and bisexual strains of the viviparous fish *Poeciliopsis*. Evolution, 15:302-325.

——— 1966. Hybridization experiments with an all-female fish of the genus *Poeciliopsis*. Biol. Bull. Woods Hole, 130:415-429.

——— 1967. Gynogenesis and triploidy in the viviparous fish *Poeciliopsis*. Science, 157:1564-1567.

SEILER, J. 1946. Die Verbreitungsgebiete der verschiedenen Rassen von *Solenobia triquetrella* (Psychidae) in der Schweiz. Rev. Suisse Zool., 53:529-533.

——— 1947. Die Zytologie eines parthenogenetischen Rüsselkäfers, *Otiorrhynchus sulcatus* F. Chromosoma, 3:88-109.

——— 1961. Untersuchungen über die Entstehung der Parthenogenese bei *Solenobia triquetrella* F. R. (Lepidoptera, Psychidae) III. Die geographische Verbreitung der drei Rassen von *Solenobia triquetrella* (bisexuell, diploid und tetraploid parthenogenetisch) in der Schweiz und in angrenzenden Ländern und die Beziehung zur Eiszeit. Bemerkungen über die Entstehung der Parthenogenese. Z. Vererbungsl., 92:261-316.

SMITH, M. A. 1935. The fauna of British India, Ceylon and Burma. Vol. II: Sauria.

STALKER, H. D. 1956. On the evolution of parthenogenesis in *Lonchoptera* (Diptera). Evolution, 10:345-359.

STEFANI, R. 1956. Il problema della partenogensi in *Haploembia solieri* Ramb. (Embioptera-Oligotomidae). Atti Accad. Naz. Lincei, Ser. VIII, 5 (Sez. IIIa): 127-201.

———— 1959. Aspetti zoogeografici di un problema evolutivo. Boll. Zool., 26:105-113.

———— 1964. La telitochia. Boll. Zool., 31:119-145.

SUOMALAINEN, E. 1945. Zu den Chromosomenverhältnissen und dem Artbildungsproblem bei parthenogenetischen Tieren. Sitzungsb. Finn. Akad. Wiss. pp. 181-201.

———— 1947. Parthenogenese und Polyploidie bei Rüsselkäfern (Curculionidae). Hereditas, 33:425-456.

———— 1953. Die Polyploidie bei den parthenogenetischen Rüsselkäfern. Zool. Anz., Suppl., 17:280-289.

———— 1961. On morphological differences and evolution of different polyploid parthenogenetic weevil populations. Hereditas, 47:309-341.

———— 1962. Significance of parthenogenesis in the evolution of insects. Ann. Rev. Entomol., 7:349-366.

TAKENOUCHI, Y. 1966. Tetraploid and pentaploid races of the Japanese parthenogenetic weevil, *Catapionus gracilicornis* Roelofs (Curculionidae, Coleoptera). Annot. Zool. Jap., 39:47-54.

THOMAS, R. 1965. The smaller teiid lizards (*Gymnophthalmus* and *Bachia*) of the southeastern Caribbean. Proc. Biol. Soc. Washington, 78:141-154.

TINKLE, D. W. 1959. Observations on the lizards *Cnemidophorus tigris, Cnemidophorus tesselatus* and *Crotaphytus wislezeni*. Southwestern Natural., 4:195-200.

UZZELL, T. M., JR. 1963. Natural triploidy in salamanders related to *Ambystoma jeffersonianum*. Science, 139:113-115.

———— 1964. Relations of the diploid and triploid species of the *Ambystoma jeffersonianum* complex (Amphibia, Caudata). Copeia, 1964:257-300.

———— and S. M. GOLDBLATT. 1967. Serum proteins of salamanders of the *Ambystoma jeffersonianum* complex, and the origin of the triploid species of this group. Evolution, 21:345-354.

WHITE, M. J. D. 1966. Further studies on the cytology and distribution of the Australian parthenogenetic grasshopper *Moraba virgo*. Rev. Suisse Zool., 73:383-398.

———— J. CHENEY and K. H. L. KEY. 1963. A parthenogenetic species of grasshopper with complex structural heterozygosity (Orthoptera: Acridoidea). Austral. J. Zool., 11:1-19.

———— and G. C. WEBB. 1968. Origin and evolution of parthenogenetic reproduction in the grasshopper *Moraba virgo* (Eumastacidae: Morabinae). Austral. J. Zool., 16:647-671.

WRIGHT, J. W., and C. H. LOWE. 1967a. Evolution of the alloploid parthenospecies *Cnemidophorus tesselatus* (Say). Mamm. Chrom. Newsl., 8:95-96.

———— and C. H. LOWE. 1967b. Hybridization in nature between parthenogenetic and bisexual species of whiptail lizards (genus *Cnemidophorus*). Amer. Mus. Novit., 2286:1-36.

ZWEIFEL, R. G. 1965. Variation in and distribution of the unisexual lizard, *Cnemidophorus tesselatus*. Amer. Mus. Novit., 2235:1-49.

9

Evolutionary Studies on *Maniola jurtina* (Lepidoptera, Satyridae): The "Boundary Phenomenon" in Southern England, 1961 to 1968

E. R. CREED
*The Genetic Laboratories, Department of Zoology,
The University, Oxford, England*

W. H. DOWDESWELL
*School of Education, Bath University of Technology,
Bath, England*

E. B. FORD
*The Genetic Laboratories, Department of Zoology,
The University, Oxford, England*

K. G. McWHIRTER
*Department of Genetics, The University of Alberta,
Edmonton, Canada*

Dedication	264
Introduction	264
Studies of the Southern English and East Cornish Types up to 1960	268
1961 to 1968	270
1961	270
1962	272
1963	273
1964	274
1965	276

1966 278
1967 279
1968 280
Discussion 282
Summary 285
Acknowledgements 286
References 286

Dedication

We are paying our respects to Professor Th. Dobzhansky on his seventieth birthday by publishing in his honour the long-accumulated results of a specific piece of research in ecological genetics. Indeed we feel that no better way could be found of acknowledging the achievements of one who has advanced so greatly the experimental study of evolution.

For we are here describing new developments in a subject acutely interesting to Dobzhansky: the genetic adjustments of wild populations. Not only has he shown that interest by giving us his advice and encouragement, but he has visited England to take part in the field-work here outlined (p. 279). It is conducted not on *Drosophila,* with its great advantages which he has so brilliantly exploited, but on the butterfly *Maniola jurtina,* which has proved suitable for the particular purposes we had in mind. We offer these studies to him as a record of our admiration for his work.

Introduction

The number of spots on the underside of the hindwings of the Meadow Brown butterfly, *Maniola jurtina,* has long been investigated in order to make a quantitative study of variation relative to the evolutionary adjustments of this species (Ford, 1965, chapters 4 and 5). From Devon across England to the North Sea the pattern of spotting is characteristic, being unimodal; individuals with two spots are commonest in the males, and those with no spots in the females: the *Southern English* type (Figs. 1, 2). However, in central and eastern Cornwall and for a distance into Devon, which has varied from season to season, the frequencies are also unimodal at two spots in the male, but in the females they are bimodal, with a greater mode at no spots and a lesser at two. This constitutes the *East Cornish* type (Fig. 3). It is one in which the females, with their second mode at two spots, have in general a higher spot-average than in the Southern English form with its single mode at the spotless class. Both the female spot-averages can vary annually as well as from one part of the insects' range to another. Indeed, the two extremes overlap so that Southern English values higher than East Cornish ones have been encountered. Thus during the

Fig. 1. Maniola jurtina, the Meadow Brown Butterfly, natural size, showing the spots on the underside of the hindwings. Numbered left to right, in three horizontal rows. 1-3, males with two, three, and four spots, respectively. 4-9, females: 4, no spots; 5, one spot; 6 and 7, two spots differently placed; 8, three spots; 9, four spots.

period now under consideration, the Southern English spot-averages have ranged from 0.80 to 1.26; the Cornish from 0.91 to 1.61. Other stabilized spot-distributions are found in the far west of Cornwall and in the Isles of Scilly (Creed et al., 1962), but they are not the subject of the present article.

These are "first order" distinctions, and certain minor adjustments can occur within them. The Southern English form may be relatively extreme, with 60 percent or more of the females spotless (the *Old English* type). Alternatively it may be less extreme, with fewer than 60 percent of the females spotless (the *New English* type). These modifications, though more evident in the females, affect the males also because in that sex those with more than two spots become commoner and those with less than two spots rarer in the New English than in the Old English stabilizations (McWhirter, 1957).

Up to 1954 the Southern English populations were Old English. However, in 1956 the majority, though not all of them, proved to be New English or even had a small secondary mode at two spots: the "Pseudo-

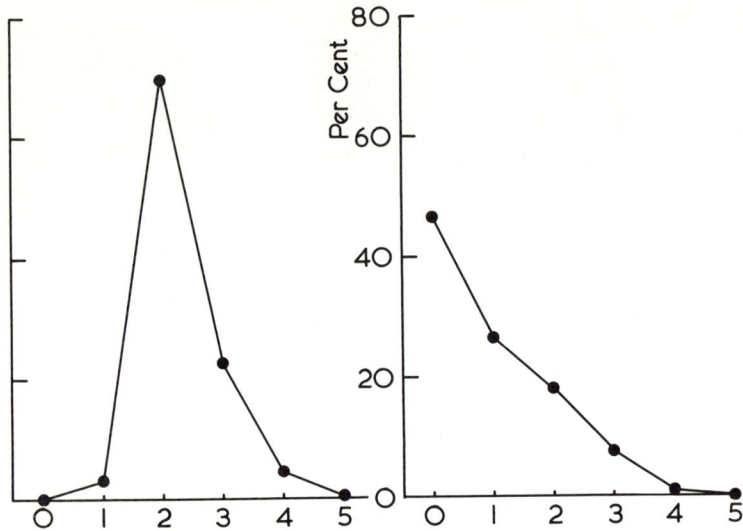

Fig. 2. Spot-distribution of the butterfly, Maniola jurtina, based on specimens from seven localities in Devon in 1960, showing the "Southern English" type with a single mode at 0 spots in the female. Males on the left; females on the right.

Cornish" form; a change possibly associated with the great abundance of the species that year, when relatively large numbers of the higher spotted individuals may have survived because of less rigorous elimination. In 1957, after an unusually mild winter and prolonged spring drought, always unfavourable to *Maniola jurtina,* the butterfly was much rarer and the proportion of spotless individuals became slightly greater. By 1958 the Old English type had in general been restored except in Devon and Somerset, where the population remained New English.

However, the main Southern English spot-distribution is powerfully buffered against environmental effects, being the same in a great diversity of habitats: those on acid or alkaline soils, on downland, in meadows, in marshes and along woodland paths. This is true also of the East Cornish type except that, in its relatively limited distribution the areas of strongly alkaline soil are of insufficient extent to determine whether or not this bimodal spot-distribution responds to them.

Although our more detailed studies of *Maniola jurtina* in the British Isles cover Southern England (with Scilly) only, the type of female spot-distribution unimodal at zero found from the North Sea to Devon seems very widespread; for it characterizes also the samples we have obtained from South Wales and the North of Scotland (Caithness, Wester Ross, and the Island of Raasay, near Skye). Moreover, in Ireland the situation is similar but more extreme, the females being nearly always spotless with a very few one-spotted specimens only. Their spot-average is about 0.2 and that of the males about 1.3.

Dowdeswell and McWhirter (1967) have also obtained extensive data on the spotting of *Maniola jurtina* throughout the greater part of its range in the Palaearctic Region. They find that the Southern English spot-distribution extends approximately from the Pyrenees and the west coast of France to Rumania and Finland. Thus the transition to a different type that occurs in the southwest of England marks the limit of an extremely widespread condition that indeed shows great stability. In its peripheral distribution, however, the situation is much less constant in spot-frequency; for the populations break up into a number of distinct types in respect of this character, each limited to a relatively small geographical area. These outlying stabilizations occur in Ireland, the Isles of Scilly, Southwest England, the Canary Isles, North Africa, Iberia, Italy, Greece, and Western Asia.

When rearing *Maniola jurtina* in the laboratory from larvae caught in the wild, the spot-frequencies may vary during the period of emergence, changing from higher to lower values. A seasonal shift in spotting also occurs sporadically in nature. When observable, it may be in either direction, but more usually towards lower values as the summer advances. Such

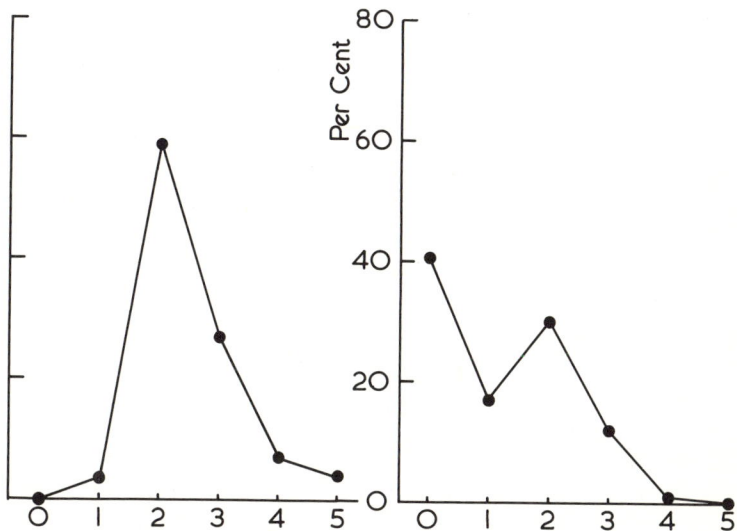

Fig. 3. Spot-distribution on the butterfly, *Maniola jurtina*, based on specimens from four localities in east Cornwall and west Devon in 1960, showing the "East Cornish" type with modes at 0 and 2 spots in the females. Males on the left; females on the right.

intraseasonal changes are particularly noticeable when considerable readjustments are taking place in the population, as in the neighbourhood of the boundary when this is about to shift or has recently done so.

In order to obtain light on this seasonal effect we have often visited the Devon-Cornwall region when the species first appears on the wing in late June and early July, in addition to conducting our principal work at the height of the emergence in late July and early August. Owing to other commitments it has not been possible for us to remain in the West Country during the whole period of about nine weeks when *Maniola jurtina* is flying.

As a result of his breeding work, McWhirter (1969), using material from the Isles of Scilly, has been able to evaluate the heritability of spotting in this species. At about 15°C, two methods of estimation gave figures of 0.63 (± 0.14) and 0.83 respectively in the females, but much less, and not significant on the numbers obtained, in the males (0.14). That is to say, the imaginal effect of the genes responsible for spotting is powerful in the females and largely sex-controlled.

Studies of the Southern English and East Cornish Types up to 1960

The distinction between the Southern English and East Cornish types was first recognized in 1952 in an east to west transect in the center of the Devon-Cornwall peninsula at the Launceston-Okehampton levels and confirmed the same season along another transect 13 miles (20 km) to the south, between Dartmoor and the sea (Dowdeswell and Ford, 1953). In 1956 Ford and McWhirter discovered that instead of merging gradually into each other, the *M. jurtina* populations changed abruptly from one type to the other at Larrick (see map. Fig. 4), approximately 2 miles (5 km) west of the River Tamar (Creed et al., 1959). At that point the two stabilizations came into contact at an ordinary field hedge across which the butterflies flew freely. Moreover, the characteristics of both forms became accentuated on approaching the interface between them: a clear indication that they are subject to powerful and opposed selection. Such a "reverse cline" effect has been observed also in the snail *Partula* on Moorea in the South Pacific (Bailey, 1956).

The boundary between the East Cornish and Southern English populations had moved eastwards to the Tamar in 1957. Moreover, it was not quite so clear-cut, since the definitive types were separated by a region, only 160 m wide, in which the population was intermediate between them. This occupied the narrow eastern flood-plain of the river. The position of the boundary remained unaltered, with the same intermediate zone of 160 m, in 1958 and 1959. In the latter year we also tested once more the southern of the two transects studied in 1952, that between Dartmoor and

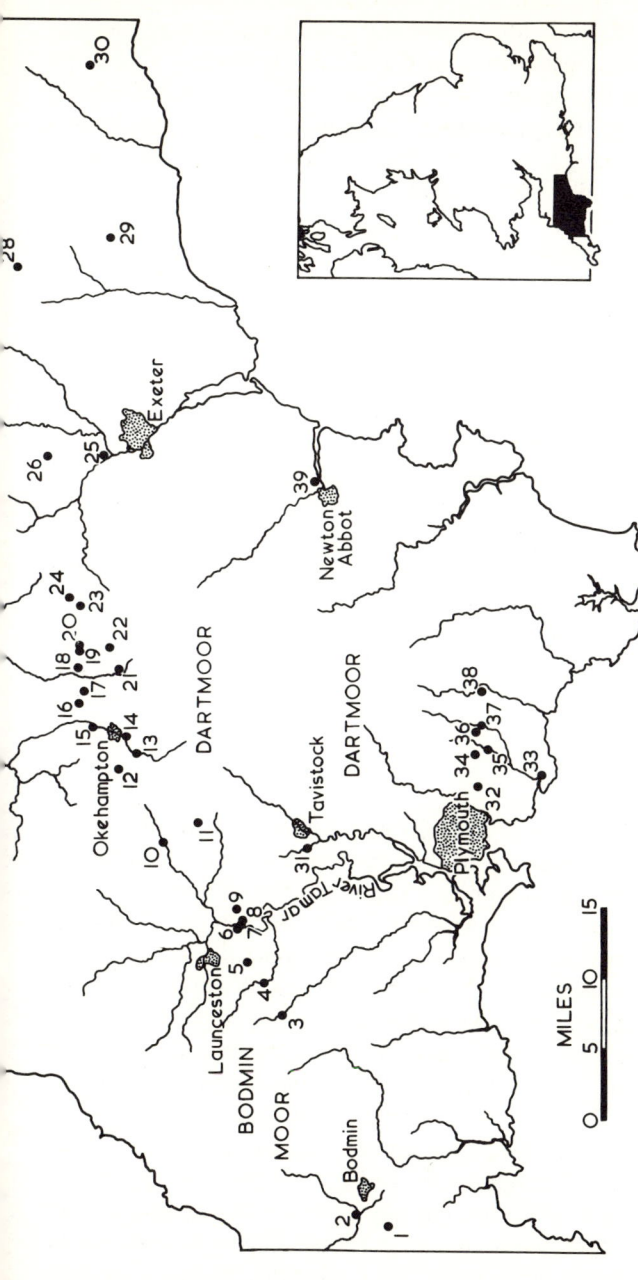

Fig. 4. Map showing the position of the collecting localities mentioned in the text; the inset gives the relationship of this map to the rest of Britain.

1. Lanivet. 2. Dunmere. 3. Beriowbridge. 4. Larrick Mill. 5. East and West Larrick 6. Tamar West Flood Plain 7. Tamar East Flood Plain. 8. Pilistreet. 9. Kelly. 10. Wrixhill Bridge. 11. Lydford. 12. Maddaford. 13. Meldon. 14. Okehampton Golf Course. 15. Abbeyford. 16. Hatherton. 17. Sampford Courtenay. 18. Taw Mill. 19. Thornes Itton. 20. Itton Moor. 21. South Tawton. 22. Powlesland. 23. Hilldown. 24. Walson Barton. 25. Upton Pyne. 26. Raddon. 27. Tiverton. 28. Broadhembury. 29. Gittisham. 30. Hawkchurch. 31. Lumburn. 32. Haye Farm. 33. Noss Mayo. 34. Tuxton. 35. Efford. 36. Lee Mill Bridge. 37. Popples Bridge. 38. Thornham. 39. Newton Abbot.

the south coast, and obtained similar evidence of a sharp boundary between the two stabilizations. We also found that the East Cornish one had extended its range 2 to 3 miles eastwards since 1952, in conformity with what had happened further to the north. The 1960 season was devoted wholly to the southern transect and we encountered the boundary in the neighbourhood of Efford (Creed et al., 1962, with maps). These facts have already been discussed in some detail (Ford, 1965).

1961 to 1968

Our results that deal with the boundary between the East Cornish and Southern English types during this period have not yet been published. They are, however, too extensive to print in full, as has been done in the past. Our complete records will, therefore, be deposited in the Natural History Museum, London. We shall here include only those portions of them which provide direct information on the subjects we are discussing. These principally involve the position and movement during successive years of the boundary between the East Cornish and Southern English spotting-types; also reference to a few enclaves within the main regions in which the alternative form exists, in general because it survives after a previous change in the area of the two stabilizations or because it anticipates a future one.

1961

As already explained, our work in 1960 was devoted wholly to investigating the boundary and its position in the southern transect, that near the coast. In 1961 we attempted to combine studies both north and south of the granite intrusion of Dartmoor, which forms a barrier to *M. jurtina*. This we had done successfully in 1959, for in that year the situation was particularly clear and uncomplicated, and it proved possible to obtain decisive data in both regions. In 1961, however, which appeared to be a season in which the site of the boundary was changing, we found it had been a mistake to reduce the size and number of our samples by endeavouring to collect data in both areas. Having established the existence of the boundary phenomenon in the south upon similar lines to that in the centre of the Devon-Cornwall peninsula (Creed et al., 1962), we decided as a result of our experience in 1961 to concentrate in the future chiefly upon the more northerly transect while obtaining occasional diagnostic information in the south, supplied principally with the kind help of Mr. and Mrs. Beaufoy.

The 1961 samples tend therefore to be defective because too small, particularly in the northern transect. We may, however, consider the southern one first. At Noss Mayo, a promontory on the coast about 4 miles east of Plymouth, a sample of 103 females proved to be East Cornish, while at Haye Farm, 3½ miles to the north, the population was Southern English (Table I):

TABLE I

Female Spotting, 1961, from Noss Mayo and Haye Farm

	Spots						Total	Spot-average	Type
	0	1	2	3	4	5			
Noss Mayo	48	23	25	7	–	–	103	0.91	C
Haye Farm	37	25	15	3	–	–	80	0.80	E

C, East Cornish.
E, Southern English.

However, the difference between them is not significant: $\chi^2_{(2)} = 2.61$. Yet Haye Farm was the beginning of the English stabilization, which was picked up in samples of 83, 81, and 104 females at three localities, Efford, Lee Mill Bridge, and Newton Abbot, respectively, passing progressively eastwards. The last of them lies beyond the eastern limit of Dartmoor, and the population in its neighbourhood joins the main Southern English one.

Tuxton, a site east of Haye Farm, gave an equivocal result (21, 9, 8, 5, 1, 0 = 44). Ideally, the region including it should have been studied in successive years, had time and personnel allowed, to determine whether or not this remains an isolated pocket of an exceptional type.

We scored old and new specimens separately, and at Efford the early emergence appeared to be East Cornish, and the later Southern English. A similar seasonal adjustment took place at Lumburn up the valley between Dartmoor to the east and Bodmin Moor to the west. This is a locality situated a mile west of Tavistock, half way between our two transects. Here again, the specimens that emerged early were East Cornish and those appearing later were Southern English.

Along the central transect, a number of small samples which, as indicated, we should have wished considerably to increase, gave a somewhat confused picture westwards from Kelly, 1½ miles east of the Tamar; a place where we were indeed able to give the time to collecting a larger number. The 91 female specimens obtained there appeared to be East Cornish in type. The only other large sample (125 females) captured in that region in 1961

TABLE II

Female *M. jurtina* Spotting, 1961, from Kelly and Lydford

	Spots						Total	Spot-average
	0	1	2	3	4	5		
Kelly	33	19	22	12	4	1	91	1.32
Lydford	43	37	32	12	–	1	125	1.14

was at Lydford, 7 miles east of Kelly, and this was Southern English of the New English form (Table II).

Yet the difference in spotting between the two localities is not significant ($\chi^2_{(3)} = 4.19$).

1962

For the reasons already given, our work during the years 1962 to 1968 was conducted principally along the route following approximately the line Bodmin, Launceston, Okehampton, Exeter. That transect is the one described in the accounts that follow, unless stated to the contrary.

Detailed results amassed from July 29th onwards, at the height of the insects' emergence, now confirmed the impression, gained in 1961, that the Cornish spot-distribution had extended its range 1½ miles eastwards to Kelly. For in 1962 four localities westwards from that locality showed the East Cornish stabilization, while three eastwards from it were Southern English. We were indeed able to pick up the boundary as passing through the Kelly collecting ground; for the female samples obtained in fields ¼ mile apart, which we designated as Kelly Main area and Kelly South, showed the following frequencies (Table III):

TABLE III

Female *M. jurtina* Spotting, 1962, from Kelly

	Spots						Total	Spot-average
	0	1	2	3	4	5		
Kelly Main	163	144	114	48	7	–	476	1.14
Kelly South	34	19	46	12	–	–	111	1.32

The difference between them is highly significant ($\chi^2_{(3)} = 16.1$), but there were other fields in the neighbourhood in which the pattern was not clear.

It will be noticed that the sample from Kelly South is indeed of an unusual and extreme East Cornish type, with a greater mode at 2 than at 0.

We had, in addition, made some early collections during the period 1st to 8th July. Those from Pilistreet and from three localities west of it (Tamar West, Larrick Mill, and Beriowbridge) were all decisively East Cornish in their modality though heterogeneous, while a single Kelly sample was of the Southern English type, being unimodal at 0. Lydford, which became typically Southern English at the end of the month, was more highly spotted (20, 15, 15) at this early date. At the same period three samples of 50 or more were captured along the southern transect. That at Efford was bimodal and two from further east (at Popples Bridge and Thornham) were unimodal, though the distinction was not significant on the numbers obtained.

1963

This year caused us great surprise as the boundary had moved 11 miles eastwards and fell between Meldon, 2½ miles west of Okehampton, and the Okehampton Golf Course lying on the southwest edge of the town. Setting aside South Tawton (see below) we obtained samples from three localities east of Meldon. They were all Southern English and proved to be homogeneous ($\chi^2_{(6)} = 3.3$). The seven populations that were studied west of Okehampton were all East Cornish and they included Kelly, sampled the year before. They began to suggest heterogeneity, with $\chi^2_{(18)} = 26.2$, for which P<0.1. However, this tendency is slight compared with the disparity obtained when the Southern English and East Cornish groups are contrasted ($\chi^2_{(3)} = 13.1$). Thus it is at least evident that a distinction of a far more considerable kind occurred between Okehampton and Meldon, separating the populations that were respectively unimodal and bimodal in the females. Indeed the second-order differences already mentioned (p. 264) makes it clear that samples drawn from either stabilization may show slight heterogeneity if large numbers are available.

The line which we studied extended from Tiverton in Devon to Lanivet in Cornwall and comprised 16 localities. Five of these are, however, not included in the calculations since they must be judged as "small samples" with a total of less than 30), although they had in fact fallen perfectly into the Southern English or East Cornish types appropriate to the areas whence they were obtained. The spot-frequencies of the two critical areas on either side of the boundary, and two others lying beyond them in either direction may be given here, passing downwards from east to west (Table IV).

A locality near South Tawton appeared to be an area of the East Cornish type within the Southern English stabilization, the female spot-frequencies (from 0 to 5) in a small sample obtained there took the values 14, 6, 11, 2,

TABLE IV

Female Spot-frequencies. Three localities on Either Side of the Boundary

	Spots						Total	Type	Spot-averages
	0	1	2	3	4	5			
Sampford Courtenay	52	25	14	8	2	–	101	E	0.84
Abbeyford	9	8	6	1	–	–	24	E	0.96
Okehampton	16	10	6	7	–	–	39	E	1.10
Meldon	11	2	17	7	1	–	38	C	1.61
Maddaford	24	14	17	11	2	1	69	C	1.36
Wrixhill Bridge	35	13	21	6	–	1	76	C	1.03

E, Southern English.
C, East Cornish.

0, 0 = 33. It is noteworthy that a major discontinuity, though in the reverse direction, was also encountered between the Sampford Courtenay population and another in the South Tawton area in 1968.

These results were obtained at the end of July and the first half of August during the main period of the emergence. We had also made a preliminary collection during the first week of July at six localities: from Lydford to Dunmere, including Kelly. All were of the East Cornish type, save that at Dunmere in which the female spotting was unimodal at 0, with values 22, 16, 11, 9, 2 = 60. Exceptional spot-distributions are common at the beginning of the emergence (p. 267,) and the population subsequently passed to the East Cornish type, normal for that area; the values, based on 134 females obtained during the main period of the emergence, were 49, 27, 35, 19, 4, 0.

A few small collections from the southern transect made at the height of the emergence tended towards higher spotting in the older specimens and lower in those which had just emerged. Mr. and Mrs. Beaufoy, however, secured a large sample for us from Noss Mayo, which reversed this tendency, passing from a Southern English distribution among the older insects to the expected East Cornish type among the newer. A large catch at Newton Abbot, which they also provided, was Southern English of the New English type.

1964

We were honoured to have Professor Ernst Mayr of Harvard working with us during the main period of the emergence this year. The collect-

TABLE V

Female Spot-frequencies: The Boundary at Itton Moor in August 1964

	Spots						Total	Type	Spot-average
	0	1	2	3	4	5			
Eastern Area	49	33	27	5	–	–	114	E	0.89
Central Area, E	45	24	18	12	–	–	99	E	0.97
Central Area, W	52	20	38	12	1	–	123	C	1.11
Western Area	68	27	40	23	–	–	158	C	1.11

E, Southern English.
C, East Cornish types.

ing began on July 30th and included eleven localities of the midcountry transect, one of which required subdivision. We wound that the boundary had moved eastwards by about 8 miles since the previous season. It was identified crossing Itton Moor (Table V). The seven samples west of that locality extended as far as Dunmere and were all decisively East Cornish.

Itton Moor itself is 1 mile (1.6 km) long, east to west, and ⅓ mile wide. The central area (300 m east to west by 200 m north to south) is demarked by the hedge of a field, which intrudes into the moor from the north. We first scored the main eastern and western sections on either side of this intermediate region. They are roughly equal in extent. The western proved to be East Cornish while the east was Southern English, so it appeared likely that the boundary passed through the central section of the Moor. This we subdivided into two parts of about the same size and we were able to obtain considerable numbers from both. The east was Southern English, and the western East Cornish, in type, but the difference between them was not significant ($\chi^2_{(3)} = 5.6$). A significant difference ($\chi^2_{(3)} = 10.6$) was, however, obtained when the two eastern and the two western groups, respectively homogeneous, were combined and contrasted. Thus the line of the boundary had been closely identified.

Although our available time was by then running short, we were able to obtain adequate samples from two sites further to the east: Hilldown and Upton Pyne, respectively 2 and 13¼ miles from Itton Moor. Both were decidedly Southern English of the New English type.

A locality at Walson Barton (2½ miles E.N.E. from Itton Moor) may represent an advanced area of East Cornish type in the neighbourhood of the boundary within the Southern English region. It was very difficult to collect. In three attempts we obtained only 38 females distributed as 15,

6, 8, 6, 2, 1, with the high spot-average of 1.4. The characteristics of this population are uncertain since we had such small numbers; but if indeed East Cornish, the sample represents the only exceptional one encountered this year.

These results received valuable confirmation from the southern transect by means of large samples collected in two areas by Mr. and Mrs. Beaufoy. In that at Noss Mayo, about 4 miles east of Plymouth and within the East Cornish region from 1961 onwards, the females are distributed as 40, 29, 33, 7, 1 = 110. Two populations from Newton Abbot occurring in close proximity can be combined as homogeneous. The total figures amounted to 84, 53, 35, 10, 0, 0 = 182.

The male spotting did not prove diagnostic this year and the numbers available were, on the whole, small.

Early collections along the more northerly transect, obtained from July 6th to 9th soon after the species had begun to emerge, were arranged to span the area of the boundary as fixed the previous season. They gave about the same frequencies as observed then, the critical region being that between Okehampton and Meldon (where, however, only 22 females could be caught). The results are shown in Table VI. They indicate that the shift of the boundary to its new position only became detectable during the main period of emergence.

1965

In our work during the main period of emergence we scored old and new specimens separately, though the distinction was not always clear. In no instance did we find any significant differences between them: either the two groups were obviously homogeneous or else the number of new individuals was so small that it provided no valid comparison. Consequently the two types have been combined throughout.

We obtained samples at seven localities west of the Itton Moor area.

TABLE VI

Female Spot-frequencies: The Boundary at the Beginning of the Emergence in 1964

	Spots						Total	Type	Spot-average
	0	1	2	3	4	5			
Sampford Courtenay	23	14	12	4	–	–	53	E	0.94
Okehampton	20	15	12	4	1	1	53	E	1.13
Meldon	6	3	10	2	–	1	22	C	1.32
Maddaford	20	10	14	9	–	–	53	C	1.23

TABLE VII

Female Spot-frequencies: the Boundary at Itton Moor in August 1965

	Spots						Total	Spot-average	Type
	0	1	2	3	4	5			
Eastern Area	42	41	17	5	1	–	106	0.89	E
Central Area, E	48	34	26	7	1	–	116	0.96	E
Central Area, W	49	21	32	17	1	1	121	1.20	C
Western Area	33	21	21	8	–	–	83	1.05	C
Thornes Itton	98	57	61	27	1	–	244	1.08	C

All were East Cornish except that at Wrixhill Bridge (24, 18, 12, 9, 0, 0 = 63). At the eastern end of the transect we collected at Upton Pyne, where the population was Southern English. We carried the work still further to the east, to Tiverton where the spotting was also of the Southern English type.

The boundary was still at Itton Moor as in the previous year, though it was somewhat less clearly defined. The main eastern section was decisively Southern English. A sample of 83 from Itton Moor West had equal numbers with one and two spots, but with the higher East Cornish spot-average. That area was separated from two adjoining fields ("Thornes Itton"), running north and south along its western boundary, only by a low hedge with some trees in it.

We divided the Central Area of Itton Moor into the same two parts as before, by an arbitrary north-to-south line crossing the middle of it (see the 1964 account). These provided an eastern Southern English sample and a western East Cornish one, identified perhaps less clearly by their modes than by their spot-averages (see Table VII). They gave, however, no valid indication that a reverse cline had yet been built up during the two seasons during which the boundary had been in the Central Area of Itton Moor. It is true that those in the West Central Area had a higher spot-average than the other Cornish samples, but the eastern section was less rather than more decisively Southern English than the main eastern area, and this was true in 1964 also.

The southern transect was exceptional this year. The Noss Mayo sample, of 110 females, was Southern English, while populations in two fields at Newton Abbot (amounting respectively to 108 and 215 specimens) were both East Cornish, except for the earlier emergences of one of them. These were Southern English, based on 62 individuals.

Early collecting was limited to the northern transect and started on July

5th. We examined seven localities from Larrick Mill to Upton Pyne. In addition, an effort was made to obtain a sample at Lydford, but only 12 females were secured there. Elsewhere the numbers ranged from 40 to 103. Excluding Lydford, which indeed suggested bimodality; Larrick Mill and Tamar West were studied westwards from Maddaford, and were clearly Cornish. Maddaford itself and three localities eastwards from it were all clearly Southern English. Upton Pyne was exceptional at this date (32, 22, 23, 8, 2). Later in the emergence, however, it became more nearly Southern English (34, 33, 17, 7, 0, 0). Two of the localities sampled at this period were the opposite ends of Itton Moor, which differed in early August this year and still more the previous one. Yet in early July they were both Southern English in type (East End: 40, 36, 17, 10 = 103; West End: 31, 17, 11, 2, 1 = 63).

1966

We were greatly surprised this season to find that we had to move 40 miles eastwards to reach the first conclusively Southern English population, at Hawkchurch on the Dorset border. We encountered it after considerable difficulty in discovering a suitable collecting site. From that point we collected at 11 localities along the transect westwards to Larrick across the Cornish border; among them Itton Moor, of course, was subdivided. At Wrixhill Bridge the old specimens appeared to be Southern English and the newer ones East Cornish, but the numbers were inadequate (33 and 30 respectively). It proved impossible to secure sufficient totals at Hilldown (24) and Walson Barton (14). Otherwise it may be said that, with the exceptions mentioned below, all the populations west of Hawkchurch were East Cornish. They had not, however, gone the whole way to becoming so since, except at Larrick, the frequency of individuals with two spots was only marginally higher than, or equal to, the frequency of those with one: though all had acquired the higher East Cornish spot-average. This is illustrated

TABLE VIII

Female Spot-frequencies: The Boundary near Hawkchurch in August 1966

| | Spots | | | | | | | Spot- | |
	0	1	2	3	4	5	Total	averages	Type
Hawkchurch	52	26	13	8	1	–	100	0.80	E
Gittisham	24	12	13	3	1	–	53	0.96	C ?
Broadhembury	39	23	25	15	1	1	104	1.22	C ?

by the next two sites westwards from Hawkchurch compared with Hawkchurch itself, in Table VIII. In fact, Gittisham and Broadhembury when combined do not differ significantly from Hawkchurch, $\chi^2_{(3)} = 6.2$.

Two of the westerly sites require special mention. At Itton Moor the main eastern area had become clearly East Cornish (49, 18, 27, 13, 0, 1 = 108, spot-average = 1.07) while the main western area was, strangely, Southern English but with a spot-average more characteristic of the East Cornish type (44, 27, 21, 10, 2, 0 = 104, spot-average = 1.03). However, the distinction between them is not significant upon the numbers obtained $\chi^2_{(3)} = 2.9$). Unfortunately, the species had become rare in the central areas of Itton Moor this year, where we could obtain only six and eight specimens respectively. The other site that requires comment, especially in view of the situation there in later years, is Sampford Courtenay. This certainly looks as if it were passing through the East Cornish to the Southern English stabilization, being unimodal at 0 but with a high spot-average (25, 19, 13, 10, 0, 0 = 67, spot-average = 1.12).

We had collected at six sites near the start of the emergence, from Larrick Mill to Itton Moor, which we subdivided. Larrick Mill and Wrixhill Bridge were East Cornish, Sampford Courtenay and the two parts of Itton Moor were Southern English; so also was Maddaford, while Okehampton was East Cornish.

1967

We were honoured to have Professor Th. Dobzhansky of the Rockefeller University, New York, working with us this year.

For the first time the boundary moved westwards, and by approximately 44 miles. The findings were clear and uncomplicated; the populations at seven sites from Hawkchurch to Sampford Courtenay, including Itton Moor East and West had become decisively Southern English. The change from the one to the other stabilization for which $\chi^2_{(3)} = 7.9$ occurred between Stampford Courtenay and the adjoining Hatherton, separated only by half a mile of rough ground and agricultural land (Table IX). Thence at

TABLE IX
Female Spot-numbers: The Boundary August 1967

| | Spots | | | | | | | Spot- | |
	0	1	2	3	4	5	Total	average	
Sampford Courtenay	35	23	11	6	–	–	75	0.84	E
Hatherton	39	18	44	11	–	–	112	1.24	C

four sites further westwards to Larrick the samples were typically East Cornish. The English spotting at the seven sites from Hawkchurch to Sampford Courtenay is homogeneous; so also at the five Cornish sites. The difference between the two groups is, of course, immensely significant.

On the southern transect, Mr. and Mrs. Beaufoy obtained samples for us at Noss Mayo (39, 22, 35, 10, 0, 0 = 106, spot-average = 1.15) and at Newton Abbot (42, 32, 25, 8, 0, 0 = 107, spot-average = 0.99). These frequencies suggest that they were respectively East Cornish and Southern English, though the difference is not significant ($\chi^2_{(3)} = 3.9$).

We also collected at seven sites along the northern transect, near the start of the emergence, during the period July 6th to 11th. These spanned the distance from Hawkchurch to Larrick. All were East Cornish, even Hawkchurch itself and both the main parts of Itton Moor, with the sole exception of Sampford Courtenay, which was Southern English (29, 18, 9, 7, 0, 0 = 63, spot-average 0.90).

1968

During the main period of the emergence this year we obtained adequate samples from Hawkchurch and Dunmere and from 13 localities between them, though that from Walson Barton amounted only to 37 females. The situation was very similar to the one found in 1967 in so far as the boundary was again well to the west, 28 to 36 miles from Hawkchurch. It had, however, moved eastwards since the previous summer, by between 6 and 16 miles, to the tract of country between Walson Barton and Raddon, in which we could find no satisfactory collecting sites. The spot-frequencies at these two places and at a locality on either side of them are given in Table X. They are, respectively, typical of what was encountered this season. As only a small sample could be obtained at Walson Barton, the districts at the boundary may be quantified by combining the samples at Broad Hembury and Raddon on the one hand, and at Walson Barton and Itton Moor East

TABLE X

Female Spotting: The Boundary During the Main Period of the Emergence in 1968

	Spots						Total	Spot-average	Type
	0	1	2	3	4	5			
Broadhembury	19	19	11	3	1	–	53	1.02	E ?
Raddon	31	26	21	12	2	1	93	1.26	E
Walson Barton	14	5	12	6	–	–	37	1.29	C
Itton Moor E	35	18	32	10	2	–	97	1.24	C

TABLE XI

Female Spotting: The "Island" of Southern English Stabilization At and Near Sampford Courtenay in 1968

	Spots						Total	Spot-average	Type
	0	1	2	3	4	5			
Hatherton	57	28	59	15	4	2	165	1.32	C
Sampford Courtenay	65	40	36	18	3	–	162	1.10	E
Taw Mill	20	21	15	5	1	–	62	1.13	E ?
Itton Moor W.	51	21	36	5	1	–	114	0.98	C
Powlesland	29	16	30	12	4	–	91	1.41	C

on the other, since each pair is homogeneous ($\chi^2_{(3)} = 1.8$ and 0.6 respectively). The difference between the two Southern English and East Cornish populations is signficant, ($\chi^2_{(3)} = 8.6$). An indication of the former extent of the East Cornish form appears to be retained in the high spot-values of the Southern English populations, manifest also at Hawkchurch (1.14).

The previous year the Sampford Courtenay population had been the most westerly of the Southern English type and in 1968 it remained as an apparently isolated pocket of the same, within the country characterized by the East Cornish spot-frequency. We attempted to discover how large that pocket might be. To the northwest, it did not extend to Hatherton, ½ mile away. Eastwards, however, it included the moor at Taw Mill, a distance of 1¼ miles, but not the next site, Itton Moor West, a further 1¼ miles off.

After some difficulty, we also discovered some suitable fields to the south of Sampford Courtenay where it was possible to obtain a considerable sample. This was Powlesland, where the spotting proved to be clearly East Cornish, so we had there passed beyond the exceptional area; the data upon which these statements are based are given in Table XI. The two isolated Southern English samples are homogeneous ($\chi^2_{(3)} = 3.3$), as are those from the three surrounding East Cornish localities, though they approach heterogeneity ($x^2_{(6)} = 10.2$). The difference between these two population-groups is significant ($\chi^2_{(3)} = 11.8$).

As will have become evident, the spot-frequencies at the beginning of the emergence differ from those to which the populations settle down as the numbers increase. These early values were more abnormal than usual this year. At Larrick and Wrixhill Bridge, female spotting had so far passed beyond the ordinary East Cornish type as to have become unimodal for individuals with two spots. The Maddaford sample of 32 females was too small to be decisive, producing equality of those with one and two spots (15, 7, 7, 3, 0, 0 = 32). That at Itton Moor East was East Cornish, while those at Raddon and Hawkchurch were both exceptional in possessing a single mode at one spot. In view of the situation found subsequently at Sampford Courtenay, it is interesting to note that this colony showed signs of the Southern English stabilization even when the specimens were just appearing (32, 29, 24, 11, 1, 0 = 97).

Discussion

The female spotting of *Maniola jurtina* is stabilized at a number of distinct frequencies, two of which are discussed in this article: the normal Southern English, very widespread also in Europe, which is unimodal at no spots, and the East Cornish type which is bimodal at no spots and at two (see Figs. 2 and 3). The situation poses certain difficulties of interpretation, in which we have to take into account the following facts.

1. Each type is strongly buffered against environmental changes of many kinds: the pH of the soil; diversity of habitat, such as open country, marshes and woodlands; a Continental compared with an Atlantic climate.

2. One stabilization in spotting can change to the other in the course of a few yards, and in the absence of any geographical or geological barrier, along a line extending for many miles and running approximately north and south across the Devon-Cornwall Peninsula, and also at the boundary of certain small enclaves of a different spotting type within one of the main stabilizations.

3. The boundary between the two forms of spotting may remain at exactly the same position for several years, while it can also shift considerable distances, 20 miles or more, from one generation to the next, of which there is one annually.

Evidently, powerful selection operating upon the highly heritable female spotting demonstrated in *Maniola jurtina* must be responsible for phenomena such as these. Indeed, when *M. jurtina* are reared in the laboratory with little elimination, spotting among the females from any one locality may be very different from that occurring in the wild population found there. This is not merely the result of the exceptional environment subsisting in laboratory conditions, as indicated by the extreme stabilization of spotting in widely differing environments in nature. Moreover, it accords with the fact that the spotting of the earlier specimens to emerge differs from that found at a later date.

The larvae of *Maniola jurtina* are subject to parasitism by Hymenoptera and Diptera and to bacterial diseases, which may destroy differentially the early stages of specimens that are destined to produce the distinct spotting-types. Yet it seems clear that the boundary between the Southern English and East Cornish forms is not controlled by the limits of distribution of such insect enemies, many of which are indeed widespread, or bacteria, the range of which would in these different species have to end abruptly and fluctuate sharply; otherwise the effect produced upon the butterfly would change gradually instead of being clearly delimited as it is.

Moreover, the clear-cut boundary between one spotting stabilization and another cannot be due to a past discontinuity allowing independent evolution followed by a subsequent extension of range as, for instance, seems to have happened in the moth *Bupalus piniarius* (Ford, 1965, p. 283). For in the present article many instances are given demonstrating that the Cornish type changes to the English, and the reverse, over wide areas of country in a single generation.

The existence of a reverse cline at Larrick in 1956 suggests that the respective forms had built up distinct and adapted gene-complexes, and that the boundary must have remained at the same place long enough for the elimination of the less well adapted intermediates to affect the population in the region of the interface between the two types. Unfortunately, however, we discovered the boundary only in 1956 and the next season it had moved 3 miles eastwards to the Tamar Valley. It remained there for only three years, nor has it persisted subsequently at any one place for more than two generations. Given time, such a reverse cline is indeed likely to develop in these circumstances under powerful selection. As already mentioned, it has been detected where two races of the snail, *Partula taeniata* come into contact on Moorea in the Society Islands.

Thus it is possible for species to adapt themselves to their environment in different ways, involving distinct genetic adjustment sharply demarked, as in the advance of the pale form of the mouse *Peromyscus polionotus,* adapted originally to the sea shore, 40 miles inland across relatively dark soil (Summer, 1930). Also, to take a polymorphic instance, we may cite the remarkable spread of the Pikes Peak (PP) inversion in *Drosophila pseudoobscura* demonstrated by the brilliant researches of Dobzhansky from 1958 onwards.

Thus, although species may resolve their genetic problems in alternative ways, the respective adaptations will break down among the intermediates where the two types meet. These, therefore, will be selectively eliminated at the interface, producing a sharp demarkation between the well-adjusted types.

This seems to be the situation encountered in *M. jurtina* in the difficult adaptations to be achieved near the edge of its range. There, as in West Cornwall and the Isles of Scilly, it may break up into a number of differently adapted gene-complexes; in other places, as in Ireland and in countries along the Mediterranean, it achieves distinct stable adaptations manifested in spotting of a different kind from that which characterizes the species over the greater part of its range. So powerful indeed are the stresses to which it must adjust itself in Scilly that it frequently readapts its spotting-type from one generation to another on the same small island when considerable seasonal or ecological changes occur there, and from one to another locality even on such a limited territory. The situation on Great Ganilly is particularly striking. This is an island of approximately 30 acres (0.12 km^2) divided, by an isthmus about 50 m wide and 90 m across, into two parts of approximately equal extent. They are occupied by differently adjusted population of *M. jurtina,* as shown by female spotting. These remain sharply distinct from one another, though no physical boundary separates them (Ford, 1965, pp. 81, 82), thus reproducing in miniature the boundary phenomenon that we have encountered between the two forms found in Southwest England. Indeed, the existence of powerful selection producing sharply marked genetic adaptation is now being widely recognized. A good example of it is provided by the adjustment of certain plants to soil containing salts of heavy metals, especially on and near the spoil-heaps of mines, a situation that is being studied with great success by Bradshaw and his colleagues (1965).

The alternative Southern English and East Cornish adjustments of *Maniola jurtina* must both be strongly buffered against the environmental diversity they normally encounter, although, at times, each appears more appropriate than the other in a single set of changing conditions along the Devon-Cornwall peninsula. These adjustments in the genes that control spotting must be of a discontinuous rather than a continuous kind. This is shown by the fact that the transition from populations with a lower to a higher spot-average is achieved not by increasing the proportion of higher spotted individuals gradually but by avoiding the frequency at which the females are unimodal at one spot. Yet that situation is not an impossible one since, though very rare, it has been encountered on a few occasions (see, for example, p. 282).

McWhirter has stressed how important such alternative genetic adaptations may be in strains of farm animals or crop plants. These adaptations would explain the disadvantageous effects sometimes encountered when what appear to be the same forms from similar environments are crossed.

The existence of distinct but alternative adjustments within different populations of *Maniola jurtina,* between which no geographical barrier exists, is in close accord with the results of Thoday and Boam (1959) in testing the deductions reached by Mather (1955) on disruptive selection. They found that stocks of *Drosophila melanogaster* selected in different directions diverged significantly even when a gene-flow of 50 percent was taking place between them.

Summary

1. The number of spots on the underside of the hindwings of *Maniola jurtina*, Satyridae, has a high female heritability and the frequency distributions of the spotting types have long been used as criteria of evolutionary adjustment in that species.

2. Different populations may be characterized by distinct spot-frequencies and two of these are studied here: (a) The *Southern English*, which is unimodal in both sexes, individuals with two spots being commonest in the males, and these with no spots being commonest in the females; (b) the *East Cornish*, which is also unimodal at two spots in the males, but bimodal in the females, with a greater mode at no spots and with a lesser at two spots.

3. The Southern English type is widespread in Europe and throughout Southern England westwards to Devon. The East Cornish is found from Central Cornwall eastwards into Devon for a distance which varies from year to year.

4. Both spotting-types are strongly buffered against environmental differences. The one generally changes to the other within a few yards and in the absence of any barrier, along a line running north and south across the Devon-Cornwall Peninsula.

5. When first discovered in 1956, a "reverse cline" had built up in the neighbourhood of the boundary between the two types.

6. Although the species has but one generation per year, this boundary had moved 3 miles eastwards in 1957. It then occupied the same position for three years. The present chapter gives an account, not previously published, of the changes to which it has been subject from 1961 to 1968.

7. In attempting to collect along the two transects in 1961, one south of Dartmoor and one (on which we later concentrated) to the north, we obtained a clear indication from spot-frequencies of the boundary between the Southern English and East Cornish forms, which, however, was not significant.

8. In 1962 we were able to identify the boundary with full significance. It had moved 1½ mile eastwards from the Tamar. As usual, the butterflies had higher spot-averages at the beginning of the emergence than subsequently.

9. The boundary between the two spotting types moved 11 miles further to the east in 1963.

10. In 1964 it had again moved eastwards and by about 8 miles. We were able to demonstrate its position within a few yards—in the middle of a continuous population where no barrier exists. It remained in the same place the year following.

11. In 1966 the boundary moved 40 miles eastwards. Except at the

western end of the transect, the population had not, however, become fully East Cornish.

12. In 1967 the boundary between the two types moved westwards for the first time in our experience, and by 44 miles. It was clearly identified between two populations separated by half a mile of agricultural land.

13. The two spotting types were very distinctly marked in 1968. The boundary between them had moved eastwards again by 6 to 16 miles. A clearly demarked isolated area of English spotting 1¼ to 2½ miles by 2 miles in extent, remained within the East Cornish stabilization near the 1967 boundary.

14. The significance of the boundary phenomenon in *Maniola jurtina* is discussed.

Acknowledgments

This work would not have been possible without the assistance of many of our friends, and to them we would like to express our great debt of gratitude: Mr. and Mrs. S. Beaufoy, Mr. D. L. Blackwell, Professor Th. Dobzhansky, Mr. L. E. Gilbert, Professor E. Mayr, Dr. and Mrs. V. Scali, and those undergraduates from the Department of Zoology at Oxford who gave up part of their vacation to help us.

We would also like to express our thanks to the following organizations for their support: E.R.C. to the Nature Conservancy for the years 1961 to 1962, W.H.D. to the Royal Society for a research grant and for the loan of a calculating machine, and K.G.McW. to the Science Research Council for the years 1965 to 1968.

References

BAILEY, D. W. 1956. Re-examination of the diversity in *Partula taeniata*. Evolution, 10:360-366.
BRADSHAW, A. D., T. S. MCNEILLY, and R. P. GREGORY. 1965. Industrialization, evolution and the development of heavy metal tolerance in plants. *In* Goodman, G. T. et al., eds. Ecology and the Industrial Society. Oxford, Blackwell. pp. 327-343.
CREED, E. R., W. H. DOWDESWELL, E B. FORD, and K. G. MCWHIRTER. 1959. Evolutionary studies on *Maniola jurtina*: the English mainland, 1956-57. Heredity, 13:363-391.
———— E. B. FORD, and K. G. MCWHIRTER. 1964. Evolutionary studies on *Maniola jurtina*: The Isles of Scilly, 1958-59. Heredity, 19:471-488.
———— W. H. DOWDESWELL, E. B. FORD, and K. G. MCWHIRTER. 1962. Evolutionary studies on *Maniola jurtina*: the English mainland, 1958-60. Heredity, 17:237-265.
DOWDESWELL, W. H., and E. B. FORD. 1953. The influence of isolation on variability in the butterfly *Maniola jurtina* L. Sympos. Soc. Exp. Biol., 7:254-273.
———— and K. G. MCWHIRTER. 1967. Stability of spot-distribution in *Maniola jurtina* throughout its range. Heredity, 22: 187-210.
FORD, E. B. 1965. Ecological Genetics. 2nd ed. London, Methuen.
MATHER, K. 1955. Polymorphism as an outcome of disruptive selection. Evolution, 9:52-61.

McWhirter, K. G. 1957. A further analysis of variability in *Maniolo jurtina* L. Heredity, 11:359-371.

———— 1969. Heritability of spot-number in Scillonian strains of the Meadow Brown butterfly (*Maniola jurtina*, L.). Heredity, 24:314-318.

Sumner, F. B. 1930. Genetic and distributional studies of three subspecies of *peromyscus*. J. Genet., 23:275-376.

Thoday, J. M., and T. B. Boam. 1959. Effects of disruptive selection. II Polymorphism and divergence without isolation. Heredity, 13:205-218.

10

The Genetic Basis of a Cell-Pattern Homology in *Drosophila* Species

T. M. RIZKI and ROSE M. RIZKI

The University of Michigan, Ann Arbor, Michigan

Introduction 289
Kynurenine Distribution in the Larval Fat Body 290
Hybrid Autofluorescent Patterns 291
Acknowledgments 297
References 298

Introduction

The authors have the good fortune of being among the students who are now honoring Professor Theodosius Dobzhansky for his distinguished leadership in the area of evolutionary biology and genetics. Most of the ideas and theoretical grounds on which we have analyzed our present work to study the problem of speciation in relation to developmental genetics are well documented in "Genetics and the Origin of Species" by Professor Dobzhansky (1951). The system with which we have worked is a specific biochemical phenotype of a cell in *Drosophila*, and we have attempted to exploit this system in order to understand the nature of the genetic changes that may have taken place in the course of the evolution of the two species, *D. melanogaster* and *D. simulans*, to stabilize a homologous cell pattern.

289

Kynurenine Distribution in the Larval Fat Body

Tryptophan pyrrolase activity is localized in the anterior region of the larval fat body of D. *melanogaster*. Assessment of in vivo tryptophan pyrolase activity in the cells of the fat body is readily accomplished by examination of the fluorescence of the cells, since kynurenine, a subsequent product in this enzyme pathway, possesses a characteristic sky-blue autofluorescence (Rizki, 1961). Mutation at the *vermilion* (*v*) locus, the structural gene for tryptophan pyrrolase, results in the loss of enzyme activity, and the anterior fat cells of *v* mutant strains lack kynurenine autofluorescence. In gynandromorphs heterozygous for the *v* allele, the fat cells behave autonomously with respect to kynurenine production, and it may be concluded that the kynurenine autofluorescence observed within an individual fat cell thus represents a synthetic product of that specific cell (Rizki and Rizki, 1968). The distribution of the autofluorescent kynurenine cells is continuous in v^+ strains and forms a specific pattern of bright sky-blue autofluorescence distinct from the remaining fat cells which are engaged in pteridine synthesis. This pattern of kynurenine synthesizing cells can be modified under the influence of various mutant genes. For example, the combination of v^{36f}, v^1, or v^2 with the *suppressor-of-sable* (su^2-*s*) gene results in altered patterns of kynurenine distribution in the fat body (Rizki, 1963; Rizki and Rizki, 1968). Furthermore, mutants at loci other than the structural gene *v* are also capable of modifying the pattern of autofluorescent cells and an example of such modification is visualized when the mutant gene, *sepia* (*se*) is introduced into the genotype. This sequence of studies on the larval fat cells of D. *melanogaster* has led to the conclusion that the sets of gene loci involved in the final expression of a specific function may differ for different cells in the body (Rizki, 1964).

Several species of *Drosophila* larvae have been examined, including D. *pseudoobscura*, D. *persimilis*, D. *virilis*, D. *ananassae*, D. *willistoni*, D. *pachea*, and D. *simulans*. The general organization of the larval fat body is similar in these species and the synthesis of kynurenine is restricted to the anterior group of fat cells forming a fluorescence pattern similar to that reported for D. *melanogaster*. The genes that influence kynurenine distribution in the fat body are known to mutate spontaneously and the so-called wild-type alleles must be under constant selection pressure in order to stabilize this species-specific trait. Since more than one gene locus is involved in the formation of the kynurenine distribution pattern in D. *melanogaster*, it is reasonable to ask whether the persistence of the autofluorescent cell pattern in a pair of closely related species shares a common genetic basis. One approach to this question involves examination of hybrids between two species, and such a study was undertaken with the closely related species, D. *melanogaster* and D. *simulans*.

Hybrid Autofluorescent Patterns

With one exception, hybrid crosses utilized *D. melanogaster* females and *D. simulans* male flies. The *D. melanogaster* stocks included a *yellow* (*y*) *vermilion* (*v*) strain and an attached-X stock carrying no free Y (♀ yvf.-/0 and ♂ yvf.Y^LY^S). A wild-type *D. simulans* strain was obtained from collections made at St. Louis, Missouri, by Professor H. L. Carson, and a *v* mutant of this species was kindly provided by Professor M. M. Green. Newly emerged females of *D. melanogaster* in groups of four or five were placed with *D. simulans* aged males that had been isolated from the females of their own species for at least five days. The eggs were collected on paper spoons containing Cream of Wheat and molasses medium smeared with Fleischmann's (baker's) yeast and honey paste. The hybrid larvae were raised in dishes with this same medium and fat bodies were dissected shortly after puparium formation when kynurenine fluorescence is intensely developed. The freshly dissected fat bodies in Ringer solution were examined with a fluorescence microscope as previously described (Rizki, 1961).

In further discussion of the hybrid crosses, the subscript *m* will designate the genes from the *melanogaster* parent and *s* will indicate the *simulans* chromosomes.

The extent of the cells displaying kynurenine autofluorescence in the fat body of the *Ore-R* wild-type strain has been described previously (Rizki, 1961) and is illustrated in a lateral half of a fat body from a *D. melanogaster* larva (Fig. 1). This pattern is similar in *D. simulans* with the exception that approximately 30 percent of the specimens examined showed some quantitative difference in fluorescence intensity in the two or three most posterior cells at the limit of the kynurenine pattern. As first noted by Sturtevant (1920) only female progeny survive from a mating of a *D. melanogaster* female with a *D. simulans* male. With use of *y* females and wild-type *D. simulans* males for this mating, the hybrid female larvae are recognizable by the black mouthparts, while the hybrid male larvae have yellow mouthparts. The latter die in the first or second instar even though they manage to survive past the time of puparium formation of the female larvae. The fat bodies of male hybrid larvae were examined at this developmental stage, and no kynurenine was present irrespective of the extended duration of development. The female hybrid larvae v_m^+/v_s^+ displayed an autofluorescence pattern different from that of either parental species, but the organization of the tissue mass is the same as in the parental species. A discontinuity in the pattern of autofluorescent cells is found such that following a short region of fat cells lacking kynureinine, a small group of fat cells is engaged in kynurenine synthesis. The latter appears either in a banded sequence of two rows of cells or stacked in a pyramidal fashion at the lateral border of the fat body. The number of cells in the discontinuous

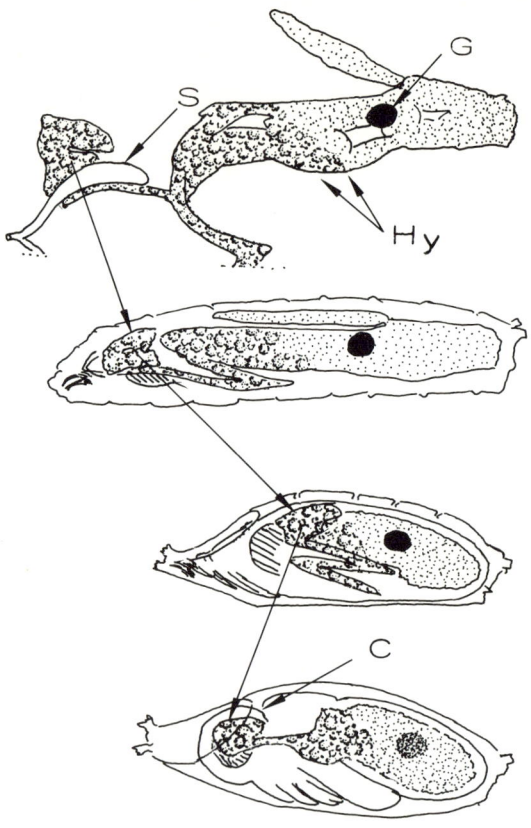

Fig. 1. Series of diagrams illustrating the organization of the larval fat body with its relative position in the larval and pupal stages. The upper drawing shows the dissected right lateral half of the fat body, with the anterior region indicated by the position of the salivary glands (S) and the posterior region indicated by the position of the gonad (G). The dark, shaded area corresponds to the distribution of the kynurenine autofluorescent cells, and the remaining fat cells show pteridine fluorescence. The region in which pattern disturbances appear in hybrid fat bodies is indicated by two arrows at HY. The subsequent drawings in this series depict the locations of the fat body regions in the larva and two puparia. The position and displacement of the fat cells overlying the salivary gland are followed by arrows; this tissue comes to rest in the head capsule, C, as the cephalic complex is everted. Soon after this stage, individual fat cells dissociate from the tissue mass.

groups ranged from 2 to 14 in various hybrid specimens, but the pattern within a single specimen was bilaterally symmetrical. There also was variability in the number of nonfluorescing cells interrupting the continuity of the kynurenine fluorescing group and, again, this feature provided bilateral symmetry of pattern. Disturbance in the kynurenine patterns was noted in hybrid larvae that were v_m/v_s^+ as well as those derived from the cross of a v_m^+ female with the v mutant male of *simulans*. This pattern modification from the various combinations of parents is illustrated in Figures 1 and 2. No kynurenine fluorescence is found in the fat bodies of hybrid v_m/v_s females and this is expected since these genes are homologous mutations.

In order to obtain male hybrid larvae, attached-X *D. melanogaster* females were crossed to *D. simulans* males. Sturtevant (1929) noted that such matings yield male progeny, and for the present study the markers in the *D. melanogaster* female were *yvf*, while this particular stock lacked a free Y chromosome. The male hybrid larvae were thus XO specimens,

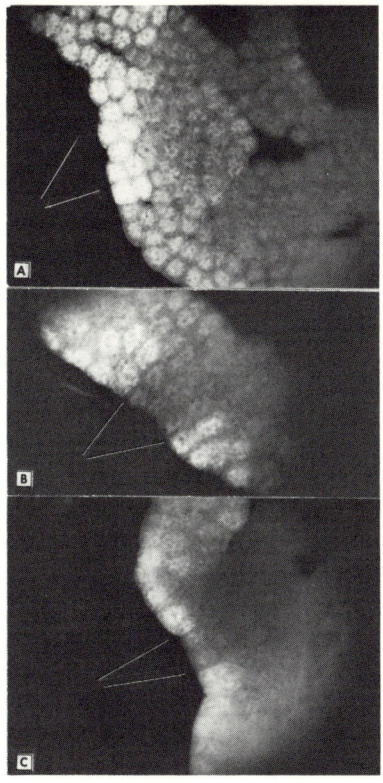

Fig. 2. Photographs of the larval fat body removed from the early puparial stage. These are freshly dissected tissues in Ringer solution viewed with the fluorescence microscope, and each corresponds to the region in Fig. 1 labeled HY. (a) Ore-R wild type; (b) and (c) v_m^+/v_s^+ hybrids demonstrating the interruption in continuity of kynurenine-fluorescing cells. The brightly fluorescent regions are the cells with sky-blue fluorescence of kynurenine, while the residual glow represents some scatter from these cells and the deep purple autofluorescence of pteridines found in the posterior cells.

with the wild-type markers clearly indicating derivation of the X from the *D. simulans* parent. Neuroblast chromosomes were examined in order to further verify the absence of a Y chromosome. The fat bodies of these male hybrids showed the pattern of discontinuity of kynurenine cells described for female hybrid larvae. In each of the crosses a few larvae showed the fluorescence pattern characteristic of the parental strains, and the combined data for these observations is presented in Table I.

It is interesting to note in retrospect Sturtevant's (1920) conclusion that the type of various bristle peculiarities found in *melanogaster-simulans* hybrids, but not in the parental species, must be due to complemental genes carried by the two species. This conclusion was expanded by the studies of Schultz and Dobzhansky (1933) on the interspecific triploid hybrid of these two species. In triploid hybrids, the normal bristle pattern was retained, indicating that the double genic complement supplied by *melanogaster* was dominant to a single dose of *simulans* genes. They concluded that a similar system of balance of sex factors must exist in these two species since hybrid individuals are clearly not intersexes and yet the types of hy-

TABLE I
Fluorescence Patterns in Hybrid Fat Bodies

Genotype of F_1 hybrid	Type of pattern		
	Normal	Disrupted	Total
$y\ v_m^+/y^+\ v_s^+$	2	18	20
$y\ v_m^+/y^+\ v_s$	1	14	15
$y\ v_m/y^+\ v_s^+$	2	13	15
$y\ v_m/y^+\ v_s$	–	–	5
$(0)/y^+\ v_s^+$	1	7	8
$a(♀)\ v_s^+/v_m^+\ (♂)$	3	2	5

[a]The geographic origin of this *simulans* strain is not known; the males were Sw-b and the F_1 hybrid larvae were raised by Miss K. Peterson, to whom we express our appreciation for this material.

brid intersexes that can be obtained in triploids mimic the range of intersexuality found in pure *D. melanogaster* intersexes. Complementary interactions of factors in *melanogaster-simulans* hybrids were also substantiated by Pontecorvo (1943). Sturtevant's prediction that the F_2 generation of a hybrid mating should yield offspring with additional new characters was fully borne out later when interspecific hybrids of cotton were obtained. *Gossypium barbadense-hirsutum* hybrids displayed a variety of blending as well as instances of characteristic expression of specific mutant phenotypes. Harland (1939) proposed that the complex intermingling obtained in these hybrid crosses resulted from the combination of background modifiers that were different in the parental species. A specific gene character would be affected in its expression by the entire complement of modifiers, and the co-adapted system of modifiers associated with the expression of major genes would be disrupted further in succeeding generations of species hybrids. Loss of genic balance in interspecific hybrids has been well documented (Stebbins, 1958), and one extremely interesting case is the report on the behavior of the satellite chromosome in interspecific hybrids of *Crepis* reported by Navashin (1934). Reexamination of such modifications would prove particularly interesting in the light of modern concepts of regulation of gene action, particularly the nucleolar genes (Ritossa et al., 1966). At the phenotypic level, studies on hybrids provide a higher resolving power of genic interaction and divergence in the course of evolution than the currently available molecular hybridization techniques. It is encouraging, however, that the DNA-DNA hybridization experiments offer additional evidence of relatedness at the level of base sequence and that the studies of Laird and McCarthy (1968) on *D. melanogaster* and *D. simulans* indicate approximately 80 percent gene homology between these two sibling species. Unfortunately, assessment of gene interaction in hybrid individuals is limited by the possibility of obtaining actual hybrid zygotes, since the disturbances in morphogenesis in hybrids are so drastic as to allow little opportunity to examine the stabilizing selection of a norm of action in preserving the patterns of differentiation and integration of cells into tissues in the development of the organism. The present study on F_1 hybrid zygotes of *D. melanogaster* and *D. simulans* is in accord with Sturtevant's observations and we would agree with his interpretation that major morphological features have been retained in the two species through the interplay and selection of a variety of modifier genes. Only when the two independently selected systems are brought together does a lack of complete genetic correspondence reveal itself.

We may now inquire more closely into the nature of the differences reflected in the hybrid autofluorescent patterns involving the expression of the gene v. The v loci in the two species are homologous since complementation at the level of eye pigment does not occur in v_m/v_s hybrids, and the absence of tryptophan pyrrolase activity is further indicated by the lack of kynurenine autofluorescence of the fat cells. The lack of kynurenine synthesis in v_m/v_s hybrids also indicates that there are no other genetic elements in the strains that can modify or suppress the mutant effect of the v locus, the structural gene for tryptophan pyrrolase. A difference in the autofluorescence pattern of the v^+/v^+ hybrids, which distinguishes them from either of parental species, now raises the question whether the new hybrid pattern is due to isoallelic differences at the structural locus in the two normal strains of the species, or whether functional disharmony in the residual genotype affects kynurenine synthesis in some of the hybrid cells. In order to discriminate between these two alternative explanations, hybrids were synthesized so that the v^+ structural gene from one of the species would be utilized for transcription. Hybrid gentoypes v_m^+/v_s and v_m/v_s^+ gave an equivalent disturbance of pattern, so that it seems that isoallelic differences at the v^+ locus in the two species do not provide a likely explanation for the observed effect. Findings in the hybrid XO males are also consistent with the sufficiency of only one X chromosome in the absence of the Y for the transcription of the v^+ gene. The pattern disturbances in these males as well as in the hybrid females can not be explained on the basis of interaction of subunit products of the v_s^+ and v_m^+ genes at the cell level. Pattern disturbances do not appear to be the result of a maternal influence since the disrupted pattern was obtained in the hybrid progeny of both *melanogaster* and *simulans* mothers; the number of specimens obtained from the latter mating was limited, but the appearance of the interrupted pattern in even the few individuals examined is important to our analysis. This series of observations on kynurenine synthesis in the fat body cells reveals that the pattern of cell involvement has been retained in the two closely related species, but the gene complexes that stabilized this pattern are presumably not the same, since disturbances are evident in interspecific combination.

That the cells within the fluorescence pattern act as autonomous units has been demonstrated by producing gynandromorph mosaics of v/v^+ fat body of *D. melanogaster* (Rizki and Rizki, 1968). Since the changes appearing in the hybrid fat body are bilaterally symmetrical, we may conclude that these modifications represent alterations of the autofluorescent cell pattern. There must be a number of cell modifiers for which the two species are polymorphic, so that when different modifiers or polygenic systems are brought together, the quantitative differences among hybrid zygotes are established. Yet these polygenic systems within the species must exist as equipotential genotypes with respect to the kynurenine cell pattern formation. It is this property of the system which leads to the conclusion that the position and function of each cell must be specified through the selection of unique sets of genetic elements in the course of evolution so that the higher levels of cell-cell integration begin to emerge as patterns. Both Stern's (1956) and Sturtevant's (1961) investigations on bristle pattern formation indicate this feature, and it is noteworthy that while a bristle is a morphological trait and tryptophan pyrrolase activity as visualized is a biochemical trait, both are single-cell phenotypes and are equally susceptible to disturbing effects of imbalance of modifiers. This comparison is particularly gratifying since model experiments in quantitative genetics in which bristle traits are used may not be as for from reality as recently feared by Robertson (1967). In multicellular organisms, various degrees of hybrid inviabilities and breakdown may be the reflection of failure in cell-cell functional harmony in the tissue-organ patterns for which modifier systems have been selected. On the other hand, this adaptive cell-cell harmony may not be necessary for survival of an individual cell, since experimentally it has been possible to produce individual cell hybrids between widely separated species (Ephrussi and Weiss, 1965).

It is, of course, not possible to determine whether the organization of the kynurenine-producing cells into various tissue masses has any significance to the developmental physiology of *Drosophila*. We may examine various aspects of tryptophan metabolism with its relationship to a physiological role. The biochemical pathways leading to the brown component of the eye pigment are well known and some of the related compounds other than intermediate precursors of brown pigment have been isolated from *Drosophila* (Kikkawa, 1953). Kikkawa (1934a, b) explored one aspect of the adaptive significance of the ommochrome pigments by experimentally evaluating the relationship between these pigments and the response of

Drosophila to light. By using normal, *brown* (*bw*), *v bw* mutants, and *v bw* raised on kynurenine (phenocopy *bw*), he demonstrated that the response to light of these genotypes was in the following order: normal > *bw* = phenocopy *bw* > *v bw*. If response to light is of importance in the ecology of *Drosophila*, then the loss of ommochrome pigments can serve to deprive the zygotes of a normal response to an important ecological factor. On the other hand, ommochromes are phenoxazones, the chromophore group which includes actinomycin, cinnabarinic acid, and anthramycin (Weissbach and Katz, 1961; Rao and Vaidyanathan, 1966; Morgan et al., 1967). No growth regulatory effects of these ommochromes have been demonstrated in *Drosophila* thus far, but the carcinogenic effects of a number of tryptophan metabolites found in *Drosophila*, including 3-hydroxy-L-kynurenine and xanthurenic acid, have been demonstrated by Bryan et al. (1964) in mice. These metabolites may therefore affect the adaptive value of *Drosophila* zygotes at various levels, and from the viewpoint of adaptogenesis (Schmalhausen, 1949) it may be worthwhile to consider the organization of the kynuerenine-producing cells in a definite pattern. The region of the fat body that overlies the salivary gland as a pair of triangular masses of tissue has been designated the head fat mass (Beadle, 1937). This mass constitutes approximately 40 percent of the kynurenine cell population. When the head capsule is everted this mass of fat cells enters the cavity of the head capsule, and, following dissociation of the fat tissue, the individual fat cells of this region come to lie in close proximity to the developing eye. This specific association is striking for it appears to place the eye-pigment precursors precisely where they are needed (Fig. 1). That this relationship may not be coincidental is further suggested by the manner in which suppression of the *v* mutant alleles expresses itself with respect to the morphology of the fat body. In all cases of combination of the suppressible *v* alleles with the su^2-*s* suppressor gene, kynurenine synthesis is always restored to the cells of the head fat, and variation is noted only in the more posterior regions (Rizki and Rizki, 1968). In the examination of the hybrid patterns we clearly see that discontinuity of pattern arises only in the posterior domain of the cell pattern and not in the anterior. In the course of evolution, the cell pattern emerging has resulted through the selection of modifiers that have conferred a high degree of developmental homeostasis to the region of the larval fat body where its functional biochemistry is coupled with the developmental processes of eye-pigment formation.

Acknowledgments

This research was supported by Grant GB-6110 from the National Science Foundation. We would like to express our gratitude to Professor Irwin Oster of Bowling Green State University for providing strains of *Drosophila* used in this study.

References

BEADLE, G. W. 1937. Development of eye colors in *Drosophila*: Fat bodies and malpighian tubes as sources of diffusible substances. Proc. Nat. Acad. Sci. (U.S.A.), 23:146-152.

BRYAN, G. T., R. R. BROWN, and J. M. PRICE. 1964. Mouse bladder carcinogenicity of certain tryptophan metabolites and other aromatic nitrogen compounds suspended in cholesterol. Cancer Res., 24:596-602.

DOBZHANSKY, TH. 1951. Genetics and the Origin of Species, 3rd ed., New York, Columbia University Press.

EPHRUSSI, B., and M. C. WEISS. 1965 Interspecific hybridization of somatic cells. Proc. Nat. Acad. Sci. (U.S.A.), 53:1040-1042.

HARLAND, S. C. 1939. The Genetics of Cotton. London, Jonathan Cape Ltd.

KIKKAWA, H. 1943a. Problems of tryptophan metabolism in insects. Kagaku, 13:282-285.

——— 1943b. Problems of tryptophan metabolism in insects. Kagaku, 13:319-325.

——— 1953. Biochemical genetics of *Bombyx mori* (Silkworm). Advances Genet. 5:107-140.

LAIRD, C. D., and B. J. MCCARTHY. 1968. Magnitude of interspecific nucleotide sequence variability in *Drosophila*. Genetics, 60:303-322.

MORGAN, L. R., R. SINGH, V. Sylvest, and D. WEIMORTS. 1967. Oxidation of o-aminophenols by mouse and human melanoma dihydroxyphenylalanine oxidase and dihydroxyphenylalanine. Cancer Res., 27:2395-2407.

NAVASHIN, M. 1934. Chromosome alterations caused by hybridization and their bearing upon certain general genetic problems. Cytologia, (Tokyo) 5:169-203.

PONTECORVO, G. 1943. Viability interactions between chromosomes of Drosophila melanogaster and Drosophila simulans. J. Genet., 45:51-66.

RAO, P. V. SUBBA, and C. S. VAIDYANATHAN. 1966. Enzymic conversion of 3-hydroxyanthranilic acid into cinnabarinic acid. Biochem. J., 99:317-322.

RITOSSA, F. M., K. C. ATWOOD, and S. SPIEGELMAN. 1966. On the redundancy of DNA complementary to amino acid transfer RNA and its absence from the nucleolar organizer region of *Drosophila melanogaster*. Genetics, 54:663-676.

RIZKI, T. M. 1961. Intracellular localization of kynurenine in the fat body of Drosophila. J. Biophys. Biochem. Cytol., 9:567-572.

——— 1963. Genetic control of cytodifferentiation. J. Cell Biol., 16:513-520.

——— 1964. Mutant genes regulating the inducibility of kynurenine synthesis. J. Cell Biol., 21:203-211.

——— and R. M. RIZKI. 1968. Allele specific patterns of suppression of the vermilion locus in *Drosophila melanogaster*. Genetics, 59:477-485.

ROBERTSON, A. 1967. Animal breeding. Ann. Rev. Genet. 1:295-312.

SCHMALHAUSEN, I. I. 1949. Factors of Evolution. Philadelphia, The Blakiston Co.

SCHULTZ, J., and TH. DOBZHANSKY. 1933. Triploid hybrids between *Drosophila melanogaster* and *Drosophila simulans*. J. Exp. Zool., 65:73-82.

STEBBINS, G. L. 1958. The inviability, weakness, and sterility of interspecific hybrids. Advances Genet., 9:147-215.

STERN, C. 1956. Genetic mechanisms in the localized initiation of differentiation. Sympos. Quant. Biol., 21:375-382.

STURTEVANT, A. H. 1920. Genetic studies on *Drosophila simulans*. I. Introduction. Hybrids with *Drosophila melanogaster*, Genetics, 5:488-500.

——— 1929. Contributions to the genetics of *Drosophila simulans* and *Drosophila melanogaster*. I. The genetics of *Drosophila simulans*. Publ. Carnegie Inst., 399:1-62.

——— 1961. Bristle pattern of *Drosophila*. Science, 134:1436.

WEISSBACH, H., and E. KATZ. 1961. Studies on the biosynthesis of actinomycin. Enzymic synthesis of the phenoxazone chromophore. J. Biol. Chem., 236:16-18.

11

Ecological Factors and the Variability of Gene-Pools in *Drosophila*

JOHN BEARDMORE

Department of Genetics, University College of Swansea, Swansea, U. K.

Introduction 299
Relations Between Ecological Heterogeneity, Genetic Variability, and
Fitness in *Drosophila* Populations 301
Conclusions 312
Summary 313
Acknowlegments 313
References 313

Introduction

In 1950 H. J. Muller wrote a paper in which he defined and discussed the concept of the genetic load. The main purposes of the paper were to point out the comparatively large fraction of deaths in man that might reasonably be attributed to genetic causes and to consider the frequencies, mutation rates, dominance relations, and other characteristics of the genes involved together with the effect of cultural factors upon the total burden or load of such genes. In addition to these considerations, however, Muller provided a conceptual framework for the structure of the gene pool in outbreeding organisms such as man. The crux of the argument is contained in pages 135 and 136 of his article. To Muller it appeared that the greater part of the genetic variability carried within populations of diploid outbreeding organisms must be due mainly to recurrent mutation from dominant wild type to deleterious recessive genes, and the frequency of such genes would be determined by mutation frequency and simple selection against the individuals possessing such mutant genes. This view has been contested by many, most notably Dobzhansky, who have regarded the major portion of such genetic variability as being due to various types of selection of which the most obvious is heterosis at individual loci. Few examples of such heterozygote advantage at a single locus are known, but this may mean that insufficient attention has been paid to such cases. There is no doubt that part of the intellectual attraction of heterosis as the primary mechanism for maintaining polymorphisms lies in the fact that the persistence of the polymorphism is better guaranteed than with some other mechanisms, such as those dependent upon alternating selective advantages and alternative niches (Ludwig, 1950; Levene, 1953; Li, 1955; Dempster, 1955). However, as I have suggested elsewhere (Beardmore, 1963), if single-gene heterosis or super-gene heterosis were really so important it would seem likely that, at least in those organisms, such as *Drosophila*, with a large reproductive potential, balanced lethal systems acting early in development would be evolved. These are relatively uncommon and I am led to the view that it is often the polymorphism as a whole which is adaptively important, in relation to heterogeneity of habitat in space and time (Dobzhansky, 1951; Cain and Sheppard, 1954a). Selection favouring two or more phenotypes in the same population as a result of spatial or temporal variety in the environment has been termed disruptive (Mather, 1955) or diversifying (Dobzhansky, 1968), and experimental evidence supporting Mather's suggestion that disruptive selection acting on quantitative characters would be effective in building up polymorphism was provided by the experiments of Thoday and his collaborators (Thoday and Boam, 1959) and by Clarke and Sheppard (1960). The results of these workers showed clearly that this type of selection is capable of forming polymorphic arrays even with considerable gene exchange between the two subpopulations with different optimum phenotypes.

Despite the elegant theoretical formulations of Levins (1962, 1963, 1964) few attempts of a quantitative kind have been made to relate genetic variability in the form of polymorphism to measurable ecological heterogeneity, although from the work of Cain and Sheppard (1954b) it may be inferred that the degree of polymorphism for shell color in *Cepaea* is relatable to variety of the habitat. Dubinin and Tiniakov (1947) provided evidence of a correlation between the frequency of inversions and urbanisation of the habitat (possibly leading to greater ecological variety) in populations of *Drosophila funebris,* and da Cunha et al. (1959 and earlier) showed that a positive correlation existed between the frequency of gene arrangements in, and ecological variety of the habitat of, populations of *Drosophila willistoni.* Indirect evidence for the possible ecological role of gene arrangements in *D. pseudoobscura* comes from the work of Lewontin (1958), who observed that a polymorphism for the Arrowhead and Pikes Peak gene arrangements broke down after a period of quasistability to yield an essentially monomorphic situation in a laboratory population. Lewontin concluded that under his laboratory conditions a relatively uniform environment was created in which this particular type of variability was not favoured as it would be in nature. However, evidence that such breakdown of chromosomal polymorphisms is very uncommon is summarized by Dobzhansky and Pavlovsky (1960). Because of the peculiar nature of gene arrangements the thesis that genetic variability of this type is related to ecological heterogeneity has been disputed on the grounds that the geographical distribution of gene arrangements lies rather in the interplay of selective forces on the amount of recombination which is, of course, partly determined by inversions (Carson, 1955).

Relations Between Ecological Heterogeneity, Genetic Variability, and Fitness in Drosophila Populations

For some years my colleagues and I have been interested in the relations between ecological heterogeneity, genetic variability, and fitness of laboratory populations of *Drosophila*. Accurately defining any of them poses very considerable problems, and in practice one simply isolates one or more factors of each and measures these factors in the hope that the answer obtained bears some correspondence to reality. How real any relationships discovered are, can, of course, be tested only by their predictive value in future work.

Van Delden and Beardmore (1968) have shown that in relatively inbred laboratory populations of *Drosophila* a significant positive correlation exists between fitness measured as productivity, and genetic variability measured as heritability for a bristle character. It was clearly important to try to see if this relationship was of a general nature and to extend the factors considered to include the heterogeneity of the environment. To this end two experiments were carried out. The first of these involved collecting wild inseminated *D. melanogaster* females from widely separated localities in the United States and assaying genetic variability and fitness in the same way as previously. Eight localities ranging from New York in the East to Iowa in the Midwest were sampled.

Flies were collected by using fermenting banana bait in large glass jars placed in suitable spots for 12 to 24 hours. Single inseminated females were isolated to establish individual strains.

Tests of productivity were carried out by placing single pairs of flies in 4-inch by 1-inch vials and counting the number of offspring produced in each of 20 replicates per strain in each of four temperatures, 15°C, 20°C,

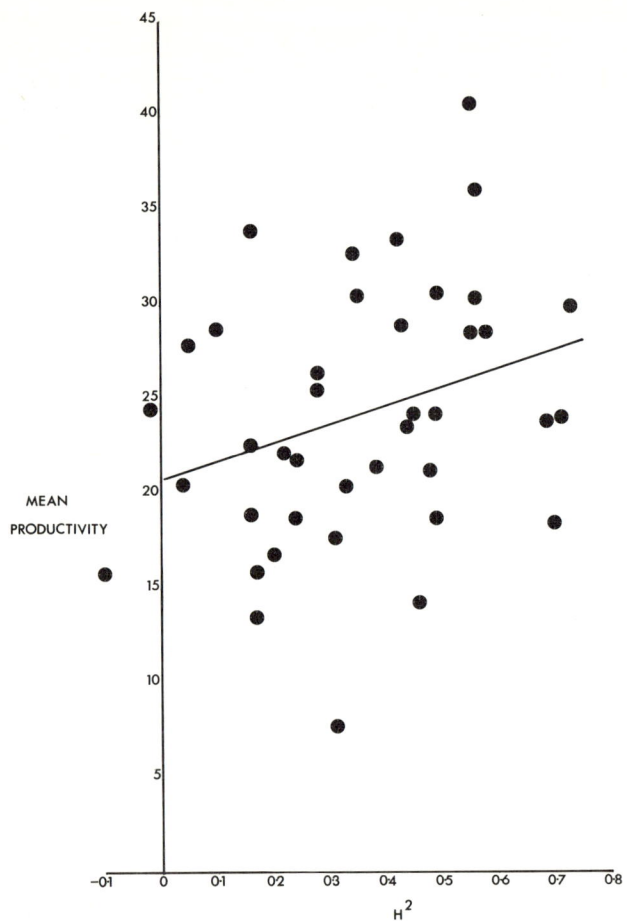

Fig. 1. Productivity (y) and heritability of sternopleural chaeta number (x) in geographic strains of Drosophila melanogaster.

25°C, and 30°C. Within each population (locality), between four and six strains were assayed and the mean productivity per strain measured over the four environments was calculated.

Incubators of identical pattern were used for each temperature, and the number of days allowed for emergence was 30, 22, 14, and 12, respectively, for the four test temperatures. Assays of genetic variability depended upon estimates of heritability of sternopleural chaeta number carried out as described previously (Beardmore and Levine, 1963). Five males and five females counted in each of 10 cultures raised at 25°C provided the estimates of progeny bristle number for each strain. In Figure 1 the regression of

TABLE I

Analysis of Variance of the Regression of Productivity on Heritability of Sternopleural Number in Geographic Strains of *D. melanogaster*

Item	M.S.	N	P
Regression	163.84	1	<0.02
Between populations	98.6	7	<0.02
Residual	30.8	31	
Total		39	

mean productivity on heritability is shown for these strains and the analysis of variance is given in Table I. The regression is seen to be positive and statistically significant.

Other workers (Lerner, 1953; Lewontin, 1956) have pointed to the relationship often found between high fitness and low variation in fitness characters. This relationship is also found here. Table II shows the coefficient of correlation between mean productivity and the coefficient of variation based on between-replicate variance in productivity. Under all temperature conditions the more productive populations tend also to be the least variable in this component of fitness. As more productive populations are also genetically richer an inverse relationship between the homeostatically significant property of productivity variability and that of genetic variability is indicated. It is probably also of significance that the value of r increases with extreme temperature in both directions.

Not surprisingly there are also significant differences in mean viability

TABLE II

The Correlation Coefficient Between Mean Viability and Coefficient of Variation of Viability for Geographic Strains Tested at Four Temperatures

Temperature °C	r	t_{38}	P
15	−0.74	5.16	<0.001
20	−0.71	6.24	<0.001
25	−0.49	3.94	<0.01
30	−0.96	21.9	<0.001

(Table I) between populations. These arise partly no doubt from true large-scale factors of ecology much as Timofeef-Ressovsky (1935) found for *D. funebris* from Asia and Europe. I have recalculated Timofeef-Ressovsky's data (Beardmore, 1966) and shown that a striking correlation exists between the mean viability, tested in a range of conditions, of his geographic strains and the annual temperature variation to which the natural populations from which they came were exposed. It is probable that *D. melanogaster* populations are influenced in much the same way, but it soon became evident in this experiment that other factors would make it difficult to compare geographic populations of *D. melanogaster* on this basis alone. In particular, the widespread use of insecticides means that shifts of population size, which are large and erratic in nature, frequently occur. Several samples were collected in the vicinity of fruit markets, whose owners informed me that spraying with insecticides was frequently carried out. Clearly a proper comparison of natural populations demands a careful choice of geographic areas comparable in as many respects as possible, and this was certainly not the case in this study. Ideally one would wish to be able to derive estimates of ecological variability both in time and space for natural populations, but as this is obviously an immensely difficult task it was decided to focus attention for the present upon laboratory populations exposed to temporal variations in temperature. Earlier work had shown that populations exposed to well-regulated short-term variations of temperature tend to develop or retain greater levels of genetic variability than populations kept in constant temperature conditions (Beardmore and Levine, 1963). In addition, such variable temperature populations are of greater fitness as determined by larval-adult survival.

Dr. T. C. Long has kindly allowed me to describe briefly some of the results he has obtained in our laboratory studying four groups of such populations, all initially similar and derived from newly captured wild flies from Groningen, Holland. The four groups were kept in conditons of constant temperature, daily cycle, 32-day cycle, and 96-day cycle, respectively. The cycle is from 30°C to 20°C and back in all cases.

In Figure 2 the productivity of these populations is plotted against the root of additive genetic variance for sternopleural chaeta number, both parameters being determined after a considerable period estimated at about 42 generations of adaptation. Genetic variance was estimated by determining the heritability of 5th sternite bristle number using parents that had developed at 25°C, and progeny raised at the same temperature. Productivity was measured by counting the number of progeny produced under crowded conditions at 25°C by 100 females given an egg-laying period of 24 hours.

Both parameters differ somewhat from those used by Beardmore and Levine (1963). The measure of genetic variability used by Long is the

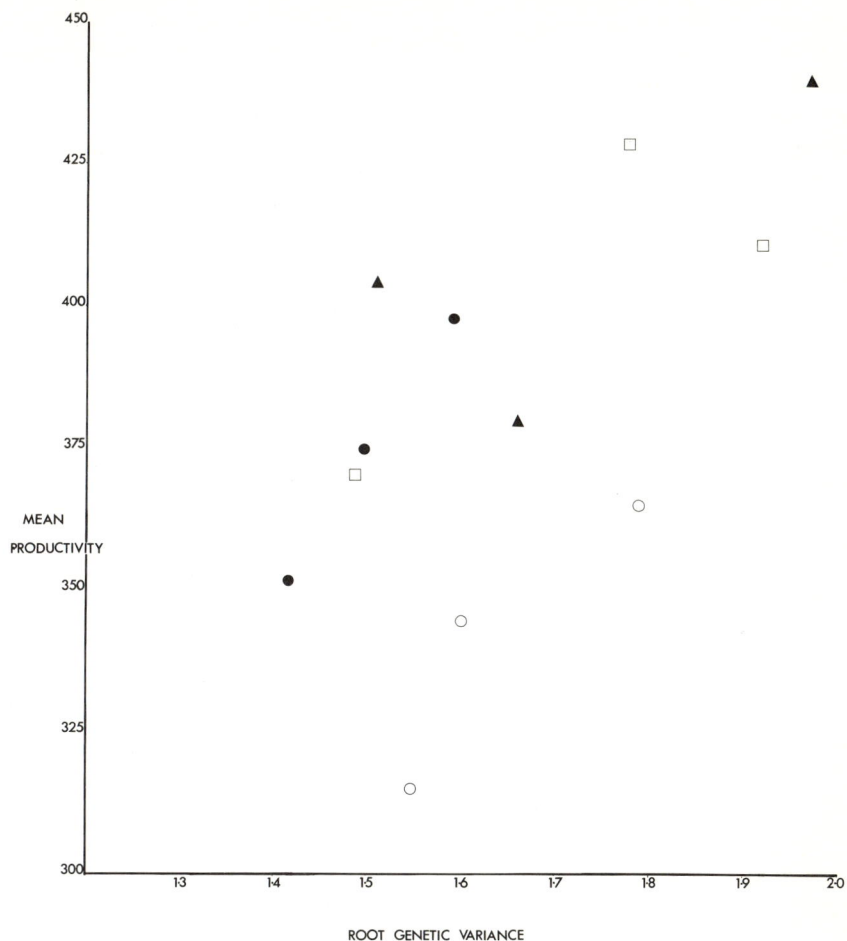

Fig. 2. Productivity (y) and root-additive genetic variance for sternopleural chaeta number (x) in laboratory populations of Drosophila melanogaster. (● constant, ☐ 24 hour, ▲ 32 day, ○ 98 day.)

root of additive genetic variance, but it appears to make little difference in an experiment of this type whether h^2, $\sigma^2 a$ or σa is used, provided that the $\sigma^2 p$ of all populations is similar. In some cases, such as the experiment with the geographic strains, there are large differences in $\sigma^2 p$ from strain to strain, and h^2 is probably the best measure to use. The productivity as a measure of fitness is a more restricted measure than that used for the geographic strains, or by Beardmore and Levine (1963), for it is derived from observations in only one set of environmental conditions. Productivity as such has been shown by van Delden (1968) to correlate very

TABLE III

Mean Fitness Index Based on Three Characters at Three Points in Time in the Temperature Populations. Constant Temperature Population Set at 1.00 in Each Sample, Each Figure Is the Mean of Three Populations (from T. C. Long, 1969).

Population sampled in	Populations			
	Constant	24 hour	32 day	96 day
1966	1.00	4.58	2.68	1.89
1967	1.00	3.46	2.63	1.15
1968	1.00	3.93	2.45	−0.35

highly with a number of other characters involved in total fitness, although Long (1969) has indicated that caution should be exercised in its use. Accepting these qualifications we may note that the regression of fitness on genetic variance is positive and statistically signficant ($P < 0.05$).

Another feature of Figure 2 is that the variable temperature populations tend also to have the higher productivity, although this is not true for the 96-day populations. A fitness index based upon a combination of productivity, competitive ability, and survival in new environments, indicates (Table III) that on the whole, V-type populations enjoy a distinct superiority over K populations. Only in the 1968 sample is the mean 96-day population productivity inferior to that of the K populations, and this corresponds in time with the data presented in Figure 2.

When the values for genetic variance are examined in relation to the environment inhabited by the population it can be seen (Table IV) that the mean values for the variable populations are, with one exception, above

TABLE IV

Mean-root Genetic Variances of Sternopleural Chaeta Number for Populations Maintained in Constant and Varying Environments at Three Points in Time; Each Figure Is the Mean of 3 Populations. (from T. C. Long, 1969).

Sampled at generation (approx.)	Constant	24 hour	32 day monthly	96 day
10	1.63	1.77	1.68	1.63
32	1.57	1.61	1.43[a]	1.70
42	1.50	1.70	1.72	1.64

[a]Incubator failure decimated these populations shortly before test.

those for the "constant" populations. The exception constitutes an assay carried out on the 32-day populations, shortly after a temperature malfunction of the incubator killed most of the individuals in all populations. Whether this is the reason for this low value cannot be established but it would seem likely to have had an effect of this sort.

A statistical comparison of all estimates for genetic variance in the constant population with all those from the variable populations, except the second sample from the 32-day population, shows that the variable populations as a group possess significantly greater genetic variability (White's test, $T = 6.5$, $P \simeq 0.02$). It seems likely therefore that the temporal heterogeneity both of the type within the life cycle and that between life cycles favours a greater assortment of genes than conditions of temperature constancy. It might be argued that this difference could be a result of an increased mutational load since it is known that temperature influences mutation rate. Samples of second and third chromosomes from all populations were assayed jointly by a II-III translocation technique a few generations prior to the last test of genetic variability. The results shown in Table V indicate that if differences between environments do exist they are in the direction that constant temperature populations possess a *greater* frequency of lethals than variable temperature populations. A mutational origin for the greater genetic variability found in the variable environments therefore seems unlikely if lethals be taken as representative of mutations in general.

The finding that populations exposed to greater ecological heterogeneity in time possess greater genetic variability is now extended from those living in conditions of diurnal variation (Beardmore and Levine, 1963) to regular fluctuations of period greater than the generation time of *Drosophila*. It also seems reasonable to conclude that selective processes are responsible for this. However, the test of genetic variability is, if not crude, one of little resolving power, for it says nothing of gene frequencies. It seemed

TABLE V

Percentage Frequency of Second and Third Chromosomes Bearing Recessive Lethal Genes at Generation 36 in the Temperature Populations (each sample based on about 130 genomes, from T. C. Long, 1969).

	25°C	24 hour	32 day	96 day
Population 1	38.9	39.6	31.9	30.4
2	38.9	29.5	33.9	37.0
3	38.6	32.9	39.1	32.7
Environment mean	38.8	34.0	35.0	33.4

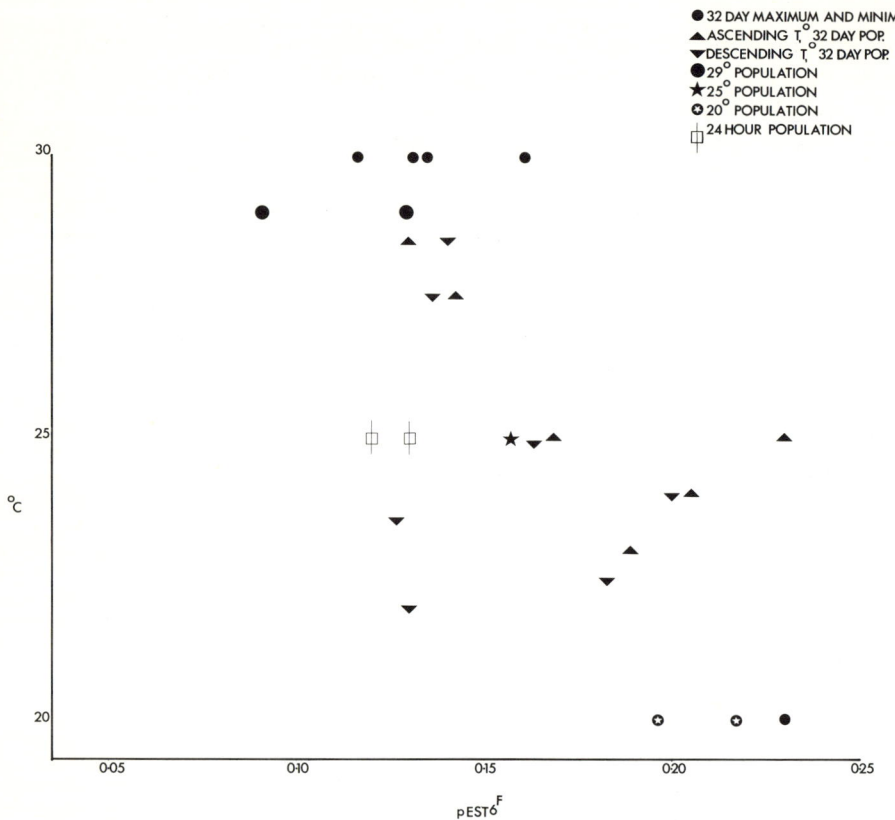

Fig. 3. Allele frequencies of Esterase-6^F in populations at different temperatures.

desirable to examine variability in systems that could be more precisely defined both as to the individual genes and to their effects.

The first system chosen was the Esterase-6 system first discovered by Wright (1963). All the temperature populations are polymorphic for this system with two electrophoretically distinguishable isozymes controlled by alleles designated Fast and Slow. There are characteristic differences between different populations but they are of a kind that, while they do not indicate that variable temperature populations have more genetic variability for this system than constant-temperature populations, do show large, ecologically significant changes in gene frequency mediated by the temperature changes. Reference to Figure 3 shows an example of this in the changes in the frequency of the Esterase-6^F allele observed in one of the 32-day group of populations. Samples of flies taken at random from the cage were typed by polyacrylamide disc electrophoresis, and the allele frequencies were calculated.

In Figure 3, the triangular symbols show the frequency of the fast allele plotted against the temperature at the point in the cycle at which the sample was taken. Upright symbols indicate those samples taken during the rising half, and inverted symbols those taken during the descending half of the cycle. Samples taken at the temperature extremes are represented by small, closed circles.

The data extend over a period of a year. They are not entirely randomly distributed within this period, and there is reason to believe that some shifts in mean gene frequencies in some populations may have occurred as a result of changes in culture conditions. Nevertheless, the conclusion that gene frequencies respond to temperature oscillation in a moderately regular way is inescapable. Most samples consist of about 100 flies, some of half this number, so that a rough standard error, which is certainly not an underestimate, would be 0.03 on the allele frequency scale. The inverse correlation between allele frequency and temperature noted in this 32-day population might suggest that populations in different constant-temperature conditions would show gene frequencies corresponding to the appropriate point in the 32-day cycle. In order to test this, data from constant-temperature populations of the same origin are also shown in Figure 3. It has not been possible to keep populations at 30°C, probably because of sterility problems, so that populations at 29°C must serve for comparison at the top end of the temperature scale. The allele frequencies for the constant temperature populations agree very well with what would be predicted from the values for the 32-day population. Sister populations of this population, for which less data exist, show a similar pattern of change of allele frequencies correlated with temperature, and 96-day populations also display a broadly similar picture.

There is some similarity between the temporal sequence shown by the Esterase-6 alleles and the seasonal change of gene-arrangement frequencies in natural populations of *D. pseudoobscura* demonstrated by Dobzhansky (1948). Temperature, though perhaps the most obvious, is only one of the changes exhibited during the march of the seasons, and Dobzhansky and others have in fact shown that other ecological factors could be important in determining the relative frequency of Arrowhead, Chiricahua, and Standard gene arrangements. Likewise, here we cannot be sure that it is temperature which is the selective factor, and preliminary investigations suggest that density may be of some importance in this connection. In Figure 4 the frequency of Est-6^F plotted against population size is shown. The data in this figure are drawn from populations in both constant and varying environments. The correlation coefficient between these two parameters (after angular transformation of gene frequency) is 0.62, with a probability of being due to chance of less than 0.02. It seems therefore that there is a relationship between gene frequency and population size; and as population size in the variable temperature populations appears to vary in a more or

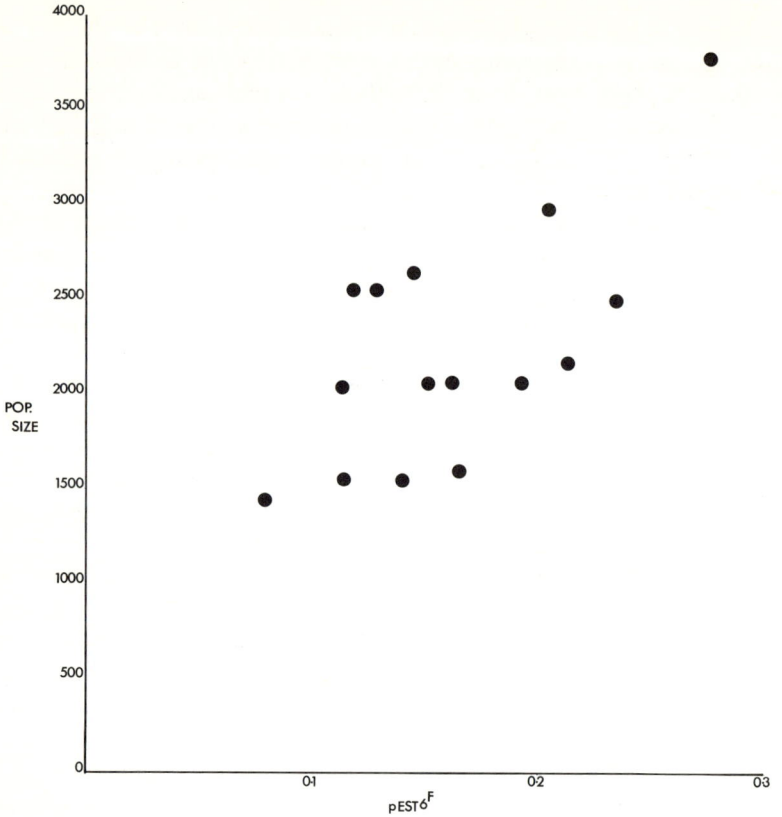

Fig. 4. Population size (y) and frequency of Esterase-6^F (x) in laboratory populations of D. melanogaster.

less regular way as a result of the temperature change, it could well be that crowding, for example, is the agent bringing about the regular changes in genetic composition. Only one small experiment to test this has so far been carried out. In this test, progeny from a series of replicate cultures started with 20 pairs of parents were compared with progeny resulting from cultures with 120 pairs of parents. The results based on randomised samples are shown in Table VI. The frequency of Est.-6^F is indeed significantly higher in the more crowded conditions.

As much of the selection in *Drosophila* is thought to occur in the larval stages, it might be argued that it would be more useful to relate gene frequency to the temperature at the time of larval development, rather than the time of sampling. This certainly will be looked at but, because of the continuous variation in development time in the varying populations a good deal of exploratory work needs to be done first.

TABLE VI
The Frequency of Est-6^F In Relation to Crowding

No. of parents/culture	Genotypes of progeny			
	F/F	F/S	S/S	pF
20 pairs	0	20	79	0.101
120 pairs	1	34	69	0.173

$\chi_1^2 = 4.43, P = <0.05$

A simple calculation shows that the change in fitnesses that must be invoked to account for the shift in the frequency of Est-6^F in the 32-day cycle populations on the simplest assumption of additive action of genes must involve differences of adaptive value of the order of 15 to 20 percent. These differences cannot easily be smaller, and may well be much larger depending upon the precise mode of selection operating.

Since the demonstration by several workers (Lewontin and Hubby, 1966; Harris, 1966) that the normal composition of the gene-pool of diploid outbreeding species is one of marked biochemical polymorphism, much thought has been given to the problem of how such widespread genetic variabilty is maintained. From some quarters it is suggested that this variability is random in the sense that many genotypes are selectively effectively neutral; e.g., Shaw (1965) and Kimura (1968) subscribe to this view, although for different reasons and with different assumptions. Recently King and Jukes (1969) have suggested that 10 to 15 percent of all mutations may fall into this category, with important consequences for evolutionary thought. I do not wish to pursue this point in depth here but would simply point out that if this measure of selective neutrality really exists one would expect to find more alleles per polymorphism than seems to be generally the case. It is true that since electrophoresis is used so widely, and as electrophoretic techniques do not pick up all variants but only those which involve a change in charge or effective molecular form, such techniques may underestimate genetic variability. Nevertheless, what is striking is not that, as King and Jukes point out, over 80 known mutants for α- and β-chains of human hemoglobin are known to be viable, but that only a very small number, perhaps three or four, of these are known to exist in appreciable frequencies of, say, greater than 1 percent.

In the Esterase-6 system Wright (1963) has described three alleles, but most wild populations examined by us, whilst polymorphic, seem to contain only two alleles, although not always the same two. This combination of ubiquitous polymorphism with a very small number of different alleles represented in each population does not support the idea of selective neutrality, although the modes of selection that may be involved in maintaining such polymorphisms are still largely unknown.

Kojima and Yarborough (1967) argued that as the fitnesses of the genotypes in the Esterase-6 polymorphism of *Drosophila melanogaster* are dependent on their relative frequency, then the genetic load would be nonexistent or minimal at the point of equilibrium. The argument is a little difficult because it is expressed in terms of the effects of selection directly upon genes; and whilst the effects of selection can be expressed as change in gene frequencies, selection is better thought of as operating on phenotypes. They concluded that it is possible that different genotypes can exploit different parts of the environment. The differential exploitation of the environment will follow if the products of the two alleles are physiologically different. Wright (1963) has shown that the temperature stability and degree of inhibition by an organophosphate is different for the two molecular forms of the Esterase-6 enzyme. It seems to me likely that physicochemical differences affecting biological factors will be found in other isozyme systems, and Dr. John Gibson and I have evidence that such differences exist in the alcohol-dehydrogenase system in *Drosophila melanogaster*. In this polymorphism large differences in fitness between genotypes exist, and these can be modified dramatically by simple techniques like putting alcohol into the culture medium. We suspect further that under conditions likely to be ecologically more complex the polymorphism is retained, whilst monomorphism tends to supervene in conditions of less ecological variability.

Conclusions

The data described in this paper are scanty and need to be greatly expanded. However, if they are accepted as representative of a more general situation, they would seem to suggest that the degree of heterogeneity of habitat will influence profoundly the amount of variability in gene-pools. Further, under such conditions the magnitude of differences in fitness between different morphs is likely to change markedly from time to time and from place to place. In the kinds of spatial and temporal heterogeneity characteristic of natural populations the ecological load may then be appreciable, although this will clearly depend on the degree to which each genotype "matches" the environment in which it finds itself. In any case, however, if polymorphisms are in general adaptive in nature (Dobzhansky, (1968), such a load is the interest paid to secure capital for future activities.

Summary

Evidence for an ecological component of genetic variability is briefly reviewed. It seems that measurable and predictable differences in genetic variability and fitness can be produced in otherwise similar populations of *Drosophila* exposed to simple temporal variations of temperature. On this basis it would be expected that a considerable and perhaps major fraction of the variability of the gene-pools in outbreeding organisms is due to selection operating through ecological factors.

In the Esterase-6 polymorphism in *Drosophila melanogaster* large differences in selective values of the genotypes can be inferred to exist since shifts in ecological factors, such as temperature and density, strongly influence gene frequencies. If this system is typical of protein polymorphisms in general, it seems unlikely that "effective neutrality" of genotypes is a meaningful notion.

Acknowledgments

This work was started during the author's tenure of a National Science Foundation Senior Foreign Scientist Fellowship at Pennsylvania State University, where the hospitality and facilities provided by Professor James Wright are gratefully acknowledged. To the Trustees of Groningen University go my thanks for their generosity in granting leave of absence. The collaboration of Dr. W. van Delden, Dr. T. C. Long, Miss G. de Jong, and Mr. Andrew Birley is acknowledged with thanks.

References

BEARDMORE, J. A. 1963. Mutual facilitation and the fitness of polymorphic populations. Amer. Natural., 97:69-74.
———— 1966. Genetic information in populations. Advance. Sci., 23:128-132.
———— and L. Levine. 1963. Fitness and environmental variation. 1. A study of some polymorphic populations of *Drosophila pseudoobscura*. Evolution, 17:121-129.
CAIN, A. J., and P. M. SHEPPARD. 1954a. The theory of adaptive polymorphisms. Amer. Natural., 88:321-326.
———— and P. M. SHEPPARD. 1954b. Natural selection in *Cepaea*. Genetics, 39:89-116.
CARSON, H. L. 1955. The genetic characteristics of marginal populations. Sympos. Quant. Biol., 20:276-287.
CLARKE, C. A., and P. M. SHEPPARD. 1960. The evolution of mimicry in the butterfly *Papilio dardanus*. Heredity (London), 14:163-173.
CUNHA, A. B. DA, TH. DOBZHANSKY, O. PAVLOVSKY, and B. SPASSKY. 1959. Genetics of natural populations. XXVII. Supplementary data on the chromosomal polymorphism in *Drosophila willistoni* in its relation to the environment. Evolution, 13:389-404.
DELDEN, W. VAN. 1968. Doctoral Thesis, Groningen.

────── and J. A. BEARDMORE. 1968. Effects of small increments of genetic variability in inbred populations of *Drosophila melanogaster*. Mutat. Res., 6:117-127.
DEMPSTER, E. R. 1955. Maintenance of genetic heterogeneity. Sympos. Quant. Biol., 20:25-32.
DOBZHANSKY, TH. 1948. Genetics of natural populations. XVI. Altitudinal and seasonal changes produced by natural selection in certain populations of *Drosophila pseudoobscura*. Genetics, 33:158-176.
────── 1951. Genetics and the origin of species. 3rd ed. New York, Columbia University Press.
────── 1968. On some fundamental concepts of Darwinian Biology. *In* Dobzhansky, Th., Hecht, M. K., and Steere, W. C., eds. Evolutionary Biology, Vol. 2. New York, Appleton-Century-Crofts. Pages 1-32.
────── and PAVLOVSKY, O. 1960. How stable is balanced polymorphism? Proc. Nat. Acad. Sci. (U.S.A.), 46:41-47.
DUBININ, N. P., and G. G. TINIAKOV. 1947. Inversion gradients and selection in ecological races of *Drosophila funebris*. Amer. Natural., 81:148-153.
HARRIS, H. 1966. Enzyme polymorphism in man. Proc. Roy. Soc. [B], 164:298-310.
KING, J. L., and T. H. JUKES. 1969. Non-Darwinian evolution. Science, 164:788-798.
KIMURA, M. 1968. Evolutionary rate at the molecular level. Nature (London), 217:624-626.
KOJIMA, K. I., and K. M. YARBOROUGH. 1967. Frequency dependent selection at the esterase locus in *Drosophila melanogaster*. Proc. Nat. Acad. Sci. (U.S.A.), 57:645-649.
LERNER, I. M. 1953. Genetic Homeostasis, Edinburgh, Oliver and Boyd.
LEVENE, H. 1953. Genetic equilibrium when more than one ecological niche is available. Amer. Natural., 87:331-333.
LEVINS, R. 1962. Theory of fitness in a heterogeneous environment. I. The fitness set and adaptive function. Amer. Natural., 96:361-373.
────── 1963. Theory of fitness in a heterogeneous environment. II. Developmental flexibility and niche selection. Amer. Natural., 97:75-90.
────── 1964. Theory of fitness in a heterogeneous environment. III. The response to selection. J. Theor. Biol., 7:224-240.
LEWONTIN, R. C. 1956. Studies on homeostasis and heterozygosity. I. General considerations. Abdominal bristle number in second chromosome homozygotes of *D. melanogaster*. Amer. Natural., 90:237-255.
────── 1958. Studies on heterozygosity and homeostasis. II. Loss of heterosis in a constant environment. Evolution, 12:494-503.
────── and J. L. HUBBY. 1966. A molecular approach to the study of genic heterozygosity in natural populations. II. Amount of variation and degree of heterozygosity in natural populations of *Drosphila pseudoobscura*. Genetics, 54:595-609.
LI, C. C. 1955. The stability of an equilibrium and the average fitness of a population. Amer. Natural., 89:281-296.
LONG, T. C. 1969. Ph.D. Thesis, University of Wales.
LUDWIG, W. 1950. Zur Theorie der Konkurrenz. Die Annidation (Einnischung) also fünfter Evolutionsfaktor. Neue Ergeb. Probleme Zool., Klatt-Festschrift, 1950: 516-537.
MATHER, K. 1955. Polymorphism as an outcome of disruptive selection. Evolution, 9:52-61.
MULLER, H. J. 1950. Our load of mutations. Amer. J. Hum. Genet., 2:111-176.
SHAW, C. R. 1965. Electrophoretic variation in enzymes. Science, 149:936-943.
TIMOVEEV-RESSOVSKY, N. V. 1935. Über geographische Temperaturrassen bei *Drosophila funebris* F. Arch. Naturgesch., N. F. 4:245-257.
THODAY, J. M., and T. B. BOAM. 1959. Effects of disruptive selection. II. Polymorphism and divergence without isolation. Heredity (London), 13:205-218.
WRIGHT, T. R. F. 1963. The genetics of an esterase in *Drosophila melanogaster*. Genetics, 48:787-801.

12

Mating Propensity and Its Genetic Basis in *Drosophila*

ELIOT B. SPIESS

Department of Biological Sciences, University of Illinois at Chicago Circle, Chicago, Illinois

Introduction	316
Mating Propensity as a Genetic Character in *Drosophila*	318
Associations with Specific Mutants	318
Strain Differences Without Genetic Markers	324
Genetic Analysis by Selection Techniques	328
Quantitative Genetic Analysis of Wild-Type Strains	337
Chromosomal Polymorphism and Mating Propensity	343
Sex Ratio in *Drosophila pseudoobscura*	344
Third Chromosome Arrangements in the *Drosophila psuedoobscura* Subgroup	346
Chromosome Polymorphism and Mating Propensity in Other Species	354
Factors Modifying Mating Propensity	356
Extrinsic Factors	357
Temperature	357
Food Volume, Crowding, and Yeast Availability	358
Light	358
Density Between Mating Participants	359
Sex Ratio	359
Age of Parents	361
Intrinsic Factors	361
Frequency Dependency	364
Summary	373
Acknowledgments	375
References	375

Introduction

Sexual behavior must have evolved as an efficient mechanism to propagate the species. As a property contributing to Darwinian fitness it must be one of the most critical in the life cycle, because throughout the preadult stages numbers of individuals succumb to the exigencies of their environment so that at the mature stage when reproduction takes place the minimum for the life cycle is attained. A diversity of genotypes acting on the efficiency of sexual behavior at this focal stage will no doubt exert selective differences which have considerable value in perfecting the efficiency of the sexual process. As a result, at least three major phenomena may evolve: (1) increasing fertility, or higher fitness for the population or the species, (2) increasing sexual dimorphism owing to intrasexual competition (Darwin's sexual selection) and to preferences for mates, and (3) increasing assortative mating leading to sexual isolation and speciation (Bateman, 1948; Smith, 1958a, b; O'Donald, 1962; Manning, 1965; Caspari, 1967, for example).

A useful strategy for a population in improving its fitness or greater contribution to the natural rate of increase (Malthusian parameter), might be to reduce the time to oviposition (Lewontin, 1965). Fast and efficient mating may insure quicker egg laying as one feature controlled by selection. For males, selection would tend to increase persistent courtship activity to insure a continuous supply of sperm and widespread insemination. For females, however, fast receptivity or lowered threshold of acceptance must be compromised with an ability to discriminate between males. Recognition between mates may develop also, especially in species with territorial and/or aggressive behavior. Selection must be aimed at the different properties in the two sexes, focused at the end point of high fertility to establish the new generation rapidly. Synchronization of male and female rhythms will doubtlessly be selected for in consequence.

The second and third phenomena, that is, Darwin's sexual selection and assortative mating, may easily follow, as has been pointed out by Smith (1958a, b). Increased sexual dimorphism with display behavior and associated secondary sexual characters can be selected for if combined with higher fitness; and preferential or discriminative behavior may lead to isolation between genetically distinct populations if the choosing ability leads to higher fitness. Given the genetic potential for behavioral differences and the possibilities that male displays and female preferences may confer a fitness advantage, genotypes determining these sexual behaviors may rapidly be established (O'Donald, 1962). At our present stage of understanding for both the genetic determination of behavior and genetic architecture of populations, it seems that our most important job is to discover and describe the genetic diversity controlling such traits. "Dissection" of these complex characters with genetic markers and with artificial selection or other quantitative techniques is essential. We wish to know the distribution of effective factors among the chromosomes, the extent of interactions between such factors, and the conditions under which they come to expression in mating behavior. Accompanying or preceding the genetic analysis is the understanding of behavioral traits involved. In fact, what to measure and how to measure the significant features of sexual activity are perhaps among the most important tasks at present.

This discussion will first summarize mating propensity studies with the use of mutant markers, as well as the evidence for genetic control in a few *Drosophila* species. If the sexes display contrasting properties on which selection acts in separate ways, it is important to describe the distinct contributions of genetic variation to the performance of each sex in terms of its relative control over the outcome of mating and how selection might act on each sex to perfect the function of increased fertility via male courtship and female receptivity. Evidence of preferential, or discriminative, tendencies will be examined to some degree; but the aspect of this subject that leads to sexual isolation is so extensive as to be beyond the scope of this chapter, though those apparent tendencies of incipient isolation which are convenient to this discussion will be mentioned.

If we postpone the question of whether any genetic variant that persists in a population might have some fitness value and also an effect on behavior (which might be asking too much of pleiotropism), it will be seen in the following pages that many of the genetic variants in *Drosophila* have marked influence on mating ability. Also the frequencies of chromosomal variants in natural populations are correlated with mating propensity, pointing to the vast store of genetic potential available for selection but linked in semipermanent blocks of loci. As a fitness trait, mating speed is less complex than traits like natural rate of increase of a population or competitive ability of genotypes, and is much more amenable to a physiological and ultimately to a direct gene-action study, providing that the mating behavioral elements can be specified along with the conditions that evoke the behavior. Therefore the genotype controlling mating speed is likely to be genetically dissectable and at the same time certain to be contributory to adaptive functions of genetic variants in natural populations. *Mating success,* however, is far more complex than such elements of behavior as response to light, gravity, wing-beat frequency, and similar quantitative orientation movements. How much of mating success may be the result of "vigor," "discrimination," "preference," "sex drive," "sensitivity," along with many other ill-defined yet meaningful qualities on the part of either or both sexes in behavior, we are hardly beginning to understand. Whatever achievements have been made are due to the careful efforts of ethologists and the observations of those intent on understanding how mating behavior influences, or is influenced by, the genotypes in populations of interbreeding organisms such as *Drosophila*. The methods of quantitative genetics as applied to morphological or physiological traits can be used much more in the future to analyze the hereditary particulars of sexual activity for each sex and the distinguishing qualities necessary for mating success.

Mating Propensity as a Genetic Character in Drosophila

Associations with Specific Mutants

We cannot in most cases associate mating propensity with prominent morphological effects by the action of any specific locus, although the evidence supports a very close association with certain loci upon which a large number of investigations have been made (for example, *yellow body, white eye, ebony body* mutants). Unfortunately, with many investigations of this sort there has been a minimum of genetic control; specific alleles are often unmentioned; both mutant and wild-type stocks used for testing have not been isogenized nor are they often even inbred or characterized with sufficient care for further good genetic analysis. Many studies have tacitly assumed that the mating propensity of the predominant mutants being observed is the indirect pleiotropic effect of those mutant loci. Recombination to attempt possible separation of behavioral control elements from mutant loci has rarely been tried. Lastly, techniques have been so varied that it is small wonder any consistency emerges at all unless some prominent mutants do have pleiotropic effects.

In spite of these shortcomings, however, the use of marker mutants has proved helpful and is thus worth reporting for the following reasons: (1) employing a mutant that disturbs the mating behavior sufficiently to make the elements of behavior observable so that the interplay of stimulus-response between individuals participating in mating may be ascertained; (2) criteria of "rivalry" between genotypes, interpretations of "preference," "sex drive," "vigor," and other factors in sexual selection can be more refined after careful observation with simple markers; and (3) direction can be given to genetic analysis of the behavioral elements that we wish to watch.

The question of whether either sex in *Drosophila melanogaster* is selective with regard to mutants or abnormalities of the opposite sex in mating activity was first investigated by Sturtevant (1915). His technique was simply to give one sex a "choice" of two forms, one normal and one abnormal, and watch until mating occurred. While the results cannot be taken as the outcome attributable to particular genic loci (since coisogenicity had not been really achieved at that time) a few significant observations were made that are worth noting: (1) when individual males were given two types of females to mate with, the matings were equally distributed between wild-type and mutant irrespective of mutants used (*white eyes, yellow body, vermilion eyes,* and *curved wings*); (2) when individual females were paired with two kinds of males, three of the mutant types of males (*white, yellow,* and *curved*) mated less frequently than their wild-type rivals. Although Sturtevant did not describe the interactivity of males or females in detail, so that we cannot determine, for example, whether the lowered success of those mutant males was a result of greater female rejection, refusal to accept them, or due to less persistence of the males in courtship, it is still evident that the males' differences rather than the females' were selectively significant. Sturtevant (1915) attributed the result to "the difference in vigor" among males and no evidence of any "choice" or preference appeared: "A female, in the greater majority of cases, seems to allow the first active, amorous, male that comes along to pair with her; or if she is disinclined to mate, resists all males indifferently. When a male is sexually excited he pairs with the first female he finds that will allow him to. As a result, the more active and vigorous males are likely to win their contests, and the greater the difference in vigor, the greater the proportion of times the better male wins." There is no indication then that females were selectively discriminating against these mutant males from this account, yet because no quantitative record was made of opportunities to mate, which the females may have had but did not take, it is difficult to rule out female control of mating and the exercise of female preference. While the point was not clear then, it can be admitted that technical problems of measuring female control and discrimination as a major element in mating success is still not easily solved. Yet the recent literature indicates that females of *Drosophila* are the major controlling sex for speed of mating, as will be discussed.

Since 1915, genetic changes of various sorts, single mutants, inbred lines, and strains differing in several ways have been analyzed in varying degrees of genetic or behavioral detail. The summaries of Bastock (1967), Manning (1965, 1966, 1967a), Ewing and Manning (1967) and Spieth (1952, 1968) have emphasized the behavioral elements and evolution in *Drosophila* studies while those of Parsons (1967) and Petit and Ehrman (1969) have stressed the genetic control of mating behavior.

Table I lists most of the studies recorded in the literature that achieved some significant results with mutant markers helping to develop our ideas about the interactions between the participants in mating of *Drosophila*. Throughout much of the earlier literature expressions such as "preference" and "choice" are made (especially with regard to incipient isolation) as conventional terminology, which must be realized do not indicate final

TABLE I

Mating Propensity Tests With Mutants

Reference	*Drosophila* sp. Mutants and Techniques	Author's Interpretation
Sturtevant (1915)	*melanogaster* yellow (y); white (w); curved (c); vermilion (v). Observed matings. Mutant X wild type, either 1 ♂ X 2 ♂♂ or vice versa. Interpretation: ♂♂ less "vigorous" than + (wild) in case of y, w, and c; ♀♀ not different from + in any case. v ♂♂ not different from + ♂♂.	
Rendel (1945)	*subobscura* yellow (y); withered (wi); eyeless and white. Multiple pairs mated in vials for 24 hours, usually 20 pairs per vial, ♀♀ dissected to verify mating. "No choice". y ♂♂ X ♀♀ > y ♂♂ X y/+ ♀♀ > y ♂♂ X +/+ ♀♀, in percent inseminated. Since courtship of y ♂♂ was observed to be "normal," conclusion: y ♂♂ repelled by non-y ♀♀. y ♀♀ normally receptive and not discriminated against by either type ♂♂ but they tolerate y ♂♂ more than non-y ♀♀ do. Selection of wi ♀♀ for 4 generations increased non-y ♀♀ receptivity towards y ♂♂. Eyeless and white-eyed ♂♂ do not court since light is essential for mating in this species.	
Tan (1946)	*pseudoobscura* yellow (y); aristopedia (ss^a); Bare-Curley (Ba-Cy). Technique: "male choice." yellow similar to melanogaster and subobscura in that ♂♂ are less successful than + ♂♂. ss^a reduced the receptivity of ♀♀ possibly by interfering with ability to detect courtship of males. Ba-Cy enhanced ♀♀ receptivity or lowered avoidance, or resistance, of ♀♀.	
Merrell (1949)	*melanogaster* yellow (y); cut (ct^6); raspberry (ras^2); forked (f). In various linkage combinations, and vs. wild type (Lausanne-Special). "Female choice" technique tested rivalry of males. Usually used 5 ♀♀ X 5 mutant ♂♂ + 5 wild ♂♂. "Male choice" tested female receptivity. Flies, either recently emerged or aged 7 days, were left together until mating; ♀♀ heterozygous were separated and progeny checked. Results: y ♂♂ similar to above being less successful than + ♂♂ and better X y ♀♀ than X non-y ♀♀. ct ♂♂ also less successful than + ♂♂; ct ♀♀ were fertilized less than any other type of ♀. ras ♂♂ half as successful as + ♂♂. f ♂♂ = + ♂♂. y ct ♂♂ more detrimental than either mutant separately. ct ras improved ♂♂. All combinations with y detrimental to success of ♂♂.	
Reed and Reed (1950)	*melanogaster* white (w); + (Lausanne-Sp.). (+/w ♀ and ww ♀) X (+ ♂ and w ♂) "multiple choice" for 24 hours. Propensity for ♀♀ equal but + ♂ : w ♂ about 1: 0.75. (progeny checked from individual ♀♀.)	
Rendel (1951)	*melanogaster* ebony (e); vestigial (vg). Technique: similar to previous Rendel study: ♀♀ dissected after 2 hours or after 24 hours. Either 1 type ♂ X both types ♀♀ or reciprocal. e ♂♂ much less successful than vg ♂♂ in light but better than vg in dark.	
Merrell and Underhill (1956)	*melanogaster* glass (gl); vestigial (vg). Technique: "♀ choice": 10 ♀♀ X 10 mutant ♂♂ + 10 mutant/+ ♂♂. Progeny checked. Both heterozygous and mutant homozygous ♀♀ mate X heterozygous ♂♂ : mutant homozygous ♂♂ in ratio of > 10: 1. Experimental populations of mutants including the sex-linked loci above plus a series of white-eye alleles showed decline in mutant frequencies as predicted from selective mating outcome.	

TABLE I continued

Reference	Drosophila sp. Mutants and Techniques	Author's Interpretation
Tebb and Thoday (1956)	melanogaster white (w); white-apricot (w^a). One ♀ X (w ♂ and w^a ♂) "female choice" with progeny classified. w/w ♀♀ or w^a/w^a ♀♀ "prefer" w^a ♂♂, but w/w^a ♀♀ "prefer" w ♂♂.	
Bastock (1956)	melanogaster yellow (y). Intercrossed y X wild type for 7 generations before behavior tests to randomize genetic background. Detailed behavioral analysis especially in courtship wing display: y ♂♂ have shorter bouts at longer intervals than + ♂♂. Females do not discriminate against y ♂♂ because of their color or scent, since discrimination is just as effective in the dark and also with antennaless ♀♀. y ♀♀ at first in the study accepted y ♂♂ quite readily because of higher receptivity, but after outcrossing to wild type the higher receptivity was acquired by the wild type ♀♀ so that feature was not a characteristic of the y phenotype in ♀♀.	
Elens (1957, 1958)	melanogaster ebony (e). Lowered sexual activity of e/e ♂♂ compared with wild type. Heterosis for activity of heterozygotes depends on maternal effect.	
Petit (1951, 1954, 1958, 1959)	melanogaster white (w); Bar (B); forked (f). Bf vs. B^+f ♂♂ and Bf/Bf, B^+f/B^+f ♀♀ (isogenic since B^+ was reversion within B stock). Females isolated from populations gave progeny indicating females had mated with non-B ♂♂ preferentially but showed no preferences for females. With white vs. wild type (Oregon RC isogenized), (w/w and w/+ ♀♀) X (w and + ♂♂) with frequency of male types varied (progeny checked from individual ♀♀). When two types of ♂♂ equally frequent, w ♂♂ are at disadvantage, but when either rare or common, w ♂♂ mate more frequently than + ♂♂. Excision of antennae and aristae from ♀♀, then observed behavior with w and + ♂♂. w ♂♂ have less courtship persistence and different wing vibration, which is detected by the ♀ mostly from her antennal basal segment (Johnston's organ). Numerous conditions affecting mating propensity were studied, especially the frequency-dependency phenomenon in Bf vs. B^+f and in w vs. + ♂♂.	
Jacobs (1960, 1961)	melanogaster ebony (e). Collected from wild population, but made "isogenic" (dominant marker technique) with wild type; allelic with e^{11}. Technique: 10 e ♀♀ X (10 e ♂♂ and 10 +/e ♂♂), or 10 e ♀♀ X (10 + ♂♂ and 10 +/e ♂♂); etherized flies before mating tests, a technical error probably affecting behavior outcome. Results agree with Rendel (1951) in that e mates in greater numbers in diminished light than in full light while wild type were opposite. e ♂♂ had reduced courtship and wild ♂♂ often displaced them. In "female choice" tests only total progeny was counted, so that individual effect was lost. Heterozygote ♂♂ produced more progeny than expected about the same in light and dark.	

TABLE I continued

Reference	*Drosophila* sp. Mutants and Techniques	Author's Interpretation
Geer and Green (1962)	*melanogaster* white *(w)*; white-apricot *(w^a)*, $w^aR1,2$ (reversions towards wild type from apricot), all in coisogenic background. (25 *w/w* ♀♀ and 25 *w^a/w^a* ♀♀) × (25 *w/w* ♂♂ and 25 *w^a* ♂♂) for 2 days either under light-dark cycle, total darkness, or total light. ♀♀ isolated and progeny identified. In light conditions, darker pigmented allele always had more mating success, while in darkness mating was random. Light may be important in first orientation of male toward female, but actual courtship and matings were not observed. Used mutant *vestigial* to remove wings and reported no difference from normal-winged mating success, but this result contradicts all other investigators working with *vg* or reduced wing area strains (see Ewing, 1961). Use of *vg* mutant has greater effect on mating than the excision of wings done artificially and does not disprove Petit's suggestion that the lower mating of *w* is due to different wing vibrations.	
Hildreth (1962)	*melanogaster* white *(w)*; + *(wild Samarkand)*. Matings observed with (*w/w* ♀ and +/+ ♀) × + ♂. First copulations were mostly × *w/w* ♀♀. Time to courtship shorter to that mating than to +/+ ♀ mating. Also longer period between copulations if first ♀ was *white*. Wild ♀♀ were more active than *white* ♀♀, avoiding ♂♂, running, kicking, and decamping. *w/w* ♀♀ more quiet permitting copulation after briefer courtship. Author records that when copulation had occurred in a vial previously +/+ ♀♀ copulated more frequently than in freshly cleaned vial. This observation agrees with Sturtevant but is one of the few recorded agreements implying a chemical or odor stimulating mating. Also no attempt was made to isogenize the *w* and wild type stocks, so that female receptivity may not be comparable with other studies with *white-eye* mutant females.	
Hildreth and Becker (1962)	*melanogaster* "Basc" (= *In (1) scute, white-apricot, Bar* = "Muller-5") and *"Bv"* (= *yellow, Bar, vermilion*) ♀♀ × Samarkand wild type ♂♂. Single ♂ × 2 ♀♀ ("male choice"). Flies only 24-32 hours old. Copulations observed. No preference in first matings, though second copulation was much more if the first mating was × *Basc* than by *Bv* ♀♀. Possible chemofactors transferred from *yellow (Bv)* ♀♀ to ♂♂ thus lowering their acceptance by second, although a scent involvement is contrary to the observations of Bastock. Positive correlation between length of 1st and 2nd copulation may indicate sperm-volume constancy.	
Barker (1962)	*melanogaster* *yellow (y)*; + wild (Oregon R-C). Several techniques employed: pair matings, male choice, female choice, and multiple choice with sex ratio equal or in 3:1 for ♂ choice or ♀ choice, and with flies aged either 4 hours or 7 days, left together either 23 hours or 5 hours for the two ages, respectively. Progeny inspected from individual females. Wild ♂♂ far more successful with wild ♀♀ than *y* ♂♂ in all tests, but only slightly better with *y* ♀♀. Success with +/*y* ♀♀ was indistinguishable from +/+ ♀♀. Isolation estimates indicate greater isolation with female and multiple-choice than with	

TABLE I continued

Reference	Drosophila sp. Mutants and Techniques	Author's Interpretation
Barker (1962) (continued)	male-choice method, but it is a spurious isolation since it results entirely from the failure of y ♂♂ to fertilize wild type ♀♀. With young flies, more y/y ♀♀ were fertilized by + ♂♂ than by y ♂♂, but with aged flies the difference was not significant. In multiple choice, any advantage of + ♂♂ would be accentuated if y ♂♂ waste much of their time courting wild type ♀♀. Interaction between courtship activity of ♂♂ and receptivity of ♀♀ rather than "preferences," determine mating outcome.	
Bösiger (1962, 1967)	*melanogaster* vermilion (v); cinnabar (cn); sepia (se); forked (f). In v vs. cn studies, 2 types ♂♂ X one type ♀ (with sex ratio equal and between approximately 100 to 250 ♀♀) into population cage for 48 hours. ♀♀ isolated and progeny checked. cn ♂♂ mated more than v ♂♂ by about 9:1 throughout. In a second study, v (B allele) and wild (Oregon R-C) ♂♂ X v (B) ♀♀. Proportion of v among males and sex ratio were varied. Mating was random when ♂♂ were 20%-40% of total, although v ♂♂ mated more than wild type ♂♂ with sex ratio outside those limits. Author attributed advantage to male "vigor" and speed rather than to ♀ choice. Aging flies for 2 days did not change the outcome. Courtship of cn ♂♂ starts before v ♂♂, with more persistence and intensity. Both cn and v stocks outcrossed X Or-R-C, with F_2 cn and v reconstituted: cn ♂♂ were then highly variable. Isogenation of cn and v stocks for 40 generations produced equal mating abilities in the two mutants. Both se and f ♂♂ inseminate about 25% of wild ♀♀, while these mutant ♀♀ are also less receptive than hybrid or wild ♀♀. Freshly caught wild strains have higher mating ability than laboratory strains.	
Crossley (1963)	*melanogaster* ebony (e); vestigial (vg). e ♂♂ show low intensity courtship with reduced frequency of vibration and licking; furthermore it lacks persistence, shows intermittent displays, apparently poor visual reactions, so that e ♂♂ tend to lose ♀♀ by allowing them to escape without following.	
Mainardi and Mainardi (1966)	*melanogaster* yellow (y). To test discrimination of ♂♂ between + and y ♀♀, single ♂♂ X (+ ♀ and y ♀) tested for courtship and copulation. Courtships were toward + ♀ nearly twice those toward y ♀, but copulations favored y ♀. Oregon (+) ♀♀ have higher receptivity threshold than y ♀♀.	

determinations of which sex controls the final outcome or how mating success is achieved. Nevertheless, many accounts document the fact that genetic differences between strains do have considerable effect on mating behavior and propensity. How much real discrimination by either sex actually occurs and affects the mating outcome has only been recently established (for one of the best examples see Bastock, 1956). Evidence in favor of female discrimination and against male discriminating ability was summarized by Bateman (1948) on the basis of several investigators' tests. The standard technique was to enclose one kind of male with two kinds of female and then to observe the relative frequency of insemination for the two kinds of female. As Bateman points out "It is easy to fall into error in assuming that it is the male that 'chooses' between the kinds of female; the results could be equally well interpreted on the assumption that two kinds of females discriminate with specific strengths against the same males." Merrell (1949a) similarly produced evidence strongly in favor of female control over mating success as the most likely explanation of male "choice" experiments. While the selection seems to be performed almost exclusively by females, it is of course directed toward differences among males. In other words, genotypes of males may be selectively different and females are the selective agents in the process, a concept that is more or less in conformity with Darwin's sexual selection.

It is noteworthy that for the vast majority of mutants (see Table I) the males differ from wild-type in their aberrant courtship and consequent lowered success, while female mutants are most often indiscriminately courted by nearly all types of males. Exceptions to the latter may be the *cut* females by Merrell and *aristopedia* females reported by Tan, although the lowered success could easily be due to their greater avoidance of males as much as to the less desirability by the males. In those cases where female relative receptivity could be tested by direct observation, it was found that mutant females had lowered thresholds, were less likely to avoid males, and consequently were mated more often than were wild-type females. For example *yellow* females (in studies by Bastock, and the Mainardis) and *white* females (Hildreth) had less resistance to males in most tests; nevertheless at least in one case (Bastock) the lowered resistance of females to males was incorporated into the genome of wild-type females by suitable outcrossing: derived wild-type females mated in higher frequency, so that the mutant locus was not so likely responsible for the *yellow* females' receptivity as was a polygenic set of determiners. Finally, it is clear that the outcome of mating success depends far more on the female's control by her acceptance of particular males than by the male's control over choice of his mate. To rephrase the point, males provide the diversity from which the females do the selecting. If the female's selection is accompanied by higher fitness, the female's discrimination, or high threshold (except toward males with "correct signals") will be maintained or increased provided it has a genetic basis.

Strain Differences Without Genetic Markers

Not only have genetic changes occurred apparently "spontaneously" in various inbred or mass-mated lines kept in laboratories, but, through artificial selection, fast or slow mating response has been demonstrated, indicating that considerable multigenic or polygenic variation is available in most *Drosophila* populations. Such genetic differences have far more significance for natural selection of mating propensities than have the pleiotropic effects on mating which may accompany the major marker mutants discussed above. Much of the literature is concerned with establishing which sex predominantly controls the speed of mating and whether preferential, or isolation, tendencies exist. However, in more recent years with clear descriptions of *Drosophila* mating behavior such as those of Manning, Bastock, and Brown, in which courtship elements and female response can be analyzed, there is more insight into what to look for, so that genetic variation affecting either sex can be described more carefully and its magnitude measured. The nature of genetic systems in each sex and the connections or separateness (linkage or pleiotropic relationships) of such systems determining sexual behavior is of considerable interest. The following account describes those experiments which significantly contributed to a genetic analysis of mating propensity.

Merrell (1949b) observed single pair matings of two inbred lines (Lausanne-Special and Oregon-R) of *D. melanogaster* and recorded time before courtship, before copulation, and the duration of copulation for the four ways of mating these strains. Time to courtship and mating was apparently characterized (controlled?) by the female line while duration of copulation was characterized by the male line. Assuming the males were equally active, it was the female which determined when mating was to occur since she has effective rejection ability.

In their classic work on the effect of artificial selection on genetic variation, Mather and Harrison (1949) described relative mating speeds of Oregon (OR) and Samarkand (SK) parent lines and three mass-selected lines (for high chaeta number). They used a "male choice" technique with single male $\times$ 5 OR females + SK females; their object was to test "discriminative" and insemination abilities of the males as well as the propensities of the females. Flies were aged 1, 4, 7, and 10 days. "D" (discriminative ability) was measured as a ratio similar to the normal deviate as follows:

$D = \dfrac{d}{\sqrt{pq}}$, where d = mated OR ♀♀ − mated SK ♀♀ = virgin SK ♀♀ − virgin OR ♀♀; p = (1 − q) = (number of ♀♀ mated)/10.

SK females mated more often than OR females irrespective of the male type, while SK males were also more successful in mating than OR males. Males improved in "discriminative" ability with age from one to four days while the improvement of females with age was irregular; however, since such an ability for males is probably a result of female differences in receptivity, the discriminative index increase with age was no doubt due to female changes. Three selected lines were mated similarly in rivalry one at a time with the parent lines; when coupled with *SK* females, the females of two lines were less receptive, indicating similarity to the *OR* parent females, and their males were also alike. The third line (with highest chaeta number) resembled *SK* in that males lacked discriminating ability (or females were more receptive to them). These data from Mather and Harrison, then, measured departure from random mating ("discriminative ability") as well as the amount of mating by which the parent stocks and selected lines differed. There was a suggestion that these differences in mating behavior depended on more than one genic locus since there had been a correlated response in discriminative ability (with likely determination on chromosome 2) accompanying the selection for chaeta number. However, its expression in one or both sexes was not established, and the data could only be considered as suggestive and preliminary.

D. subobscura is a species with much more ritual in mating than *D. melanogaster;* that is, in the details of courtship movements and female response where considerable action and interaction apparently require visual cues, since light is necessary for mating. Smith (1956) documented the discriminative ability on the part of females and mating success for inbred and outbred strains of *D. subobscura* in some detail, a study which may serve as a model of observation for current studies in any *Drosophila* species. Females perform a side-stepping courtship dance and accept those males which can keep up. At the start of courtship the male orients towards the female, and he circles to face her head to head. When the female is receptive her reaction to a courting male is to perform a side-stepping dance moving rapidly from side to side, maneuvers which the male must follow keeping his facing orientation if he is to be successful. Mating often occurs with a single dance, although several repeated dances may be necessary before the female stops, stands still, and permits the male to mount her. All females used in Smith's study were hybrids between two inbred lines (B/K) while the males were either their sibs (B/K) or from a separate inbred line (O). Hybrid males had about 90 percent success in mating, usually in the first 15 minutes. Maximum success was attained when males were stored singly before the mating test; however, if either B/K males were stored together in lots of five before mating or if they had been paired with already inseminated females (which refuse them), they had a lower capacity for mating—either not courting, or courting much less, a "discouragement" which may be due to lack of stimuli from the female partner. In contrast, the inbred (O) males had only about 50 percent success when mated to the B/K females. Reasons for lack of success of these inbred males were different from those of "discouraged" B/K males, namely, that O type males in following the side-stepping dance of the females could not keep up and did not keep facing the female as B/K males did. Such dances ended up by the female turning away. If the female did remain still and the male attempted to mount he seemed clumsy, and the female would reject him by kicking him off. Such lower mating success was then not so much the result of lower intensity of courtship as inaccuracy and insensitivity to the female and lack of rapid reactivity to her movements, which Smith summarizes in the expression: "the spirit is willing but the flesh is weak." If females then can be said to discriminate between males that court "properly" and those which court "clumsily," there is some evidence that Darwin's sexual selection may be important in the evolution of the courtship ritual, provided that such sexual rivalry and fertility as a parent are correlated. In Smith's outbred and inbred lines there was such a correlation since outbred males produced four times as many progeny as inbred males (mated to the same B/K females). Just why the quantity and quality of sperm should be correlated with the lowered mating success is not clear: it could be simply that the inbred males were homozygous for numerous deleterious factors, but the experiments are clearly a case of females being stimulated to accept males that do have higher fitness value.

The parallel study by Bastock (1956) has already been mentioned in Table I, but the *yellow* mutant males in *D. melanogaster* have a courtship which differs from wild-type in having shorter bouts at longer intervals, details which the wild-type females recognize and discriminate against. It seems logical that most of the mutants mentioned previously that affect male success may act substantially the same way, namely, by producing less stimulation for female receptivity. Correlation with fertility of such males has not been reported, however. The apparent effect of *yellow* on the females, raising their receptivity to both types of males, was found not to be a pleiotropic feature of the mutant locus since on backcrossing to wild type for a few generation, wild-type females became more receptive and *yellow* females less so. Possibly selections in *yellow* stock cultures had favored low-threshold females, i.e., requiring less stimulation to mate.

A further instance of spontaneous differences between strains is to be found in the reports of Hoenigsberg and Koref-Santibañez (1960) in their search for evidence of preferential mating or partial isolation between strains of *D. melanogaster* (*SK* outbred, *OR* outbred, and three inbred strains). Their technique was unfortunately only that of "male choice." Although they observed single male × two types of female in courtship details, their interpretation seems unduly misleading, largely because of the technique's shortcomings. Their results, which they interpreted as evidence of preferential courtship and mating by males, could just as well be shown (as they were by Merrell, 1960) to be due to the following: greater receptivity of *SK* females than *OR* females (similar to Mather and Harrison's observation [1949]) and outbred *OR* females greater than inbred line *OR* females. As for males, there was a homogamic tendency for *OR* males to "tap" females of their own strain, but *SK* males tapped only one third as often as *OR* males. Male preference was not really involved in the apparently preferential copulations, since the inbred females mated much less frequently than outbred females with either type of male. As Merrell points out, "In view of the fact that the outbred males spent equal or greater amounts of time courting the inbred females than they did courting their own females, these data cannot indicate incipient sexual isolation." Inbred males spent more time courting their own inbred females simply because those females seemed to require longer courtship before mating. With all their courtship directed towards their own females, they still copulated more with the outbred *OR* females. All those facts agree with previous studies in showing lowered mating success of inbred members of both sexes, as well as the control of mating success by females.

In these cases inbred females differ from many of the mutant strain females mentioned in the previous section in that more often the mutant females (like *yellow* [Bastock]; *white* [Hildreth]; and *vermilion* [Bösiger]) had lower thresholds than the outbred or wild-type strains, while inbred females tended to have higher. In a few cases (*cut* [Merrell] and *aristopedia* [Tan]) mutant females were less likely to mate. It is difficult to generalize, however, since most experiments have aimed at testing males rather than females. It is evident, nevertheless, that for any single genetic strain the sexual activity of males has little predictive value for the threshold of their sibling females. Males with aberrant courtship, which are usually discriminated against by wild-type females, may have females with either lower or higher thresholds than wild type, a fact that implies sex-limited or sex-influenced polygenic systems with different loci likely to be involved in either sex.

Genetic Analysis by Selection Techniques

One of the common observations made by mixing virgin flies for mating tests is the variation in time taken before mating. Flies taken from recently sampled natural populations characteristically mate rapidly provided that they are cultured under near optimal conditions. For example, *D. melanogaster*, *D. pseudoobscura*, and other members of the subgenus *Sophophora* will mate within the first five to ten minutes with a probability of between 0.50 and 0.80. After inbreeding or merely maintaining mass cultures in the laboratory for a few generations, however, strains typically differ greatly in mating propensity, and usually a considerable fraction of nonmaters remain even after observation for several hours. Demonstration that a large fraction of this variation is genetic has been done by artificial selection and analysis-of-variance techniques. The following discussion includes the

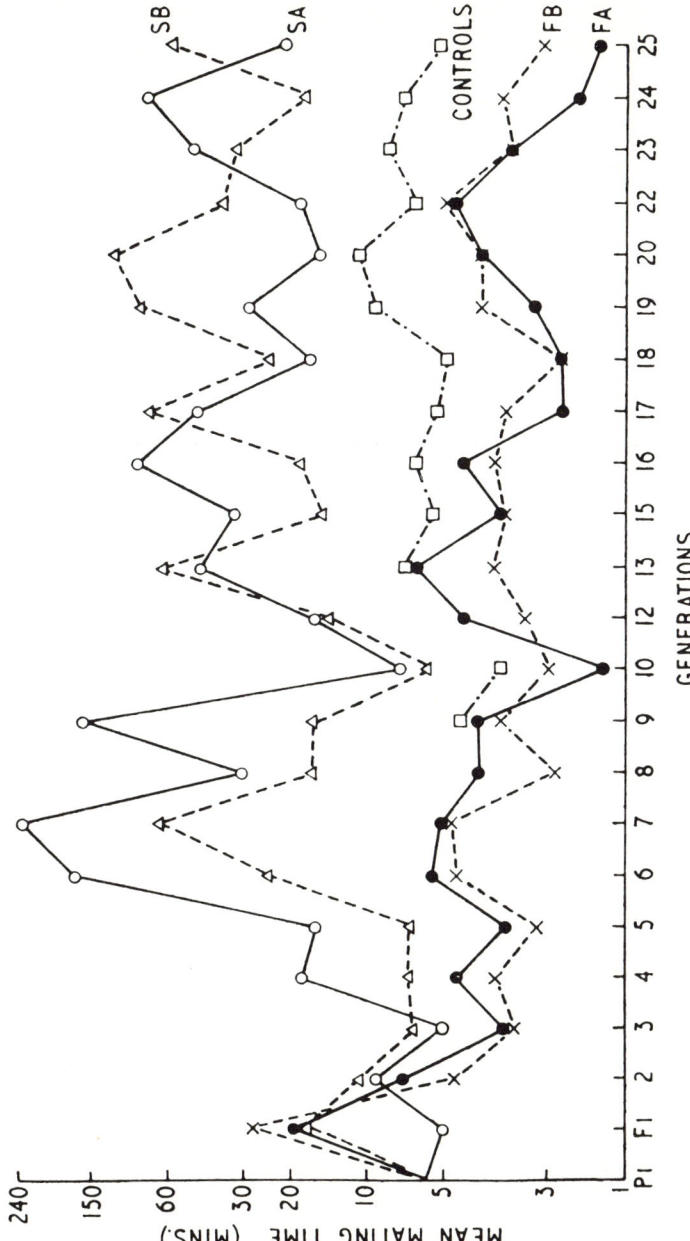

Fig. 1. Response to selection for fast (FA, FB) and slow (SA, SB) mating speed in Drosophila melanogaster. Mean mating time is plotted on a logarithmic scale. Selection was relaxed in generations 11, 14, and 21. Controls are unselected. (From Manning, 1961. Courtesy of Baillière, Tindall & Cassell Ltd.)

results of selective breeding, a technique that is especially suited to separating the behavioral elements of courtship and mating, the genetic determiners of those elements, and the organization of the wild type's genetic system. Variance analysis more accurately "dissects" the genetic system allocating portions of additivity, dominance and interaction. Of course the two techniques are largely reinforcing.

The experiments of Manning (1961) are most valuable not only for the genetic and behavioral information they produced but for their utility as a prototype. Beginning with a large population of *D. melanogaster*, collected originally by Dr. Forbes Robertson, Manning selected the first 10 (fast line) or last 10 (slow line) pairs to mate from a set of 50 pairs within each line. All pairs were timed as they mated and were aspirated out of the mating bottle while an unselected control stock was measured for comparison. The response was almost immediate from about the F_3 to F_7 (Fig. 1), with rapid divergence in mean mating time in both replicates of

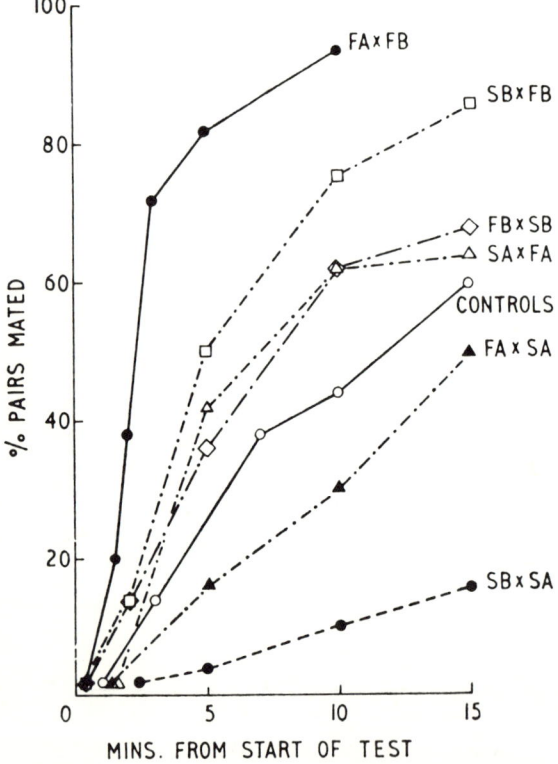

Fig. 2. Mating speeds of F_1 hybrids from generation 17 parents. The male parent's line is given first. (From Manning, 1961. Courtesy of Baillière, Tindall, & Cassell Ltd.)

fast and slow lines. (Mean time was measured by transformation of percent mating into probits and plotted against log time in the mating test.) The fluctuations of some magnitude between generations are conspicuous in being parallel and thus due to some common environmental factors, "the nature of which are still obscure," according to Manning; and this observation is characteristic of the trait measured, assuredly, and at present has no further explanation. The divergence of selected strains was quite constant nevertheless as long as they were kept contemporaneous. For example, at the 18th generation, fast line matings averaged about 85 percent in 5 minutes, the controls about 60 percent, and the slow lines less than 10 percent. The results are consistent with the accumulation in the selected lines of many genes with nearly additive effect whose heterozygous expression is about intermediate in hybrids between lines (Fig. 2); the hybrids between the two fast and two slow replicate lines were very similar to their parents and thus probably accumulated genes with very much the same actions. There is also a possible maternal effect, although significance of differences between the cumulative percentage curves is not given. While mating speed is a complex trait and depends on

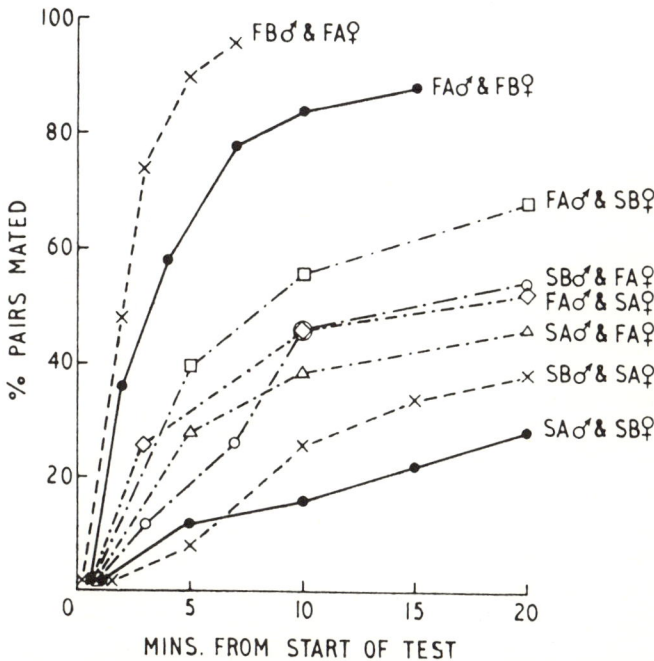

Fig. 3. Mating speeds of flies from generation 7 when tested by pairing flies between the selected lines, omitting FB × slow lines. (From Manning, 1961. Courtesy of Baillière, Tindall, & Cassell Ltd.)

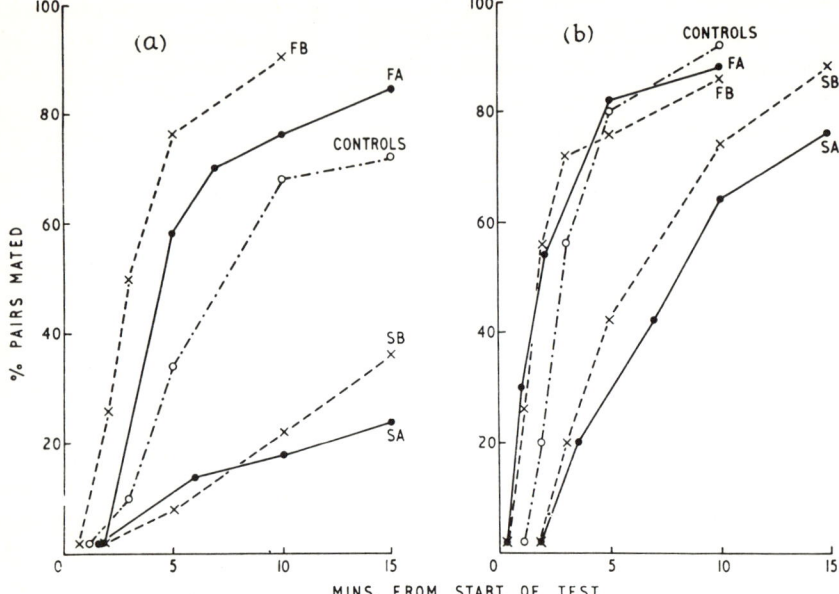

Fig. 4. The mating speeds of (a) selected line and control males with unrelated stock females, and (b) unrelated stock males with selected line and control females. Selected flies are from generation 18. (From Manning, 1961. Courtesy of Baillière, Tindall, & Cassell Ltd.)

the interaction of two individuals, a rough heritability was estimated at approximately 0.30, the larger part being due to response in the slow rather than in the fast lines.

Two sets of experiments were then done to ascertain which of the sexes had been affected by selection: (1) one sex from fast by the other sex from slow lines were tested in mating chambers (Fig. 3), and (2) each line and the controls were tested by using a foreign wild-type stock as the independent other member of the test (Fig. 4). From the first set of percentage curves it is evident that both sexes had been affected, since interchange of either sex altered the mating speed about equally. When outcrossed to the independent stock, however, males showed good divergence, but slow line females, although less than fast females and the controls, did accept the foreign males rather readily. Courtship activity of males and receptivity of females then were highly correlated in the selected lines.

Whenever a response to selection is observed, there is always a question of whether the organism is responding to the investigator's efforts or to some unconscious change of which the observer is not aware. Manning noticed at an early stage in the selection process that fast-mating flies were sluggish while in contrast the slow-mating ones were very disturbed by merely shaking them into the bottle and did not settle down for several minutes. To get an objective criterion of activity differences, flies were tested in an open "arena" with a grid of squares to quantify their speed of movement. Slow lines were indeed much higher in activity than fast lines, and in that characteristic they were similar to the controls. Manning summarizes these behavioral changes as follows:

> Briefly then, whilst the S-lines show high activity and low sex, the F-lines show low activity and high sex, but the controls show both high activity and high sex. It appears that selection has been most successful in directions based upon raising the thresholds of performance. The most conspicuous effects are the raising of the activity thresholds in the F-lines and the courtship of the S-lines. This fact indicates in turn that under normal domesticated conditions, in a population cage or stock bottle, natural selection keeps the levels of general activity and sexual behaviour near to the possible maximum . . . It is not so much the level of activity that matters as the speed with which the fly can make the change over from activity responses to sexual ones.

Such flexibility undoubtedly is an expression of what can be considered high fitness for the wild-type. As Manning points out, "Clearly, a simple concept of 'vigour' is not sufficient," since the artificial selection has separated what were two positively correlated components of behavior: vigorous "general activity" with low courtship and vigorous courtship with low activity. Genes accumulated in these selection lines must be affecting neutral thresholds and/or hormonal mechanisms controlling these behavioral components.

In selecting for one sex at a time with the same intensity as before but without selecting for mating speed of the opposite sex (and omitting female fast lines which are technically difficult), Manning (1963) was unsuccessful in getting a response for fast male or slow female lines, and of the two slow male lines one was just slightly slower than the controls. Certainly part of this reduced success was due to dilution of selection by incorporation of unselected genotypes from the opposite sex. Nevertheless, from the slow male line that did respond well, behavioral tests were made which contrasted sharply with the previous experiment (both sexes selected). To find out whether the change was limited to males, females were mated to control males (at generations 11 and 21); at the earlier time when the males were clearly slow maters the females were unaffected but later did show reduction in mating speed. While these results could be derived from sex-limited genes that at first only had effects on males but with further accumulation extended to females, a more likely interpretation was possible after testing both sexes in the "open field arena" for general activity. Both sexes from the slow line turned out to be sluggish, slow movers in contrast with the earlier experiments (Manning, 1961), in which the slow-moving strains were easily disturbed and hyperactive. Males tested in single pairs for "licking" frequency as a measure of courtship intensity had a lowered value also. Sluggish males in this case had a reduced mating speed merely because males need a modicum of activity coupled with low sexual threshold (as with the fast mating males to the 1961 experiments) in order to accomplish mating to control females. Females, on the other hand, are mated with more frequently if they are slow to move away when courted. In the earlier generations of testing, these females' sluggishness may have stimulated extra courtship from control males with the net result of no lowering of mating speed, while at the later generation sexual responsiveness must have been affected so much as to overbear this compensation. Finally, the possibility of sluggishness being the outcome of extreme inbreeding and high homozygosity was ruled out by the success of reverse selection (for fast mating) at the 25th generation in the main slow line, and lack of hybrid vigor when the two slow lines were intercrossed (at the 24th generation).

These two sets of experiments, then, show up some of the differences in the organization of behavior in the two sexes, particularly those involving the conflicting effects of genes affecting activity levels, as well as illustrating the diverse ways that selection may change the same character. Genes affecting general activity and sexual activity then must be separate systems whose levels can be changed quite independently, provided the selection method is directed carefully toward one or the other. Nevertheless, the two activities are highly correlated. By selecting for "spontaneous activity" Ewing (1963) found his inactive lines to display greater sexual activity, as might be expected from Manning's observations. The selection techniques, however, involved placing 50 flies of one sex in the first tube of a line of interconnected tubes and selecting the first 10 flies (active) to reach the opposite end or the last 10 flies to leave (inactive). While selection was quite successful in separating flies that went through the tube in a hurry from those that remained behind, when the respective lines were tested by introducing single flies in the "open field arena" of Manning, there was no significant difference between them, so that they could not be said to be different in spontaneous activity but rather in "dispersal" activity, or reactivity to each other when placed in a group as they always were in the selection trials. In fact, the subtlety of the responses both in that experiment and a subsequent one of sending single flies through a series of funneled chambers and getting a response to the stimulus of the funnel walls ("claustrophobic effect"), yet no response toward spontaneous activity, emphasizes the numerous genetic systems controlling different behaviors possible in this fly. The sexual responses were correlated with "avoidance," or reactivity between adjacent flies, then, and must be considered still separably distinct from Manning's earlier correlation between sexual activity and the open arena "general" activity.

Most recently, in a physiological study of the control of receptivity in *D. melanogaster* females, Manning (1966, 1967c) indicated that female acceptance of a courting male had two distinct elements, namely, *receptivity* (an all-or-none process apparently related to juvenile hormone level following eclosion), and *courtship summation* (a critical level of stimulation from the male). He described then the behavior of a selected slow mating line of *D. simulans* (Manning, 1968). Selection technique was the same as in the earlier work (10 out of 50 mated), but in addition, tests were made on single pairs in which duration of courtship required by a female before accepting the male could be measured. As before, selection response was excellent, giving no overlap with controls after the fourth generation. Yet when the sexes were tested separately for sexual activity, the males not only were fast to court and mate with control females, but were even better than control males. It was the female sex only which was responsible for the slow mating outcome, since slow line females "performed the most vigorous repelling movements, extruding the ovipositor and twisting or lifting their abdomen beyond the reach of the courting male." Slow line males persisted in courtship when in mass matings, and possibly that feature was essential for survival of the line to counteract the extreme high threshold of the females. Such a result provides still another contrast with the outcome of slow selection in all previous experiments with *D. melanogaster*: in the earlier case both sexes were affected, and lowered mating had probably been correlated with sensitivity to disturbances, then the second achievement of slow-mating (one sex selected) produced sluggishness of activity. Finally there was a strain with the two sexes markedly differing in their sexual responses. These slow-mating *D. simulans* had normal ovarian growth, and their corpora allata complex, when implanted into normal control pupal hosts, were quite capable of inducing precocious receptivity in the freshly eclosed females. Consequently, within the framework of Manning's hypothesis of female acceptance these selected line females had presumably the hormonal concentration for receptivity but they lacked the neural target-organ action necessary to respond in courtship summation and thus bring about acceptance of the male.

In *D. pseudoobscura*, Kessler (1968, 1969) was able to select quite effectively for fast and slow mating speed from a heterogeneous foundation population obtained from an intercross of three wild-type strains (from British Columbia, California, and Guatemala). Essentially the technique was similar to that of Manning's: 10 selected mating pairs out of 50 were aspirated out of the mating tube, either the first or the last being selected. After the 12th generation of selection, tests were made in all possible combinations of fast, slow and control lines to assess the relative roles of the

TABLE II

Mean Mating Indices[a] for the Control (C), Fast (F), and Slow (S) Selected Lines of *Drosophila pseudoobscura* after 12 Generations of Selection (after Kessler, 1968)

Males	Females			Mean for Males
	C	F	S	
C	6.35 (11)[b]	9.28 (6)	2.42 (6)	6.02
F	11.04 (6)	10.57 (8)	3.83 (11)	8.48
S	6.73 (6)	8.29 (11)	1.80 (8)	5.61
Mean for Females:	8.04	9.38	2.68	6.70

[a]Mating index (Spiess, Langer, and Spiess, 1966) is average of matings weighted by the reciprocal of time in 5-minute intervals; since Kessler's index has a maximum of 20 if all matings took place in 5 minutes and a minimum of 1 if all matings took place after 30 minutes, it can be used for samples of unequal size and in the analysis of variance.
[b]numbers in parentheses indicate replicates.

two sexes in mating outcome. The nine mating combination results are given in Table II as mean mating indices. The mating index uses a method of weighting the frequencies of mating in 5-minute intervals within a period of 30 minutes observation by the reciprocal of time (see Spiess, Langer, and Spiess, 1966). As used by Kessler it is limited to values between 1 and 20; since the maximum of 20 would be achieved if all mating occurred within 5 minutes, the percent of that maximum can easily be computed by dividing any mean index by 20. In sexual activity, or success in mating, both sexes were separable from the controls but in different lines: that is, slow-mating females were definitely controlling reduced mating whenever they were involved (and fast line females were not significantly faster than control females), while fast-mating males speeded up all matings in which they were involved (and slow line males were not significantly different from the control males). As for the components of variance, females accounted for 66 percent of total variance (mostly owing to the slow line's effects) while the males accounted for about 12 percent. From examining the reciprocals of Table II, it is evident that the speed of mating is mostly derived from female control.

It is noteworthy at this point to mention that in data of Spiess, Langer, and Spiess (1966) on mating differences of karyotypes in the same species (discussed in the section on chromosomal polymorphism), female determination of mating speed is also indicated. (Kessler did not analyze his selection strains for chromosomal arrangements, however.) These results contrast with Manning's earlier selection experiments, which were technically very similar to Kessler's in that Manning's selected lines showed both sexes affected about equally. Kessler points out that the performance of the unselected females resembles that of fast line females, whereas the unselected males' performance resembles that of slow line males. "When placed together, the unselected flies exhibit a relatively efficient speed of mating only slightly less than that shown by the fast line, or a compromise between the complementary mating tendencies of the two sexes, skewed towards the fast side as a result of the relatively greater influence of the female" (Kessler, 1968).

Quantitative Genetic Analysis of Wild-Type Strains

While it is abundantly clear that genetic factors affecting mating speed are commonly available to produce response to selection and to account for a considerable portion of the observed variation in sexual activity, it is particularly difficult as a trait to analyze quantitatively, in contrast with morphological traits, for at least two reasons: (1) as a character, mating speed depends on performance of two separate individuals that may have separate sets of genes controlling their sexual activities, and (2) it is not a normally distributed character in time of action but is usually extremely skewed, so that data transformations, which may lack biological reality or which may differ in degree of skewness between populations, have to be used. A considerable part of current analysis then is devoted to exploring better descriptive criteria and methods of measurement.

Since values of heritability and variance components due to genetic factors should be estimated from testing each sex separately, two fairly efficient types of test have been used: (1) diallele crossing between strains so that each strain can be averaged for either sex's "additive" or "nonadditive" effect; or (2) testing either sex from many strains separately by using some standard genotype of the opposite sex for uniformity of comparisons. The former technique has been employed by Parsons (1964, 1965, 1967) and his colleagues. Using six inbred lines derived by sib-mating from an Oregon stock background of *D. melanogaster*, Parsons tested flies in single pairs noting the time of copulation and the number of mating out of 7 pairs tested per trial (2 trials per cross and 36 crosses). Each line was tested by mating of intraline pairs and also following hybridization between lines by mating of interline hybrids. A summary of the data from Parson's studies is given in Table III. In the upper portion of the table hybrid mating speeds are compared with their inbred parent strains; and it is evident that hybrid matings are greater than inbreds for all lines except $N2$. Parsons treated the two sets of time period data (10 minutes versus 40 minutes) as if they were independent statistical quantities, and made the statement after finding that the two sets gave different estimates of additivity, dominance, and heritability that "either the dominance relations of genes controlling

TABLE III

Mean Number of Successful Matings[a] for Hybrids and Inbred Lines
(after Parsons, 1964, 1965, 1967)

A. Flies aged 6 days; F_1 Hybrids and Inbred Lines

	Time to Mating					
	≤ 10 Minutes			≤ 40 Minutes		
Line	♀ Parent[b]	♂ Parent[c]	Inbred[d]	♀ Parent[b]	♂ Parent[c]	Inbred[d]
N1	2.4	3.5	1.5	4.9	5.5	4.5
N2	2.9	3.5	4.5	5.9	5.9	6.5
Y1	4.1	2.6	2.0	6.1	5.5	2.5
Y2	3.9	2.4	0.5	5.5	4.1	0.5
G5	2.7	3.6	1.5	5.1	5.9	6.5
OR	5.2	5.6	1.0	6.3	6.9	3.5
Means:	3.53		1.83	5.63		4.00

B. Flies aged 3 days; Inbred Lines Only

Line	e	f	g	e	f	g
N1	2.7	0.9	0.5	4.6	3.4	4.5
N2	2.4	1.7	3.0	5.2	4.5	7.0
Y2	2.1	2.1	3.0	4.4	5.9	3.0
G5	2.1	3.9	2.5	6.0	6.6	5.5
D5	2.9	3.6	2.0	6.4	6.2	3.5

[a] Out of 7 pairs tested per trial
[b] Means for hybrids from ♀ parent crossed with remaining lines
[c] Means for hybrids from ♂ parent crossed with remaining lines
[d] Means for inbred lines, intra se
[e] Means for ♀ inbred lines tested by remaining lines
[f] Means for ♂ inbred lines tested by remaining lines
[g] Means for inbred lines tested, intra se

mating speeds change with time or different loci control this character at different times" (Parsons, 1964). However, he failed to consider that the results at 40 minutes include the results at 10 minutes plus the additional matings in the interval between. Besides that fact, experience in timing mating speed does not give such confidence in the results of any particular time period of such short duration as to merit generalizing such major conclusions. Finally, since the 40-minute data approach the maximum value, the difference between inbreds and hybrids will become less simply because of that fact. Consequently there is no justified indication that there are separate genic loci or different dominant relations responsible for what is observed at the end of the time period (total matings) in contrast with the early time period (matings in first 10 minutes).

Parsons (1967) discussed his analysis of variance for these diallele hybrids, which may be summarized as follows: *general combining abilities,* expressed as deviations from the overall hybrid mean mating speed, are the average effects of given strains in hybrid combinations, and *special combining abilities* are the departures from the additivity of the general combining expectations, that is, the unpredictabitity of hybrid mating outcome. Highly significant general combining abilities indicated that the lines differed greatly in their success in hybrid combination. For example, the *OR* line was very positive, raising the value of the hybrids in every combination, while the *N1* was almost consistently negative, lowering hybrids below the overall mean. Reciprocal (maternal effects) and special combining abilities were also significant; these facts can mean only that a considerable interaction between the sexes had been the case with either males or females or both determining the mating outcome. Unfortunately Parsons did not test the sexes separately to ascertain which sex accounted for the hybrid advantage.

In the lower half of Table III, mating speed of the inbred lines (not their hybrids) is given (Parsons, 1965). For each time interval there was less difference between the intraline results and interline tests than when hybrids were used, of course; however, at a glance it can be seen that the variation between lines is greater for the males than for the females (rows "e" versus "f"). In both time intervals variation among effects in the males is highly significant in contrast with lack of significance between mean values for females. Whether mating was rapid or slow, then, depends on the male's genotype for these inbred lines. Such a result recalls the work of Bateman (1948), who demonstrated that by counting genetic dominant marker progeny from males versus females, the males produced the greater contribution to the variance of the next generation irrespective of environmental and heritability factors, and that males are inherently more subject to intrasexual selection than females owing to the fact that their progeny contributions are dependent on the frequencies of insemination. Females may control whether a mating occurs, but the differences between males from these inbred lines of Parsons, probably expressed via courtship activity, have produced the major portion of variation.

In most recent studies Parsons and others (Hosgood and Parsons, 1967; Parsons, Hosgood, and Lee, 1967) have measured mating speed and duration of copulation in certain strains derived from single wild females (*D. melanogaster*) that differ also in chaeta number and which must then have come from a highly polymorphic and genetically diverse population. By testing these derived strains in single pair matings (50 pair matings per test per generation), Parsons and colleagues again found that the strain of male accounted for the largest amount of variance, the strain of female also for a significant but smaller portion, as well as some interaction between the sexes. In the diallele analysis, most of the variance was additive. The duration of copulation, which averaged about 20 minutes for these strains, was controlled almost entirely by the male strains and was additive predominately with considerable reciprocal (maternal) effects.

A more efficient technique for quantitative analysis has been used by Fulker (1966) in *D. melanogaster*: by focusing on the performance of the males alone, a reliable measure of genetic variance, heterosis, and heritability was achieved. At first, tests were done to ascertain the relation between speed of mating, time of copulation, the number of copulations resulting in fertilization, and the number of progeny produced. Single males (from each of six inbred lines and from a sample of F_1 hybrids made from these lines) were tested with six virgin females (one female from each inbred line) over a period of 12 hours. These data are given in Table IV. Since each male was given the same array of females to mate with, the female can be considered the "standard" testing set. Speed of mating was much faster and more uniform for hybrid males than inbred, and the longer the time taken before copulation, the less the number of

TABLE IV

Mean Scores for 12 Male Genotypes and Intercorrelations[a]
(after Fulker, 1966)

Genotype of males	(Item 1)[a] Time to first copulation (min.)[b]	(Item 2)[a] Observed number of copulations	(Item 3)[a] Number of copulations resulting in fertilization (maximum 6)	(Item 4)[a] Number of offspring produced	
Edinburgh (Ed)	9.2[c]	4.6[c]	4.0[c]	147[c]	
6 CL	34.6	1.6	1.0	33	Inbred
Samarkand (S)	26.4	2.8	1.8	98	Lines
Wellington (W)	16.4	4.2	3.6	236	
Oregon (Or)	18.2	2.8	2.4	158	
Florida (F)	14.8	3.6	2.8	118	
F × Ed	3.6	6.2	5.4	302	
F × W	6.8	5.8	5.4	340	
Or × W	5.6	5.2	5.0	270	Hybrids
6 CL × Ed	5.6	5.0	4.6	234	
6 CL × S	7.6	3.2	1.4	63	
Or × Ed	5.2	5.4	4.2	225	

[a]Intercorrelations: Item 1 and item 2, -0.87 ($p < 0.001$); item 4 and item 3, -0.78 ($p < 0.01$); item 1 and item 4, -0.69 ($p < 0.02$); item 2 and item 3, 0.96 ($p < 0.001$); item 2 and item 4, 0.90 ($p < 0.001$); item 3 and item 4, 0.95 ($p < 0.001$)
[b]Those not mating by 40 minutes were assigned a value of 41
[c]Each mean is based on 5 ♂♂, each tested × 6 ♀♀

TABLE V

Mean Number of Females Fertilized Out of 6 Maximum from Replicated Diallele Cross F_1 and Inbred Line Males in 12 Hours (after Fulker, 1966)

Line	Hybrid males from father's line crossed to remaining lines	Hybrid males from mother's line crossed to remaining lines	Inbred males
6 CL	3.02	3.04	1.30
Edinburgh	3.63	3.74	3.40
Oregon	2.99	3.03	1.30
Wellington	3.48	3.68	2.60
Samarkand	3.06	2.76	1.20
Florida	3.49	3.42	2.30

copulations and the fertility; thus, these four measures appear to be general characteristics of male mating behavior. Furthermore, the high correlation between time of mating and the number of fertilized females gave to the latter a measure of utility for analysis of mating speed without the time-consuming task of watching the flies over a long period. A summary of the mean number of fertilized females in the diallele set of hybrid (F_1) males and inbred line parent males is given in Table V. Again the heterosis of hybrids is evident. From the analysis of variance Fulker indicated the significance of "additive" variation, or general combining ability, and dominance to be "strongly directional towards higher speed." There was no significant reciprocal effect, in contrast with Parsons' 1964 data. If there is any consistency between the two sets of experiments (Fulker's and Parsons') it may be that the reciprocal effects in the latter's results could have been expressed in females only, but since Parsons did not separate the effects of the two sexes in mating tests it is difficult to be certain. Finally, there was no evidence of complex interaction, so that the genetic analysis was relatively simple: genes for high mating speed were dominant over low mating, with the Edinburgh and Wellington lines carrying mostly dominant alleles and Oregon carrying mostly recessive. "Narrow" heritability was estimated at 0.36 and the number of effective independent genic blocks approximately 5, which is certainly an underestimate.

As Fulker points out, since dominance is for high mating speed, most progress in selection would be expected in the direction for low mating speed; indeed, Manning (1963), when selecting in males alone, achieved response for low mating speed but not for fast mating speeds. It is then in natural populations where most gene combinations are likely to be heterozygous that directionally dominant or slightly overdominant genes ensure a high proportion of fast-mating males. Natural selection is likely to be aimed towards the maximum expression of this trait, as well as high fitness in terms of progeny produced is consequently closely correlated, as is shown by the independent evidence of these experiments.

More recently a genetic quantitative analysis has been made by Kessler (1969) on his selected strains of *D. pseudoobscura* as an extension of his earlier work. Following 15 generations of selection for fast and slow mating speed in which response had been symmetrical (in spite of greater intensity in the fast direction) and effective for the first five generations mainly, Kessler crossed the two lines, measuring mating speeds in the F_1, F_2, and backcrosses to both parent lines. In addition he reversed selection as well as initiating relaxed selection lines to ascertain the amount of genetic variation available to those lines. The problem of choosing a proper measurement for the analysis was solved in part by attempting to fit the mass mating data to the additivity scales of Mather (1949). First the mating frequencies gave a poor fit; by using the mating index (Spiess, Langer, and Spiess, 1966), which gave a better fit, and finally taking the log of the mean mating index, he achieved the best fit and estimated realized heritability at about 20 percent in the fast and 6 percent in the slow line. From the genetic analysis (F_2's and backcrosses) he estimated 25 percent heritability with a large number of genetic factors and considerable dominance from the faster parent. As for relaxed and reversed selection lines, from the fast line a tendency to return to control level suggested that selection for heterozygote combinations may have occurred in that line, while from the slow line changes were not significant even though there was a net increase in mating speed towards the control. Unfortunately the sexes were not tested separately in the genetic analysis, so that it was not certain whether the dominance of fast mating was really female- or male-based, though Kessler's earlier account (1968; discussed in the previous section, and see Table II) implies predominately a female determination, in contrast with Fulker's experimental results. Fulker had only tested males, however. Nevertheless, in the light of Manning's more extensive work, where in three different selection experiments slow mating speed was achieved by three different behavioral changes (see previous section), it could easily be the case that either sex is capable of control and that the outcome of selection for mating speed may depend on very subtle distinctions between the aims of selection applied towards different components of sexual behavior. In Kessler's view the responses to selection have been aimed at speeding up the males in the fast lines (because the control males are more like the slow males) while slowing down the females (because the control females resembled fast line females). In consequence it would appear that in natural populations of *D. pseudoobscura* dominant genes controlling fast mating in females have been selected, while in males there may be a dampening effect on maximal expression of mating speed. Such conclusions must be qualified and preliminary, however; Kessler's unselected stock was a triple hybrid from very diverse widely separated original populations, possibly quite uncharacteristic of any single population's mating speed in either sex. Fulker's inbred lines and their F_1 male responses may not be typical of natural populations of *D. melanogaster* either. Yet the methods and results of these authors are important for comparative purposes of making speed determination by the two sexes from natural populations.

Chromosomal Polymorphism and Mating Propensity

It has been more than 35 years since Dobzhansky and Sturtevant began describing the extensive chromosomal polymorphism in *Drosophila pseudoobscura* and its relatives in western North America (Dobzhansky and Sturtevant, 1938; Sturtevant and Dobzhansky, 1936). We still do not know precisely what these arrangements do for their possessors in nature, though we can be fairly certain that those arrangements with stable frequencies in natural populations (either displaying gradients in relative frequencies over wide geographic areas and/or fluctuating with the seasons, whether "flexible" or "rigid" in their occurrence) contain genotypes whose functions are adaptive. Whether there are any functions characterizing particular arrangements, with uniform properties from locality to locality, is still a matter of conjecture. While the number of adaptive functions that have been associated with chromosomal arrangement polymorphism is quite large (see Dobzhansky, 1961; Dobzhansky, Lewontin, and Pavlovsky, 1964; Spiess, 1968b), the function of mating propensity and sexual selection is a highly significant one.

While it is conceivably possible that a single genic locus might have sex-limited expression for persistent courtship in males, and in females "preference" for persistence of her mate, it is more likely that separate genic systems in each sex will determine that specific behavioral set. Or, at least, behavioral traits are more commonly polygenically determined, and the set of polygenes necessary for efficient mating in one sex is likely to be different from the set in the opposite sex if either different nerve pathways or polypeptide-hormonal-type substances are involved. Linkage between such separate sets is likely to be selectively advantageous. As stated by O'Donald (1962), "Any mutation that enhances the female preference and that is also linked to a sexually advantageous gene will immediately be picked out by selection; the closer the linkage the more rapidly will the advantageous combination advance; and any factor that tightens the linkage will also be selected. Favorable combinations of closely-linked factors, or supergenes, should therefore evolve under sexual selection. Thus we should expect to find that one or more supergenes control both male display and female response to it." These well-known chromosomal rearrangements (inversions) of many *Drosophila* species can be considered as prime suspects to combine in close linkage advantageous sex-limiting genotypes controlling mating propensity with expression of synchrony between mates.

Sex Ratio in Drosophila pseudoobscura

The earliest report of the control by gene arrangement over sexual activity is that by Wallace (1948) of the "sex ratio" set of triple inversions (*SR*) in the right arm of the X chromosome of *D. pseudoobscura*. Among the several fitness traits that he measured which might have been influencing the outcome of changes in relative frequency of the *SR* and *ST* (Standard) arrangements, was the insemination frequency of females by *SR* and *ST* males in "male choice" experiments, that is, 10 males of one karyotype in a single vial with 20 females, 10 females of each of two karyotypes at a time. At the end of two hours the females were dissected to ascertain which had mated. Six sets of tests were run at 25°C and six sets at 16°; the data are presented in Table VI. At both temperatures, according to Wallace's relative fitness calculations, the *SR* males inseminated about 80 percent of the females compared with the number inseminated by *ST* males. After noting the details of Table VI, however, it becomes evident that the two karyotypes of males are about equal except for two sets of tests, that is set #3 (25° with both homokaryotype females) and set #5 (16° with *ST/SR* and *SR/SR* females). In the former case, both *SR* males and *SR/SR* females showed less sexual activity than their counterparts (*ST* and *ST/ST*) so that on superficial examination a "preference" of standard × standard might be supposed. However, *SR* males simply mated less with both types of female, and *SR/SR* females also mated less with both types of male, so that a "preferential" interpretation is not needed (as invoked by Wallace, 1968, pp. 124-125, for example). As a matter of fact the tests were not "multiple choice" in which females would have had an opportunity to accept or reject the two kinds of male. The results can more simply be a matter of relative sexual activity: *SR* males are nearly equal to *ST* males at 25° except in the special set #3 with homokaryotype females, while at 16° with all matings more frequent there is only a significantly lower value for SR males in one set (#5). "Male choice" tests like these probably measure the females' relative acceptance, especially if females are the major controlling sex for mating. From the female totals and averages it is evident

TABLE VI

Number of *D. pseudoobscura* Females Inseminated (karyotypes with sex ratio [*SR*] and standard [*ST*] X-chromosomes) When Mated to *SR* or *ST* Males (after Wallace, 1948)

A. Stored at 25°:	Female Set #1			Female Set #2			Female Set #3			Overall
	ST/ST	*ST/SR*	Av% ♂♂	*ST/ST*	*SR/SR*	Av% ♂♂	*ST/ST*	*SR/SR*	Av% ♂♂	Average for ♂♂
Males: *SR*	39/87	59/96	54%	26/78	42/66	47%	38/85	28/86	39%	232/498 = 46.6%
ST	51/91	45/92	51%	28/75	32/74	40%	78/91	49/85	72%	283/508 = 55.7%
Average for ♀♀:	51%		55%	35%		53%	66%		46%	

B. Stored at 16°:	Female Set #4			Female Set #5			Female Set #6			Overall
	ST/ST	*ST/SR*	Av% ♂♂	*ST/ST*	*SR/SR*	Av% ♂♂	*ST/ST*	*SR/SR*	Av% ♂♂	Average for ♂♂
Males: *SR*	61/97	83/108	70%	31/81	41/79	45%	44/88	58/88	58%	318/541 = 58.8%
ST	82/113	85/108	76%	66/85	56/83	73%	47/88	71/84	69%	407/561 = 72.5%
Average for ♀♀:	68%		78%	58%		60%	46%		75%	

Female Overall Averages:	*ST/ST*	*ST/SR*	*SR/SR*
At 25°	206/354 = 58.2%	158/341 = 46.3%	152/311 = 48.9%
At 16°	224/386 = 58.0%	265/382 = 69.4%	226/334 = 67.7%

that there were fewer differences between females than between males, that ST/ST females were nearly equal at the two temperatures, while the two karyotypes with SR mated also at amounts close to each other, although at about 20 percent lower at warm temperature (47 percent) than at cool (68 percent). In summary, then, the SR arrangement compared with ST had a slight lowering effect on male sexual activity at cool temperature but was not consistently different from ST at warm temperature, while females with SR arrangement accepted males more than ST/ST females at cool temperature throughout at warm.

This information, of course, is not sufficient to account for the relative changes in Wallace's experimental populations—not only because of the special interactions in particular combinations of these karyotypes, and because of the unpredictability of what might be the outcome with all karyotypes mating together, but also because mating activity is only one of a large number of fitness traits that these karyotypes might control. Nevertheless, it is interesting that SR was not so disadvantageous at 16° as at 25° from the experimental populations: SR reached an apparent equilibrium at 90 to 95 percent at 16° after about ten generations, compared with complete elimination of SR at 25°. How much of that difference was due to improvement in SR female mating activity at 16° would be difficult to ascertain, but the data are suggestive (although Wallace, 1948, did not enter female activity in his summary of fitness values, presumably because he considered his tests only as referring to male mating abilities).

Third Chromosome Arrangements in the Drosophila pseudoobscura Subgroup

In searching for fitness properties by which two arrangements (Whitney and Klamath) of chromosome 3 in *D. persimilis* might differ and thus help to account for changes in their relative frequencies within populations of that species, Spiess and Langer (1961) found that in every laboratory population initiated with homokaryotypes the WT arrangement (which was expected to increase and therefore was begun at low frequencies) always increased relative to the KL arrangement between the introduction of adult virgin flies and the deposition of eggs. Such an increase indicated unequal mating propensities of the two karyotypes introduced into the experimental populations. In subsequent mating tests, these authors found the WT arrangement consistently was associated with the higher sexual activity: first, with single males × 10 females of one karyotype only, WT males achieved insemination 36 percent more than KL males; second, with 5 males of each karyotype × 10 females of each in "multiple choice" sets (sex ratio of 1 male to 2 females), WT males mated two-and-one-half times as frequently as KL males while WT females were also more receptive than KL females; and third, in a similar "multiple choice" set with sex ratio equal, again *WT* males mated more than twice as often as *KL* males while there was a slight but not significantly higher value for *WT* females as compared with *KL*. Such results were especially important in view of the fact that the *WT* is common and the *KL* rare in the natural population from which they had been taken (White Wolf area in Yosemite National Park at an elevation of 8000 feet).

All those tests had been made over a 24-hour period so that they were not informative as to the details of mating speed and behavior. A technique modified from Manning's (1961) and Ehrman's (personal communication) was designed simply to facilitate observing all corners of a mating chamber and to remove mating pairs at the time of copulation. The karyotypes containing *WT* and *KL* arrangements were cultured in various ways at cool temperature (16°) and flies were tested by introducing 10 pairs at a time into plexiglass mating chambers and aspirating out copulating pairs. Only one combination at a time was tested—that is, "no choice matings." An overall average of percentage of matings is given in Table VII for the main experiments. The results can be summarized as follows: *WT/WT* matings were most rapid, half of them being completed in five minutes and more than 80 percent of them in an hour's time, while *KL/KL* matings were very slow with less than 20 percent in an hour. Substitution of *KL/KL* males for *WT/WT* males lowered the *WT/WT* rate slightly, while the reverse substitution increased the *KL/KL* rate by about the same amount. Apparently, then, females were the principal determiners of

TABLE VII

Percent Mating in 60 Minutes of *D. persimilis* Strains from Yosemite Locality (Karyotypes with Whitney and Klamath Arrangements[a] (after Spiess and Langer, 1964a)

Males	Females			Average for Males
	WT/WT	*WT/KL*[b]	*KL/KL*	
WT/WT	84 (120)[c]	75 (140)	30 (60)	63.0
WT/KL	74 (130)	57 (160)	16 (170)	49.0
KL/KL	63 (60)	26 (180)	18 (120)	35.7
Average for Females:	73.7	52.7	21.3	

[a]No choice matings averaged over all tests on F_1 progeny from strain crosses, raised in vial cultures, and aged for 8 days.
[b]Heterokaryotypes include reciprocals
[c]Numbers in parentheses = pairs tested

mating speed; and indeed observations of their behavior indicated that *KL/KL* females refused males continually while *WT/WT* (and heterokaryotype) females accepted males readily. Males paralleled females of the same karyotype, with *WT* males courting more actively than *KL/KL* males. Heterokaryotypes when mated *inter se* were intermediate in propensity, but when tested by homokaryotypes their performance was determined by the speed of their mate irrespective of which sex was the heterokaryotype: from Table VII it is evident that the upper left three corner and the lower right three corner values demonstrate that fact. These *WT/KL* flies, then, showed "additivity" when only tested to each other, but were quite flexible as to their reactivity to the other two karyotypes. Finally, an aging study was done to ascertain how soon after emergence from the pupa the two homokaryotypes would mate: briefly, no matings took place (in 24 hours of testing) until the flies were at least 48 hours old, at which time *WT/WT* began mating (18 percent in 24 hour test), while *KL/KL* did not begin until the third day (about 10 percent in 24 hours). *WT/WT* rose in mating propensity with age to a maximum at about 8 days of age at which a plateau was reached characteristic for that karyotype, while *KL/KL* continued to rise with age, but it never achieved as high a propensity as *WT/WT* even after 21 days of age.

It appears, then, that the genic contents of the most frequent chromosomal arrangement in the natural population (*WT* = 80 percent approximately at White Wolf) control the mating outcome by influencing the responses of both sexes which lead to rapid insemination as compared with the rare arrangement (*KL* less than 10 percent), which slows mating speed as well as maturation rate. Chromosomal polymorphism is so widespread in *Drosophila* that it was worthwhile testing mating speed of the arrangements in *D. pseudoobscura* (as was suggested to the author by Professor Dobzhansky in 1963). Strains from the Mather, California, population containing the Standard (*ST*), Arrowhead (*AR*), Tree Line (*TL*), Chiricahua (*CH*), and Pikes Peak (*PP*) arrangements (10 strains each) were then tested using the same technique. The observed proportions of these arrangements in the natural population are in descending order, respectively, and their mating frequencies (homokaryotypes only) paralleled remarkably as follows (from Spiess and Langer, 1964b): *AR* × *AR*, 81 percent; *ST* × *ST*, 75 percent; *CH* × *CH*, 45 percent; *TL* × *TL*, 44 percent; and *PP* × *PP*, 24 percent mating in an hour of observation. In all cases most of the matings that did occur took place in the first 10 minutes (about 90 percent of all matings, whether fast *AR* or slow *PP*).

This work was then extended (Spiess, Langer, and Spiess, 1966) to include heterokaryotypes and various mating combinations omitting rare types and restricting the possible combinations (15 karyotypes and 225 mating combinations) in various ways for economy. Observation time was shortened to 30 minutes. The objectives were to achieve enough of the range to ascertain some general conclusions about sex differences in control, heterokaryotype values (whether additive, dominant, or heterotic), and the likelihood that chromosomal polymorphism might be sustained in some degree by these sexual activity traits. Because of the extent of these data and the need to make significance comparisons a mating index was devised as the weighted average of 5-minute intervals, (using the reciprocal of time for the weights: 5 minutes = 20, 10 minutes = 10, 15 = 7, and so on, with those not mating in 30 minutes having a value of 1). The average index for a given mating then could be compared with a maximum value of 200 (if 10 matings had all taken place in 5 minutes). In Table VIII average mating indices for each set of karyotypes (AR versus ST, CH, TL, and PP; ST versus CH, TL, and PP) are given in percent of this maximum value.

A striking sex difference is found by comparing heterokaryotype performance in males with that in females: in every case the value for heterokaryotype males is superior to both homokaryotypes, although in only three cases out of seven is it significantly so within the set of competing karyotypes. In females there is no consistent superiority (that is, greater receptivity), so that any heterosis displayed is clearly due to greater activity of the males, persistence in courtship, or to greater female acceptance of

TABLE VIII

Mating Propensities of *D. pseudoobscura* Karyotypes in Percentage of Maximum Value of Mating Index (after Spiess, Langer, and Spiess, 1966)

Combination of Arrangements	Female Averages			Male Averages		
	HOK_1[a]	HTK[b]	HOK_2[a]	HOK_1	HTK	HOK_2
AR_1 vs. ST_2	56.7	58.6	52.0	53.5	62.2*	51.8
AR vs. CH	61.5	67.4	52.3	65.4	66.0	49.8
AR vs. TL	61.5	57.9	44.8	54.3	63.1*	46.7
AR vs. PP	49.9	42.5	22.3	42.1	43.6	29.0
ST vs. CH	44.9	48.3	35.3	45.8	48.4	34.3
ST vs. TL	59.1	57.2	45.9	51.0	63.1**	47.7
ST vs. PP	46.4	49.1	33.4	47.6	52.9	28.4

[a]HOK_1 = homokaryotype for first arrangement, HOK_2 = homokaryotype for second arrangement
[b]HTK = heterokaryotype
*Statistical significance to heterokaryotype superiority over AR/AR or ST/ST.

heterokaryotype males—probably because of their increased sexual activity. With the slower tendencies of the rarer homokaryotypes (*CH, TL,* and *PP*), however, both males and females were usually similar in that they both showed more reluctance to mate than the common *AR* and *ST* karyotypes. Outstanding single exceptions were the following: *PP* females were especially reluctant to accept males (in the *AR* combinations), while *PP* males and *CH* males were both low in courtship. Kessler (1968) has recently pointed out that in three out of the seven combinations homokaryotype control of the mating outcome is mostly by females (*AR* with *TL, AR* with *PP,* and *ST* with *TL*), while in one combination (*ST* with *PP*) males are more influential and in the other three there is no clear single sex determination. Of course in natural populations these rare homokaryotypes constitute less than one percent of the flies, yet if mating propensity is a fitness trait of high value the question of whether retention in the population exists at all for such deleterious karyotypes must be asked. Undoubtedly the heterosis in males is an important factor, and there may be frequency dependency (see the next section on conditions affecting mating propensity) in some cases that give the rarer form an advantage in mating. It is clear, nevertheless, that mating speeds must be important in controlling the frequencies of these chromosomal arrangements in natural populations and that the deleterious effects are mostly attributable to rare karotypes in females.

Kaul and Parsons (1965, 1966), using a single pair techinque to measure copulation time as well as mating speed, studied *ST* and *CH* (3 strains each) and found time to mating longer for *CH/CH* than for *ST* karyotypes in agreement with Spiess and colleagues. Both mating speed and duration of copulation (which was shortest for *CH/CH*) were found to be male-determined with greater heterogeneity within female karyotypes than within male. However, later Parsons and Krul (1966) who used *AR* versus *PP* in single pair matings, did not find the low mating values for *PP* as Spiess and Langer had, though they did agree that with those karyotypes females were more influential than males in the final outcome. (The discrepancies between *PP* results in the two laboratories can be attributed to a number of differences in techniques, temperature, changes in strains with time, and fewer strains tested by Parsons and Kaul [5 compared with 10 in the Spiess laboratory]).

Parsons (1967) has summarized these contributions and further suggested that differences between strains of gene arrangements may be so great that genetic backgrounds may give contrasting results from separate laboratories even though all strains may have originated from the same natural population and locality. Indeed, much larger samples of wild populations and genetic analyses for determination of behavioral traits must be completed before final generalities can be made on these inversion systems and their control of mating activity. Evidence for changes either spontaneous or purposely selected tend to imply considerable genetic modification of mating propensity by a "polygenic" background: Parsons and Kaul (1967) as well as Spiess, Langer, and Spiess (1966) imply that some strains do not remain constant over long periods of time. Also, strains selected for other behavioral traits of AR and CH homokaryotypes, such as geotaxis (Dobzhansky and Spassky, 1962), were tested in multiple-choice matings with use of the Elens-Wattiaux observation chamber (Ehrman, Spassky, Pavlovsky, and Dobzhansky, 1965) and found to be substantially equal, a fact which could be interpreted as a change in genetic background in the selected lines since the original strains were the same as those used by Parsons and Spiess; but unselected lines from two experimental populations of Dobzhansky and Spassky's also displayed no difference between karyotypes. In contrast, Spiess and Spiess (1966), testing for mating speed populations of *D. persimilis* that had been selected in the laboratory for fast or slow development rate and in which the WT and KL arrangements were at the same frequency, found that the populations displayed fast or slow mating speed as well as development rate. Such a result could imply that considerable genetic variation had acquired control over mating propensity apart from the inversion blocks on the third chromosome. It must be surmised then that the large linkage blocks retained by inversions represent a major control system which may respond quickly and efficiently to selective forces, while genetic modifiers certainly exist that may be concentrated by selection to influence the expression of those major blocks.

The fact that these chromosomal variations affect mating propensity in parallel with their frequency in natural populations strongly suggests that with the origin of new arrangements selective differences in behavioral traits were conferred. In the course of local selective forces during increased distribution of newly arisen arrangements, specific features, or norm of reaction to local conditions, may have been altered through polygenic modification, but the general feature of mating propensity control must have been retained. (Of course this statement should not be construed to mean that many other fitness properties may not also be controlled as well by chromosomal polymorphism.) For the sake of argument, if a new arrangement increased in frequency and spread, its advantage may well have been modified in other localities to suit conditions of those environments encountered. Certainly, flexibility of control to meet local exigencies found in courtship and mating is likely to accompany such cytogenetic mechanisms. How genetically uniform each karyotype is for determination of this trait within any local population, and how flexible it is homeostatically, or how it may respond to selection, are questions yet to be answered with larger samples from natural populations plus a linkage analysis of chromosomal factors determining the components of mating behavior.

As for the question of which sex exerts the greater control over the mating outcome, it must be kept clearly in mind that the answer depends entirely on the karyotypes (or genotypes) being tested: for example, with a nonreceptive type of female present, such as *PP D. pseudoobscura* or *KL D. persimilis* (raised in the cold), with all the persistence of courting males from *AR* or *WT* of these two species, respectively, few matings will take place, since the females exert the greater control. However, if the type of female is receptive, variation in male courtship may be more influential. Furthermore, the type of female may modify the male's courtship initiation or persistence, although from present experiments the indication is that the initial, or primary, sexual activity of the male is not markedly influenced by the type of female as long as she is receptive. Experiments (Spiess, 1968c) designed to elucidate some of the variation observed within sets of karyotype matings in *D. pseudoobscura* employed a "standard" strain that had been presumed to be more reliable from previous tests in each mating to an "experimental" strain that had been thought to be highly variable before. Time to courtship, mating speed, and time in copulation were recorded using a single-pair technique similar to that of Parsons and his colleagues. The F_1 progeny of two strains of *AR* constituted the "experimental," while heterokaryotype *AR/PP* was the "standard." By re-

TABLE IX

Time to Courtship in Minutes and Variances Between Parental Pairs[a] (from Spiess, 1968c)

Type of male	Number of courtships Total	Average time ± standard error	Variance between parent exp. pairs
AR/AR (F_1) (X stand.)	265/300	4.64 ± 0.36	2.58
AR/AR (F_2) (X stand.)	169/200	4.18 ± 0.41	3.33
AR/PP (X exp. F_1)	200/300	7.76 ± 0.82	12.71
AR/PP (X exp. F_2)	112/200	8.72 ± 0.91	16.67
Control: *AR/AR* (F_1)	52/60	6.54 ± 0.85	14.35
AR/AR (F_2)	23/26	7.59 ± 1.36	29.51
Control *AR/PP*	24/31	8.38 ± 1.54	56.85

[a] 20 pairs per weekly set for 3 weeks were combined.

TABLE X

Percentage of Females Courted and Mated[a] **(from Spiess, 1968c)**

Type of female	Number tested	% courted ± standard error	% mated	% females accepting
AR/PP (X exp. F_1)	300	88.33 ± 2.46	77.67 ± 3.50	87.92
AR/PP (X exp. F_2)	200	84.50 ± 3.20	77.00 ± 4.71	79.30
AR/AR F_1 (X stand.)	300	66.67 ± 3.87	57.00 ± 4.31	85.50
AR/AR F_2 (X stand.)	200	56.00 ± 4.32	46.00 ± 4.44	82.14
Controls:				
AR/AR (F_1)	60	86.67	76.67	88.46
AR/AR (F_2)	26	88.46	73.08	82.61
AR/PP	31	77.42	64.52	83.33

[a]Data parallel to Table 9.

ciprocal test differences it was possible to determine which sex was more responsible for sexual activity. The "experimental" flies (AR/AR) were controlled by their emergence order from their culture bottles, while the "standard" flies (AR/PP) were not. In Tables IX and X, male courtship times and females mating, respectively, are presented, and in Table XI copulation time is given. Male activity in courtship time is strain-specific, being faster in AR/AR, but it was not significantly affected either by age of parents or by inbreeding one generation (F_2). Female response was relatively unvarying, in contrast, as can be seen from the right-hand column in Table X: all strains and generations of females were about equal in accepting males once courted (80 to 90 percent). Consequently the percentage of total mating here depended on male propensity. The lower value for AR/PP then must have been due to less copulatory tendency for those males. In view of the well-known indiscriminate nature of the males' courtship activity towards both males and females, they are probably not greatly affected by the females' "desirability" for courtship initiation, although it is undoubtedly important for continuation and final mating. The female role appears to be passive here with these particular karyotypes in view of their equal acceptance once courted. Nevertheless, courtship initiation is not completely determined by the male since AR/AR males × AR/PP females courted significantly faster than AR/AR males to their own females (con-

TABLE XI

Time in Copulation[a] (from Spiess, 1968c)

Type of mating[b]	Number mated	Time (minutes) in copulation
AR/PP × $F_1 AR/AR$	233	5.51 ± 0.14
AR/PP × $F_2 AR/AR$	134	5.52 ± 0.18
$F_1 AR/AR$ × AR/PP	171	7.67 ± 0.18
$F_2 AR/AR$ × AR/PP	92	7.45 ± 0.26
Controls:		
$AR/AR\ F_1$	46	5.82 ± 0.24
$AR/AR\ F_2$	19	5.71 ± 0.30
AR/PP	20	7.55 ± 0.28

[a]Data parallel to Tables 9 and 10.
[b]Female's Karyotype is first.

trols), as can be seen in Table IX, second column. Stimulation to males by particular females needs considerable further testing.

Finally, duration of copulation was the most uniform part of male mating activity, quite strain-specific and male-determined as evidence from reciprocal matings (Table XI); the data are in agreement with those of Parsons and his colleagues. The faster mating AR/AR spent less time in copulation than the slower-mating AR/PP, but there was no general relationship between courtship speed and duration of copulation (correlation coefficients were nonsignificant). MacBean and Parsons (1967) selected for duration of copulation, and they presented evidence that indicated the duration of copulation to be an expression of the rate of sperm transfer; at least their selected lines had responded to short duration by reducing the time for sperm transfer, Then courtship time and duration of copulation may each prove to be under different genetic control.

Chromosomal Polymorphism and Mating Propensity in Other Species

In conformity with the fact that outbred lines exhibit greater and more mating activity than inbred ones (see section on spontaneous differences between strains), some interesting positive correlations have been demonstrated between chromosomal polymorphism and the mating activity of samples from natural populations when tested in the laboratory. Brncic and Koref-Santibañez (1964) tested males of *D. pavani* from two Chilean populations (Copiapó and Bellavista) for their ability to court and copulate with females of a sibling species, *D. gaucha,* which is not sexually isolated from *D. pavani.* Males were classified into groups as to (1) those which mated within 30 minutes, (2) those which courted but did not copulate, and (3) those sexually inactive during the observation period. A salivary-gland chromosome analysis of males tested indicated that the frequency of heterokaryotypes was higher by 20 to 30 percent among the mating and courting group than the inactive.

Moreover, positive correlation has been shown between natural populations having different levels of chromosomal polymorphism and the mating activity of samples from those populations. Sperlich (1966) compared males of *D. subobscura* from northern Europe (with low polymorphism) with those from more southern areas (with higher polymorphism) by direct 2-hour observations for speed of mating. There was greater success in mating for males from the southern (more polymorphic) than from the northern localities.

An extensive study of mating speed and chromosomal variation was carried out by Prakash (1967a, b; 1969) in *D. robusta* with hetero- and homokaryotypes of chromosomes XR versus XR-1, $2L$ versus $2L$-1 and $3R$ versus $3R$-1 (collected at Creve Coeur, Missouri). Progeny of both sexes from wild females were tested for mating propensity and fertility (number of offspring produced). Single males were presented with three nonsib random females for a period of 2 hours, after which karyotypes were determined by outcrossing the males to standard laboratory females for larval salivary identification. When data for males were pooled for the karyotypes of each chromosome (2 and 3 separately) there was no evidence of heterokaryotype superiority in mating frequency over homokaryotypes. However the males with X-chromosomal rearrangement XR-1 inseminated more females (average, 2.2 females) than did XR males (1.9 females) in the 2-hour period; but more interesting was the heterokaryotype greater mating frequency for males with particular chromosomal associations: first, the XR-1 males which were also heterokaryotypic on the second chromosome inseminated 2.4 females compared with 1.7 females for XR males with the same second chromosome karyotypes. A similar association was found later on derived strains between the $2L$ and $3R$ chromosomes: that is, while $2L/2L$-1, $3R/3R$ showed no faster male mating frequency than homokaryotype 2nd chromosomes, the $2L/2L$-1, $3R/3R$-1 did show strong heterosis.

Females to be tested were determined by Prakash to be homokaryotypes for the two X-chromosome arrangements (XR and XR-1) and the two $2L$-arrangements ($2L$ and $2L$-1). Hour-long tests were then run on 10 pairs of virgin flies varying in age from 6 days to 30 days; mating pairs were aspirated out by a technique essentially the same as that of the Spiess group. None of the homokaryotype mating frequencies were significally different from each other. However, when females of XR/XR-1 (both $2L/2L$ and $2L/2L$-1) were mated to homokaryotype males, their mating frequency was consistently higher than homokaryotype XR/XR or XR-1/XR-1 females at all ages tested, by a factor of 6 to 13 percent.

These results contrast with those mentioned previously for *D. pseudoobscura* and *D. pavani,* in which males with two arrangements in heterokaryotype were most often heterotic, but they also contrast with the *D. persimilis* experiments vis-à-vis slow versus fast performances of the homokaryotypes. Chromosomal interactions were more predominant with rearrangements on all three chromosomes in *D. robusta.*

In view of the fact that *D. robusta* differs from the *pseudoobscura* group of species in having rearrangements on all three of its chromosomes, while in the latter they are mostly confined to the third chromosome, it may well be expected, if the chromosomal polymorphism is maintained in these species to any degree by mating speed control in one or both sexes, that interchromosomal associations should be selected as well as linkage blocks within chromosomes. Mating speed as a component of fitness has been shown by Prakash (1967b) and by Fulker (1966) to be exceedingly effective because not only do the most rapidly mating males inseminate females sooner and therefore produce offspring ahead of slower-mating males, but also they inseminate more females because they mate repeatedly in a given period of time. For females, however, mating repeatedly is less likely since in *Drosophila* a certain refractory period always follows mating; a disinclination to mate exists, and repeated mating for females does not increase their productivity if the second mating comes before any sperm have been utilized from the first mating (Prakash, 1967b). Consequently the action of selection on the two sexes is quite different and the sets of genes controlling their responses are likely to be independent. Thus high fitness in natural populations containing chromosomal polymorphs that are common enough to occur in considerable heterokaryotype frequencies will couple together sets of such linkage blocks in single individuals even though their behavioral activity expression may be sex-limited. Such a mechanism must have substantial fitness value.

Factors Modifying Mating Propensity

The control of courtship and mating, while cytogenetically based, is at the same time very delicately conditioned by environmental factors including those which affect the preadult stages as well as interactions among mating adults. Environmental variables affect not only the behavioral expression of genotypes via developmental changes, but also their selective value in the struggle for existence. Consequently, an analysis of mating behavior is likely to encounter "genotype-environmental interactions" of considerable degree and complexity. While measuring such behavior in a uniform set of conditions is a primary rule for achieving information of its genetic determination, it is by testing the behavioral repertoires of genotypes through a spectrum of conditions that we learn the "norm of reaction," the reactions and responses of the genotype to those conditions which it may have encountered in natural populations and to which it may have become adapted.

An exhaustive list or consideration of factors that influence mating behavior is well beyond the scope of this paper. In the first place, it may be not far from the truth to state that every conceivable extrinsic factor that has an effect on growth and development of Drosophila must have some influence on its behavior in mating; and second, our ignorance of hormonal and neurophysiological mechanisms in Drosophila plus ecological factors in its natural environment is still too great to permit any but tentative conclusions about adaptive reactions and responses to external conditions. Nevertheless, some fundamental consideration of these factors will be helpful to keep in mind throughout any analysis of mating propensity. (More extensive reviews bearing on this subject may be found in Manning, 1965, 1966, 1967a, b, c; Bastock, 1967; Parsons, 1967; Petit and Ehrman, 1969.)

Extrinsic Factors

TEMPERATURE. Sensitivity of mating propensity, as with most fitness traits, depends for expression on rather sharp temperature conditions. Differences between mating tests due to this factor have probably been important in the results obtained by Parsons and Kaul (1966) versus those of Spiess, Langer, and Spiess (1966) with karyotypes of *AR* and *PP* in *D. pseudoobscura*. When cultured at cool temperature (15°) there is much less difference between the performances of *AR* and *PP* than at 25°, since the latter arrangement was much more increased in speed than the former with the temperature change. Also in *D. persimilis*, *WT* and *KL* karyotypes differ in their optimal temperatures (Spiess, unpublished) with *WT* mating faster at cool and *KL* faster at warm temperatures than they do in the reverse conditions.

Heterosis may be expressed more at "stress" temperatures than at optimal (Spiess and Spiess, 1967) as in the experiments with *D. persmilis* from Humboldt, California, in which heterokaryotypes of *KL/MD* were faster in mating speed than homokaryotypes at cool temperature (which lowers mating speed of all karyotypes from that locality compared with warm temperature speeds). The greater heterosis at cooler temperature contrasted with the hypothesis of Langridge (1962), which had proposed ubiquity of temperature-sensitive recessive alleles and had given considerable evidence for heterotic expression at higher than optimal temperatures (see also discussions by Van Valen, 1967, and Spiess, 1967). Furthermore, heterokaryotypes may display behavioral homeostasis as in Parsons and Kaul's experiments with *AR* and *PP* (1966). As pointed out by Parsons (1967), these results are analogous to those found for many fitness characters—including the fundamental study of Dobzhansky and Levene (1955) on viabilities of second chromosome homozygotes and heterozygotes in which the latter were much more uniform (developmentally homeostatic).

FOOD VOLUME, CROWDING, AND YEAST AVAILABILITY. Conditions that limit growth in larval stages reduce mating activity of males and receptivity of females (see Roberston, 1963, 1966, for limiting conditions of larval growth). Crowding and density of larvae at "stress" levels have effects on delaying matings (male courtship) and lowering receptivity in females (Kaul and Parsons, 1965; Spiess and Spiess, 1969). Perhaps the most critical factor is the availability of yeast both during growth and during maturation of adults immediately after eclosion, apparently for ovarian growth and female receptivity (Manning, 1967c; L. D. Spiess and Spiess, 1968), as well as a conditioning factor without which adults will not be sexually active.

LIGHT. While many species of *Drosophila* will mate in the dark, those which will are slowed down usually by complete darkness (Spieth, 1952; Manning, 1965) while those with elaborate courtship displays like *D. subobscura* will not mate at all in the dark. Other species like *D. auraria* and *D. simulans* do not have complex displays but are markedly reduced in mating by darkness. *D. pseudoobscura* on the other hand is much slowed up by very bright illumination, while *D. persimilis* is relatively unaffected by the intensity that stops its sibling species from activity. In addition, various mutants such as *ebony* and *white* in *D. melanogaster* are light-dependent (Rendel, 1951; Jacobs, 1961; Geer and Green, 1962). A very extensive set of experiments to determine the relative influence of light (vision) and wing display in courtship and mating of several species and the dependence of these elements on illumination has been described by Grossfield (1966, 1968). While many species seem to depend on sound produced by wing vibrations, those species dependent on light use the wings apparently for visual information, and considerable reactivity between the sexes results; for example, *D. auraria* females signal receptivity to males with their wings while *D. suboobscura* males need the wings in response to the female's "dance."

DENSITY BETWEEN MATING PARTICIPANTS. The number of adults in an observation chamber may be critical to the amount of mating that takes place. Spiess (1968b) and Spiess and Spiess (1969) have compared either *AR/AR* with *AR/PP D. pseudoobscura* or *KL persimilis* at varying densities in mating chambers. As one might expect if male courtship depends on frequency of encounter, the less sexually active forms such as *AR/PP* and *KL/KL* of these sibling species are density-dependent, in the sense that single and double pairs mate less than when multiple pairs (greater than five) are in the chamber. The higher propensity *AR/AR* karyotype, however, mated equally at all densities. The density effect was simply due to increased numbers and not to some "preferences" which could conceivably become established more easily in multiple pairs than in single pairs. Spiess and Spiess showed that by a sequential change of partners in single pairs no increase in total mating took place; at least there was no evidence that, by giving each individual more than one choice of partner, mating was in any way speeded up or facilitated. These facts conform to the hypothesis of Manning (1967c) that receptive females require a "courtship summation" before mating and that those which are slower to mate may simply need more contact from courting males before their threshold is lowered.

SEX RATIO. In some studies of mutant markers (see Merrell, 1949; Rendel, 1951; Barker, 1962) male "choice" and female "choice" experiments employed unequal sex ratios, but little effect of the sex ratio difference can be evaluated since matings took place over several hours in those experiments, and the investigators were not analyzing their data for sex ratio particularly. It is rather interesting, however, in Barker's experiments with the *yellow* mutant in *D. melanogaster* to find that in a comparison of male choice (5 males × 15 females) with female choice (10 males × 5 females) *yellow* males and aged wild-type males always mated in greater frequency in the former than in the latter condition—that is, when females were more in abundance.

When *ST* and *CH* karyotypes of *D. pseudoobscura* were tested by Kaul and Parsons (1966) in sex ratios of three to one (either with three of the same karyotype or of different karyotypes), there was consistent excess of matings when males had a "choice" of females (1 male × 3 females), as compared with the case in which the sex ratio was in the opposite direction. Certainly, males court each other as well as females, so that when they are in excess, they waste much time before mating takes place since each male is more likely to encounter another male rather than a female. Both these facts were interpreted by Kaul and Parsons as a likely male interference where that sex is in excess. Ehrman (personal communication) and Spiess (unpublished) have recently confirmed these observations with *D.*

TABLE XII

Sex Ratio Effects on Mating Propensity in *D. pseudoobscura*

A. Using *AR* and *PP* homokaryotypes in ratios of
(1) *AR* × *AR*:[a]

	1 ♀ × 2 ♂♂	2 ♀♀ × 1 ♂
Time to Courtship	6.08 ± 1.08	7.64 ± 1.19
Courting Frequency	72%	66%
Time to Mating	7.19 ± 1.26	7.75 ± 1.22
Mating Frequency	64%	64%
Time in Copulation	6.34 ± 0.28	5.72 ± 0.23

(2) *AR* × *PP*:[a]

	1 *AR* ♀ × 2 *PP* ♂♂	2 *PP* ♀♀ × 1 *AR* ♂
Time to Courtship	13.50 ± 4.80	4.19 ± 0.59
Courting Frequency	8%	72%
Time to Mating	12.67 ± 6.69	10.33 ± 2.43
Mating Frequency	6%	24%
Time in Copulation	5.33 ± 0.88	6.00 ± 0.33

(3) *AR* × *PP* and *AR*:[b]

	1 *AR* ♀ × 1 *AR* ♂ and 1 *PP* ♂	1 *AR* ♀ and 1 *PP* ♀ × 1 *AR* ♂
Time to Courtship	7.24 ± 0.99	5.47 ± 0.86
Courting Frequency	70%	90%
Time to Mating	7.84 ± 1.13	5.87 ± 1.02
Mating Frequency	60% (AR), 4% (PP)	78%
Time in Copulation	5.78 ± 0.24	5.95 ± 0.22

B. Using *AR* only in ratios of

	2 ♀♀ × 8 ♂♂[c]	8 ♀♀ × 2 ♂♂[d]
Time to first Mating	6.80 ± 2.77	1.33 ± 0.14
Time to second Mating	8.33 ± 2.84	4.17 ± 0.69
Mating Frequency	60%	100%

[a] 50 single × 100 opposite sex
[b] 50 *AR* × 50 *AR* and 50 *PP*
[c] 20 chambers run
[d] 18 chambers run

pseudoobscura: Table XII presents some data which indicate such a sex-ratio effect. When a single male or female *AR* is mated by two of opposite sex, there is no significant difference in any mating activity (line 1). In line 3 of Table XII, however, there is faster mating and in higher frequency when there are two females present to one male, even though one of the females is the unreceptive *PP* type. (Note in line 2 of Table XII that *AR* males court *PP* females but only one-third are accepted by those females; and *PP* males have very low courtship activity even though mated to highly receptive *AR* females.) With a sex ratio of 4:1 (section B), however, it is evident that a very significant excess of sexual activity takes place when females are in excess, a phenomenon that is in agreement with Kaul and Parsons' data. Whether a true male × male interference is the explanation is yet to be determined. Extra females could induce greater activity in the males as an alternative explanation.

In *D. persimilis,* however, Spiess and Spiess (1969) did not find that an excess of males delayed the average time to mating (ratios of 8:2), although in 30 minutes of observation more matings did occur in the ratio that favored females (94 percent) compared with the opposite ratio with males in excess (72 percent). After observing for a period of 3 hours, these authors found that all receptive females had mated in the 8 female to 2 male ratio—a value of 72 percent, identical with that of the opposite ratio. Thus interpretations of mating totals may easily depend on the time of observation, as pointed out by Kaul and Parsons.

AGE OF PARENTS. Lines of *D. subobscura* selected for age of parents were found by Wattiaux (1968) to differ in male sexual activity, increasing in the line with young parents (3 to 9 days old) but decreasing in that with old parents (6 to 8 weeks old). Since reciprocal hybrids were not studied it is difficult to ascertain what the mechanism of the age effect was, but it was cumulative in the selected lines and could easily have been due to selection of genotypes favoring fast or slow speeds of mating, although a cumulative cytoplasmic factor is not ruled out. In a similar set of experiments with fast and slow development time being selected, Spiess and Spiess (1964, 1966) also showed a delay in mating for those flies from slow selected populations in which parental age at each generation was older in the slow than in the fast populations—on the average, by about two weeks.

Intrinsic Factors

Flies to be tested for mating may differ in a wide variety of attributes, which can be controlled by the investigator by treating them during growth in particular ways. While the following brief summary serves merely to provide the reader with some introduction to a very voluminous literature, it should be kept in mind that we are far from a complete understanding of those morphological and physiological features which lead to sexual activity, For example, a glance at Figure 1 (Manning's selected lines' performance), noting the parallelism in spite of enormous fluctuations in contemporaneous mating propensity, demonstrates how unpredictable flies' activity may be in spite of genetic and environmental control. Any investigator who follows the performance of *Drosophila* lines for several generations, either wild unselected or selected strains, will experience occasional sudden changes in performance in spite of careful control over external conditions. While much of this irregularity is often frustrating to the observer, it is nevertheless an expression of flexibility on the part of the flies' behavioral repertoire, presumably for homeostatic functioning under the variety of conditions these flies encounter in nature.

At least in the subgenus *Sophophora* (the *D. obscura* and *melanogaster* subgroups), in which visual cues are probably not very critical (see Grossfield, 1968), it has been emphasized by Spieth, Manning, Brown, Ewing, Bennet-Clark, and others that chemical and tactile stimuli are probably most important at the start of courtship: males begin to orient toward females following their "tapping" with foretarsi. Olfactory and tactile receptors are thus stimulated in both sexes. Males continuing to court and then vibrate one wing in the direction of the female with a highly specific pulse of sound of very short (2 to 3 milliseconds) duration produced at intervals of about 30 to 50 milliseconds, depending on the species (Bennet-Clark and Ewing, 1968; Waldron, 1964; Shorey, 1962), in amplitude about one-fourth of the normal flight wing-beat frequency, and producing a tone at about A, B, or C on the musical scale. Females are then stimulated via their antennal auditory receptors predominantly, although olfactory sensitivity is not ruled out. Receptive females need a minimal amount of courtship (summation) before they will accept the male. A detailed behavioral description is beyond the scope of this chapter. There are, of course, many other minor elements in the sophophoran sexual activity, which are described in the references given at the end of this chapter. It is important to realize, however, that variation in wing size and body size, for example, accounts for much of the differences in mating success of males, while variation in antennal receptors as well as their tactile and olfactory sensitivity must be critical in females.

In addition the stage of maturity is equally important. In the experience of most observers, males age faster than females and begin their courtship several hours before their sibling females of contemporary age become receptive. For example single males of *D. melanogaster* of different ages given 10 virgin females 3 days old were observed for 12 hours by Kvelland (1965): when less than 2 hours from pupal emergence, males mated with 1 female on the average, then in successive daily ages increased linearly to an average of 5 females mated at 7 days, plateaued at that point, and at ages greater than 12 days dropped in ability to mate. Three-day-old virgin males, in contrast, mated with an average of 6 females initially, dropped to approximately the same plateau, and paralleled the other set of males in senility. Female receptivity (studied by Manning, 1967), in the same species and in *D. simulans* was shown to begin just about at the start of the second day after emergence from the pupa. Receptivity or lack of it can be ascertained within the first 15 minutes of a mating test, since more than 90 percent of flies clearly either mate right away or do not mate at all for more than an hour, a fact that suggests that the change from unreceptivity to receptivity is very rapid indeed. That change which takes place between days 1 and 2 was determined to be an all-or-none process, which male courtship does not hasten since, as Manning stated, young unreceptive females have been observed for more than 7 hours not to accept the male. On the other hand, courtship summation is necessary for those females which are receptive; that most of the stimuli are received from wing vibration was shown by using wingless males, which were accepted but after an increased courtship time of about 20 minutes. Finally, females tend to become less receptive in old age, a parallel situation with males (being about 50 percent receptive when older than 16 days in both species). Manning presented evidence indicating that the corpus allatum and ovaries show a growth cycle parallel to receptivity and that increase in juvenile hormone may be responsible for the "switch-on" of receptivity at 2 days of age, which in turn defines the female as "accessible" to stimulation from the courtship of a male. Virgin females remain receptive for many days but after mating become immediately unreceptive, an inhibition that seems to have two components: (1) an effect of copulation itself, which wears off after 48 hours (ascertained by using sterile males with immobile sperm), and (2) an effect of live sperm in the seminal receptacles, which wears off after about 8 to 10 days when the sperm supply is exhausted.

In summary, genetic and environmental factors may act via numerous mechanisms to affect mating propensity: variation in the speed, sequence, and quality of courtship elements; sensitivity of both sexes and their reactivity to these elements; and the physiological state of the participants may account for the diversity of sexual activity observed. Perhaps a simplified working diagram to express in summary form the tendencies and expressed qualties of each sex will serve to condense these factors, which may be looked for in genetic comparative studies (based upon ideas of Manning, Brown, and Spieth):

	Primary "Drives"	Sexual Expression	Quality
Male	Copulation and Avoidance	Courtship	Speed Persistence Discrimination
Female	Avoidance and Receptivity	Acceptance of male	Reactivity to male Resistance Discrimination

The final outcome of mating propensity must be the resultant of variation in these elements via the qualities of expression, which may be intensified or lessened by genetic and environmental factors.

Frequency Dependency

Just how delicately conditioned may be the interactions between mating adults has been explored simply by varying the relative frequency of genotypes, karyotypes, or environmentally altered types among participants of a mating group. The increase in mating success, especially among rarer types, has considerable significance as a phenomenon both for behavioral analysis and for population genetics. Not only can the change in mating propensity give us clues for interpreting the behavioral stimuli and responses between flies, but also it provides an interesting case of "frequency-stat," or frequency-regulated, selection when the fitness of genotype may be a function of its frequency (Li, 1962, 1967). When a form possesses an advantage in the minority that it does not enjoy in the majority, an equilibrium is possible with polymorphism maintained and a lowering or elimination of the genetic load at the point of equilibrium (Clarke and O'Donald, 1964; Lewontin, 1958).

Petit (1951) first observed in multiple-choice matings between *Bar* and wild-type *D. melanogaster* that *Bar* males mated with relatively greater success when they were rare than when common. Later (1958) she described in considerable detail (with a "coefficient of sexual selection" for *Bar* versus wild and for *white eye* versus wild) that there was a "minority advantage" for both mutants plus a high frequency advantage for *white eye*. In 1964, Spiess observed a minority advantage for mating propensities of *WT* versus *KL* of *D. persimilis* by varying the frequency of the two karyotypes among males only (reported in Ehrman, 1966); and since 1965 Ehrman and her colleagues have run several series of tests with various karyotypes and mutant strains of *D. pseudoobscura* plus certain species strains from the *D. willistoni* group (Ehrman, Spassky, Pavlovsky, and Dobzhansky, 1965; Ehrman, 1966, 1967, 1968, 1969; Ehrman and Petit, 1968) to demonstrate almost consistent minority advantage. In *D. pseudoobscura*, the Ehrman group has found the mating advantage of rare genotypes not only for mutants but also for gene arrangement strains of different geographic origin and even for flies of the same strain grown at two different temperatures. In *D. tropicalis*, *D. willistoni*, and *D. equinoxialis* minority males from geographic strains some of which exhibit a tendency for ethological isolation displayed greater mating success than the majority, and to a lesser extent the same was true of minority females; between completely isolated strains (incipient species) of *D. paulistorum*, however, there was no minority advantage. Spiess and Spiess extended their earlier work to include a confirmation of minority advantage between *AR* and *PP* arrangements of *D. pseudoobscura* (Spiess, 1968a) and between geographically separate strains of *D. persimilis* (L. D. Spiess and Spiess, 1969).

Most of these experiments have been summarized by Petit and Ehrman (1969), but some details of later progress are pertinent. Though the different laboratories have used slightly varied techniques, similar net results have been produced and a working behavioral hypothesis of the minority-advantage mechanism can be formulated. Both Ehrman and Petit have consistently employed the technique of marking flies sufficiently (either with visible mutants or by clipping wings) to distinguish them under low-power magnification at mating in observation chambers (Elens-Wattiaux). No flies are withdrawn from the mating chamber and are then undisturbed for the period of observation. Males can, of course, mate more than once, and it is technically difficult to distinguish males that have mated from those which have not. Spiess and colleagues aspirate out pairs that have copulated in order to prevent males from a second mating. Under these circumstances males that enjoy a minority advantage may have the opportunity for increasing that advantage in the Petit-Ehrman technique, while Spiess's technique allows recording only such advantage as is due to a first-mating excess. There remains the possibility that aspirating disturbs the remaining flies, although there is no evidence that a significant alteration in their behavior results; most matings take place in the first few minutes of a trial and the lowering of density by removal does not substantially affect the main result. Finally, one other principal technical difference lies in the fact that Ehrman usually has tested mating success by altering ratios simultaneously in both sexes, while Spiess has tried one sex at a time (leaving the other at equality) as well as varying both sexes together. These technical differences arising from consideration of observational advantages in the two laboratories have proved to be reinforcing in the long run, and the following summary with illustrations of certain results (from Spiess) will serve as a background example for a working hypothesis of the minority-advantage mechanism.

It can reasonably be assumed that a minority advantage must imply some dissimilarity between rival types, which must be "recognizable" as a difference among the participants in a mating box. Lack of an advantage, of course, can imply either lack of difference in behavioral, morphological, or physiological properties, or independence of behaviors among the participants. In making comparisons between karyotypes or genotypes, which might be identical, employing a minority test then may be a useful technique to ascertain behavioral contrasts. L. D. Spiess and Spiess (1969) have attempted such a comparison with strains of *D. persimilis* with *KL* arrangement from two widely separated areas: Yosemite region (White Wolf locality) and Humboldt State Park, California. That arrangement is the predominant one in northern coastal populations but is rare in the Sierra Nevada, and it is of considerable interest to ascertain what the extent of similarity may be between flies of identical karyotype from the two regions. The reader may be reminded that when cultured at cool temperature (15°) *KL* flies from either locality mate at low propensity (about 10 to 20 percent in 30 minutes) while at warm (25°) they average 60 to 70 percent in the same time. Multiple-choice mating tests for flies raised at two temperatures and from the two separate localities were run as follows: F_1 progeny from intralocality-strain crosses were isolated at emergence from the pupa and stored in lots of usually 10, then marked with a pinhole through one wing to distinquish types. Twenty pairs of flies (equal sex ratio) were introduced into plexiglass chambers, and one or both sexes were varied by changing the ratio of types (populational or temperature-cultured) from 2:18, 4:16, 6:14, through 18:2, or in later studies (for economy only) the extreme ratios (2:18, 18:2) and the control (10:10) were used. Either 30 (in

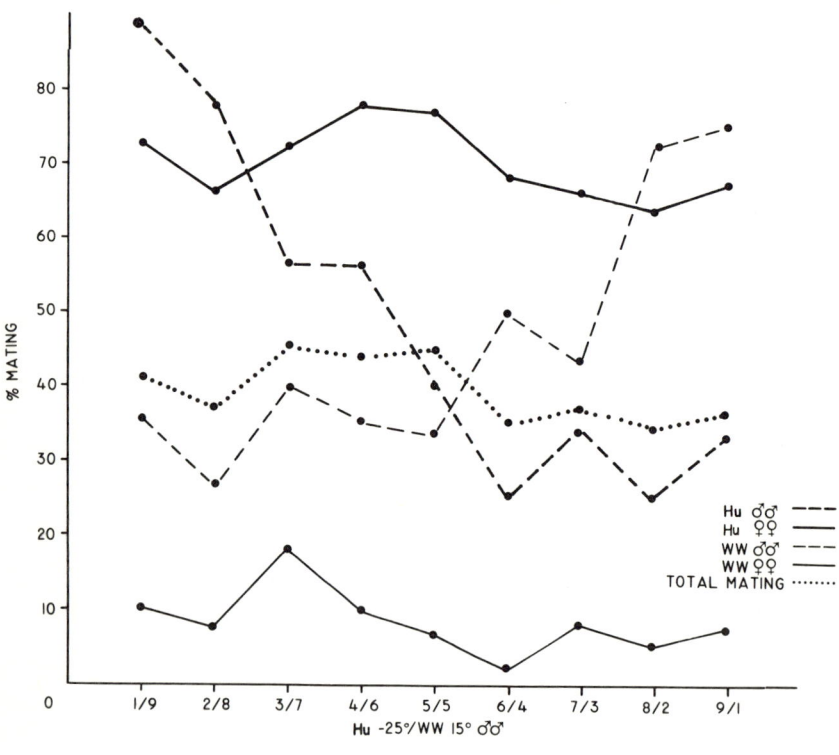

Fig. 5. *Percentage of males, females, and total matings out of those introduced into chambers at different ratios of Humboldt (25°): White Wolf (15°) males with the two kinds of females equal (10 each per chamber). (From Spiess and Spiess, 1969. Courtesy The American Naturalist.)*

former) or 15 (in latter) runs were done to insure sufficient numbers at the extreme ratios for reasonable significance.

Figure 5 illustrates the main effects when one population (Humboldt) is cultured at a warm temperature, and the other (White Wolf) at a cool one, and the two types of male are varied. As shown at the left end of the graph, "*HU*" males mate more than twice as much as they do when they are equal to the "*WW*" males or are in the majority. From the opposite end the *WW* males show an essentially reciprocal tendency. Since all mating pairs are removed from the chamber this advantage is expressed by allowing each male just one mating. Females, on the other hand, are little affected by the change in ratio among the males: *HU* (warm) females mate constantly about six to seven times more readily than *WW* (cool) females, and the total mating roughly reflects this constancy at about 40 percent. Consequently with the male propensities being nearly mirror images of each other, it can be concluded that the two types of males cultured at either temperature have about the same tendencies towards mating. Further tests with females showed that for both populations females were equally

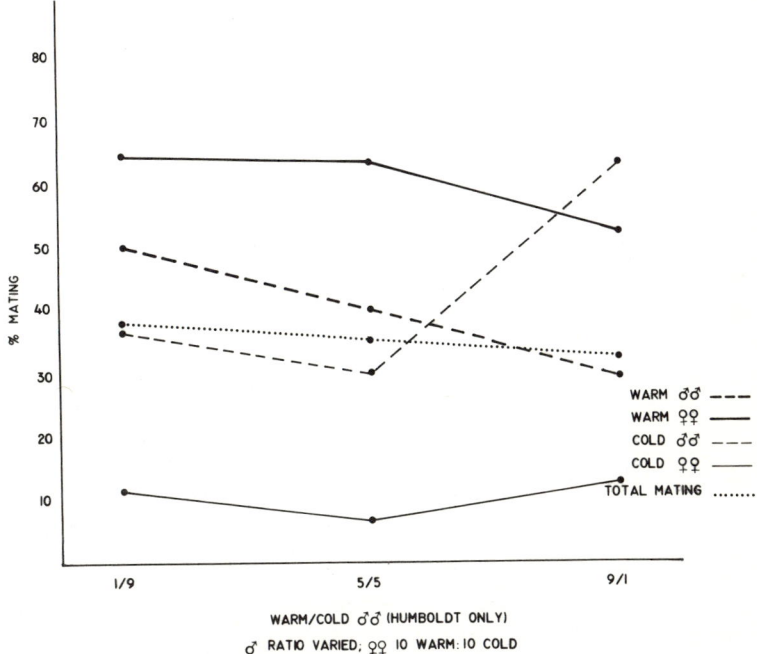

Fig. 6. Percentage of HU matings of both sexes raised at 15° or 25° while the ratio of warm:cold was varied among males with females from the two temperatures equal. (From Spiess and Spiess, 1969. Courtesy The American Naturalist.)

temperature-sensitive: *HU* females were just as unreceptive at cool as *WW* females (see Fig. 6), while it can be seen from Figure 7 that *WW* females at warm temperature are nearly identical with *HU* females.

When both populations are cultured under identical conditions (25°), males still show a significant minority advantage (Fig. 7), although to a lesser degree than when the populations are cultured at different temperatures. If the ratio of populations is varied among females, however, no frequency effect occurs (Fig. 8). Finally, when strains from a single population (Humboldt) are cultured at the two temperatures and males are varied, an increase in mating for males occurs that is statistically significant for cool (Fig. 6, right) but not significant for warm (left) males. This latter observation is in agreement with Ehrman's finding (1966) that *D. pseudoobscura* (*AR*) cultured at contrasting temperatures displayed a minority advantage. Certainly, low temperature slows development and increases size, so that while these changes do not seem to affect the overall mating propensity of males very much, there must be a difference detectable to the mating participants. The greatest minority advantage

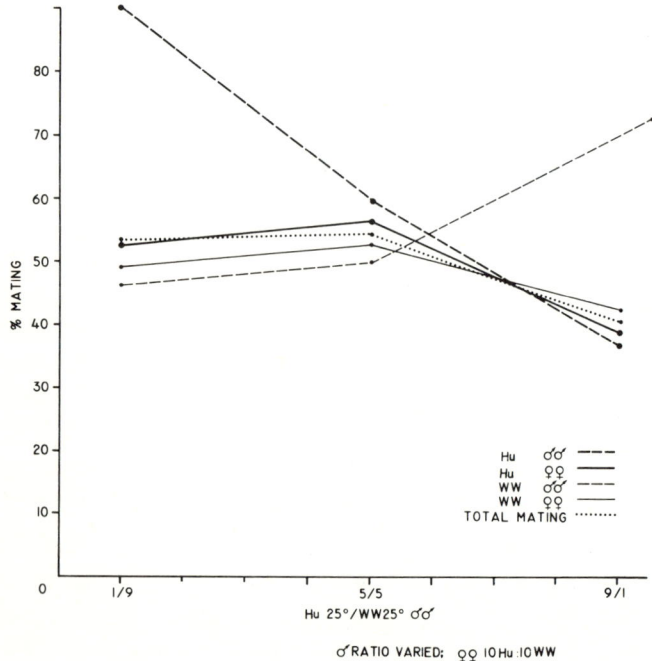

Fig. 7. Percentage of HU and WW of both sexes mating while the ratio of HU:WW was varied among the males with females of the two populations equal; both populations raised at 25°. (From Spiess and Spiess, 1969. Courtesy The American Naturalist.)

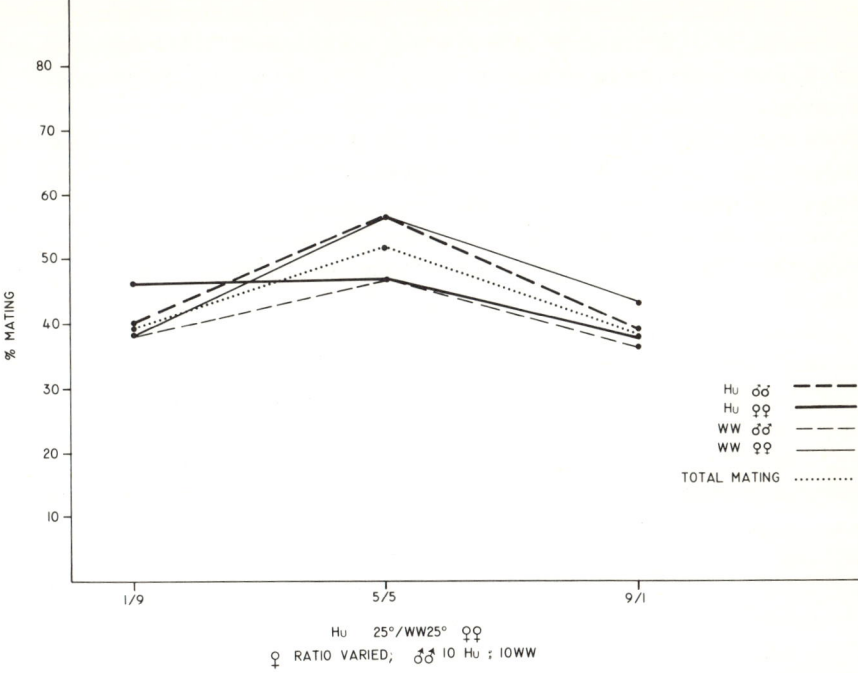

Fig. 8. Percentage of HU and WW of both sexes mating while the ratio of HU:WW was varied among the females with males of the two populations equal. Both populations raised at 25°. (From Spiess and Spiess, 1969. Courtesy The American Naturalist.)

in this *D. persimilis* study was expressed, then, when both population and temperature difference were combined in a reinforcing, or "additive," performance, and it was significant only for males in the example considered here.

To effect this frequency-dependency change in mating propensity, there must be change in stimuli distinguishable by the individuals present when rivals occur at varying frequencies. The phenomenon is not detectable in every instance (as in the last experiment with *HU* flies raised at two temperatures, the warm-cultured males did not display a significant effect), and the fact that the advantage is less with only one variable (population or temperature) than when both variables reinforce each other, indicates that if no variables are applied, the minority advantage should become null. We feel justified therefore in treating the minority advantage as a tool in determining behavioral differences between individuals of the same sex.

It is important to discuss possible mechanisms for this minority effect in *D. persimilis* males in the light of observations made in *D. pseudoobscura*

(Spiess, 1968b; Ehrman, 1966, 1967, 1968). Particularly with regard to our previous work with *AR* and *PP* arrangements, *AR* males mate about four times more frequently than *PP* males within 30 minutes of observation. Both types displayed a minority advantage that was tentatively accounted for by some sort of male × male interaction at opposite ends of the varied ratio: one could postulate different behavioral interactions at either end of the scale. For example, when *AR* males are common (*PP* rare), they might "interfere" by courting so much, including each other, that opportunity to mate with females is afforded the less active, rare *PP* males. One could as easily postulate under those conditions that the increased activity of *AR* males lowers the females' threshold to all courtship so that the total mating should be increased, and *PP* males could then benefit by finding the females more receptive than they otherwise would be. At the opposite end of the scale, with *AR* rare (*PP* common), one cannot postulate the same mechanism(s), since *PP* males are not active in courtship. One could then visualize either (1) that lack of courtship between most males eliminates the "interference" effect, thus letting the rare *AR* males have greater access to the females than when they are common, or (2) that the threshold of *AR* males is lowered by the excess of females available (that is, with *PP* males being discounted, effectively the sex ratio is altered). While either interpretation may be correct, none of these interpretations seems sufficient for the cases here reported in Klamath *D. persimilis* homokaryotypes, in which both ends of the scale (male ratio varied) display about *equal* minority advantage and essentially equal central propensities. One can only assume at this stage of our knowledge that males that usually behave equally towards each other and towards the females when at control or majority frequencies, become distinguishable from each other when rare. In other words their efficiency at mating is equal but they must possess recognizably different properties (perhaps chemically detectable pheromones or behavioral characteristics), which either bring about male × male interactions of a different sort or change the female thresholds towards them when rare.

The male × male interactions are quite unlikely as an explanation for the mechanism when the results of Ehrman's double-chamber tests are considered (1967, 1969; Ehrman and Spiess, 1969). Ehrman has succeeded in nullifying the minority advantage by separating a "minority" ratio set of flies (usually *AR* versus *CH* with 20 pairs of one type to 5 pairs of the other in the upper section of the double chamber) from a compensating group added to the lower section below cheesecloth, so that the two sections combined have equal numbers of both types. Until recently in all such tests Ehrman had added the compensating group in pairs (both sexes). Under these circumstances the minority advantage is nullified in the top section. In other words, whatever cue the flies above were receiving as to frequency, they are "aware of the change even though the upper and lower flies are not in physical contact." More recently, Ehrman (1969) has pumped air through the two adjoining chambers and has shown that the cues are definitely airborne—that is, either chemical (odor to the flies) or sound (vibration).

In his 1968a paper, Spiess listed four possible ways in which, when both sexes were varied simultaneously, some kind of interaction might take place: (1) females "recognize" that there are two types of males present, and they change their receptivity in favor of the minority males; (2) females "recognize" that their sex occurs in two kinds, and the minority female type increases its receptivity; (3) males "recognize" that they occur in two kinds, and the rarer increases its courtship activity; and (4), opposite to the first, males "recognize" that there are two kinds of females present and court the rare type female. Models (2) and (4) were eliminated by the experimental evidence: when two types of *Drosophila* female occur together the rarer type is only seldom observed to mate in excess of expectation, and males do not court rare females more than common ones. Spiess tentatively did not favor the first model, but he concluded that the critical experiments indicating a decision between models (1) and (3) have not been done yet. In view of the fact that all of her experiments were done by varying the frequency of both sexes simulatneously together, Ehrman (Ehrman and Spiess, 1969) has undertaken further experiments with the double chamber using a tightly stretched cheesecloth partition of two layers between upper and lower sections and varying just *one* sex at a time. Table XIII gives the results of experiments using such double chambers in which only females or males were varied (rows 1 and 2), while the third row is a repeat of the previous experiment performed with CH (neg.) versus AR (neg.) strains, which had given the more significant minority advantage before (Ehrman, 1966, 1967, 1969). Rows 4 through 6 are parallel with CH (pos.) versus AR (pos.) strains. These results indicate that the minority advantage of males rare above the partition is eliminated by the compensation of either additional couples (rows 3 and 6) or simply *males alone* (rows 2 and 5) but not by females alone (rows 1 and 4) to the section below the partition. With the addition, the rare type above is no longer rare in totality, but it is "recognized" as a change in frequency only when males are added below.

These results parallel those of the Spiesses (1968, 1969) with AR versus PP *D. pseudoobscura* and interpopulational matings with *D. persimilis* in which minority advantage was usually significant when male ratios were varied but not when female ratios were. They consequently confirm that males principally provide the cues in these species for the minority advantage observed so far.

In view of these results it is our hypothesis, therefore, that from the

TABLE XIII

Numbers of *Drosophila pseudoobscura* Males Mating Recorded in Double Chambers with Rare Type Arrowhead Individuals Confined to the Lower Section to Equalize $AR=CH$ (from Ehrman and Spiess, 1969)

	Runs	All Matings			Early Matings		
		AR	CH	χ^2	AR	CH	χ^2
A. 20 $Ch-$: 5 $Ar-$ (per run)							
$AR-$ ♀♀ BELOW	5	40	70	18.4[a]	16	41	2.3
$AR-$ ♂♂ BELOW	5	20	86	0.1	14	41	1.0
$AR-$ ♀♀ and ♀♀ BELOW	4	21	79	0.1	15	37	2.5
B. 20 $CH+$: 5 AR (per run)							
$AR+$ ♀♀ BELOW	5	48	54	46.7[a]	28	24	37.2[a]
$AR+$ ♂♂ BELOW	5	20	82	0.01	11	42	0.02
$AR+$ ♀♀ and ♀♀ BELOW	5	22	80	0.16	17	36	4.8[b]

Note: Female mating results omitted since they are not significantly different from expected proportions (except "early matings" of $AR+$ females BELOW in which AR mated 20: CH 32, with $\chi^2 = 11.1$)
[a] $.01 > P$
[b] $.05 > p > .01$

four models proposed the first is most likely. We may suppose that with two kinds of males present, one being rare with respect to the other, when a female is courted she is stimulated by some cue from the male (either chemical or auditory). Her first reaction is more often "reluctance," and consequently she receives more than one stimulus since she will encounter probably more than one male. If the majority produce a set of stimuli that bring about a sensory "adaptation," it may be the difference in stimulus brought about by the minority which induces her to accept the courting male, simply because his cue is different. Some of the Spiess's data when examined for the order of mating for the males that were in the minority indicated that it was those males rather than the majority type males which were accepted first by the females. There is no question that females determine the mating frequency in most of these cases; their receptivity varies considerably but their wariness is evident especially in slower-mating strains, although even in fast-mating strains of *D. persimilis*, for example, the first reaction of the female is reluctance to mate, and she will need more than one encounter with a male before accepting him. The well-known feature of nervous saturation (sensory "adaptation") produced by the majority followed by a detectably different stimulation from the minority might be a physiological mechanism effective in these cases. (See, for example, Dethier, 1964, on sense organ "adaptation" or lowering of impulses with continued stimulation in the case of the blow-fly, *Phormia*, and its chemoreceptors for sugar.) At least this hypothesis does account for the double-chamber results without direct contact between the flies. If this hypothesis is born out by future evidence, we shall necessarily come much closer to appreciation of the extreme sensitivity of these flies to slight differences in their mating behavior cues, stimuli, and responses.

Summary

From techniques designed to measure the significant features of sexual activity in *Drosophila* species (predominantly in the *melanogaster* and *obscura* groups), it is clear that for the most part males provide a diversity of courtship from which females do the selecting. Whether one uses morphological mutant markers, genetically distinct wild strains, or cytologically marked wild strains, most observations point to variations in male courtship toward which females are discriminating. On the other hand, genetically differing females are most often indiscriminatingly courted by nearly all types of males. Female receptivity, then, is the usual deciding factor in mating success provided a minimal amount of courtship by the males is available. In Darwinian terms, the female's control by her acceptance of particular males constitutes the principal selective agent in developing "correct signals" for efficiency of intraspecific propagation.

Usually both mutant and inbred males suffer from aberrant courtship, which tends to rejection by wild-type outcrossed females, while mutant females tend to have lower thresholds (less rejection), and inbred wild types higher thresholds, than outbred wild-type females. Since mutant strains are usually maintained in homozygous condition for several generations, lowered female thresholds may have developed as responses to the aberrant courtship of males in those strains, while opposite tendencies of inbred females may be explained as a result of homozygosity for recessives of poor fitness quality. It is evident, however, that for any single genetic strain the sexual activity of one sex has little predictive value for the activity of the siblings of the opposite sex, a fact that implies sex-limited or sex-influenced polygenic systems with different loci likely to control the mating behavior of either sex.

The diversity of genetic and behavioral avenues open for achieving a particular mating propensity is illustrated by artificially selecting for mating speed. In particular, the experiments of Manning for slow mating speed exemplify three distinct ways of attaining that result: (1) both sexes may be affected probably by a response correlated with sensitivity to external disturbances; (2) sluggishness of males in courtship; or (3) high threshold of female neural centers against male acceptance. Other responses to selection via juvenile hormone levels in controlling receptivity or other controlling mechanisms may yet be demonstrated. While rapid selection responses must imply considerable additivity in gene action, there is also evidence that in natural populations directionally dominant or overdominant genes guarantee high proportions of fast-mating males, so that selection in the laboratory would be expected to be more efficient in slowing up mating speed. However, the fact that increased speed can also be achieved easily in some lines at the expense of "general activity" indicates the outbred wild-type is not at its maximum for this character, but is more likely at a compromise position between high mating speed and awareness or sensitivity for external disturbances.

The frequency of chromosomal polymorphs from natural populations is positively correlated with mating speed, indicating that genic systems that may be quite sex-limited when closely linked, may gear the sexes for greater efficiency of reproduction, including both male display and female response to it. Heterokaryotype superiority is often displayed in faster mating speed and plasticity of response to the opposite sex. Total time from courtship through the end of copulation is often most brief for certain heterokaryotypes as compared with homokaryotypes of common gene arrangements. It is to be expected, then, that if specific loci controlling each sex's mating activity were operative, linkage blocks preserved through inversion systems would probably couple together such loci within heterokaryotypic individuals in greater degree than in homokaryotypes. Such preservation would be expected to evolve in populations with considerable degree of cytogenetic heterogeneity. Possibly populations with less heterokaryotypy might be derived secondarily if such controlling mechanisms had been "built into" the more common arrangements during initial stages.

In view of the fact that nearly every extrinsic factor affecting development and growth in *Drosophila* probably has an effect on behavior and speed in its mating, genotypes that affect mating would be expected to show marked environmental interactions. But in addition, interactions between adults in expression of sexual activity, in terms of qualities such as speed, persistence, and discrimination, bring a level of complexity with which the geneticist is at first "uncomfortable" but with which the behaviorist may learn to appreciate the sensitivity of these flies to their operational cues and stimuli for sexual activity. Especially in working out the phenomenon of frequency dependency in which two or more genotypes or environmentally induced types are interacting, detection of extremely delicate cues by the mating participants (possibly following "sensory adaptation" by the majority males so that females accept the minority by their difference in courtship) leads us to realize how subtle the operation of selection may be. As a likely mechanism for preservation of genetic variation (polymorphism) in the population, the organism then ceases to be merely passive but operates as its own selective agent in the process.

Acknowledgments

The work from the Spiess laboratory was supported by Contract AT(11-1)-1652, U. S. Atomic Energy Commission. The author wishes to express appreciation to his wife, Luretta D. Spiess, Dr. Bozena Langer (University of Pittsburgh), and to Mr. Richard N. Sherwin (Chicago) for all their assistance in past years.

References

BARKER, J. S. F. 1962. Studies of selective mating using the yellow mutant of *Drosophila melanogaster*. Genetics, 47:623-640.
BASTOCK, M. 1956. A gene mutation that changes a behavior pattern. Evolution, 10:421-439.
——— 1967. Courtship: an Ethological Study. Chicago, Aldine Publishing Company.
——— and A. Manning. 1955. The courtship of *Drosophila melanogaster*. Behaviour, 8:85-111.
BATEMAN, A. J. 1948. Intra-sexual selection in *Drosophila*. Heredity, 2:349-368.
BENNET-CLARK, H. C., and A. W. EWING. 1968. The wing mechanism involved in the courtship of *Drosophila*. J. Exp. Biol., 49:117-128.
BÖSIGER, E. 1962. Sur le degré d'hétérozygotie des populations naturelles de *Drosophila melanogaster* et son maintien par la sélection sexuelle. Bull. Biologique, 96:3-122.
——— 1967. La signification évolutive de la sélection sexuelle chez les animaux. Scientia, 102:207-223.
BRNCIC, D., and S. KOREF-SANTIBAÑEZ. 1964. Mating activity of homo- and heterokaryotypes in *Drosophila pavani*. Genetics, 49:585-591.
BROWN, R. G. B. 1964. Courtship behaviour in the *Drosophila obscura* group. I. *Drosophila pseudoobscura*. Behaviour, 23:61-106.
——— 1966. Courtship behaviour in the *Drosophila obscura* group. Part II. Comparative studies. Behaviour, 25:281-323.
CASPARI, E. 1967. Remarks on evolutionary aspects of behavior. *In* Hirsch, J., ed. Behavior-Genetic Analysis. New York, McGraw-Hill Book Company. pp. 3-9.
CLARKE, B., and P. O'DONALD. 1964. Frequency-dependent selection. Heredity, 19:201-206.
CROSSLEY, S. 1963. An experimental study of sexual isolation within a species of *Drosophila*. Doctor of Philosophy Thesis, University of Oxford, England.
DETHIER, V. 1964. Microscopic brains. Science, 143:1138-1145.
DOBZHANSKY, TH. 1961. On the dynamics of chromosomal polymorphism in *Drosophila*. *In* Kennedy, J. S., ed. Insect Polymorphism. London, Royal Entomological Society. pp. 30-42.

―― and H. LEVENE. 1955. Genetics of natural populations. XXIV. Developmental homeostasis in natural populations of *D. pseudoobscura*. Genetics, 40: 797-808.

―― R. C. LEWONTIN, and O. PAVLOVSKY. 1964. The capacity for increase in chromosomally polymorphic and monomorphic populations of *Drosophila pseudoobscura*. Heredity, 19:597-614.

―― and B. SPASSKY. 1962. Selection for geotaxis in monomorphic and polymorphic populations of *Drosophila pseudoobscura*. Proc. Nat. Acad. Sci. (U.S.A.), 48:1704-1712.

―― and A. H. STURTEVANT. 1938. Inversions in the chromosomes of *Drosophila pseudoobscura*. Genetics, 23:28-64.

EHRMAN, L. 1966. Mating success and genotype frequency in *Drosophila*. Anim. Behav., 14:332-339.

―― 1967. Further studies on genotype frequency and mating success in *Drosophila*. Amer. Natural., 101:415-424.

―― 1968. Frequency dependence of mating success in *Drosophila pseudoobscura*. Genet. Res., 11:135-140.

―― 1969. The sensory basis of mate selection in *Drosophila*. Evolution (in press).

―― and C. PETIT. 1968. Genotype frequency and mating success in the *willistoni* species group of *Drosophila*. Evolution, 22:649-658.

―― B. SPASSKY, O. PAVLOSKY, and TH. DOBZHANSKY. 1965. Sexual selection, geotaxis, and chromosomal polymorphism in experimental populations of *Drosophila pseudoobscura*. Evolution, 19:337-346.

―― and E. B. SPIESS. 1969. Rare type mating advantage in *Drosophila*. Amer. Natural. (in press).

ELENS, A. A. 1957. Importance sélective des différences d'activité entre mâles *ebony* et *sauvage*, dans les populations artificielles de *Drosophila melanogaster*. Experientia, 13:293-294.

―― 1958. Le rôle de l'hétérosis dans la compétition entre *ebony* et son allèle normal. Experientia, 14:274-276.

―― and J. M. WATTIAUX. 1964. Direct observation of sexual isolation. *Drosophila* Information Service, 39:118-119.

EWING, A. W. 1961. Body size and courtship behaviour in *Drosophila melanogaster*. Anim. Behav., 9:93-99.

―― 1963. Attempts to select for spontaneous activity in *Drosophila melanogaster*. Anim. Behav., 11:369-378.

―― 1964. The influence of wing area on the courtship behaviour. Anim. Behav., 12:316-320.

―― and A. MANNING. 1963. The effect of exogenous scent on the mating of *Drosophila melanogaster*. Anim. Behav., 11:596-598.

―― and A. MANNING. 1967. The evolution and genetics of insect behavior. Ann. Rev. Entom., 12:471-494.

FULKER, D. W. 1966. Mating speed in male *Drosophila melanogaster*: A psychogenetic analysis. Science, 153:203-295.

GEER, B. W., and M. M. GREEN. 1962. Genotype, phenotype, and mating behavior in *Drosophila melanogaster*. Amer. Natural., 96-175-181.

GROSSFIELD, J. 1966. The influence of light on the mating behavior of *Drosophila*. Studies in Genetics: III. Univ. Texas Publ., 6615:147-176.

―― 1968. The relative importance of wing utilization in light dependent courtship in *Drosophila*. Studies in Genetics: IV. Univ. Texas Publ., 6818:147-156.

HILDRETH, P. E. 1962. Quantitative aspects of mating behavior in *Drosophila*. Behaviour, 19:57-73.

―― and G. C. BECKER. 1962. Genetic influences on mating behavior in *Drosophila melanogaster*. Behaviour, 19:219-238.

HOENIGSBERG, H. F., and S. KOREF-SANTIBAÑEZ. 1960. Courtship and sensory preferences in inbred lines of *Drosophila melanogaster*. Evolution, 14:1-7.

HOSGOOD, S. M. W., and P. A. PARSONS. 1967. Genetic heterogeneity among the founders of laboratory populations of *Drosophila melanogaster.* II. Mating behaviour. Aust. J. Biol. Sci., 20:1193-1203.

JACOBS, M. E. 1960. Influence of light on mating of *Drosophila melanogaster.* Ecology, 41:182-188.

——— 1961. The influence of light on gene frequency changes in laboratory populations of ebony. Genetics, 46:1089-1095.

KAUL, D., and P. A. PARSONS. 1965. The genotypic control of mating speed and duration of copulation in *Drosophila pseudoobscura.* Heredity, 20:381-392.

——— 1966. Competition between males in the determination of mating speed in *Drosophila pseudoobscura.* Aust. J. Biol. Sci., 19:945-947.

KESSLER, S. 1966. Selection for and against ethological isolation between *Drosophila pseudoobscura* and *Drosophila persimilis.* Evolution, 20:634-645.

——— 1968. The genetics of *Drosophila* mating behaviour. I. Organization of mating speed in *Drosophila pseudoobscura.* Anim. Behav., 16:485-491.

——— 1969. The genetics of *Drosophila* mating behavior. II. The genetic architecture of mating speed in *Drosophila pseudoobscura.* Genetics (in press).

KVELLAND, I. 1965. Some observations on the mating activity and fertility of *D. melanogaster* males. Hereditas, 53:281-306.

LANGRIDGE, J. 1962. A genetic and molecular basis for heterosis in *Arabidopsis* and *Drosophila.* Amer. Natural., 96:5-27.

LEWONTIN, R. C. 1958. A general method for investigating the equilibrium of a gene frequency in a population. Genetics, 43:419-434.

——— 1965. Selection for colonizing ability. *In* Baker, H. G., and Stebbins, G. L. eds. The Genetics of Colonizing Species. New York, Academic Press, Inc. pp. 77-94.

LI, C. C. 1962. On "reflexive selection." Science, 136:1055-1056.

——— 1967. Genetic equilibrium under selection. Biometrics, 23:397-484.

MACBEAN, I. T., and P. A. PARSONS. 1967. Directional selection for duration of copulation in *Drosophila melanogaster.* Genetics, 56:233-239.

MAINARDI, D., and M. MAINARDI. 1966. Sexual selection in *Drosophila melanogaster.* The interaction between preferential courtship of males and differential receptivity of females. Atti Soc. Ital. Sci. Nat. Museo Civ. St. Nat. Milano, 105:284-286.

MANNING, A. 1961. The effects of artificial selection for mating speed in *Drosophila melanogaster.* Anim. Behav., 9:82-92.

——— 1963. Selection for mating speed in *Drosophila melanogaster* based on the behaviour of one sex. Anim. Behav., 11:116-120.

——— 1965. *Drosophila* and the evolution of behaving. *In* Carthy, J. D., and Duddington, C. L. eds. Viewpoints in Biology. London, Butterworths. pp. 125-169.

——— 1966. Sexual Behavior. *In* Haskell, P. T., ed. Insect Behaviour. London, Royal Entomological Society. pp. 59-68.

——— 1967a. Genes and the evolution of insect behavior. *In* Hirsch, J., ed. Behavior-Genetic Analysis. New York, McGraw-Hill Book Company. pp. 44-60.

——— 1967b. An Introduction to Animal Behavior. Reading, Mass., Addison-Wesley Publishing Co., Inc.

——— 1967c. Control of sexual receptivity in female *Drosophila.* Anim. Behav., 15:239-250.

——— 1967d. Antennae and sexual receptivity in *Drosophila melanogaster* females. Science, 158:136-137.

——— 1968. The effects of artificial selection for slow mating in *Drosophila simulans.* Anim. Behav., 16:108-113.

MATHER, K. 1949. Biometrical Genetics. London, Methuen.

———and B. J. HARRISON. 1949. The manifold effect of selection. Part II. Heredity, 3:131-162.

MERRELL, D. J. 1949a. Selective mating in *Drosophila melanogaster.* Genetics, 34:370-389.

——— 1949b. Mating between two strains of *Drosophila melanogaster.* Evolution, 3:266-268.

—— 1960. Mating preferences in *Drosophila*. Evolution, 14:525-526.
—— and J. C. UNDERHILL. 1956. Competition between mutants in experimental populations of *Drosophila melanogaster*. Genetics, 41:469-485.
O'DONALD, P. 1962. The theory of sexual selection. Heredity, 17:541-552.
PARSONS, P. A. 1964. A diallele cross for mating speeds in *Drosophila melanogaster*. Genetica, 35:141-151.
—— 1965. The determination of mating speeds in *Drosophila melanogaster* for various combinations of inbred lines. Experientia, 21:478.
—— 1967. The Genetic Basis of Behaviour. London, Methuen.
—— S. M. W. HOSGOOD, and B. T. O. LEE. 1967. Polygenes and polymorphism. Molec. Gener. Genetics, 99:165-170.
—— and D. KAUL. 1966. Mating speed and duration of copulation in *Drosophila pseudoobscura*. Heredity, 21:219-225.
—— 1967. Variability within and between strains for mating behaviour parameters in *Drosophila pseudoobscura*. Experientia, 23:121-132.
PETIT, C. 1951. Le rôle de l'isolement sexuel dans l'évolution des populations de *Drosophila melanogaster*. Bull. Biologique, 85:392-418.
—— 1954. L'isolement sexuel chez *Drosophila melanogaster*. Étude du mutant *white* et son allélomorphe *sauvage*. Bull. Biologique, 88:435-443.
—— 1958. Le déterminisme génétique et psycho-physiologique de la compétition sexuelle chez *Drosophila melanogaster*. Bull. Biologique, 92:248-329.
—— 1959. De la nature des stimulations responsables de la sélection sexuelle chez *Drosophila melanogaster*. C. R. Acad. Sci. (Paris), 248:3484-3485.
—— and L. EHRMAN, 1969. Sexual selection in *Drosophila*. *In* Dobzhansky, Th., Hecht, M. K., and Steere, W. C., eds. Evolutionary Biology, Vol. III. New York, Appleton-Century-Crofts (in press).
PRAKASH, S. 1967a. Chromosome interactions in *Drosophila robusta*. Genetics, 57:385-400.
—— 1967b. Association between mating speed and fertility in *Drosophila robusta*. Genetics, 57:655-663.
—— 1969. Chromosome interaction affecting mating speed in *Drosophila robusta*. Genetics, 60:589-600.
REED, S. C., and E. W. REED. 1950. Natural selection in laboratory populations of *Drosophila*. II. Competition between a white-eye gene and its wild type allele. Evolution, 4: 34-42.
RENDEL, J. M. 1945. Genetics and cytology of *Drosophila subobscura*. II. Normal and selective matings in *Drosophila subobscura*. J. Genet., 46:287-302.
—— 1951. Mating of ebony, vestigial, and wild type *Drosophila melanogaster* in light and dark. Evolution, 5:226-230.
ROBERTSON, F. W. 1963. The ecological genetics of growth in *Drosophila*. 6. The genetic correlation between the duration of the larval period and body size in relation to larval diet. Genet. Res., 4:74-92.
—— 1965. The analysis and interpretation of population differences. *In* Baker, H. G., and Stebbins, G. L., eds. The Genetics of Colonizing Species. pp. 95-115.
SHOREY, H. H. 1962. Nature of the sound produced by *Drosophila melanogaster* during courtship. Science, 137:677-678.
SMITH, J. M. 1956. Fertility, mating behaviour, and sexual selection in *Drosophila subobscura*. J. Genet. 54:261-279.
—— 1958a. Sexual selection. *In* Barnett, S. A., ed., A Century of Darwin. London, Heinemann. pp. 231-244.
—— 1958b. The Theory of Evolution. Baltimore, Penguin Books, Inc.
SPERLICH, D. 1966. Unterschiedliche Paarungsaktivität innerhalb und zwischen verschiedenen geographischen Stämmen von *Drosophila subobscura*. Z. Verebungsl., 98:10-15.
SPIESS, E. B. 1967. Temperature sensitivity and heterosis. Amer. Natural., 101:93-95.
—— 1968a. Low frequency advantage in mating of *Drosophila pseudoobscura* karyotypes. Amer. Natural., 102:363-379.

────── 1968b. Experimental population genetics. Ann. Rev. Genet., 2:165-208.
────── 1968c. Courtship and mating time in *Drosophila pseudoobscura*. Anim. Behav., 16:470-479.
────── and B. LANGER. 1961. Chromosomal adaptive polymorphism in *Drosophila persimilis*. III. Mating propensity of homokaryotypes. Evolution, 15:535-544.
────── and B. LANGER. 1964a. Mating speed control by gene arrangement carriers in *Drosophila persimilis*. Evolution, 18:430-444.
────── and B. LANGER. 1964b. Mating speed control by gene arrangements in *Drosophila pseudoobscura* homokaryotypes. Proc. Nat. Acad. Sci. (U. S. A.), 51:1015-1019.
──────B. LANGER, and L. D. SPIESS. 1966. Mating control by gene arrangements in *Drosophila pseudoobscura*. Genetics, 54:1139-1149.
────── and L. D. SPIESS. 1967. Mating propensity, chromosomal polymorphism, and dependent conditions in *Drosophila persimilis*. Evolution, 21:672-678.
────── and L. D. SPIESS. 1966. Selection for rate of development and gene arrangement frequencies in *D. persimilis*. II. Fitness properties at equilibrium. Genetics, 53:695-708.
────── and L. D. SPIESS. 1964. Selections for rate of development and gene arrangement frequencies in *D. persimilis*. Genetics, 50:863-877.
SPIESS, L. D., and E. B. SPIESS. 1968. Mating frequency and conditions of yeast and water in culture of *D. persimilis*. Drosophila Inform. Serv., 43:130-131.
────── 1969. Minority advantage in interpopulational matings of *Drosophila persimilis*. Amer. Natural., 103:155-172.
────── and E. B. SPIESS. 1969. Mating propensity, chromosomal polymorphism, and dependent conditions in *D. persimilis*. II. Factors between larvae and between adults. Evolution, 23:225-236.
SPIETH, H. T. 1952. Mating behavior within the genus *Drosophila* (Diptera). Bull. Amer. Mus. Natural. Hist., 99:395-474.
────── 1968. Evolutionary implications of sexual behavior in *Drosophila*. Evol. Biol., 2:157-193.
STURTEVANT, A. H. 1915. Experiments on sex recognition and the problem of sexual selection in *Drosophila*. J. Anim. Behav., 5:351-366. *In* Lewis, E. B., ed. Selected Papers of A. H. Sturtevant: Genetics and Evolution. San Francisco, W. H. Freeman and Co. Publishers. pp. 24-37.
────── and TH. DOBZHANSKY. 1936. Inversions in the third chromosome of wild races of *Drosophila pseudoobscura*, and their use in the study of the history of the species. Proc. Nat. Acad. Sci. (U.S.A.), 22:448-450.
TAN, C. C. 1946. Genetics of sexual isolation between *Drosophila pseudoobscura* and *Drosophila persimilis*. Genetics, 31:558-573.
TEBB, G., and J. M. THODAY. 1956. Reversal of mating preference by crossing strains of *Drosophila melanogaster*. Nature (London), 177:707.
VAN VALEN, L. 1967. Heterosis and temperature. Amer. Natural., 101:92-93.
WALDRON, I. 1964. Courtship sound production in two sympatric sibling *Drosophila* species. Science, 144:191-193.
WALLACE, B. 1948. Studies on 'sex ratio' in *Drosophila pseudoobscura*. I. Selection and sex ratio. Evolution, 2:189-217.
────── 1968. Topics in Population Genetics. New York, W. W. Norton & Company, Inc.
WATTIAUX, J. M. 1968. Cumulative parental age effects in *Drosophila subobscura*. Evolution, 22:406-421.

13

Observations on the Microdispersion of Drosophila melanogaster[1,2]

BRUCE WALLACE

Cornell University, Ithaca, N.Y.

Introduction	381
The Bryant Park Experiments	382
Experimental Populations	386
The "Tropical Rainforest" Greenhouse Experiments	393
Venetian Experiments	395
Discussion	396
References	398

Introduction

Natural populations live for the most part in unbounded areas. The effective population size of a local population of any species, then, is related to the movement of individuals and to their mating habits. Either

---

[1] The work reported here was carried out under the support of Contract No. AT-(30-1)-2139, U.S. Atomic Energy Commission.
[2] This paper is dedicated to Professor Th. Dobzhansky on the occasion of his 70th birthday. In a sense it represents the completion of one full cycle; my first professional job was that of student assistant helping Dr. Dobzhansky study the dispersal of *Drosophila* at Keen Camp, California, during the summer of 1941. It is appropriate then, to express here the best wishes of the other two student assistants of that summer, the present Professors Harlan Lewis of the University of California at Los Angeles and Alexander Sokoloff of the California State College at San Bernardino.

limited dispersal or predispersal mating would tend to reduce the effective size of the population and to promote inbreeding. Dispersal over extended distances, the systematic dispersion of large segments of the population, and the postponement of mating until dispersion has occurred would tend to diminish inbreeding and to increase the effective size of the local population.

A number of studies have been carried out on the dispersal of several *Drosophila* species (*D. funebris*: Timofeeff-Ressovsky and Timofeeff-Ressovsky, 1940, and Dubinin and Tiniakov, 1946; *D. pseudoobscura*: Dobzhansky and Wright, 1943; *D. willistoni*: Burla et al., 1950). These studies have been made by the release-recapture technique, one that requires (if it is to be "efficient") the release of considerable numbers of marked flies within a (generally) already occupied area; traps for recapturing the released flies have usually been set at intervals of 10 or 20 meters for, in some instances, hundreds of meters.

A recent analysis (Wallace, 1966b) of the relation between the probability of allelism of recessive autosomal lethals and the distance between collection sites has shown that this probability decreases detectably as distance increases from 0 to 90 m. An attempt was made (Wallace, 1966a) to relate the decreasing frequency of allelism over these short distances to the movement of flies themselves; Wright (1968) has criticized this effort especially as it relates to *pattern* of dispersal but also as it bears on *distance*.

The present paper presents the results of a series of studies on the dispersal of *D. melanogaster* over small distances. These studies also deal with the mating behavior of these flies in relation to their dispersion. A variety of experimental conditions have been used and so techniques and results will be described in sequence. It may be mentioned here, though, that a common feature of the "field" experiments has been the deliberate attempt to release the marked flies gradually and in low numbers so as not to overcrowd the neighborhood immediately surrounding the point of release.

The Bryant Park Experiments

The first of the dispersal experiments were carried out on my own lot in the Bryant Park section of Ithaca, New York. The back of the lot is 60 feet (approximately 20 m) wide. Wooden stakes approximately 1 foot high were driven into the ground at 6-foot intervals across the back of the lot and for 30 feet along each side. The stakes marked 20 sites in all. The entire area involved is shaded by lilac bushes, wild grape vines, pine, sugar maple, and elder trees.

A small board 1 foot long and 3 inches (approximately 7 cm) wide was nailed to the top of each stake to make a T-shaped apparatus. These boards served to shield 1-inch-by-4-inch shell vials attached to their undersides from the rain and, if it happened to penetrate the surrounding bushes, direct sunlight.

The flies used in the Bryant Park experiments were wildtype strains from Kentucky and Georgia and a strain homozygous for *sepia* that came originally from Raleigh, North Carolina.

Vials that contained originally two or three pairs of parents were unstoppered immediately before the first adult offspring were to emerge; one vial was attached to the underside of the crossboard at each site. Vials from which homozygous *sepia* (*se/se*) flies would emerge were placed at every third site (3, 6, 9, 12, 15, and 18); the vials at the other sites contained wildtype (+/+) flies.

At the opposite end of each crossboard was fastened a freshly prepared vial, lightly yeasted and unstoppered, with its open end facing away from the old vial from which flies were about to emerge. The distance from the open end of the old vial to the open end of the fresh one was very nearly 1 foot; the distance between vials at adjacent sites was (as stated above) about 6 feet.

Roughly a week after the vials had been set out, stoppers were inserted deftly into the "old" and "fresh" vials at each site, the vials were labeled, and they and the flies captured within them were brought into the laboratory.

The results of these experiments are summarized in Tables I through V. Table I gives the number of wildtype and *sepia* flies captured within the *old* vials at +/+ and *se/se* sites (sites at which vials containing homozygous wildtype and homozygous *sepia* flies were placed). Not enough time had elapsed since setting out the original vials for offspring to hatch in the

TABLE I

The Distribution of Adult Wildtype and *sepia* Flies Among *Old* Vials[a]

	Flies	
Sites	Wildtype	*sepia*
+/+	230	1
se/se	1	102

[a]Old vials from which wildtype and *sepia* flies emerged are designated +/+ and *se/se*; the sites at which such vials were placed are also designated by these symbols.

TABLE II
The Distribution of Adult Wildtype and *sepia* Flies Among *Fresh* Vials

	Flies	
Sites	Wildtype	sepia
+/+	20	5
se/se	0	6

field; consequently, the adult flies recovered were those which had hatched from the old vials themselves. These flies were recovered almost exclusively from vials in which they hatched; only 2 of 334 flies were found at the "wrong" site. It would seem that the food in old vials is not particularly attractive to flies from other sources; on the other hand, emerging flies seemed to be in no hurry to depart.

Table II lists the number of wildtype and *sepia* flies captured within the fresh vials at +/+ and *se/se* sites. These are flies that were in these vials at the end of the experiment; they are not necessarily the only flies that visited these same vials during the entire preceding week. The data suggest that wildtype flies are found within the fresh vials only at the sites where they initially emerged (1 foot away). *Sepia* flies, on the other hand, are found both at *se/se* and +/+ sites. (The local *Drosophila* population did not contain large numbers of *D. melanogaster* at the time of these studies; this had been established by trapping immediately before the experiment and is confirmed by the absence of wildtype *D. melanogaster* in the fresh vials at *se/se* sites).

The data given in Table III refer to the offspring that hatched from the fresh vials at +/+ and *se/se* sites. Once more we see that wildtype flies are encountered only at +/+ sites; there is no evidence that any wildtype

TABLE III
The Distribution of Wildtype and *sepia* Flies Emerging as Offspring from the Fresh Vials

	Flies	
Sites	Wildtype	sepia
+/+	433	124
se/se	0	116

TABLE IV

Numbers of Wildtype and *sepia* Flies Hatching from
Fresh Vials at Those +/+ Sites Which Yielded Wildtype Flies;
the Genotypes of Tested Wildtype Males are Also Listed

Site	Wildtype	*sepia*	Wildtype males +/+	Wildtype males +/*se*
4	52	22	1	15
5	132	11	6	6
7	119	3	7	14
14	105	0	9	11
17	25	0	2	0
Totals	433	36	25	46

fly migrated to a food source 6 feet away from its birthplace. *Sepia* flies are found at both *se/se* and $+/+$ sites, presumably at the $+/+$ sites nearest their birthplaces (6 feet away).

Wildtype males that hatched from the fresh vials were tested to determine whether they were $+/+$ or $+/se$; the results are listed in Table IV. Wildtype flies hatched from five vials; altogether these flies accounted for 433 of the 469 flies hatching from these vials. Despite the paucity of homozygous *sepia* flies hatching from these five vials, two thirds of the tested wildtype males proved to be $+/se$ heterozygotes. The observed and and expected zygotic distributions of flies hatching from these five vials are given in Table V. Ostensibly, more heterozygotes were observed than were expected.

Several tentative conclusions can be stated here: Dispersion of *D. melanogaster* may be slight if the simultaneous release of thousands of flies is avoided and if the released flies are not agitated. The restricted dispersion should promote inbreeding and homozygosis. A compensatory factor has

TABLE V

A Comparison of the Observed and Expected Distributions
of Genotypes Among the Flies Listed in Table IV

	+/+	+/*se*	*se/se*
Observed	0.325	0.598	0.077
Expected	0.389	0.469	0.141

been suggested by the excess of +/se offspring in vials that yielded both wildtype and sepia offspring; this excess may reflect a preferential mating of parental flies of dissimilar origins (see Ehrman, 1966). Nevertheless, sepia females seem to have departed from se/se sites as nonvirgins because sepia offspring were obtained exclusively from several fresh vials at +/+ sites. Again, in reference to premigration mating, it should be stated that nearly all flies captured in "old" vials (Table I) were old and the females were obviously not virgins.

Experimental Populations

Two experimental populations, C-6 and C-7 (C = Cornell), were set up to investigate further the mating patterns of "migrating" flies. Migrant individuals, we may recall, tend to produce inbred offspring if they are fertilized before leaving "home." An extreme example would be the wide dispersal of seeds of an obligatorily self-fertilizing plant; the distance over which such seeds might be dispersed would not affect the homozygosity resulting from inbreeding.

Conditions were arranged in population C-6 so that about half of the adult flies were homozygous wildtype (+/+) while the remainder were homozygous sepia. These two types of flies hatched from separate vials (Fig. 1) that were placed within the regular (but empty) plastic food cups

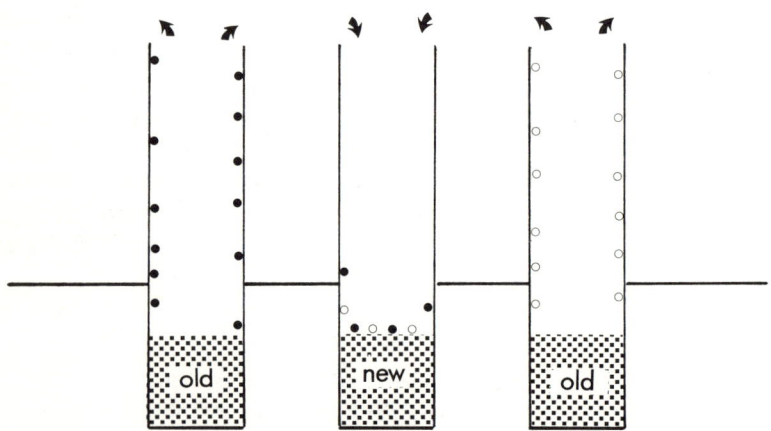

Fig.1. An attempt to illustrate the basis for expecting an excess of homozygosity following the "premigration" mating of two kinds of homozygous flies in an experimental population. The "old" vials (left and right) contain homozygous +/+ or se/se flies (open and closed circles); females already inseminated before leaving the old vial will tend to produce homozygous offspring in the "new" one (center).

TABLE VI

Proportion of *sepia* Flies Among the
Offspring Obtained by Means of Daily Egg
Samples Taken from Population C-6

Sample	Total flies	*sepia*
1	703	0.788
2	411	0.533
3	428	0.369
4	330	0.218
5	343	0.210
6	296	0.135
7	350	0.183
8	508	0.242
9	531	0.260
10	485	0.169

of the population cage. Five vials produced $+/+$ flies and six produced *se/se*. The plastic population cage has places for 15 food cups; the remaining four were used to expose four fresh vials to the population of flies each day for 10 days. The sample vials were retained until the eggs within them had given rise to adults; these were sorted and counted. A number of wildtype males of each sample were crossed to virgin *se/se* females in order to determine whether their genotypes were $+/+$ or $+/se$.

The relative proportion of *sepia* flies in each of the 10 daily samples is listed in Table VI. The proportion of *sepia* declines steadily over the period of 10 days. Presumably the *sepia* flies in the population cage emerged somewhat sooner from their vials than did the wildtype and, hence, the high frequency of *sepia* homozygotes at the beginning. If the frequencies of $+/+$ and *se/se* flies were about 50:50 and if matings were to occur at random, we would expect about 25 percent *sepia* offspring in the sample vials.

Wildtype males hatching from the daily samples were mated with *se/se* females in order to determine the genotypes of the males; $+/+$ males left only wildtype offspring while $+/se$ heterozygotes produced offspring half of which were *sepia*. The numbers of males tested together with their genotypes are listed in Table VII.

Table VIII lists the observed and expected distributions of zygotic fre-

TABLE VII

Numbers of +/+ and +/se Flies Among Phenotypically Wildtype Offspring Obtained by Means of Egg Samples from Population C-6

Sample	# +/+	# +/se	Total
1	4	33	37
2	3	28	31
3	13	63	76
4	6	26	32
5	10	47	57
6	12	43	55
7	14	71	85
8	14	52	66
9	29	43	72
10	22	60	82

quencies for the 10 samples. The observed distributions are calculated directly from the data of Tables VI and VII: Table VI gives the observed frequency of *sepia* homozygotes; the remainder is apportioned between $+/+$ and $+/se$ according to the results shown in Table VII. The expected distributions are calculated according to Hardy-Weinberg expectations where gene frequencies are estimated by summing the frequency of the appropriate homozygote and one half that of heterozygotes. The data shown in Table VIII do not support the notion outlined at the start of this section—namely, that mating within the parental vials might lead to an excess of homozygous offspring. The majority of samples tested reveal the opposite relationship; fewer homozygotes are observed than are expected.

The experimental technique used in analyzing population C-6 contained a flaw that was remedied in the study of a second experimental population, C-7. The egg samples from population C-6 were collected in vials and then the larvae were allowed to develop within the same vials. Because of the number of eggs laid, the developing larvae were severely overcrowded. Consequently, the surplus of heterozygotes observed in most of the egg samples may have resulted from differential survival. Indeed, homozygous wildtype and *sepia* zygotes may have been in considerable excess only to be reduced to an apparent deficit by excessive larval mortality. The ambiguity can be removed by subdividing the egg sample among a num-

TABLE VIII

Observed (O) and Expected (E) Distributions of Zygotic Frequencies in Successive Samples from Population C-6

Sample		+/+	+/se	se/se	Homozygotes O/E
1	O	0.023	0.189	0.788	
	E	0.014	0.207	0.780	1.02
2	O	0.045	0.422	0.533	
	E	0.066	0.381	0.554	0.93
3	O	0.108	0.523	0.369	
	E	0.136	0.466	0.398	0.89
4	O	0.147	0.635	0.218	
	E	0.215	0.497	0.287	0.73
5	O	0.139	0.651	0.210	
	E	0.216	0.498	0.286	0.70
6	O	0.189	0.676	0.135	
	E	0.278	0.499	0.224	0.65
7	O	0.183	0.682	0.134	
	E	0.275	0.498	0.226	0.63
8	O	0.242	0.597	0.161	
	E	0.293	0.498	0.212	0.80
9	O	0.260	0.442	0.298	
	E	0.231	0.499	0.269	1.11
10	O	0.169	0.608	0.223	
	E	0.224	0.499	0.278	0.78

ber of culture bottles so that larvae develop under near-optimal conditions.

Population C-7 was set up by placing 12 1-inch-by-4-inch shell vials into otherwise empty food cups of a standard population cage. Homozygous wildtype flies were developing in six of these cultures, homozygous *sepia* in the others. Adult flies had not yet appeared in the vials although parental flies had been mated in them nine days previously.

Three days after the population had been set up, cups of food were inserted in the remaining three spaces in order to obtain an egg sample from the flies that were now in the cage. These cups were removed (November 21, 1967) and replaced by three others; the eggs of the first sample were distributed among five half-pint (approximately quarter-liter) culture bottles.

Part of the second egg sample was taken in a specially prepared food cup

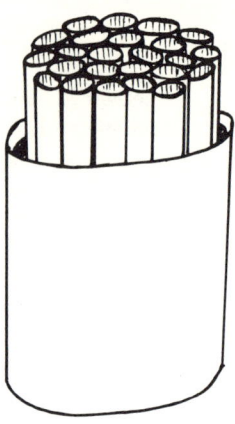

Fig. 2. A food cup into which numerous short pieces of plastic soda straws have been inserted. This technique was used to disrupt the food surface and to increase the probability that at least some females (+/+ or se/se) that had been inseminated before leaving their "home" vials would fail to encounter and mate with males of the other homozygous genotype.

(see Fig. 2). Instead of an uninterrupted food surface, numerous 1-inch lengths of plastic soda straws were inserted into the food contained in this cup. These straws were supposed to impede the movement of flies on the food and, hence, to exaggerate the effects of premigrational mating—that is, to bias the distribution of zygotic frequencies in favor of homozygous +/+ and *se/se* offspring.

The second egg sample was beset with troubles; holiday travel and illness postponed the removal of this sample until November 26, 1967. By the time the sample cups were removed from the cage, the developing larvae had worked the food considerably; the sample, then, was by no means an *egg* sample. The third egg sample, one which involved exclusively a food cup equipped with the honeycomb of short soda straws, was taken on November 27th; each length of soda straw with its adherent plug of food was placed in a vial where the eggs and larvae developed under uncrowded conditions.

The data obtained from the egg samples of population C-7 are listed in Tables IX through XIV. Table IX lists the numbers of *sepia* and wildtype flies hatching from the five culture bottles containing the first egg sample. The proportion of *sepia* obviously varies from bottle to bottle but the overall frequency of *sepia* is about 0.36. The genotypes of wildtype flies hatching in these cultures were not determined by test crosses.

Table X lists the numbers of wildtype and *sepia* flies hatching in half-pint cultures containing the second egg sample; the frequency of *sepia* in these cultures was about 0.29. The same egg sample taken in a food cup studded with soda straws gave considerably different results (Table XI); in this case the frequency of *sepia* flies was only 0.13, less than half that observed for the sample raised in bottles. It seems that this difference is related to the extended period during which the sample cups were left in the population cage. Test crosses with *se/se* females revealed that 29 of

TABLE IX
Flies Hatching (in half-pint bottles) from the First Egg Sample of Cage C-7

Bottle	Wildtype	sepia[a]
a	63	31
b	107	82
c	98	22
d	74	30
e	106	89
Totals	448	254

[a] Frequency of *sepia* = 0.362.

TABLE X
Flies Hatching (in half-pint bottles) from the Second Egg Sample of Cage C-7

Bottle	Wildtype	sepia[a]
a	478	196
b	513	193
c	480	202
Totals	1471	591

[a] Frequency of *sepia* = 0.287.

TABLE XI
Flies Hatching (in shell vials) from the Second Egg Sample of Cage C-7

Date counted	Wildtype	sepia[a]
12/5/67	639	101
12/8/67	980	146
Totals	1619	247

[a] Frequency of *sepia* = 0.132.

TABLE XII

Flies Hatching (in shell vials) from the Third Egg Sample of Cage C-7

Vial	Wildtype	sepia	Vial	Wildtype	sepia
1[a]	105	20	9[a]	37	6
2	101	38	10	82	36
3[a]	99	20	11	43	2
4	44	3	12	100	18
5[a]	40	15	13[a]	107	9
6[a]	56	42	14	63	18
7[a]	41	13	15[a]	69	59
8	67	18	16	51	14
			Totals	1105	331

[a] Frequency of *sepia* in these vials = 0.249; frequency of *sepia* in remaining vials = 0.231.

31 males hatching in bottles were $+/se$ heterozygotes while all 22 of the males tested from the sample grown in vials were heterozygotes.

Table XII lists for the third egg sample the numbers of wildtype and *sepia* flies hatching in 16 vials each of which was seeded by the small food plug at the base of a single soda straw. The frequency of *sepia* flies varies from vial to vial; the overall mean is approximately 0.23. Wildtype males whose genotypes were determined by test crosses were taken from the marked

TABLE XIII

Genotypes of Wildtype Males Obtained from the Third Egg Sample of Cage C-7

Vial	+/+	+/se
1	0	29
3	5	8
5	1	13
6	0	10
7	1	4
9	0	14
13	2	16
15	7	26
Totals	16	120

TABLE XIV

Summary of Observed (O) and Expected (E) Proportions of the Three Genotypes in Flies Hatching from the Second and Third Egg Samples of Cage C-7; the Third Sample Yielded the Most Extensive Data

		+/+	+/se	se/se
Second sample (Bottles)	O	0.046	0.667	0.287
	E	0.143	0.471	0.386
Second sample (Vials)	O	0	0.868	0.132
	E	0.188	0.491	0.320
Third sample (Vials)	O	0.089	0.662	0.249
	E	0.176	0.487	0.336

(^{a}) vials of Table XII; the frequency of *sepia* in these vials was about 0.25, not significantly different from the entire array of vials counted.

The numbers of $+/+$ and $+/se$ males among those tested of the third egg sample are listed in Table XIII; as in the previous samples, the proportion of $+/se$ heterozygotes in unexpectedly high. The observed and expected arrays of $+/+$, $+/se$, and se/se genotypes of the tested males for the second (bottles and vials) and third (vials only) egg samples are given in Table XIV. In each instance the frequency of $+/se$ heterobygotes is greater than expected. These results agree with those observed in both the Bryant Park experiments (Table V) and the analysis of population C-6 (Table VIII). Despite the expectation that females hatching in homozygous cultures might be inseminated before departing and, hence, might produce an excess of homozygous offspring (or, in even stronger terms, despite our knowledge that premigration mating does in fact occur), the actual distribution of zygotes seems to be quite the opposite. In each of the experiments performed, the heterozygous offspring are too numerous. The observed excess of heterozygotes exists despite the bias caused by premigrational matings that *must* operate to favor homozygotes.

The "Tropical Rainforest" Greenhouse Experiments

The experiments described in this section were carried out in a small (8-feet × 17½-feet) greenhouse at Cornell University that is maintained under the direction of Wm. L. Brown, Jr. for raising tropical ant species. Plants found in the greenhouse include *Monstera, Ficus, Croton, Poinsettia,* and *Mimosa; Tetramorium oceanicum, Pheidole* (two species), *Meranoplus dichrous,* and the tailor ant, *Oecophylla smaragdina* are among the ant species. A rather large "natural" *Drosophila* population is maintained by placing overripe bananas and melon rinds on the greenhouse benches; these flies serve as food for the predacious ants. The greenhouse is sprayed liberally and violently with water nearly every day to maintain its rainforest characteristics.

The first experiment on the dispersal of flies within the tropical greenhouse consisted of labeling the flies already present in the population instead of introducing additional flies as is the usual procedure in release-recapture experiments. A coffee tin containing nearly 1 quart (approximately 1 liter) of fermenting bananas stained with blue food coloring was placed on the greenhouse bench. Preliminary experiments had shown that all flies kept in vials for 24 hours on such food were clearly labeled by virtue of their blue abdomens. Furthermore, half of these flies were clearly labeled after being removed from the colored food for 24 hours; at the end of 40 hours about 20 percent of the flies could still be identified as being blue.

Table XV gives the consolidated data of five days' collections within the greenhouse at various distances (½ m intervals) from the can of colored bananas. The flies with the colored abdomens are found for the most part within ½ m of the food source itself. Others have strayed up to 2½ m away. That the food source was an attractive feeding place within the greenhouse is suggested by the large number of wildtype flies captured nearby. A second attractive spot existed at a distance of 1½ to 2 m, however, but this spot did not lure a great many flies from the colored food once they had visited this seemingly more delicious repast.

Another experiment carried out in the same greenhouse consisted merely

TABLE XV

Consolidation of Five Days' Collections of Flies (*D. melanogaster*) that were Marked by Eating Fermenting Bananas Stained with Commercial Food Coloring; the Experiment was Carried Out in a "Tropical Rain Forest" Greenhouse

Distance (m)	Stained	Unstained	Total
½	31	185	216
1	5	69	74
1½	7	89	96
2	5	119	124
2½	3	57	60
3	0	27	27

TABLE XVI

Sepia and Wildtype Flies Captured in the "Tropical Rain Forest" Greenhouse After Two Days of Dispersal of the sepia Flies from Unplugged Culture Bottles

Distance (m)	sepia	Wildtype	$\frac{sepia}{Wildtype}$
½	92	72	1.28
1	32	20	1.60
1½	trap filled with water		
2	74	94	0.79
2½	37	51	0.73
3	4	8	0.50

of placing three unstoppered cultures of homozygous *sepia* flies on the greenhouse bench. Two days later a collection of flies at ½-m intervals yielded the data listed in Table XVI. A clear gradient exists in the relative proportion of *sepia* to wildtype flies captured over a total distance of only 3 m.

Venetian Experiments

The final experiments on dispersal were carried out at the Istituto di Biologia del Mare, Venice, where I was a guest during the summer of 1968. Cultures of homozygous *sepia* flies (first one, then two, and finally three cultures) in half-pint bottles (each containing a small segment of ripe banana) were exposed unstoppered in a small shady garden beside the Institute's "villa." These bottles were protected from the rather frequent showers by a waterproofed box from which one side had been removed. Traps, 1-inch-by-4-inch shell vials baited with fermenting peaches or bananas, were set at intervals of ½ or 1 m more or less daily for 15 days; usually a single collection was made in the late afternoon after the traps had been out since early morning but occasionally two collections (noon and late afternoon) were made in a single day. The captured flies of each collection were stupefied in the freezing compartment of a refrigerator, classified under a hand lens, and destroyed.

The results of this experiment on dispersal are listed in Table XVII. Of 36 *sepia* flies recaptured over the 15-day period, 27 were captured within 1 m of the source of these flies; none were captured beyond 6 m. The variable numbers of wildtype flies collected at vials as closely spaced as these were

TABLE XVII

Sepia and Wildtype Flies Captured Over a 15-day Period at Various Distances from Unplugged Culture Bottles that Served as a Continuous Source of *sepia* flies

Distance (m)	Total *sepia*	Total wildtype	Wildtype 8-10 days	Wildtype 11-15 days
½	13	128	33	82
1	14	106	50	39
1½	5	56	22	34
2	1	43	16	23
2½	0	37	23	14
3	1	31	10	13
4	0	37	18	12
5	0	54	22	16
6	2	48	25	10
7	0	98	38	32
8	no trap at this distance			
9[a]	0	(9)	–	–
10[a]	0	(11)	–	–
11[a]	0	(15)	–	–

[a] Traps at distances 9, 10, and 11 were used only from 9/6/68 until 9/11/68 (2nd to the 7th days of the experiment).

suggest that the "native" flies, too, do not disperse rapidly. A final suggestion of this same point (see the rightmost column of Table XVII) comes from the large number of wildtype flies captured ½ m from the *sepia* culture bottles once sufficient time had elapsed (11 to 15 days) to permit the hatching within the bottles themselves of offspring of the native wildtype flies.

Discussion

The studies described above, as well as earlier ones carried out with Professor John Beardmore (see Wallace, 1968), were made in order to learn about the distances that flies move in populations. Patterns of dispersal—whether dispersing flies form normal, leptokurtic, or exponential distributions—have not concerned us in this work. If the experiments appear to be simple in design and execution, it is not entirely accidental. There is reason to suspect that "efficient" experiments, those in which thousands of flies are released in order that hundreds might be recaptured, reveal the dispersal of overcrowded and agitated flies, not of flies in a normal, everyday environment.

The data we have gathered resemble those obtained originally by Timofeeff-Ressovsky (Timofeeff-Ressovsky and Timofeeff-Ressovsky, 1940); his studies, too, emphasized the slight extent to which flies move. In his studies Timofeeff made extensive use of mutant strains; it is difficult to decide how much the choice of these strains affected his results. In our own experiments the mutant *sepia* was used to a great extent. This particular mutant strain is a relatively recent aquisition from a natural population, however. Furthermore, the data offer two opportunities to compare the dispersal of these flies with that of wildtype strains; in both instances (Tables II and III; Table XVII) it appears that *sepia* flies move from place to place at least as readily as do wildtype individuals.

The data obtained in the studies reported here do not exclude the possibility that *D. melanogaster* might migrate considerable distances, perhaps with considerable speed. Experiments similar to those carried out in the "rainforest" greenhouse were attempted in a complementary "desert" greenhouse. These experiments failed dismally; upon leaving the unstoppered culture bottles, the flies flew immediately to the glass walls of the greenhouse. Literally, they dispersed throughout the small room in minutes. Yerington and Warner (1961) found that individual flies (*D. melanogaster*) can travel four or five miles (approximately 8 km) within 24 hours; 40,000 adult flies were released in the experiment that gave this result.

The experiments carried out with Professor Beardmore had shown that *D. subobscura* when placed in one of a series of population cages will move into the second and even into the third. Cages beyond the third were visited by very few individuals despite the presence of hundreds in the first cage, mere inches away. The near immobility of the flies in that experiment encouraged an attempt to study the dispersal of flies by means of a long, narrow plastic tube. Although Oregon-R flies, wildtype flies virtually domesticated after years of laboratory use, moved slowly within the tube, individuals of a recently captured wildtype strain (EV; New York) when introduced into one end of the tube ran immediately and as a group to the opposite end.

Work by T. M. Rizki (personal communication) has suggested that *Drosophila* flies locate food much as the male gypsy moth locates the female; the response elicited by the odor of food is a tendency for the fly to move into the wind. (See, too, Michelbacher and Middlekauff, 1954, p. 918.) A response to an actual odor gradient presumably occurs only in the immediate vicinity of the food source. The extent to which individuals of *D. melanogaster* can be induced to move, despite the data that emphasize their sedentary nature, is illustrated by those flies which located the fermenting bananas in a third-story apartment in Venice.

Wright (1968) points out that data on dispersal and on the allelism of lethals in *D. pseudoobscura* lead in both instances to an estimate of about 250 m as the radius of the "neighborhood" which the parents of individuals at a given place might be supposed to have been drawn at random. This is very much larger than the microdispersal data for *D. melanogaster* would suggest. Of the two species, *D. pseudoobscura* and *D. melanogaster*, the latter has been shown in previous experiments (Dobzhansky and Wright, 1943; Timofeeff-Ressovsky and Timofeeff-Ressovsky, 1940) to be the more sedentary by far; that *melanogaster* may be as sedentary as the present results suggest had not been suspected before the sharp decline in the allelism frequency of lethals over very small distances had been demonstrated (Wallace, 1966b; Paik, 1968).

The excessively high proportions of heterozygous individuals ($+/se$) that were seen repeatedly in experiments that were deliberately designed to emphasize the relative *paucity* of these individuals deserves a final comment in this concluding section. If this excess heterozygosity is based upon a mating preference between individuals of dissimilar origins, then a population of *D. melanogaster* (as Bateman, 1950, has already suggested) is a curious mixture of greater-than-expected frequencies of hybridity and, simultaneously, greater-than-expected frequencies of homozygosity. Given, for illustrative purposes, an environment with small, equally attractive food sources spaced at intervals of 3 or 4 m, most adult flies upon emergence would remain at their place of origin; presumably, mating within these small colonies would lead to excessive inbreeding. On the other hand, chance migrants, according to our observations, would mate more successfully than natives; an unexpectedly large proportion (unexpected in respect to the number of migrants) of intercolony would arise as a result. It is conceivable, as well, that mortality during larval development would fall for the most part upon the inbred portion of the larval population so that surviving adults would be largely of hybrid origin. The latter, though, would be subjected to the inbreeding pressures of the small colony in the following generation. This picture, I believe, reflects reasonably well the observations on allelism, dispersal, and mating patterns observed in studies on *D. melanogaster*.

References

BATEMAN, A. J. 1950. Is gene dispersion normal? Heredity, 4:353-363.
BURLA, H., A. B. DA CUNHA, A. G. F. CAVALCANTI, TH. DOBZHANSKY, and C. PAVAN. 1950. Population density and dispersal rates in Brazilian *Drosophila willistoni*. Ecology, 31:393-404.

DOBZHANSKY, TH., and S. WRIGHT. 1943. Genetics of natural populations. X. Dispersion rates in *Drosophila pseudoobscura*. Genetics, 28:304-340.
DUBININ, N. P., and G. G. TINIAKOV. 1946. Inversion gradients and natural selection in ecological races of *Drosophila funebris*. Genetics, 31:537-545.
EHRMAN, L. 1966. Mating success and genotype frequency in *Drosophila*. Anim. Behav., 14:332-339.
MICHELBACHER, A. E., and W. W. MIDDLEKAUFF. 1954. Vinegar fly investigations in northern California. J. Econ. Entom., 47:917-922.
PAIK, Y. K. 1968. Behavior of lethals in *Drosophila melanogaster* populations. Proc. XII Int. Congr. Genet., 2:164-165.
TIMOFEEFF-RESSOVSKY, N. W., and H. A. TIMOFEEFF-RESSOVSKY. 1940. Populationsgenetische Versuche an *Drosophila*. II. Aktionsbereiche von *D. funebris* und *D. melanogaster*. Z. indukt. Abstamm. Vererbungsl., 79:39-43.
WALLACE, B. 1966a. On the dispersal of *Drosophila*. Amer. Natural., 100:551-563.
―――― 1966b. Distance and the allelism of lethals in a tropical population of *Drosophila melanogaster*. Amer. Natural., 100:565-578.
―――― 1968. On the dispersal of *Drosophila*. Amer. Natural., 102:85-87.
WRIGHT, S. 1968. Dispersion of *Drosophila pseudoobscura*. Amer. Natural., 102: 81-84.
YERINGTON, A. P., and R. M. WARNER. 1961. Flight distances of *Drosophila* determined with radioactive phosphorus. J. Econ. Entom., 54:425-428.

14

Studies on the Evolutionary Biology of Chilean Species of *Drosophila*[1]

DANKO BRNCIC[2]

Department. de Genética, Facultad de Medicina y Facultad de Ciencias, Universidad de Chile, Santiago, Chile

Introduction 401
Widespread Species of *Drosophila* 405
    Evolutionary Biology of *Drosophila funebris* and
    Drosophila immigrans 408
Endemic and Ecologically Restricted Species 411
    Evolutionary Biology of *Drosophila flavopilosa* 412
Endemic and Ecologically Versatile Species 418
    Evolutionary Biology of *Drosophila pavani* 419
    Evolutionary Biology of *Drosophila gasici* 426
Isolation and Chromosomal Structure 428
Acknowledgments 432
References 432

Introduction

Chile is situated in the southwestern part of South America, between parallels 17°32′ S. and 56° S., and between meridians 75°40′ W. and

---

[1] This paper is dedicated to Professor Theodosius Dobzhansky as an homage to his 70th birthday.
[2] The work reported in this article has been partially supported by grants from the School of Medicine, University of Chile, and the Consejo Nacional de Investigación Científica y Tecnológica, Santiago, Chile.

66°33′ W. The country is isolated from the rest of the neotropical zone by strong geographic and bioclimatic barriers: in the north, the Peruvian-Chilean desert, one of the most arid expanses in the world; in the east it is separated from Argentina and Bolivia by the Andes mountains, which prevent the migration of many living organisms. Only south of the lake region (39° S.) is there a zone that allows some species to pass from one side of the Andes to the other. Even so, the climate of this part of the country is unfavorable for the survival of many insects, among them those of the genus *Drosophila* to which this paper refers. The Andean areas of Chile and Argentina, south of the lake region down to the southern tip of the Hemisphere, constitute a zoogeographic region which is separated from the rest of South America by the cold and arid barrier of the Patagonian "pampa."

This particular geographic configuration of Chile determines a subdivision into regions with strongly contrasting climates and vegetation: from the rigorous conditions in the subtropical northern desert, to the Antarctic oceanic climate in the extreme south, passing through temperate environments in the central region. Moreover, some of these regions may be subdivided into still more isolated zones, especially in the north and north central part, where there are green valleys and oases separated from one another by mountains or deserts. Finally, the whole Chilean territory has a variety of subclimates, which depend on altitudinal gradients going from sea level to an altitude of 5000 or more meters.

The isolation of Chile from the rest of the neotropical region has determined the fundamentally endemic nature of its flora and fauna (Reiche, 1907; Fuenzalida, 1950). The Drosophilidae of Chile represent a good example of this phenomenon. Of the 33 species described (Brncic, 1957a, 1962a), 60 percent represent local forms, practically restricted to Chile, and at the most, to the neighboring Andean regions. It should be pointed out that the Andes represent a very special type of barrier, as in themselves they constitute a well-defined biogeographic zone, many of whose animal or plant species are distributed along both the eastern and western slopes.

From an evolutionary point of view, the endemic nature of the animal and plant species in Chile is of particular interest. Chile represents not only a geographical isolate, but also the southern and western limit of distribution of a large number of neotropical species. Wagner (1868), Jordan (1896), and more recently Dobzhansky (1951), Rensch (1960), and Mayr (1963) emphasized the importance of geographical isolation in the process of race formation and speciation, in particular along the margins of major areas of distribution. Jordan (1896) wrote: "The study of localized varieties is of the greatest importance in respect to the theory of evolution: the study of geographical races or subspecies or incipient species is a study of the origin of species." Mayr (1963), furthermore, pointed out:

"Geographic speciation is the almost exclusive mode of speciation among animals, and most likely the prevailing mode even in plants."

The genus *Drosophila* comprises forms with strongly contrasting population structures. There exist nearly cosmopolitan forms like *D. melanogaster* and others, whose ecology is strongly influenced by human activities. Other species are adjusted to wild habitats, and their distributional areas are confined to certain definite territories. Some of these species form small isolated populations; small populations interconnected through restricted migration; large but thin populations that may be isolated or interconnected; or finally, large populations. While some species can be collected only during some seasons, others are available throughout the whole year. There are some ecologically diversified species, and others that are strictly adjusted to rare and specialized niches. Although no species is strictly comparable to any other in relation to the type of populations it forms, and therefore all classifications are rather arbitrary, from a practical point of view, in the present paper, the Chilean species of *Drosophila* have been divided into three groups according to distribution and ecology (Table I):

(A) *widespread species,* which include the eight cosmopolitan forms and those which are amply distributed in the New World;

(B) *endemic and ecologically restricted species,* which include the highly specialized forms;

(C) *endemic and ecologically versatile species,* which include certain localized species that are nevertheless abundant in several environments.

Since many of the species belonging to those three groups have been studied cytologically, an assay has been made to correlate the data obtained from the cytogenetic analysis with certain characteristics of their distribution and ecology.

The study of the chromosomal structure of the populations gives valuable information on the evolutionary status of the species. In the first place, it may lead to the knowledge of the sequence of the chromosomal changes that historically have occurred in the process of race formation or speciation. A detailed analysis of this type of investigation in many species of the genus *Drosophila* may be found in the books of Dobzhansky (1951), and Patterson and Stone (1952). Nevertheless, besides the aforementioned facts, the cytogenetic approach may be valuable for other aspects of their evolutionary biology. Many species of the genus *Drosophila* are polymorphic respecting the gene orders in their chromosomes that result from inversions of their segments. This polymorphism, in the better-studied cases, has proven to be of the "balanced" type, in which the heterozygotes for chromosomes with different gene sequences (structural heterozygotes or heterokaryotypes) show selective advantage. This superiority assures the maintenance of the different chromosomal alternatives that constitute the

TABLE I

Chilean Species of *Drosophila*

Species	Distribution	Chromosomal arrangements
A. Geographically widespread species		
A-1. Cosmopolitan species:		
D. busckii Coquillett	Northern to central Chile	polymorphic
D. melanogaster Meigen	From north to 43° S.	polymorphic
D. simulans Sturtevant	From north to 43° S.	monomorphic
D. ananassae Doleschall	Central Chile	–
D. funebris Fabricius	Central and southern Chile (unique species found south of 43° S.)	polymorphic
D. immigrans Sturtevant	From north to 43° S.	polymorphic
D. hydei Sturtevant	Northern to central Chile	polymorphic
D. repleta Wollaston	Central Chile	monomorphic
A-2. Widespread but not cosmopolitan species:		
D. virilis Sturtevant	Central Chile	monomorphic
D. mercatorum Patterson and Wheeler	Northern to central Chile	polymorphic
D. nigricruria Patterson and Mainland	Northern Chile	monomorphic
D. cardini Sturtevant	Northern to central Chile	monomorphic
B. Endemic and ecologcially restricted species		
D. flavopilosa Frey	Central and southern Chile	polymorphic
D. amplipennis Malloch	Southern Chile (occasionally Central Chile)	unknown
D. appendiculata Malloch	Southern Chile	unknown
D. huilliche Brncic	Southern Chile	unknown
D. osornina Brncic	Southern Chile	unknown
D. kuscheli Brncic	Juan Fernández Islands	unknown
D. allei Brncic	Northern Chile	unknown
C. Endemic and ecologically versatile species		
D. araucana Brncic	Central and southern Chile	monomorphic
D. camaronensis Brncic	Northern Chile	monomorphic
D. serenensis Brncic	Central Chile	polymorphic
D. pavani Brncic	Central Chile	polymorphic
D. gasici Brncic	Northern Chile	polymorphic

polymorphism found in their natural populations. The products that result from crossing-over between chromatids that are bearers of different gene orders are eliminated at female meiosis without any decrease of fertility. This latter circumstance determines that inversions constitute a powerful means for preserving together local gene complexes that have adaptive significance for the population (Dobzhansky, 1951). Therefore, analysis of the peculiarities of chromosomal polymorphism, of its variations within the same or of closely related species, allows valuable inferences on the genetic mechanisms that operate in the particular populational models of each species.

In light of the aforementioned viewpoint I believe it is of interest to summarize in the present paper the characteristics of the chromosomal polymorphism found in the local populations of the cosmopolitan and widespread species and of the different kinds of endemic species living in Chile. This analysis has revealed many similarities and differences between these diverse types of populations that may contribute to a better comprehension of the relationships between genetic structure and ecology, and also of the real significance of geographic isolation in the evolutionary process. From the Chilean species of *Drosophila* indicated in Table I different sorts of data have been taken. These are not always strictly comparable, as not all the species allow the same possibilities of experimental analysis. Some of the data refer to geographic distribution and ecological requirements, which are well known in some species but not in others. From the cytogenetic point of view, in some species it is known only whether they are monomorphic or polymorphic, while in others there is quantitative information that yields knowledge about the adaptive nature of the polymorphism, based on cyclic seasonal changes of the different chromosomal alternatives, their distributional gradients in correspondence with geographical clines, or their variations under experimental laboratory conditions. In other species the balanced nature of chromosomal polymorphism has been established by means of the study of some of the physiological properties of structurally homo- or heterozygous individuals, such as longevity, mating activity, rate of development, and other components of fitness. All these data have been used in the present discussion on the evolution of the Chilean species of the genus *Drosophila*.

Widespread Species of Drosophila

Patterson and Stone (1952) recorded eight species of the genus *Drosophila* that have become established in the six continental regions (Darlington, 1957). These cosmopolitan species are: *D. busckii* Coquillett, *D. melanogaster* Meigen, *D. simulans* Sturtevant, *D. ananassae* Doleschall, *D. funebris* Fabricius, *D. hydei* Sturtevant, *D. repleta* Wollaston, and *D. immigrans* Sturtevant. All of them have also been found in Chile in different regions and in different frequencies. In Chile there live other species that, although not cosmopolitan, are also wide-ranging ones and exist in the neotropical as well as elsewhere in the six great zoogeographic regions. One of these is *D. virilis* Sturtevant, found in four of these regions according to Patterson and Stone (1952), and the other is *D. mercatorum* Patterson and Wheeler, found in the neoarctic and neotropical regions, and in Hawaii. Finally, there exist species that have a tendency to spread over all the neotropical zone, from Mexico, covering all of South America; *D. nigricruria* Patterson and Mainland, and *D. cardini* Sturtevant belong to this group. These 12 species constitute the counterpart of those which are endemic for Chile and exceptionally spread to the neighboring zones of the South Andean region.

The reason for the widespread character of the above-mentioned species of *Drosophila* in some cases may reside in their great ecological versatility, which allows them to live and reproduce in many different environments. In other species it may be due to the fact that they are adapted to a few specific niches that are very widely distributed, some of them associated with human habitat. According to recent reviews of Mayr (1963) and Dobzhansky (1965), the populations of common and widespread species of *Drosophila* may contain high levels of genetic variability. When they become isolated from their main distribution range, they may retain a large amount of this variability, not differentiating themselves greatly from the neighboring populations. On the other hand, they may evolve into well-defined races as a genetic adaptive response to the local environmental conditions. As Mayr (1963) has pointed out, "the most distinct isolates of a species are nearly always situated along the periphery of the species range." Due to its geography and ecology, Chile represents the southern and western limit of distribution of many neotropical and cosmopolitan species of *Drosophila*. This fact justifies a discussion of the genetic and cytological structure of the Chilean populations of this group of widespread species.

Of the eight cosmopolitan species found in Chile, at least one, *D. ananassae,* seems to be an occasional foreign visitor that has not been able to establish permanent populations. Only two individuals belonging to this species have been collected, both in a fruit market that imports foreign fruits such as pineapples and bananas. *D. melanogaster* is also relatively scarce in Chile; it has been found only within houses and in large numbers in wine cellars and fruit markets. No systematic investigations of its chromosomal structure have been undertaken. On the other hand, its sibling species, *D. simulans,* constructs large populations during the summer months in the northern, central, and southern parts of the country where there is vegetation, even in regions distant from the artificial environments created by man. Its populations, just as has been described for other parts of the world (Carson, 1965), have proven to be monomorphic for the gene arrangements in its chromosomes.

Drosophila busckii has been found in many places, especially where rotten vegetables are deposited. This species exhibits a certain degree of chromosomal polymorphism; unfortunately no comparison has been made with the polymorphism found elsewhere (Krivshenko, 1963) and no quantitative studies have been undertaken. *D. virilis* and *D. repleta,* found in small numbers in many parts of Chile, appear to be strictly monomorphic, as also in other regions of the world (Patterson and Stone, 1962; Carson, 1955). Chilean populations of *D. hydei* are polymorphic for an inversion in the second chromosome, identical to that denominated $2a^2$ by Warters (1944), and, according to Wasserman (1962a), present in moderate or low frequency in many populations from all parts of the world.

Drosophila nigricruria deserves special consideration. This species was described from Mexico to South America. Chilean specimens were erroneously considered a new species, designated as *D. hoeckeri* (Brncic, 1957a). Several strains of *D. hoeckeri,* including the one from which the type specimens were taken, were cytologicaly analyzed by Wasserman (1962b). He found their gene arrangements to be identical to the "Standard" sequence of *D. nigricruria,* which is monomorphic over all its area of distribution. All crosses between *D. hoeckeri* and other *D. nigricruria* strains resulted in fully fertile offspring. According to Wasserman *D. hoeckeri* is therefore a synonym of *D. nigricruria,* or a subspecies of the latter, since the Chilean populations are easily distinguished from those from other parts by their highly contrasting external color pattern. This fact is interesting since it indicates the existence of a certain degree of genetic divergence respecting the center of the distribution area of the species.

The other species of the *repleta* group, included within the category of widespread species, which has also been found in Chile, is *D. mercatorum.* According to Wasserman (1962c) this species shows frequent and extensive chromosomal polymorphism in some populations and a reduced one in others. The same author thinks that *D. mercatorum* can be divided into two subspecies: *D. m. mercatorum,* an essentially homozygous form, very wide-ranging, extending from the north of the United States to the Peruvian and Bolivian Andean region, with a tendency to spread to other zones such as Hawaii; the other is a polymorphic form, designated as *D. mercatorum pararepleta,* found in the Brazilian lowlands. The Chilean populations, which are isolated geographically from both major populations, are polymorphic for the $2s^3$ genetic arrangements present only in Brazil and Bolivia. According to the observations of Wasserman, the Chilean *mercatorum* represents a colonizing population of *pararepleta* that came westward across the Andes.

The chromosomal structure of *D. cardini* has not been extensively studied. Nevertheless, Heed (1962) has found that the Chilean strains of this species give fully fertile offspring when crossed with strains from Mexico, the West Indies, and South America.

Evolutionary Biology of D. funebris and D. immigrans

Brncic and Dobzhansky (1957) reported that *Drosophila funebris* Fabr., 1787, is the only species of the genus *Drosophila* known to live in Chile south of latitude 43° to 44° S. According to Basden (1956), it is also the northernmost member of the genus. He recorded the species in arctic Finland, Norway, Iceland, and Greenland. Consequently, *D. funebris* represents the most widely distributed species of *Drosophila*, and extends its area farthest north as well as farthest south. Like other cosmopolitan species, it occurs mostly in close association with man, and has been introduced into Chile probably by human agencies, but it is apparently the most cold adapted of the "domestic" species, and it is rare or absent in the tropics.

Dubinin and his school (1937, 1946) have established that *D. funebris* is polymorphic with respect to the gene orders in the chromosomes due to the presence of inversions. These authors also detected some correlation between the frequencies of the different chromosomal types and the nature of the ecological niches exploited by the species. A report of the chromosomal structure of the populations of *D. funebris* living in Chile seems therefore of interest. Brncic and Sánchez (1958) have analyzed many stocks and natural populations of *D. funebris* from different places extending from La Serena in the north central part of Chile (Lat. 29°53′ S) to Tierra del Fuego in the extreme south (Lat. 53°40′ S.). The samples were mostly heterozygous for a long inversion in the second chromosome, corresponding to the *C-II-M* inversion of Dubinin et al. (1957), which was found also in Sweden and France by Perje (1954). In the most northern population (La Serena) some of the flies were heterozygous for an inversion, in addition to the *C-II-M* one, located in the middle part of the third chromosome; it is probably the same one described as *IV-1* by Dubinin in Russian populations.

The finding that the Chilean populations of *D. funebris* are polymorphic with respect to the gene orders in their chromosomes, and that at least one of the inversions is the same as found in northern Europe, indicates that this genetic arrangement is very old in the evolutionary history of the fly. Such an old and widely distributed kind of polymorphism may be regarded as adaptively important for the species. Since *D. funebris* seems to be the most widely distributed species of the genus in the world, inversion *C-II-M* is actually the most widely distributed inversion ever discovered.

The adaptive nature of the inversions in the genus *Drosophila* has been demonstrated in many ways. In natural populations of "wild" species, not associated with human habitats, the different gene arrangements usually show quantitative and qualitative geographic variations (Dobzhansky, 1951). "Domestic" species have, on the contrary, chromosomal variants that are widely distributed over the range of the species, and do not normally show differences in the concentration of the inversions from one place to another. Nevertheless, it would be wrong to conclude that the chromosomal variability present in species associated with man is not adaptively important. As Dubinin and Tiniakov (1946) pointed out "human activities create distinct ecological complexes which form an integral part of the environment in which the evolution of the species is being enacted." The observations of these authors have shown that the populations of *D. funebris* that inhabit urban places are distinct with respect to the concentration of inversions from those living in rural localities, and Brncic (1955) has found that in another "domestic" species, *D. immigrans,* there are some relationships between the type of the habitat and the chromosomal polymorphism, as will be discussed further on.

In *D. funebris,* some experimental data have indicated that selection can operate at different stages of the life cycle of the fly. Dubinin and Tiniakov (1946, 1947) have shown that the homozygotes and the heterozygotes for some gene orders exhibit differential longevity at low temperatures. In Chilean stocks of the same species, Brncic and del Solar (1961) have demonstrated that there is a higher frequency of inversion heterozygotes among adult flies than among larvae. Moreover, 100-day-old flies contain a greater frequency of heterozygotes than younger, ten-day-old flies. These last results indicate that there is differential mortality favoring the inversion heterozygotes. The selective pressures that confer a higher fitness to these heterozygotes seem to act both at the preadult and adult stages of life.

Finally, it is important to discuss the situation offered by the Chilean populations of the cosmopolitan species *D. immigrans.* Freire-Maia et al. (1953) have published drawings of the salivary gland chromosomes, as well as some data on the genetic arrangements found in several Brazilian populations of this species. These authors found that about 20 percent of the individuals are heterozygous for an inversion located in the middle part of the left arm of the second chromosome.

In Chile, *D. immigrans* is a very widespread and common species. It has been found breeding on overripe fruit near towns, but also in natural habitats, such as woods, zones of xerophytic bushes, and even under the stringent ecological conditions of the desert. The distribution extends vertically from near sea level to altitudes above 1500 m in the Andes. It seems that *D. immigrans* tolerates low temperatures, since it is one of the commonest species in spring and autumn collections, but rare during the summer. Samples collected in different parts of Chile were analyzed cytologically (Brncic, 1955). Some of them were taken in the north desert regions, such as Azapa (Arica) where the average rainfall is under 0.1

mm per year. At La Serena, Paihuano, and Rapel in the north central part of Chile it rains about 100 to 120 mm per year; at Bellavista and San Vicente in the central part, around 300 to 600 mm. In the southern rainforests, Valdivia, Ensenada, Puerto Varas, and Puerto Montt, the rainfall amounts to about 2000 mm yearly.

No inversions have been found in the first (X), third, or fourth chromosomes of the Chilean populations of *D. immigrans*. In the left arm of the second chromosome one inversion has been observed, and in the right arm two more were seen. The inversion in the middle part of the left arm was designated as *Inversion A* and is the only inversion reported by Freire-Maia in Brazilian populations. The gene arrangement *B* gives a relatively small subterminal inversion at the distal end of the right arm of the second chromosome. The small inversion *C* lies in the proximal part of the right arm of the same euchromatic strand. Pictures of these three inversions have been given in the above-mentioned publication (Brncic, 1955).

Inversion B is the only one, besides the "Standard" gene arrangement, that is distributed throughout Chile. *Inversion A* is restricted to southern Chile from Bellavista to Puerto Montt. *Inversion C* seems to be endemic in the population of Valdivia. The quantitative data indicated that the mean frequency of heterozygous inversions per individual in the Chilean populations was about 0.20, exactly the same as found in Brazil by Freire-Maia et al. (1953). However, only one kind of inversion is found in Brazil, while in Chile three inversions are found. In Chile, the widely distributed *Inversion B* at the heterozygous state has an average frequency of 0.13 per larva. Homozygotes for the "Standard" gene arrangement are most frequent in northern Chile, while the gene arrangement *B* seems to be most abundant, or at least as abundant as "Standard" in the southern part of the country. The arrangement *C* alone, or in combination with *B*, is frequent at Valdivia.

Dobzhansky et al. (1950) and DaCunha et al. (1953) have shown that in several species of *Drosophila* there is a correlation between the amount of the chromosomal polymorphism and the diversity of the habitats in which a population lives. The greater the ecological versatility of a population, the greater the chromosomal polymorphism, and vice versa. The distribution of the gene arrangements in *D. immigrans* is, then, very instructive. In the populations from northern Chile, which live under the stringent ecological conditions of the desert, only the "Standard"- and *B*-gene arrangements are found. On the other hand, in the south, where the species occupies the rich habitat of the rainforest, there are at least three different genetic arrangements beside "Standard." It should be pointed out that *D. immigrans* is not a strictly domestic species like *D. melanogaster*.

The above facts clearly indicate the existence of racial differences between the different Chilean populations of *D. immigrans*, and between the Chilean populations and those from other parts of South America. These racial differences are evidently in response to the ecological conditions existing in each place, and are maintained by the geographic isolation between the populations.

Endemic and Ecologically Restricted Species

Within the genus *Drosophila* there are species that are geographically very localized and ecologically highly specialized. Their distribution and abundance depend on the abundance and distribution of the ecological niche that they exploit. Typical representatives of such species are those which are found associated with a single plant host. Since these species are rarely or never collected by net sweeping over fallen plant parts or in the usual fruit-baited traps, their discovery is frequently accidental. Examples of this type of species in Chile are *D. flavopilosa* Frey (Wheeler et al., 1962), closely associated with *Cestrum parqui* L'Her (Solanaceae); *D. allei* Brncic 1962, which until now was found associated only with *Datura arbustiva;* and *D. appendiculata* Malloch, associated with *Chusquea* (Bambuseae). From an experimental point of view, the main characteristic of all these species is that it is not possible to breed them in the laboratory with the usual methods employed for cosmopolitan species. Their maintenance requires particular procedures for each, which are frequently so tedious that the species are discarded as experimental material for genetics or cytology. This feature is a consequence of their ecological restriction.

Although their association with any known plant has not yet been discovered, there exist in Chile other species that could be included within the group of ecologically restricted ones, due to their scarcity, to the fact that they are not attracted by fermented-fruit baits, and that they do not mate or oviposit under laboratory conditions. Some of these species are: *D. amplipennis* Malloch, which lives in the rainforests of southern Chile and has occasionally been found in the central zone, near the coast, and *D. huilliche* Brncic and *D. osornina* Brncic, both very rare and also found in the southern part of the country.

The study of the genetics of these ecologically and geographically restricted species is important for the knowledge of the complexities of the evolutionary process in the genus *Drosophila*, since they represent the most common kinds of species. Stone et al. (1960) estimate the probable number of living species of the genus *Drosophila* to be about 1500 to 2000. Most of these species, over 95 percent, are rare in that they are geographically very localized and ecologically highly specialized. On the other hand, since the breeding and feeding sites of some of these species are relatively well known, sometimes measurable both qualitatively and quantitatively, it may be possible to establish certain interrelationships between their genetic and their ecological structure. The joint project of the Universities of Hawaii and Texas on the evolution and genetics of the Drosophilidae living in Hawaii, where of the 400 species of the family 18 are immigrant and 383 are endemic and ecologically restricted (Hardy, 1965, 1966), represents the most valuable work under way in this field. Of the ecologically restricted Chilean species, the genetic structure of only one has been studied in some detail due to certain peculiarities of its biology. This species is *D. flavopilosa*.

Evolutionary Biology of Drosophila flavopilosa

Drosophila flavopilosa Frey belongs to the *flavopilosa* group of the subgenus *Drosophila* (*Drosophila*), established by Wheeler et al. (1962) for a group of 14 neotropical species. *Drosophila flavopilosa*, the type species of the group, has been recorded for Chile, Argentina, Uruguay, Bolivia, and Peru. In Chile, its exclusive natural breeding sites are the flowers of the solanaceous shrub *Cestrum parqui* L'Her. In Argentina, it was also found associated with *Cestrum euanthes* Schlecht. All members of the *flavopilosa* group seem to be flower-breeding species. Larvae of two other species of the group have been found by D. A. Hunter and the present author to be living in flowers of *Cestrum tomentosum* Sandwith, near Bogotá (Colombia). Pipkin et al. (1966) have found other species of the *flavopilosa* group breeding in flowers of *Solanum rubidum, Aphelandra micans, Calathea allouia,* and *Heliconia* sp., all growing at 800 m in Panama.

In Chile, *Cestrum parqui* is very abundant from sea level to an altitude of about 1600 m. In the central part of the country the flowering period of the plant extends from early spring to late autumn, but it is possible to find a few flowers in particularly protected places during wintertime. *D. flavopilosa* females lay their eggs in the lower part of the flowers, probably perforating the corolla tube with their strong, spined ovipositor. Larvae develop inside the flowers and are pollen feeders, while pupae can be seen fixed on the inner walls of the corolla tube. Very seldom does a flower contain more than one preadult; the reasons for this are unknown. Although nothing is known about the feeding habits of the adults, it is probable that they also depend on the same plant, for both females and males have been observed fluttering over the flowers, and they have never been attracted by the different kinds of baits utilized for collecting other species of *Drosophila*. The whole developmental cycle of the fly, from egg to adult, takes about 20 to 21 days.

The number of *Cestrum* flowers containing preadult forms varies according to the season and geographic zone. Early in spring, at the beginning of the flowering period, the preadults are very scarce inside the flowers, but they increase quickly, reaching a maximum in summer. There are occasions in which more than 80 percent of the flowers were found to be parasitized by the flies. Until now in no place have *Cestrum* plants been found without *D. flavopilsoa*. Therefore the distributional area of the fly coincides very strictly with that of its plant host.

So far it has not been possible to breed *D. flavopilosa* on the standard *Drosophila* food medium, but the chromosomal structure of natural populations of the fly has been analyzed by observing the salivary gland cells of larvae directly taken from their natural breeding sites. These studies have shown that natural populations of the fly are polymorphic for the gene arrangements in their chromosomes, due to the presence of four independent inversions, all of which are located in the right arm of the fifth chromosome. The descriptions and frequencies of the different gene arrangements in various populations have been given in previous papers (Brncic, 1962b, 1966). Quantitative analysis has shown that some of the inversions are relatively rare: *Inversion C* was found in low frequency in samples from every locality, while *Inversion D* appears only in certain places. The other two gene arrangements (*Inversions A* and *B*) are abundant and widely distributed. The comparison of different populations has shown that there are geographic variations of the frequencies of some of the chromosomal orders, and that the *A* and *B* heterokaryotypes exhibit altitudinal gradients and seasonal fluctuations. Table II shows some of these variations in certain selected zones of central Chile. Heterozygotes for

TABLE II

Frequencies (percent) of Inversions Found in Heterozygous Condition in the V-R Chromosome of *Drosophila flavopilosa* in Five Populations in the Maipo Valley (pooled data from several collections made from 1962 to 1964) (from Brncic, 1966)

Heterozygous inversions	El Tabo 6 m	Vizcachas 900 m	San Alfonso 1200 m	Queltehues 1400 m	El Volcán 1800 m
none	56.3 ± 1.6	49.0 ± 1.5	51.1 ± 3.8	50.6 ± 1.5	54.3 ± 2.3
A	10.2 ± 0.9	35.8 ± 1.4	37.0 ± 3.7	40.3 ± 1.4	41.4 ± 2.3
B	27.4 ± 1.4	10.8 ± 0.9	9.4 ± 2.2	6.2 ± 0.5	2.7 ± 0.7
C	2.9 ± 0.9	2.4 ± 0.4	1.7 ± 1.0	0.9 ± 0.3	0.6 ± 0.3
D	0.5 ± 0.2	0.4 ± 0.2	–	0.2 ± 0.1	0.2 ± 0.2
A + B	1.0 ± 0.3	1.1 ± 0.3	0.5 ± 0.5	1.3 ± 0.3	0.2 ± 0.2
A + C	–	–	–	0.1 ± 0.1	0.4 ± 0.3
B + C	0.8 ± 0.3	0.2 ± 0.1	–	0.1 ± 0.1	–
B + D	0.2 ± 0.1	0.1 ± 0.1	–	0.1 ± 0.1	–
Larvae tested	969	1052	170	1089	466

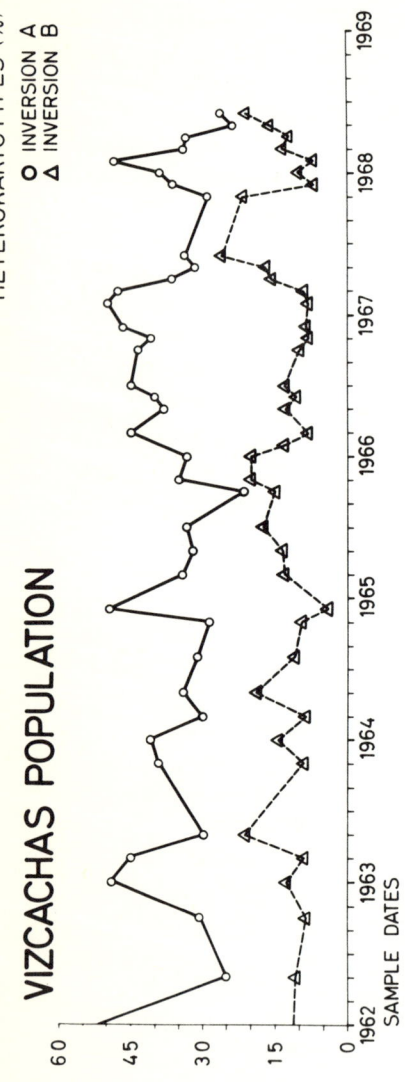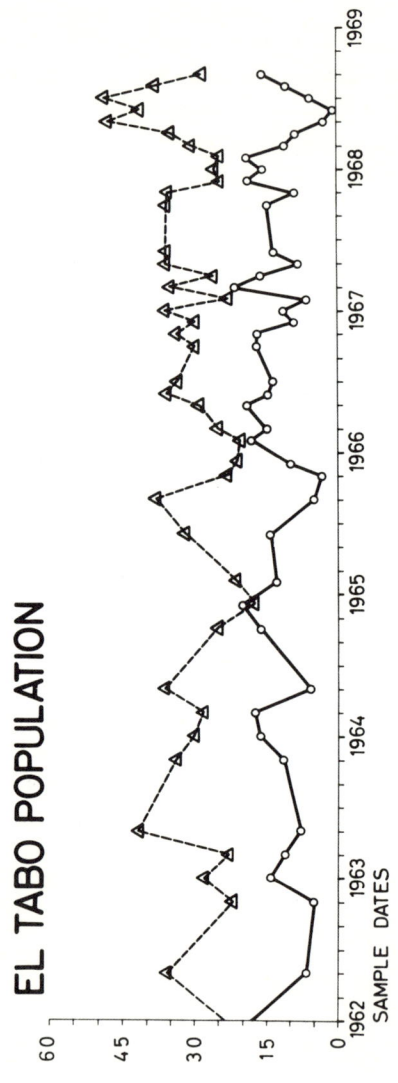

Fig. 1. Fluctuations in the frequencies of heterokaryotypes for inversion A and Inversion B from 1962 to 1968 in two natural populations of D. flavopilosa.

Inversion A are more frequent at high altitudes than at sea level. However, heterozygotes for *Inversion B* are abundant at sea level, but practically disappear at high altitudes. Figure 1, which summarizes the data collected from the summer of 1962 until the autumn of 1968 in two populations near Santiago (El Tabo at sea level, Lat. 33°31′ S., and Vizcachas, Lat. 33°37′ S) at 900 m, shows that the frequency of *A* heterokaryotypes increases during the summer months (December, January, and February) and tends to decrease in autumn and winter. An inverse relationship exists with respect to the *B* heterokaryotypes. The variations in the *B* arrangement are more noteworthy in the El Tabo population, where its frequency is higher.

It is of interest to comment that, recently, samples of *D. flavopilosa* larvae from three localities in Argentina (Mendoza, San Luis, and Buenos Aires) were examined. In these populations none of the Chilean inversions could be identified, but three new rearrangements in chromosome 5 were found. This finding is important, since it indicates that the Andes mountains constitute a real barrier, and that Chilean *D. flavopilosa* is a subspecies representing the populations on the western side of the Andes.

As a result of the field work on the biology of *D. flavopilosa* in Chile and of subsequent analyses of the information gained, a number of interesting facts about the ecology, relevant to the understanding of the population dynamics of the fly, have emerged. Counting the number of flowers of *C. parqui* containing preadult forms of the fly in different geographic regions, an estimation was made of the abundance of the species in relation to climate (temperature, humidity, and rainfall), available food, competitors, predators, and parasites. A species of Thysanoptera, *Thrips tabaci* (competitor), two species of parasitic hymenopterans, *Opius trimaculatus* Spinola (Braconide°), and a cynipid of the genus *Ganaspis,* seem to constitute, together with climate, the main regulating factors of *D. flavopilosa* population density (Brncic, 1966).

The two species of parasitic wasps, *Opius trimaculatus* Spinola (Braconidae) and a *Ganaspis* sp. (Cynipidae), constitute the most conspicuous parasites of *D. flavopilosa* in all the zones investigated. Parasitic hymenopterans lay their eggs in the very small larvae. These eggs develop mainly during the pupal stage of the fly, the pupae being devoured completely. The total developmental cycle of the parasite exceeds that of the flies by several days (about 30 days for *O. trimaculatus* and 40 to 50 days for *Ganaspis* in pupae maintained under laboratory conditions). From 1963 to the present, in each collection of flowers of *C. parqui* made in different periods of the year, about 100 pupae were left to develop until the emergence of *D. flavopilosa* adults as well as of wasps. By this method it was possible to determine the number of deaths due to parasites. There are great differences between one population and another, and the percentage of wasp parasitism also exhibits seasonal fluctuations. In El Tabo and Vizcachas populations near Santiago the fraction of pupae that are parasitized by wasps, tends to increase during the summer months, reaching values of over 60 percent. An attempt was made to establish some correlations between the seasonal variations of A and B heterokaryotypes and the fluctuations of the percentages of parasitized individuals (Brncic, 1966, 1967). A heterozygotes tend to increase with the increase of parasitism.

It is known that some climatic components determine variations of the genetic make-up under laboratory conditions in many species of *Drosophila* (references in Dobzhansky, 1951; daCunha, 1960; Spiess, 1962). Population density can also induce changes in the genetic composition (Dobzhansky and Spassky, 1944; Lewontin and Matsuo, 1963). Furthermore, as Birch (1960) has pointed out, there are selective differences between genotypes that may be a function of density. In uncrowded populations of *Drosophila pseudoobscura,* the inversion *Chiricahua* is favored over "Standard," but when larvae are crowded at a high density, there is a selective mortality favoring the "Standard" arrangement over *Chiricahua* (Birch, 1955). These observations may help to explain the causes of the fluctuations in the number of individuals and in the frequencies of certain genotypes in natural populations. Furthermore, they indicate that changes in genotypic composition may be of importance in maintaining the population size within certain critical values. Parasites may also influence the genetic make-up of populations. There are some indications that wasp parasitism in experimental populations of *D. melanogaster* may act as a selective agent in developing certain genetic resistance to the parasites. On the other hand, populations of wasps are capable of developing genetic resistance to the defense reaction of the host. The experiments of Walker (1959, 1961, 1962) on *D. melanogaster* and the parasitic wasp *Pseudeucoila bochei* Weld represent a good example of such an interaction. Of the same sort are the results reported by Pimentel (1961) and Pimentel et al. (1963), who found in laboratory populations of *Musca domestica* and their parasitic hymenopteran *Nasonia vitripennis* that each species acts as selective pressure for the other, modifying the genetic composition of the respective population.

A better understanding of the mechanisms and causal connections between the aforementioned factors acting in $D.$ *flavopilosa* was obtained by means of some experiments carried out under controlled laboratory conditions. Some of these experiments lead to the conclusion that the temperature at which larvae develop represents a critical factor for the viability of the A and B heterokaryotypes in *D. flavopilosa* (Brncic, 1968). A heterokaryotypes seem to be favored at 25°C, while B heterokaryotypes seem advantageous at 16°C (Table III). These data are in accord with

TABLE III

Influence of Temperature on Chromosomal Polymorphism of *D. flavopilosa* Larvae (adapted from Brncic, 1968)

Temperature	No. of individuals	Heterokaryotypes A	B	others	Homokaryotypes
El Tabo samples					
16°C	400	36	156	16	192
25°C	400	66	106	22	206
			$\chi^2 = 19.80$		P (d.f. 3) <0.001
Vizcachas samples					
16°C	400	134	60	12	194
25°C	400	305	49	8	172
			$\chi^2 = 7.72$		P (d.f. 3) 0.05-0.07

what is observed in the natural populations of the species, where the *A* inversion is more abundant during the summer months, while the *B* inversion tends to increase during winter (Brncic, 1966). As a result of the increase of *A* concomitant to the decrease of *B* at 25°C, and the reverse tendency at 16°C, the total heterozygosity for inversions is maintained at a more or less constant level. The data also concur with the altitudinal gradients found in the distribution of the inversions observed in natural populations. El Tabo, at sea level, has a lower mean temperature than Vizcachas during summer and autumn, when the samples for the experiments were collected. The El Tabo population has a much lower frequency of *A* heterokaryotypes than Vizcachas, and the reverse occurs for the *B* inversion. In *D. flavopilosa,* as in most polymorphic species of *Drosophila,* the different inversions may have become coadapted to each other in the course of evolution. Therefore, changes in the environment may determine modifications of the entire genetic make-up of the population and not only of a single inversion.

Chromosomal polymorphism in certain *Drosophila* species is very sensitive to relatively small environmental changes. Wright and Dobzhansky (1946) showed, under laboratory conditions, that in *D. pseudoobscura* the adaptive values of the inversions depended, among other factors, on temperature. At 25°C, the inversion heterozygotes were superior, but at 16°C heterozygotes and homozygotes were equal. In *D. persimilis,* Spiess (1950) has demonstrated a reverse situation. In *D. funebris* (Dubinin and Tiniakov, 1947), and in *D. robusta* (Levitan, 1951), the different gene arrangements were also temperature-sensitive. None of these studies indicates the stage of the life cycle at which temperature affects the adaptive value of the karyotypes. Only Hartmann-Goldstein and Sperlich (1963) found that in *D. subobscura* high temperature favored the structural heterozygotes for certain inversions during the preadult stage.

The finding that in *D. flavopilosa* the inversion polymorphism can be modified by breeding the larvae at different temperatures contributes to the understanding of the genetic structure in relation to the population dynamics of this species. The flexible character of the polymorphism in *D. flavopilosa* has led to the conclusion that the different chromosomal types possess specific selective values in relation to some environmental factors.

Endemic and Ecologically Versatile Species

In contrast to the ecologically highly specialized species of *Drosophila*, such as *D. flavopilosa*, found associated with a single plant host, there live in Chile another kind of endemic species that are undoubtedly more versatile, although up to the present their breeding sites are not known exactly. They are attracted by fermented-fruit baits placed in different types of environments, they feed and reproduce easily on the usual *Drosophila* food medium, and can be maintained indefinitely under laboratory conditions. Their endemic character may be deduced from the fact that until now they have been found only in Chile and, exceptionally, in neighboring regions in the south Andean system of mountains. But, although these species are highly localized, they share certain characteristics with the "domestic" species of *Drosophila*, since many of them may be collected by net-sweeping over fallen overripe fruit in orchards, gardens, or other sites closely associated with man, where they coexist with domestic species, such as *melanogaster, immigrans,* or *hydei*. These endemic species living in Chile closely resemble in their habits members of the *obscura* group found in North America and Europe. According to Dobzhansky (1965) these could be species in the process of colonizing new habitats and spreading into man-made environments. As Dobzhansky suggests, "these latter species are the most interesting of all for an eventual understanding of the characteristics of the genetic systems which make a species qualify for a career of a domestic and colonizing species."

The following species found in Chile can be included within this group: *D. camaronensis* Brncic, a highly localized species living only in certain narrow green valleys and oases in the northern deserts; *D. araucana* Brncic, found in small numbers in central and southern Chile; *D. serenensis* Brncic, a member of the *repleta* group of species, which lives in central Chile; and two species of the *mesophragmatica* group: *D. pavani* Brncic and *D. gasici* Brncic.

Although no quantitative investigations have been undertaken on the chromosomal structure of *D. camaronensis* and *D. araucana,* small samples of the offspring of females inseminated in nature revealed them to be monomorphic for the genetic arrangements in their chromosomes. On the other hand, *D. serenensis* is polymorphic, and the small samples of individuals collected in different places were mostly heterozygous for at least three different inversions. *D. pavani* and *D. gasici* belong to the *mesophragmatica* group of species (Brncic and Koref-Santibañez, 1957). These two species have been studied extensively in our laboratory during the last 15 years, and the data obtained are especially valuable for the present discussion.

The *mesophragmatica* group of species was established by Brncic and Koref-Santibañez (1957) in order to include a cluster of South American species that form a phyletic unit. At the present time, the group comprises eight species: *D. mesophragmatica* Duda, 1927; *D. gaucha* Jaeger and Salzano, 1953; *D. pavani* Brncic, 1957; *D. gasici* Brncic, 1957; *D. altiplanica* Brncic and Koref-Santibañez, 1957; *D. orkui* Brncic and Koref-Santibañez, 1957; *D. viracochi* Brncic and Koref-Santibañez, 1957; and *D. brncici* Hunter and Hunter, 1964. With the exception of *D. gaucha,* which lives in southern Brazil, Argentina, and Bolivia, the other species of this group seem to be fundamentally Andean. Its members are abundant in nature, and in some places they constitute the dominant species of the genus. All are very similar in their morphology and some of them, such as *D. pavani* and *D. gaucha,* are considered sibling forms. The eight species have varying degrees of reproductive isolation, and when hybrids are produced, they are completely sterile. In spite of the fact that some of these species are sympatric in certain regions, there is good chromosomal and genetic evidence to show that no gene exchange takes place in the natural populations (Brncic and Koref-Santibañez, 1957; Koref-Santibañez, 1963, 1964). The evolution of the chromosomal pattern in the *mesophragmatica* group has been given in previous papers (Brncic and Koref-Santibañez, 1957; Brncic, 1957b, 1957c).

Evolutionary Biology of Drosophila pavani

Drosophila pavani lives in central Chile (between Lat. 27° and 39° S) and on the eastern slope of the Andes in Argentina (San Luis, Long. 66° W). It is a chromosomally polymorphic species and the inversions are concentrated in the second chromosome and in both arms of the fourth chromosome. In the second chromosome two inversions have been discovered, one of them included in the other (*A* and *B* inversions). These two inversions have never been found separately. Both are present in all Chilean and Argentinian populations studied. The frequency of heterozygotes for these inversions ranges from 38 to 56 percent in the different populations. In all samples analyzed there are also three overlapping inversions in the right limb of the fourth chromosome (*Inversion IVR; A+B+C*); the heterozygotes for this condition occur in frequencies that range from 46 to 60 percent in different populations. Finally, in the left arm of this same chromosome, most populations studied contain heterozygotes for another complex gene arrangement which differs from "Standard" by three overlapping inversions (*Inversion IVL; A+B+C*). The frequency of heterozygotes for this gene sequence ranges from 44 to 62 percent. In none of the populations, and in none of the crosses between them, have these inversions in the right arm or those in the left arm been found isolated; they always occur together as complexes.

The presence in nature of complexes of inversions and the apparent lack of separate occurrence of the constituents of these complexes constitute an interesting fact. The number of inversions that must have occurred during the evolutionary history of *D. pavani* must have been rather large. Wallace (1953) provides a possible explanation for the absence in natural populations of the gene arrangements intermediate between the existing terminal links of the inversions series found in many species of *Drosophila*. According to the Wallace hypothesis, the adaptive role of inversions is to maintain intact coadapted complexes of polygenes that give rise to heterosis in heterozygotes. This coadaptation might be lost if gene exchange were to occur, breaking up the coadapted gene complexes. Such a situation is likely to occur if three links in the chain of overlapping inversions (the "triads" of Wallace) were to coexist in the same population. Crossing-over then might result in transfer of blocks of genes from one chromosome with one gene arrangement to another chromosome with a different arrangement, and this might lead to disintegration of the coadapted gene systems. As Wallace has pointed out, this mechanism might explain the fact that in many populations of polymorphic species of *Drosophila* only two members of the "triads" are found with high frequencies. The selective pressures that control the frequency and the distribution of the chromosomal orders would, according to Wallace, maintain the adaptive gene combinations guarded by the inversions.

After 15 years of cytogenetic studies, both in natural populations as well as in laboratory stocks of *D. pavani*, it has been possible to establish that there are no appreciable geographic, seasonal, altitudinal, or other environmental fluctuations of the different chromosomal variants. In all the samples analyzed, the frequency of the structural heterozygotes for the genetic arrangements in the fourth chromosome is always around 50 percent. In the stocks maintained in the laboratory, the different chromosomal arrangements conserve in general the same frequencies found in nature. In other words, *D. pavani* is a good example of a species bearing "rigid" or "stable" polymorphism, according to the definition given by Dobzhansky (1962), in contrast to other species in which polymorphism is of the "flexible" type, where the frequencies of the different arrangements are modified by environmental changes both in nature and under laboratory conditions (references in: Dobzhansky, 1961; Carson, 1965; Sperlich, 1967). These facts have led to the conclusion that the chromosomal polymorphism actually present in *D. pavani* represents the end-product of a long and continuous selective process, that has mainly favored the heterokaryotypes in most of the environments exploited by the species ("hetero-selection" according to Carson, 1959). As a result, the heterokaryotypes may be expected to be heterotic, that is, superior in adaptive value.

In searching for the heterotic properties of the heterozygotes for inversions in *D. pavani*, it was found that they exhibit a greater longevity, a superior mating activity, and a faster rate of development. Brncic and del Solar (1961) have shown that the frequency of heterokaryotypes with respect to some gene orders is significantly higher in 100-day-old flies than in larvae, or in young 10-day-old adults. This finding suggests that the selective pressures that confer a higher fitness to these heterokaryotypes may act both during the preadult and the adult life stages (Fig. 2). Brncic and Koref-Santibañez (1964) and Koref-Santibañez and Brncic (1965) studied the mating activity in two very heterozygous Chilean populations, one from Copiapó in northern Chile, the other from Bellavista in the

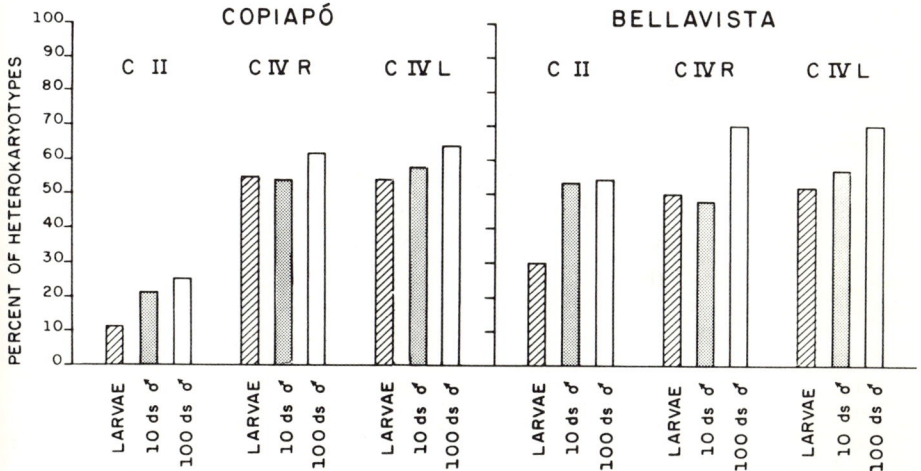

Fig. 2. Frequencies (in percent) of larvae (10-day-old males and 100-day-old males) that were heterozygous for the gene arrangements in the second chromosome (C-II), and in the right and left arm of the fourth chromosome (C-IV-R and C-IV-L) in two stocks of D. pavani (Copiapó and Bellavista). (After Brncic and del Solar, 1961)

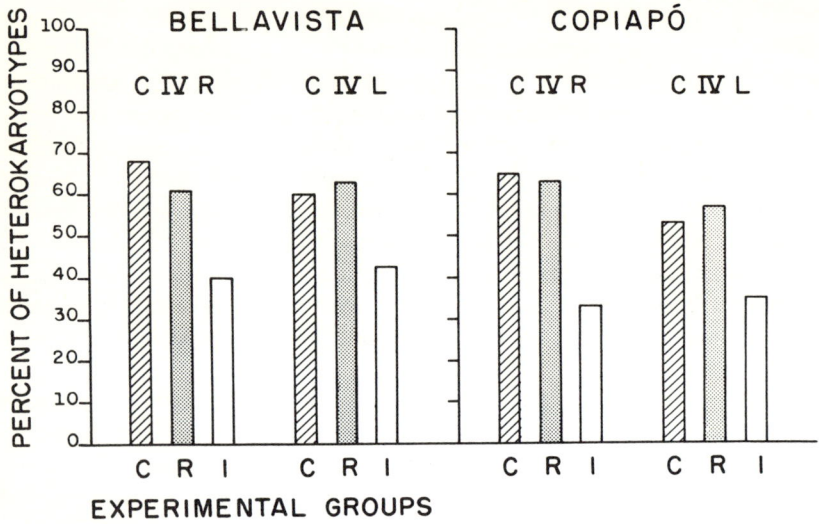

Fig. 3. Frequencies (in percent) of D. pavani males (Bellavista and Copiapó stocks) that were heterozygous for the gene arrangements in the right and left arms of the fourth chromosome, among groups which copulate (C), court (R), or are inactive (I) during the first 30-minute observation periods after being confronted with virgin females. (After Brncic and Koref-Santibañez, 1964)

central region. The results indicated that the frequency of heterokaryotypes was significantly higher among the males and females that mated and courted within a period of 30 minutes than in those which remained inactive during the time of observation (Fig. 3). Brncic, Koref-Santibañez, Budnik, and Lamborot (1969) studied the rate of development as another physiological property of the heterokaryotypes in *D. pavani,* which may also be a good indicator of heterosis. In these experiments there was a significantly higher number of heterokaryotypes for the gene order in the fourth chromosome among the flies that developed rapidly (30 to 35 days) in relation to those which took longer (44 to 65 days; Fig. 4). The fact that the structural heterozygous individuals survive longer, mate more, and have a faster rate of development, indicates that they have a greater reproductive efficiency.

Since the pioneer work of Wright and Dobzhansky (1946), it is known that chromosomal polymorphism in *Drosophila* can be maintained by the superior fitness of the structural heterozygotes. Following the classic experiments in *D. pseudoobscura,* heterotic properties of the inversion heterozygotes have been reported in many species of the genus, such as *D. melanogaster, D. funebris, D. persimilis, D. robusta, D. willistoni, D. subobscura, D. ananassae, D. pavani, D. flavopilosa,* and others. In most of

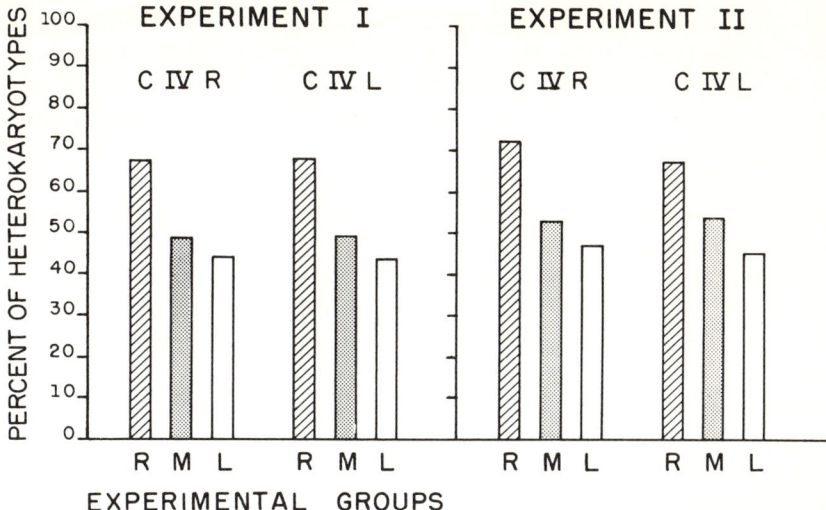

Fig. 4. Frequencies (in percent) of D. pavani males (Bellavista stock) that were heterozygous for the gene arrangements in the right and left arm of the fourth chromosome among groups of rapid (R), medium (M), and slow (L) rate of development. (After Brncic, Koref-Santibañez, Budnik, and Lamborot, 1969)

them, the criteria for heterosis were referred to the superiority in fitness components such as viability, rate of development, longevity, sexual activity, hatchability, and homeostatic adjustment. Complete references of the work done on this subject may be found in the reviews of DaCunha (1960), Dobzhansky (1961), Spiess (1962), and Sperlich (1967). For example, in *D. funebris*, Dubinin and Tiniakov (1946, 1947) have shown that the homozygotes and the heterozygotes for some gene orders exhibit differential longevity at low temperatures. Dobzhansky (1947) and Dobzhansky and Levene (1948) have demonstrated that in some natural as well as in laboratory populations of *D. pseudoobscura* differences in mortality of the carriers of certain gene arrangements occur between the egg and the adult period, and that the frequency of heterozygotes is higher in adults than in egg samples coming from females of the same populations. In *D. persimilis,* Spiess et al. (1952) have shown that the heterozygotes for some gene orders have a greater longevity, and in *D. tropicalis* Dobzhansky and Pavlovsky (1955) have discovered that during the preadult period there is a high mortality of homozygotes for certain gene arrangements. Nevertheless, in other species, such as *D. robusta* (Levitan, 1951), differential survival between the egg and adult period of the carriers of different chromosomal arrangements has not been demonstrated.

Research on sexual behavior in *Drosophila* has shown that the carriers of different genotypes may differ in their mating ability. Bastock (1956),

Petit (1958), Bösiger (1960), and others, found that in *D. melanogaster* some genotypes are more efficient than others in mating. Rendel (1944) and Maynard Smith (1956) have observed similar situations in *D. subobscura*. In *D. persimilis*, Spiess and Langer (1961), while searching for elements of fitness conferred by various homokaryotypes within a population, found significant differences in mating propensity. In several species, namely, *D. melanogaster* (Bösiger, 1960; Hoenigsberg and Koref-Santibañez, 1959) and *D. subobscura* (Maynard Smith, 1956), it has been observed that outbred lines exhibit greater and more efficient mating activity than inbred ones. This would be an expression of the heterotic properties of the heterozygotes in relation to the corresponding homozygotes.

The experiments on *D. pavani* show that the heterokaryotypes for gene arrangements in their fourth chromosome are superior in mating ability to the corresponding homokaryotypes within the same population. They court and mate more rapidly. This has interesting evolutionary implications. Darwinian fitness is a measure of the genetic contribution of the carriers of a genotype to the next generation. Any selective pressure that modifies the courtship or mating efficiency, as Spurway (1955) and others have emphasized, may introduce a nonrandom element into the process of fertilization. If a greater mating ability is a property of a certain genetic constitution, and is correlated with characteristics that confer higher fitness, one may rightly suppose that it would contribute towards the maintenance of well balanced genotypes in the offspring.

The statement that chromosomal polymorphism in *D. pavani* is of "rigid" nature does not imply lack of variation in the genetic contents within the inverted zones of the chromosomes. When individuals of the species coming from different geographic populations are crossed, the frequency of heterokaryotypes decreases drastically after a few generations (Brncic, 1961a; Fig. 5). The adaptive superiority of the heterozygotes that carry chromosomes with different gene arrangements is determined by the polygene complexes contained in the inverted sections of the chromosomes. Natural selection tends to act in such a way as to coadapt the gene complexes present in each Mendelian population. The role of the inversions is to maintain these balanced gene complexes by suppressing recombination.

In *D. pavani*, the drastic decrease of the incidence of heterozygotes for inversions in the mixed populations indicates that the same inversions maintain different balanced gene systems in different populations. This finding is of special significance, since in this species no clear qualitative and quantitative variations of the chromosomal polymorphism in different Mendelian populations has been found. Nevertheless, an opportunity for evolutionary divergence at the mutational level is preserved. As a consequence, each geographic race of *D. pavani* has acquired, under the selective pressures, its own genetic endowment. Dobzhansky (1957) and Dobzhansky and Pavlovsky (1958) have also found that in *D. paulistorum* and *D. willistoni* the fitness of homozygotes and heterozygotes for inversions depends on the geographic origin of the chromosomes they carry.

There are other arguments to support the thesis that there are certain racial differences between the different geographic populations of *D. pavani*. After an analysis of courtship and sexual isolation in various populations of *D. pavani*, Koref-Santibañez and del Solar (1961) concluded that in this species variations have arisen that express themselves as a courtship preference of males toward females of their own population, although this preference does not reach the mating level. This is specially clear in the behavior of individual populations: the more distant ones, as for instance Copiapó, in the northern part of Chile, and Chillán, in the south, separated by strong mountain barriers, show the highest discrimination. Therefore, these populations, geographically isolated, with different genetic constitutions, have evolved, together with the processes of divergence of their genetic systems, incipient barriers to genetic exchange.

Besides the aforementioned facts, there is a third type of evidence that suggests racial differentiation in *D. pavani*. Brncic (1961b, 1969) has observed that in *D. pavani*, although the different geographic populations and laboratory stocks conserve similar frequencies of the same inversions, they may be differentiated in the degree of linkage between the genetic arrangements present in the right and left arms of the fourth chromosome. In some populations these are randomly associated, while in others they are not. Furthermore, there is a direct relationship between the number of genera-

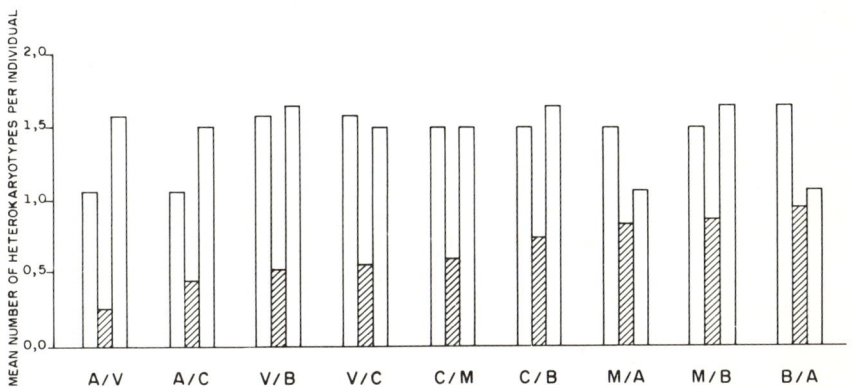

Fig. 5. Mean number of heterokaryotypes found in different populations of D. pavani and in their hybrids after 10 generations. The white columns indicate the parental populations, the hatched ones represent the hybrid populations. (A) Arrayan, (B) Bellavista, (C) Copiapó, (M) Mendoza, (V) Vallenar. (After Brncic, 1961)

tions for which a stock has been cultured and the degree of association of the inversions. In stocks maintained for 10 to 15 years in the laboratory, the tendency of nonrandom association has increased to such an extent that in many of the older stocks the linkage between both "Standard" or both inversion arrangements in the same chromosome is complete, and thus recombinants do not appear.

There are several records of similar nonrandom association of inversions in *Drosophila* (reviewed in Levitan, 1958). Of interest are cases such as *D. robusta* (Carson, 1958; Levitan, 1958), *D. paramelanica* (Stalker, 1960), *D. euronotus* (Stalker, 1964), and *D. subobscura* (Krimbas, 1964), in which there are clear relationships between inversion association and geographic distribution. In other organisms, such as the midge *Chironomus intertinctus* (Martin, 1965), there is also evidence of nonrandom association of some of the gene arrangements in certain populations. In areas with higher rainfall the association seems to be advantageous. In *D. pavani* it seems probable that the main factor involved in the establishment of nonrandom association of gene sequences in some natural populations, and in the increase of the tendency in the course of time, under laboratory conditions, should be related to selective pressures. In each geographic locality, natural selection will favor the fittest combination of gene arrangements. The greater homogeneity observed in the data from laboratory stocks, in contrast to the heterogeneity found in different natural populations, may be a result of more intensive and uniform selective pressures. As Fisher (1930) has pointed out, tight linkage is an advantage in all the heterotic genes concerned in balanced polymorphism. Furthermore, the theoretical models involving the interaction of selection and linkage show that changes in linkage relationships can be very fast under some conditions (Kimura, 1956; Bodmer and Parsons, 1962; Lewontin and Hull, 1967).

Summarizing, although the stocks of *D. pavani* maintained in the laboratory for many generations preserve a considerable amount of their chromosomal polymorphism, thus indicating the "rigid" or "stable" nature of the latter, some changes have occurred both in the number and frequency of the different gene arrangements (Brncic, 1969), as well as in the linkage relationships between the complex orders in the right and left arms of the fourth chromosome. One can hardly expect a rigid polymorphism to be absolute, as by definition it represents an adaptive condition subject to evolutionary pressures.

Evolutionary Biology of Drosophila gasici

Drosophila gasici was first described on the basis of some individuals collected in 1956 in the Azapa and Camarones valleys in the district of Arica in the northern part of Chile. Since then it has been found in San Luis

(Argentina), Cochabamba (Bolivia), and near Bogotá in Colombia. Like the other members of the *mesophragmatica* group, it seems to be relatively abundant in some localities in the Andes mountain system, such as Bogotá (Hunter and Hunter, 1964). The comparative study of all these populations has disclosed that the species has split into well-defined geographic races. The Colombian and Chilean flies differ from those living in Bolivia and Argentina by three independent inversions in the first chromosome (the sexual pair). The only polymorphic populations seem to be the Chilean ones, which exhibit two inversions in the second chromosome besides the "Standard" gene arrangement. All the other populations are homozygous for all their chromosomal sequences. Studies on reproductive isolation have demonstrated that there is some sexual discrimination of the Colombian and Chilean flies with respect to the Bolivian and Argentinean ones (Brncic and Koref-Santibañez, 1965).

In a comparative study of chromosomal variation in the species of the *mesophragmatica* group (Brncic, 1957c), it was observed that in addition to *D. gasici* three other species exhibit polymorphism with respect to their gene sequences. In *D. pavani* the inversions are concentrated in the second chromosome and in both arms of the fourth chromosome; *D. mesophragmatica* is polymorphic with respect to the gene orders in the third and in the left arm of the fourth chromosome; *D. orkui* has inversions in the second and third chromosomes. Nevertheless, there are some interesting differences in the polymorphism in *D. gasici* as compared to that shown by the other species of the group.

According to Brncic (1957b, 1957c, 1958), in *D. pavani*, in *D. mesophragmatica*, and probably also in the less-investigated *D. orkui*, there is little evidence for qualitative and quantitative geographic variations of the different gene sequences. In the better studied *D. pavani* there is no evidence either for geographic, seasonal, or altitudinal gradients in the distribution of the chromosomal arrangements. In this respect polymorphism in these three species is rather rigid. In contrast, in *D. gasici* the cytological structure exhibits very clear populational differences.

The existence of variability in the gene sequences in the sex chromosomes gives *D. gasici* a rather distinctive position within the *mesophragmatica* group. It is known that changes due to inversions in this chromosome are important in the phylogenetic history of the group (Brncic and Koref-Santibañez, 1957). The crosses between *D. pavani* and its sibling species *D. gaucha* indicate that the only chromosomal differences are given by two overlapping inversions in the X-chromosome. The analysis of F_1 hybrids between *D. mesophragmatica* and *D. pavani* or *D. gaucha* shows that changes related to speciation have involved many paracentric inversions in this same chromosome. The differences between three independent inversions in the X-chromosome of the Chilean and Colombian populations of *D. gasici* as contrasted with the Bolivian and Argentinean ones could be interpreted as an important evolutionary step that has given rise to well-defined geographic races. If it is kept in mind that each inversion has appeared independently in time, the homozygous condition for the inversions found in the Bolivian and Argentinean flies probably represents an end-product of a long evolutionary process in which the intermediate chromosomal arrangements, differing from the "Standard" by only one or two of the inversions, have been lost.

Summarizing, the populational structure of *D. gasici* indicates that the species has split into several isolates, each of which has acquired a well-defined genetic endowment, revealed both by differences in their cytological structure and in their sexual behavior. The establishment of these geographic isolates is a consequence of the distribution area of the species. *D. gasici* is essentially Andean, and the discontinuity of the Andes mountain system is specially favorable for the formation of local races, prevented from free gene exchange with other populations of the species.

Isolation and Chromosomal Structure

In the preceding pages an analysis was made of the chromosomal structure of the Chilean populations of different kinds of species of the genus *Drosophila*: widespread, localized, ecologically restricted, and ecologically versatile ones. An attempt was also made to compare the structure of Chilean populations to that exhibited by other populations of the same species, or of closely related species from other parts of the world. The question arises whether geographic barriers, such as the Andes mountains or the northern desert, which separate Chile from the rest of South America, really constitute isolating barriers for the species of *Drosophila* that could be revealed by differences in their genetic make-up. The answer is evidently affirmative. Among the group of cosmopolitan species *D. immigrans* probably represents the best example, since its populations are different cytogenetically from the populations of other parts of the neotropical region, such as Brazil (Freire-Maia et al., 1953). In other wide-ranging but not cosmopolitan species, such as *D. nigricruria* and *D. mercatorum,* some evidence also exists that could indicate a certain degree of genetic divergence of the Chilean populations with respect to those from other parts. The only local and ecologically specialized species that has been studied with intensity, *D. flavopilosa,* clearly shows that the Chilean populations differ cytogenetically from the Argentinean ones. Among the endemic and ecologically versatile species the Chilean populations of species of the *mesophragmatica* group, particularly *D. gasici,* represent a good example of local races.

Nevertheless, there are no good arguments to indicate that the geographic barriers are a major influence on the genetic structure of certain cosmopolitan species such as *D. melanogaster, D. hydei, D. busckii,* or *D. funebris.* It is remarkable to note that in this latter the same inversions found in the north of Europe are also present in Chile. The situation of these species must be analyzed within the context of domestic species, whose distribution and abundance strictly depends on the activities of man. They are species that are transported from one place to another together with fruits and vegetables, and their populations must suffer periodic, massive invasions of new colonizers. In consequence, it is not to be expected that the local populations of different parts of the world should differ greatly in their genotypes. In the same sense, *D. ananassae,* found only twice in Chile on imported bananas and pineapples, may represent an occasional immigrant that does not form colonies; *D. melanogaster* and other domestic species normally found on imported fruits may also be interpreted as recent colonizers whose genotypes can be mixed with the local populations.

The second type of facts that may be of interest to discuss in relation to the different kinds of species living in Chile bears reference to whether it could be possible to establish certain correlations between their endemic or widespread nature and their chromosomal structure. It is noteworthy that in all the groups of species, be they cosmopolitan, widespread, endemic and ecologically restricted, or ecologically versatile, there exist chromosomally monomorphic and chromosomally polymorphic forms. In the group of cosmopolitan and widespread species, six out of eleven analyzed are polymorphic. Among the endemic species analyzed, considering both the ecologically restricted as well as the versatile ones, four out of six prove to be polymorphic. In none of the Chilean species has such an exuberant polymorphism been found as exists in certain tropical species as *D. willistoni* and its relatives (daCunha, 1960) or in some widespread neoarctic forms as *D. pseudoobscura* (Dobzhansky, 1951) or *D. robusta* (Carson, 1958). The Chilean species that apparently shows the most abundant polymorphism is *D. pavani,* with eight inversions. Nevertheless, due to the absence of the intermediate steps in the complex rearrangements found in this species, this polymorphism is reduced to only one alternative besides "Standard" for the second chromosome, and to two besides "Standard" for the fourth chromosome.

Dobzhansky, Burla, and daCunha (1950), and daCunha et al. (1953, 1954) suggested that restricted species are likely to be ecologically less versatile, and therefore may be expected to be less polymorphic in comparison with the more versatile and wide-ranging species. These same authors explain the greater amount of chromosomal polymorphism in the center of distribution of widespread species, as compared with that found in marginal populations of the same or of closely related species, in the sense that in the center of the range the species are able to dispose of a greater number of ecological niches, which may be occupied by the different adaptive alternatives offered by the polymorphic condition. Carson (1955) offers an alternative hypothesis. For him, the difference in the frequencies of gene arrangements between the central and marginal populations is due to selection for increased or decreased amount of recombination. In the center of the distribution range, inversions are favored because they contribute to maintain linked complexes of internally balanced genes that are usually heterotic in heterozygous state. But, in the distributional margin, where the ecological resources are less diversified, and suffer fluctuations that may sometimes be rather drastic, a smaller number of inversions would be advantageous, since it allows the release of the genetic variability stored by the inversions, thus giving a possibility for new potentially adaptive genetic combinations.

The lower frequency of inversions in marginal populations has been observed in widespread species such as *D. willistoni,* in the tropics (DaCunha and Dobzhansky, 1954), and *D. robusta* in North America (Carson, 1955), but not in other species, such as *D. subobscura* (Goldschmidt, 1956). Mayr (1963) pointed out that isolation seems to be a more important factor for the reduction of genetic variability in the marginal populations, due to the reduction of gene flow from the center of the range. In the isolated marginal populations there may also be a more severe selection pressure that permits only a limited number of genotypes to survive.

Summarizing, it may be affirmed that differences are not always found between the frequency of polymorphism observed in marginal and isolated populations, and that observed in the central populations of the same or of closely related species. When these differences are found, they are not always easy to explain. No arguments exist that indicate that they are always due to similar evolutionary factors. The well-studied cases are few, and the exceptions are numerous.

There are species of *Drosophila* in which the polymorphism is "flexible," i.e., there are geographic and seasonal fluctuations of some of the different chromosomal types maintained in most populations. *D. flavopilosa* in Chile, some populations of such well-studied species as *D. pseudoobscura* (Dobzhansky, 1956), and *D. robusta* (Carson, 1958) in North America, *D. willistoni* in tropical America (DaCunha, 1960), and *D. subobscura* (Sperlich, 1961) in the Old World, represent examples of such types. In other species, on the contrary, the polymorphism is more "rigid," i.e., there is no good evidence for geographical, seasonal, or altitudinal fluctuations of the frequency of the gene arrangements, as in *D. pavani* in Chile, and in some populations of *D. robusta* or *D. willistoni* (Dobzhansky, 1962). The reasons why the polymorphism for gene sequences in *Drosophila* may be more "rigid" or more "flexible" are not well known; in general, this may depend on the selective values of the genes maintained by the polymorphic conditions within each population. Selection favoring homo- or heterokaryotypes according to the environment will give a "flexible" polymorphism. But, if selection favors the heterokaryotypes in most of the environments mastered by the species or populations, a more "rigid" polymorphism will be established ("hetero-selection" according to Carson, 1959). It is difficult to try to correlate "flexible" or "rigid" polymorphism with the natural history and the ecology of the species, because of the lack of precise information about the breeding and feeding sites of the majority of the members of the genus *Drosophila*. Nevertheless, some efforts have been made. Carson (1965), among others, has studied the genetic characteristics of widespread species of *Drosophila*, comparing them with endemic ones. The data presented by this author tend to indicate that, while many endemic species of the genus show chromosomal polymorphism of the "flexible" type (that is, resembling in pattern that of *D. flavopilosa* in Chile), only some of the widespread species show this type of variability. In these latter species most of the chromosomal polymorphisms are of a "rigid" character. It is important to stress that fact that the widespread or endemic character of a species, in absence of geographical barriers, is a function of the abundance and distribution of the ecological resources that the species may utilize, but is not a function of the number of different ecological resources exploited. The existence of endemism may be the expression of ecological restriction, but a wide distribution does not always need to be related to ecological versatility. This is the case for the wide-ranging species *D. buzzatii*, narrowly associated with the cactus genus *Opuntia* (Patterson and Stone, 1952), and probably for some of the domestic species of the genus strictly adjusted to some human-made habitat. In general terms, it is useful to consider that the genetic constitution of species or populations that are ecologically more versatile, i.e., the ones that may utilize many qualitatively different foods and breeding sites is different from that exhibited by species that are narrowly adjusted to few, or sometimes to a unique ecological resource. Ecological versatility may depend on genetic constitutions that confer greater phenotypic plasticity, i.e., a capacity to react to environmental changes by individual adaptability, and not by genetic specialization (Carson, 1965). Ecological restriction, on the contrary, means species having genotypes that confer less phenotypic plasticity. For their survival, the latter type of species must respond by changing their genetic make-up. A chromosomal polymorphic system of the "flexible" type, such as that observed in *D. flavopilosa*, represents, in this sense, a mechanism of genetic response to the modifications of the unique ecological niche occupied by the species (cyclic changes of climate, parasitism, predation, competition). This may be the reason why endemic species of *Drosophila* (if endemism is due to ecological restriction) exhibit "flexible" polymorphism. Consequently, in widespread species, only in those which are restricted to a few ecological niches, although they may be quantitatively well distributed and abundant, can a "flexible" polymorphism be expected.

It is also possible to establish other obvious relationships between chromosomal polymorphism and the natural history of the species. Some observed variations of the polymorphism, such as seasonal fluctuations, may be observed only in species that have rapid developmental cycles and reproduce over many months of the year. *D. funebris* and *D. flavopilosa* represent good examples of this type of species. On the other hand, those species of ubiquitous polymorphism may be present only in some parts of the year: for example, *D. pavani* (Brncic, 1957b) among the endemic species living in Chile.

Acknowledgments

The author wishes to acknowledge the many fruitful discussions about the manuscript with Dr. Susi Koref-Santibañez of the University of Chile, and Dr. Antonio Brito DaCunha of the University of São Paulo, Brazil; and the very able assistance of his secretary Mrs. Eliana Rojas de Arce.

References

BASDEN, E. B. 1956. Drosophilidae (Diptera) within the Arctic circle. Trans. Roy. Entom. Soc. London, 108:1-20.
BASTOCK, M. 1956. A gene mutation which changes a behavior pattern. Evolution, 10:421-439.
BODMER, W. F., and P. A. PARSONS. 1962. Linkage and recombination in evolution. Advances Genet., 11:1-100.
BIRCH, L. C. 1955. Selection in *Drosophila pseudoobscura* in relation to crowding. Evolution, 9:389-399.
——— 1960. The genetic factor in population ecology. Amer. Natural., 94:5-24.
BÖSIGER, E. 1960. Sur le role de la séléction sexuelle dans l'évolution. Experientia, 16:27-273.
BRNCIC, D. 1955. Chromosomal variation in Chilean populations of *Drosophila immigrans*. J. Hered., 46:59-63.
——— 1957a. Las especies chilenas de Drosophilidae. Col. Monografías Biol., Univ. Chile. Santiago (Chile), Imprenta Stanley.
——— 1957b. Chromosomal polymorphism in natural populations of *Drosophila pavani*. Chromosoma, 8:699-708.
——— 1957c. A comparative study of chromosomal variation in species of the *mesophragmatica* group of *Drosophila*. Genetics, 42:798-805.
——— 1958. Evolución en el groupo mesophragmatica del género *Drosophila*. Biológica (Santiago), 26:3-46.
——— 1961a. Integration of the genotype in geographic populations of *Drosophila pavani*. Evolution, 15:92-97.

———— 1961b. Non random association of inversions in *Drosophila pavani*. Genetics, 46:401-406.
———— 1962a. New Chilean species of the genus *Drosophila*. Biológica (Santiago), 33:3-6.
———— 1962b. Chromosomal structure of populations of *Drosophila flavopilosa* studied in larvae collected in their natural breeding sites. Chromosoma, 13:183-195.
———— 1966. Ecological and cytogenetic studies of *Drosophial flavopilosa*, a neotropical species living in *Cestrum* flowers. Evolution, 20:16-29.
———— 1967. Chromosomal polymorphism in an ecologically restricted species of *Drosophila* living in Chile. Ciencia e Cultura (Brazil), 19:45-53.
———— 1968. The effects of temperature on chromosomal polymorphism of *Drosophila flavopilosa* larvae. Genetics, 59:427-432.
———— 1969. Long-term changes in chromosomally polymorphic laboratory stocks of *Drosophila pavani*. Evolution, 23:502-508.
———— and E. DEL SOLAR. 1961. Life cycle and the expression of heterosis in inversion heterozygotes in *Drosophila funebris* and *Drosophila pavani*. Amer. Natural., 95:211-216.
———— and TH. DOBZHANSKY. 1957. The southernmost Drosophilidae. Amer. Natural., 91:127-128.
———— and S. KOREF-SANTIBAÑEZ. 1957. The *mesophragmatica* group of species of *Drosophila*. Evolution, 11:300-310.
————and S. KOREF-SANTIBAÑEZ. 1964. Mating activity of homo- and heterokaryotypes in *Drosophila pavani*. Genetics, 49:585-591.
———— and S. KOREF-SANTIBAÑEZ. 1965. Geographical variation of chromosomal structure in *Drosophila gasici*. Chromosoma, 16:47-57.
———— S. KOREF-SANTIBAÑEZ, M. BUDNIK, and M. LAMBOROT. 1969. Rate of development and inversion polymorphism in *Drosophila pavani*. Genetics, 61:471-477.
———— and P. SÁNCHEZ. 1958. The most widely distributed inversion in the most widely distributed species of *Drosophila*. Amer. Natural., 92:61-63.
CARSON, H. L. 1955. The genetic characteristics of marginal populations of *Drosophila*. Sympos. Quant. Biol., 20:276-287.
———— 1958. The population genetics of *Drosophila robusta*. Advances Genet., 9:1-40.
———— 1959. Genetic conditions which promote or retard the formation of species. Sympos. Quant. Biol., 24:87-105.
———— 1965. Chromosomal morphism in geographically widespread species of *Drosophila*. In Baker, H. G., and Stebbins, G. L., eds. The Genetics of Colonizing Species. New York, Academic Press, Inc. pp. 503-531.
DACUNHA, A. B. 1960. Chromosomal variation and adaptation in insects. Amer. Rev. Entom., 5:85-110.
———— D. BRNCIC, and F. M. SALZANO. 1953. A comparative study of chromosomal polymorphism in certain South American species of *Drosophila*. Heredity, 7:193-202.
———— and TH. DOBZHANSKY. 1954. A further study of chromosomal polymorphism in *D. willistoni* in its relation to the environment. Evolution, 8:119-134.
DARLINGTON, P. J., Jr. 1957. Zoogeography: The Geographical Distribution of Animals. New York, John Wiley & Sons, Inc.
DOBZHANSKY, TH. 1947. Genetics of natural populations. XIV. A response of certain gene arrangements in the third chromosome of *Drosophila pseudoobscura* to natural selection. Genetics, 32:142-160.
———— 1951. Genetics and the Origin of Species. 3rd ed. New York, Columbia University Press.
———— 1956. Genetics of natural populations. XXV. Genetic changes in populations of *Drosophila pseudoobscura* and *Drosophila persimilis* in some localities in California. Evolution, 10:82-92.

―――― 1957. Mendelian populations as genetic systems. Sympos. Quant. Biol., 22: 385-393.
―――― 1961. On the dynamics of chromosomal polymorphism in *Drosophila*. Sympos. Rev. Entomol. Soc. London, 1:30-42.
―――― 1962. "Rigid" vs. "flexible" chromosomal polymorphism in *Drosophila*. Amer. Natural., 96:321-328.
―――― 1965. "Wild" and "domestic" species of *Drosophila*. *In* Baker, H. G., and Stebbins, G. L., eds. The Genetics of Colonizing Species. New York, Academic Press, pp. 533-546.
―――― H. BURLA, and A. B. DaCUNHA. 1950. A comparative study of chromosomal polymorphism in sibling species of the *willistoni* group of *Drosophila*. Amer. Natural., 84:229-246.
―――― and H. LEVENE. 1948. Genetics of natural populations. XVII. Proof of operation of natural selection in wild populations of *Drosophila pseudoobscura*. Genetics, 33:537-547.
―――― and O. PAVLOVSKY. 1955. An extreme case of heterosis in a Central American population of *Drosophila tropicalis*. Proc. Nat. Acad. Sci. (U.S.A.), 41:289-295.
―――― and O. PAVLOVSKY. 1958. Interracial hybridization and breakdown of coadapted gene complexes in *Drosophila paulistorum* and *Drosophila willistoni*. Proc. Nat. Acad. Sci. (U.S.A.), 44:622-629.
―――― and B. SPASSKY. 1944. Genetics of natural populations. XI. Manifestation of genetic variants in *Drosophila pseudoobscura* in different environments. Genetics, 29:270-290.
DUBININ, N. P., M. N. SOKOLOV, and G. G. TINIAKOV. 1937. Intraspecific chromosome variability. Biol. Zhurn., 6:1007-1054.
―――― and G. G. TINIAKOV. 1946. Structural chromosome variability in urban and rural populations of *Drosophila funebris*. Amer. Natural., 80:393-396.
―――― and G. G. TINIAKOV. 1947. Natural selection in experiments with population inversions. J. Genet., 48:11-15.
FISHER, R. A. 1930. The Genetical Theory of Natural Selection. Oxford, Oxford University Press.
FREIRE-MAIA, N., I. F. ZANARDINI, and A. FREIRE-MAIA. 1953. Chromosome variation in *Drosophila immigrans*. Dusenia, 4:303-311.
FUENZALIDA, H. 1950. Clima, 1:188. Biogeografía, 1:371. *In* Geografía Económica de Chile. Santiago (Chile), Corp. de Fomento de la Producción, Imp. Universitaria.
GOLDSCHMIDT, E. 1956. Chromosomal polymorphism in a population of *Drosophila subobscura* from Israel. J. Genet., 54:474-496.
HARDY, D. E. 1965. Diptera: Cyclorrhapha II. Family Drosophilidae. Honolulu, University of Hawaii Press.
―――― 1966. Descriptions and notes on Hawaiian Drosophilidae (Diptera). Studies in Genetics. III. Univ. Texas Publ., 6615:195-244.
HARTMANN-GOLDSTEIN, I., and D. SPERLICH. 1963. The viability of structural heterozygotes of *Drosophila subobscura* at different temperatures. Genetics, 48: 863-869.
HEED, W. B. 1962. Genetic characteristics of island populations. Univ. Texas Publ., 6205:173-206.
HOENIGSBERG, H. F., and S. KOREF-SANTIBAÑEZ. 1959. Courtship behavior in inbred and outbred lines of *Drosophila melanogaster*. 1st. Lombardo Sci. e Lett. (Clase Scienze), 93:3-6.
HUNTER, A. S., and R. A. HUNTER. 1964. The *mesophragmatica* species group of *Drosophila* in Colombia. Ann. Entom. Soc. Amer., 57:732-736.
JAEGER, C. P., and F. M. SALZANO. 1953. *Drosophila gaucha*, a new species from Brazil. Rev. Brazil. Biol., 13:205-208.
JORDAN, K. 1896. On mechanical selection and other problems. Novit. Zool., 3: 426-525.

KIMURA, M. 1956. A model of a genetic system which leads to closer linkage by natural selection. Evolution, 10:278-287.
KOREF-SANTIBAÑEZ, S. 1963. Courtship and sexual isolation in five species of the *mesophragmatica* group of the genus *Drosophila*. Evolution, 17:99-106.
——— 1964. Reproductive isolation between the sibling species *Drosophila pavani* and *Drosophila gaucha*. Evolution, 18:245-251.
——— and D. BRNCIC. 1965. Mating activity and chromosomal polymorphism in *Drosophila pavani* females. Genetics, 52:453.
——— and E. DEL SOLAR. 1961. Courtship and sexual isolation in *Drosophila pavani* Brncic and in *Drosophila gaucha* Jaeger and Salzano. Evolution, 15:401-406.
KRIMBAS, C. B. 1964. The genetics of *Drosophila subobscura* populations. II. Inversion polymorphism in a population from Holland. Z. Vererbungsl., 95:125-128.
KRIVSHENKO, J. D. 1963. The chromosomal polymorphism of *Drosophila busckii* in natural populations. Genetics, 48:1239-1258.
LEVITAN, M. 1951. Experiments on chromosomal variability in *Drosophila robusta*. Genetics, 36:285-305.
——— 1958. Non-random association of inversions. Sympos. Quant. Biol., 23:251-268.
LEWONTIN, R. C., and P. HULL. 1967. The interaction of selection and linkage. III. Synergist effect of blocks of genes. Der Züchter, 37:93-98.
——— and Y. MATSUO. 1963. Interaction of genotypes determining variability in *Drosophila busckii*. Proc. Nat. Acad. Sci. (U.S.A.), 49:270-278.
MARTIN, J. 1965. Interrelation of inversion systems in the midge *Chironomus intertinctus* (Diptera: Nematocera). II. A non-random association of linked inversions. Genetics, 52:371-383.
MAYNARD SMITH, J. 1956. Fertility, mating behavior and sexual selection in *Drosophila subobscura*. J. Genet., 54:261-279.
MAYR, E. 1963. Animal Species and Evolution. Cambridge, Mass., Harvard University Press.
PATTERSON, J. T., and W. S. STONE. 1952. Evolution in the Genus *Drosophila*. New York, The Macmillan Company.
PERJE, A. M. 1954. Genetic and cytological studies of *Drosophila funebris* Fbr. Acta Zoologica, 35:259-288.
PETIT, C. 1958. Le determinisme génétique et psychophysiologique de la competition séxuelle chez *D. melanogaster*. Bull. Biol. France Belg., 92:248-329.
PIMENTEL, D. 1961. Animal population regulation by the genetic feed-back mechanism. Amer. Natural., 95:65-79.
——— W. P. NAGEL, and J. L. MADDEN. 1963. Space-time structure and the survival of parasite-host systems. Amer. Natural., 97:141-167.
PIPKIN, S. B., R. L. RODRÍGUEZ, and J. LEÓN, 1966. Plant host specificity among flower-feeding neotropical *Drosophila* (Diptera: Drosophilidae). Amer. Natural., 100:135-156.
REICHE, K. 1907. Grundzüge der Pflanzenverbreitung in Chile (Die Vegetation der Erde, VIII). Leipzig, Engelmann.
RENDEL, J. M. 1944. The genetics and cytology of *Drosophila subobscura*. II. Normal and selective mating in *Drosophila subobscura*. J. Genet., 46:287-302.
RENSCH, B. 1960. Evolution Above the Species Level. New York, Columbia University Press.
SPERLICH, D. 1961. Untersuchungen über den chromosomalen Polymorphismus einer Population von *Drosophila subobscura* auf den Liparichen Inseln. Z. Vererbungsl., 92:74-84.
——— 1967. Populationsgenetik (Teil. I. *Drosophila*). Fortschr. Zool., 18:223-278.
SPIESS, E. B. 1950. Experimental populations of *Drosophila persimilis* from an altitudinal transect of Sierra Nevada. Evolution, 4:14-33.
——— M. KETCHEL, and B. P. KINNE. 1952. Physiological properties of gene

arrangements carriers in *Drosphila persimilis*. I. Egg-laying capacity and longevity of adults. Evolution, 6:206-215.
——— 1962. *In* Spiess, E. B., ed. Papers on Animal Population Genetics. Boston, Little, Brown and Company. Introduction, pp. xi-xxi.
——— and B. LANGER. 1961. Chromosomal adaptive polymorphism in *Drosophila persimilis*. III. Mating propensity of homokaryotypes. Evolution, 15:535-544.
SPURWAY, H. 1955. The sub-human capacities for species recognition and their correlation with reproductive isolation. Acta 11th Ornithol. Cong. (Basel): 340-349.
STALKER, H. D. 1960. Chromosomal polymorphism in *Drosophila paramelanica* Patterson. Genetics, 45:95-114.
——— 1964. Chromosomal polymorphism in *Drosophila euronotus*. Genetics, 49:669-687.
STONE, W. S., W. C. GUEST, and F. D. WILSON. 1960. The evolutionary implications of the cytological polymorphism and phylogeny of the virilis group of *Drosophila*. Proc. Nat. Acad. Sci. (U.S.A.), 46:350-361.
WAGNER, M. 1868. Die Darwin'sche Theorie und das Migrationsgesetz der Organismen. Leipzig, Duncker und Humblot.
WALKER, I. 1959. Die Abwehrreaktion des Wirtes *Drosophila melanogaster* gegen die Zoophage Cynipidae *Pseudeucoila bochei* Weld. Rev. Suisse Zool., 66:569-632.
——— 1961. *Drosophila* und *Pseudeucoila*. II. Schwierigkeiten beim Nachweis eines Selektionserfolges. Rev. Suisse Zool., 68:252-263.
——— 1962. *Drosophila* und *Pseudeucoila*. III. Selektionsversuche zur Steigerung der Resistenz des Parasiten gegen die Abwehrreaktion des Wirtes. Rev. Suisse Zool., 69:209-227.
WALLACE, B. 1953. On coadaptation in *Drosophila*. Amer. Natural., 87:343-358.
WARTERS, M. 1944. Chromosomal aberrations in wild populations of *Drosophila*. Univ. Texas Publ., 4445:129-174.
WASSERMAN, M. 1962a. Cytological studies of the *repleta* group of the genus *Drosophila*: IV. The *hydei* subgroup. Univ. Texas Publ., 6205:73-83.
——— 1962b. Cytological studies of the *repleta* group of the genus *Drosophila*: V. The *mulleri* subgroup. Univ. Texas Publ., 6205:85-117.
——— 1962c. Cytological studies of the *repleta* group of the genus *Drosophila*. III. The *mercatorum* subgroup. Univ. Texas Publ., 6205:63-71.
WHEELER, M. R., H. TAKADA, and D. BRNCIC. 1962. The *flavopilosa* species group of *Drosophila*. Univ. Texas Publ., 6205:395-413.
WRIGHT, S., and TH. DOBZHANSKY. 1946. Genetics of natural populations. XII. Experimental reproduction of some of the changes by natural selection in certain populations of *Drosophila pseudoobscura*. Genetics, 31:125-156.

15

The Evolutionary Biology of the Hawaiian Drosophilidae[1]

HAMPTON L. CARSON
Department of Biology, Washington University
St. Louis, Missouri

D. ELMO HARDY
Department of Entomology, University of Hawaii
Honolulu, Hawaii

HERMAN T. SPIETH
Department of Zoology, University of California
Davis, California

WILSON S. STONE[2]
Genetics Foundation, University of Texas
Austin, Texas

Introduction (Herman T. Spieth) 438
Habitat (Herman T. Spieth) 439
    Land Areas 439
    Rainfall 445
    The Forests 447
Drosophilid Fauna (D. Elmo Hardy) 450
Biology and Behavior (Herman T. Spieth) 469
    Life Cycle 469
    Courtship Behavior 482
    Evolutionary Implications of Courtship Behavior 490

---

[1] This paper represents Journal Series No. 1092 of the University of Hawaii Agricultural Experiment Station.

[2] Wilson S. Stone possessed a powerful intellect, fertile with ideas, and analytical in action. As one of the Principal Investigators of the Hawaiian *Drosophila* Project, he effectively utilized these abilities in the planning, encouragement, and development of the diverse research efforts involved in the project. As early as 1966, recognizing the appropriateness of such a review as is presented in the following pages, he began the preliminary organization for the writing of the manuscript. His untimely death in February 1968 prevented participation in the final achievement, but because of his incalculable contributions to the entire Hawaiian *Drosophila* Project he is by unanimous agreement of the other three authors listed as a contributor to this review.

Chromosomal and Genetic Characteristics (Hampton L. Carson)	492
Metaphase Karyotypes	492
Polytene Chromosome Sequences	496
Intraspecific Chromosomal Polymorphism	504
Fixed versus Polymorphic Inversions	504
Parallel Chromosomal Polymorphisms in Different Species	506
Chromosomal Polymorphisms that are Identical in Two Different Species	506
Quantitative Studies of Chromosomal Polymorphism in Selected Species	508
Homosequential Species	514
Sibling Species	516
Interspecific Hybridization	516
Evolution, Speciation, and Migration (Hampton L. Carson)	520
Chromosomal Evidence for the Ultimate Origin of the Picture-Winged Species from Certain Continental Forms	520
Migration and Speciation of the Picture-Winged Species within the Hawaiian Islands	523
Colonization and Speciation Patterns in *Antopocerus*	532
Conclusion	532
Summary (Hampton L. Carson)	536
Acknowledgments	540
References	540

Introduction

Grimshaw (1901, 1902) and Perkins (1910) first focused the attention of dipterists on the aberrant characteristics and abundance of the Hawaiian Drosophilidae. They described 47 species; additionally, Perkins' extensive descriptions (1913) of the biotic habitat in which the insects live alerted biologists to the unique but unknown ecology of the flies.

Following T. H. Morgan's introduction of *D. melanogaster* as a valuable organism for the study of genetics and evolution, investigators in these fields became interested in the Hawaiian drosophilid fauna. Richard Goldschmidt planned to study in Hawaii but was prevented from doing so by the outbreak of the first World War; subsequently one of his students was likewise deterred from his proposed study of the Hawaiian drosophilids by the second World War (Zimmerman, 1958). In the early 1930's Curt Stern also unsuccessfully hoped to conduct studies in Hawaii. In the meantime various taxonomists had devoted some attention to the flies (see Hardy, 1965, pp. 21-22).

At the University of Texas, J. T. Patterson began his extensive study of the biology and evolution of the Drosophilidae in 1932. Patterson's collaborator, W. S. Stone, quickly appreciated the significance of the unique Hawaiian fauna, especially after Zimmerman published his introductory volume to the *Insects of Hawaii* (Vol. I, 1948). In 1946 G. B. Mainland, a student of Patterson's, joined the staff of the University of Hawaii with the avowed intent of studying the Hawaiian flies. He was successful in collecting numerous species but was frustrated in his attempts to rear the species under laboratory conditions.

In 1948 D. E. Hardy became a member of the University of Hawaii faculty and began his taxonomic studies of the Diptera of the islands. As he acquired more intimate knowledge of the rich Hawaiian *Drosophila* fauna, he made an intensive effort to interest evolutionists and geneticists in the unique problems presented by these insects. In 1961, Stone and Hardy met and developed plans for the present Hawaiian *Drosophila* Project, which was initiated in early 1963 under the auspices of both the University of Hawaii and the University of Texas. Involved in the research efforts have been a number of senior investigators from various institutions, assisted by energetic and highly competent senior technical personnel, by graduate students, and by responsible and able undergraduates.

At the time of the initiation of the project, general skepticism was expressed by competent authorities as to the probability of any successful resolution of the problems that had prevented earlier investigators from conducting successful laboratory investigations of the Hawaiian species. In such successes as have been achieved to date, serendipity has occasionally played an important role, but the major factors have been the close cooperation and integration of the efforts of all personnel involved in the project. Be that as it may, today probably more is known about the ecology, evolutionary relationships, and cytogenetics of the Hawaiian Drosophilidae than of any other comparable complex of drosophilid species. As related below a number of new, often unexpected, findings have added significant new parameters to our understanding of the taxonomy, biology, and evolution of the family Drosophilidae.

Habitat

Land Areas

The Hawaiian Archipelago consists today of a series of atolls, reefs, islets, satellite islands, and islands oriented along a 1600-mile (2500-km) northwest to southeast axis and located between 154°41′ and 171°75′ W. longitude, 18°54′ to 28°15′ N. latitude. Numerous authors (including Zimmerman, 1948; Fosberg, 1961; Stearns, 1966) have presented detailed information containing the geological history and existing nature of the Hawaiian Islands. The material presented here is minimal and has been selected for its pertinence to the analysis of the drosophilid fauna and its evolution.

The Archipelago can be characterized as the result of a series of volcanic

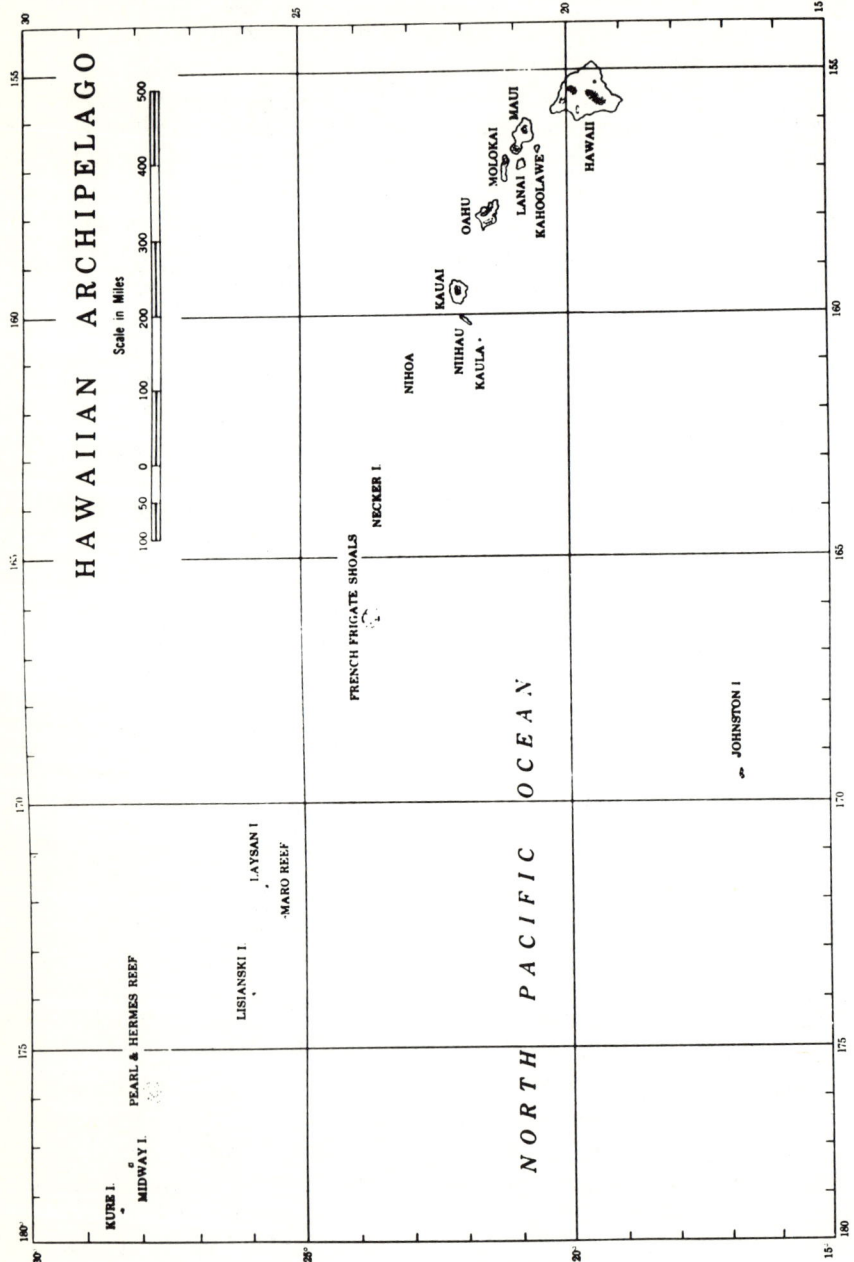

Fig. 1. The Hawaiian Archipelago.

episodes that began tens of millions of years ago at the northwest end of the chain and is still continuing at the southeast termination. Cores from drill holes on Oahu and Midway (Stearns, 1966) substantiate that the western islands are much older than the eastern ones. Further, Stearns (1966) shows that each of the Hawaiian volcanoes experienced a more or less similar historical development.

As a result of erosion, both fluvial and marine, aided by subsidence, the older, westernmost islands such as Kure, Midway, Pearl and Hermes Reef, Lisianski, and Laysan (Fig. 1) have been reduced and planed off from original high islands into reefs. French Frigate Shôals today consist of small pinnacles of volcanic rock, while Necker, Nihoa, and Kaula have been reduced to rocky islets. These islets, pinnacles, and reefs are known as the leeward chain in comparison to the eight younger islands, which form the windward or main islands of the Archipelago (Table I).

The eight windward islands are scattered over a 400-mile stretch of the ocean, separated from each other by channels of varying widths and depths (Table II). Their total area amounts to 6435 sq. mi. (16,667 km²), which is approximately half the size of the Netherlands (12,978 sq. mi.) and slightly less than that of the Fiji Islands (7070), but more than Jamaica (4411 sq. mi.). Each island consists of one or more volcanic domes. Niihau is a low remnant of an original single dome. Kauai is mainly a large symmetrical dome with a smaller dome on its southeastern flank. The potas-

TABLE I

Land Area, Maximum Elevation, and Median Rainfall of the Main Hawaiian Islands

Island	Area (sq. mi.)	Max. Alt. (ft.)	Median Annual Rainfall (inches)[a]
Hawaii	4,030	13,796	75
Maui	728	10,023	75
Oahu	604	4,046	58
Kauai	555	5,170	99
Molokai	260	4,970	42
Lanai	141	3,370	21
Niihau	72	1,281	–
Kahoolawe	45	1,477	–
	6,435		

[a] From Taliaferro (1959).

TABLE II

Approximate Distances and Minimum Depths of Channels Between the Hawaiian Islands

Islands	Channel Name	Distance (miles)	Depths (ft.)
Niihau-Kauai	Kaulakahi	17.0	2568
Kauai-Oahu	Kauai	73.0	9630
Oahu-Molokai	Kaiwi	26.0	1722
Molokai-Maui	Pailolo	8.5	564
Molokai-Lanai	Kalohi	9.0	264
Lanai-Maui	Auau	9.0	138
Maui-Kahoolawe	Alalakeiki	6.0	276
Maui-Hawaii	Alenuihaha	30.0	6180

sium-argon (K-Ar) method of aging indicates that Kauai (which has a minimal age of 5.6 to 3.8 million years) is older than the more eastern islands and as a consequence has suffered much greater erosion than have the younger islands. Oahu is composed of the remains of two domes, i.e., the older Waianae Range (3.4 to 2.7 million years of age), and the younger Koolau range (2.5 to 2.2 million years). Molokai is also formed from two domes, the western lower one having an age of 1.8 million years, the eastern one 1.5 to 1.3 million years. Maui likewise is the result of the fusion of two domes, West Maui with an age of 1.3 to 1.15 million years and East Maui less than 1 million years old. Hawaii, the most eastern and largest island of the windward group, consists of five fused volcanic domes, each being less than one million years of age. The northwest Kohala dome is the oldest, followed by Mauna Kea, Mauna Loa, Hualalei, and Kilauea (Table III).

The pairs of volcanic domes which form each of the present islands of Oahu, Molokai, and Maui, arose independently and originally formed independent islands, separated by narrow channels that were eventually obliterated by subsequent vulcanism, primarily from the eastern member of each pair, whose lavas now overlie those of the western member at the point of junction between the two domes (Stearns, 1966). In each instance the eastern dome is also 200,000 to 300,000 years younger than the western member of the pair.

The history of Molokai, Lanai, West Maui, East Maui, and Kahoolawe has been especially complex, a fact of importance to the evolution of the Drosophilidae fauna, as is shown later. This complexity was caused by three features, i.e., (1) the younger age of East Maui; (2) the low elevation of the land that joins East and West Maui; and (3) the narrow and shallow channels that exist between the four islands (see Table II). During periods of worldwide glaciation during the Pleistocene, the ocean levels were lowered due to removal of water to the extent that some or all of these islands were joined together at least twice. According to modern estimates, the duration of the Pleistocene was of such a magnitude that the entire geologic development of these islands is encompassed in this period. During interglacial periods when the ocean levels rose, not only were the present islands separated by oceanic channels, but even East and West Maui must have been separated by a shallow channel.

During periods of glaciation, accompanying the lowering of the ocean levels there was a general lowering of temperature, as a result of which the forests probably moved downward and then upward during the interglacial periods. In any case, it is clear that forests containing trees (such as *Cheirodendron*) and other plants of the same genera as are now present in the forests and upon which many of the Drosophilidae are dependent did move up and down the mountain slopes, as shown by Selling's studies (1948) of the fossil pollens.

Other channels, excepting those between the islands of the Maui-

TABLE III

Estimated Ages of Hawaiian Island Areas[a]

Island	Million Years
Kauai	5.6-3.8
West Oahu (Waianae Mts.)	3.4-2.7
East Oahu (Koolau Mts.)	2.5-2.2
West Molokai	1.8
East Molokai	1.5-1.3
West Maui (Wailuku Mts.)	1.3-1.15
East Maui (Halaekala)	0.8
Hawaii (Kohala Mts.)	<1.0
Hawaii (Mauna Kea)	0.6
Hawaii (Mauna Loa)	<0.5
Hawaii (Puu Waawaa)	0.4

[a] From Stearns (1966). The data were derived by the K-Ar technique. Although the samples are relatively few in number, the results are of significance in showing the relative ages of the islands.

Molokai complex, are of such depths and widths (Table II) that the individual islands have been separated throughout their entire history by ocean waters. Thus, the migrations of Drosophilidae from one to another of these land masses must of necessity have been accomplished by air-borne adults or by individuals that were able to tolerate movement of floating debris. Considering the nature of the developmental forms, the latter possibility (i.e., floating debris) seems highly improbable as a mechanism for safely allowing immigrants to be carried across these oceanic stretches.

The ages given for the islands represent the time when the major domes were built up above sea level, and do not signify the cessation of lava eruptions. These are still occurring on the island of Hawaii; during the period from 1850 to 1950 Mauna Loa liberated approximately 4 billion (4×10^9) cubic yards (3×10^9 m^3) of lava (Stearns, 1966); Kilauea since 1823 has extruded 5-billion cubic yards or 0.92 cubic miles of lava; Hualalei last erupted in 1800-1801.

Especially when the eruptions issue from the major vent or caldera on the top of the dome or from rifts high on the flank of the volcano, the lava flows downhill, often reaching the sea. In doing so it cuts a swath through the existing vegetation. Often the flow is such that numerous large areas of forest are surrounded and cut off, thus forming islands (or kipukas) within the river of fresh lava (Fig. 2). Isolating barriers are thereby created, which may at least partially if not completely stop gene flow between adjacent populations for considerable periods of time.

Fig. 2. Kipuka on the saddle between Mauna Loa and Mauna Kea, Hawaii. The picture is taken from a distance of 1 km; the trees are about 12 m high; the area of kipuka is about 25000 m² (6 acres) and is basically square in shape. Photo by M. P. Kambysellis.

All evidence indicates that the older islands experienced a similar history to that which has occurred within historical time on Hawaii. Furthermore, late in the life of a volcanic island, it may after a long period of erosion and submergence experience a rejuvenation of volcanicity with the occurrences of secondary eruptions mostly along or near the coasts. Diamond Head, Kaau Crater, Punch Bowl, Round Top, Sugar Loaf, and Tantulus on Oahu are products of this secondary vulcanism; the last three are only about 5,000 years old (Stearns, 1966). Thus the fauna and flora of the Hawaiian Islands have evolved with vulcanism as a constant companion, a companion which created barriers to gene flow and provided new sere areas of considerable size for reinvasion by the already established organisms from surrounding regions.

Rainfall

The windward islands all lie in the path of the persistent northeast trade winds. The flowing air as it strikes the islands is forced upward, rain clouds form, and the resultant rain drenches the windward-facing slopes and crests. On the leeward slopes the precipitation falls off rapidly and progressively. Thus on Kauai the Waialeale Station at 5075 foot elevation over a period of 46 years recorded a median rainfall of 465.5 inches (11.8 m), while Waimea on the coast directly southwest and only 13.4 miles (21.4 km) from Waialeale received 21.6 inches (Table IV). Likewise, on Oahu the Pauoa Flats Station at 1800 feet elevation recorded during 31 years a median of 162.4 inches, while the Honolulu Substation only 4.25 miles away had a median of 25.5 inches during 51 years. Trade-wind-engendered precipitation also decreases at elevations above 7000 feet (2100 m). Only East Maui (Haleakala) and Mauna Kea and Mauna Loa on Hawaii exceed this elevation; the uplifted trade winds do not go over their tops but rather flow around their flanks. A north-south transect of stations on East Maui, extending from Punaluu on the windward side to Kahikiniu on the leeward side of Haleakala (Table IV) shows the combined effect of elevation and rain shadow occurring over a distance of less than 20 miles. Additionally, as Fosberg (1961) points out, on these high islands the phenomenon of invection intrudes, with the result that certain leeward sites which normally would be expected to be arid do receive afternoon rains.

Finally, it is to be observed that there are periods of varying length of time when the trade winds cease and are replaced by air movements from other directions, typically by southern (Kona) winds. As an example, such a cessation of trades happened between early December 1967 and the beginning of March 1968. Associated with this period were a number of winter frontal-type storms that drenched the islands and especially the

TABLE IV
Rainfall Data Showing Effect of Elevation and Rain-shadow[a]

Station Number	Name	Elev. in Ft.	Number Years	Annual Rainfall Inches Max.	Median	Min.	Approx. Distance between Stations
Kauai							
1047	Waialeale	5075	46	624.0	465.5	218.0	} 13.4 miles
947	Waimea	20	45	42.2	21.6	7.0	
Oahu							
784	Pauoa Flats	1800	31	239.2	162.4	98.2	} 2.50 miles
706	Pacific Heights	680	29	95.2	67.3	39.8	} 1.75 miles
704	Honolulu Subst.	12	53	45.0	25.5	10.3	
							4.25
Maui							
447	Punaluu	700	51	220.6	123.3	76.9	} 1.20 miles
449	Waikamoi Gulch	1200	51	392.1	199.2	96.7	} 4.65 miles
336	Waikamoi Dam	4250	47	427.6	225.7	48.7	} 3.00 miles
341.2	Waikamoi Runoff #2	7200	6	128.1	112.1	39.3	} 3.15 miles
338.3	Haleakala	9750	8	76.4	32.6	27.5	} 7.25 miles
254	Kahikiniu	1400	32	43.9	28.4	9.7	
							19.25
Hawaii							
39	Mauna Loa Slope Observatory	11,146	5	28.11	21.85	14.28	–

[a]Data from Taliaferro (1959), except for Station 39, Mauna Loa Slope Observatory. The data for this station are from years 1963-67 inclusive, as presented in the monthly publication *Climatological Data, Hawaii*, U.S. Dept. of Commerce.

"normally" leeward or southern and southwestern portions. Thus Waimea (Station 947), with a median annual rainfall of 21.6 inches, received 17.59 inches during these three months; and the Honolulu Substation (704) received 18.58, as against its median annual amount of 25.5. Likewise the gauges at high elevations registered increased amounts. During the two months of December and February, Haleakala summit received 21.85 inches. The trade winds returned in March but were frequently interrupted in April and replaced with winter storms; Haleakala summit then received an additional 19.04 inches, some of which fell in the form of 9 inches of snow accompanied by 29°F temperatures. These winter storms may result in extraordinary amounts of precipitation within a short period of time, e.g., the maximum recorded rainfall on the islands for a 24-hour period is 38.5 inches at Kilauea, Kauai in January 1956. In January 1969, Kokee, Kauai, recorded 34 inches in a 24-hour period (personal communication, Saul Price, Regional Climatologist).

The fluctuating interactions of frontal storms and northeast trade winds are a consistent phenomenon, and their interplay appears to have been a constant phenomenon during the Pleistocene. These interactions result in great variations in the annual rainfall at any given location on the islands, as can readily be seen by studying the data presented in Table IV. An important corollary of these fluctuations is that when a period of low rainfall occurs, the high porosity of the underlying rocks quickly results in the occurrence of drought conditions.

The Forests

The endemic Hawaiian flora, derived from a few original stocks (Fosberg, 1961), evolved therefore under conditions of great diversity of physiography and climate, accentuated still further by the fluctuations caused by Pleistocene glaciation (Selling, 1948). The unique assemblage of endemic vegetation types that evolved under these conditions displays enormous diversity, occupying most available habitats and consisting of a number of forest types. It is within certain of these forest types that the great array of endemic Drosophilidae are found. Various authors have proposed classifications of the Hawaiian vegetation (Hillebrand, 1888; Rock, 1913; Hosaka, 1937; Robyns and Lamb, 1939; Ripperton and Hosaka, 1942; Selling, 1948; Krajina, 1963). Fosberg (1961) distinguishes 18 forest and scrub ecosystems. Four of these are formed by recently introduced plants and in all instances have replaced native forests. Of the remaining 14, 4 serve as the home for almost all of the drosophiloid[1] and scaptomyzoid species. These are (1) the cloud forest; (2) the dryland sclerophyll forest; (3) the *Metrosideros* forest; and (4) the mixed mesophytic forests.

METROSIDEROS FOREST. This forest is found in moderately moist to wet localities, from fairly low to middle elevations. Exhibiting considerable variation in composition, complexity, and structure at different sites, it may consist of almost pure strands of *Metrosideros* trees, especially on young lava flows and ash beds. In the older forests, species of many other trees and shrubs are to be found mixed with the *Metrosideros*. In areas of high precipitation, from 2000 to 5000 feet, these older forests become highly complex, forming a true montane rainforest (Fig. 3). In such situations other trees may share dominance with *Metrosideros* and a second tree layer is formed, while shrubs create a still lower layer. Ferns of various types, including the tree ferns of the genus *Cibotium,* occur in abundance, and climbers such as *Freycinetia* and an abundance of bryophytes and lichens grows on the trunks and limbs of the trees. Most importantly for the drosophiloid species, their prime host plants such as *Tetraplasandra,*

---

[1] I.e., *Drosophila* and closely related genera.

Fig. 3. Metrosideros forest, Kula pipeline, Waikamoi, Maui. Typical rainforest habitat of Drosophilidae. Photo by L. H. Throckmorton.

Ilex, Myrsine and especially *Cheirodendron* and the lobeliads *Clermontia* and *Cyanea* are common members of this complex assemblage.

CLOUD FOREST. Above and contiguous with the *Metrosideros* forest there often is found a cloud forest (Hosaka, 1937). This is a complex forest with an abundance of shrubs, ferns, mosses, and hepatics. In sheltered areas such as ravines, the vegetation reaches forest stature while on the exposed slopes and ridges it is dwarfed and tangled, reduced to scrub status or even boggy grass areas (Fosberg, 1961). In the sheltered areas drosophiloids are found along with their major host plants such as the lobeliads and *Cheirodendron*.

MIXED MESOPHYTIC FOREST. These are forests of very diverse composition, found at low to moderate elevations and also at higher elevations. They differ from the rain forests in receiving less precipitation, typically 50 to 100 inches, but do not suffer from an actual moisture deficit (Fosberg, 1961). *Cheirodendron* and the lobeliads are characteristically lacking from such forests, but other host plants of the drosophiloids such as *Sapindus, Osmanthus,* and *Myrsine* are common components. Furthermore, these forests have considerable numbers of fungi, sometimes large in size and abundant in number, which serve as food for the adults of numerous species and as the developmental sites for a number of species, including the entire fungus-feeding drosophiloid group. At higher elevations these forests reach their most exuberant development in the rich soils of the kipukas such as Kipuka Puaulu and Kipuka Ki on Hawaii. These kipukas with their relatively low temperatures have amazingly rich drosophiloid faunas. Few of the endemic drosophiloids are able to tolerate the temperatures that exist below 1000 feet elevation, even on the cooler windward exposures. Those species which do live at lower elevations seem typically to be most commonly found in the mixed mesophytic forest areas.

DRYLAND SCLEROPHYLL FOREST. Only a few remnants remain of this type of forest, which Rock (1913) indicated was species-wise the richest of all Hawaiian forests. Originally, large areas of the dry coastal slopes and higher rain shadows up to at least 5000 feet were covered with this scrub-type forest. The Auwahi forest on the south slopes of East Maui appears to be the sole remaining remnant on Maui. Amongst the 50 species of trees and shrubs that Rock (1913) found here are a number of host plants for drosophiloids, e.g., *Cheirodendron, Charpentiera, Tetraplasandra,* and others. Despite the declining fate that this forest, located at an elevation of 3500 to 4050 feet, is experiencing, it still is a rich drosophiloid collecting area. In its pristine state an enormously rich and varied drosophiloid fauna must have inhabited this ecosystem.

All of these four types of forest have been ravaged by the activities of man. The native forests originally covered the islands from seashore to timber line (Zimmerman, 1948). The arrival of the Polynesians introduced fire and caused a rapid retreat of the lowland and dryland forests, but the real destruction began with the arrival of Western man in 1778. Today the dry sclerophyll forest is almost extinct, and introduced plants and animals insure its eventual demise. The mixed mesophytic forest has been radically reduced and only in areas such as the National Parks is it partially protected. The *Metrosideros* and cloud forests have been the least disturbed, but even they have been seriously modified as man and his feral and domestic animals have slashed their way through the vegetation. The drosophiloids, like the honeycreepers, have suffered severely; probably many species and certainly countless populations of drosophiloids have been destroyed during the past 300 years. Fortunately most of the extant species dwell in the *Metrosideros* and cloud forests at elevations of 2500 to 5000 feet on the windward or wetter areas of the islands. These forests are relatively much more resistant to the intrusion of man than is any other area of the islands, as any *Drosophila* collector well knows, and additionally they have relatively high recuperative powers.

Drosophilid Fauna

The family Drosophilidae is most remarkably developed in the Hawaiian Islands and represents one of the most unusual faunas of any area of the entire world; no other drosophilids are so diversified morphologically and biologically. We have here a striking example of rapid, explosive evolution that is almost unparalleled in the animal kingdom, and this group is most ideally suited for evolution and genetic studies. That the islands have served as an area of spectacular adaptive radiation is exemplified in many different groups of native plants and animals, but the dipteran family Drosophilidae is the most unusual of any animal group that has been studied.

As now known, the Hawaiian drosophilid fauna is composed of almost 500 species. About 460 endemic species have been named to date and 17 introduced species are present; 96.4 percent of presently known species are endemic. Of the named endemic species, approximately 330 belong to *Drosophila* Fallén and closely related genera (drosophiloids), and 132 belong to *Scaptomyza* Hardy and related genera (scaptomyzoids). These represent approximately one third of the total number of *Drosophila* species known for the entire world, and twice the number of species of *Scaptomyza* known for the world, excluding Hawaii. In spite of the large number of known species, the taxonomy of this group is still in a preliminary state. It is now obvious that the total fauna consists of 650 to 700 species. It is probable that an additional 125 species of *Drosophila* and 75 to 100 *Scaptomyza* remain to be described.

Only the basic morphological descriptions have been given for roughly two thirds of the species in the fauna, and since it has become so obvious that it is impossible to determine relationships based on morphology alone, it has not been feasible to date to set up a detailed phylogenetic arrangement for the species. The team approach, which is now being used to study the evolution and genetics of Hawaiian Drosophilidae, gives a unique opportunity to gain a thorough understanding of one of the most complex and remarkable groups of animals known in the world. To my knowledge, this is the first time that a group of highly qualified specialists have pooled their efforts for determining paths of evolution and the factors affecting speciation rates in animal groups under certain insular conditions. Aside from the spectacular accomplishments resulting from the varied aspects of this study, the combined results are providing a sound taxonomic basis for this family. Phylogenies are being worked out as a thorough understanding of various species groups is being gained, based upon internal and external morphology, genetics, hybridization, behavior, ecology, ovarian transplants, and biochemical studies of enzyme systems. Such detailed knowledge has rarely been obtained for such a group of animals and the combined efforts will result in a classical example of the proper approach to systematics. Carson's arrangement of the picture-winged flies based upon comparisons of the gene sequences of the giant salivary gland chromosomes (p. 520ff.) is the most significant development that has occurred regarding our understanding of systematics of Hawaiian *Drosophila*. The detailed studies of the male genitalia by Takada (1966) and Kaneshiro (1968) correlate closely with the arrangement of Carson in most details. Following the work of Okada (1953, 1954, 1955), Kaneshiro has demonstrated that the character of the male aedeagus is of considerable importance in determining relationships between species.

The Hawaiian drosophilids have not only developed fantastic numbers of species, but exhibit the greatest diversity of form and habits in any known area of the world. A great many structural peculiarities are found that apparently do not occur in other drosophilid faunas. Some of these structural modifications are illustrated in Figure 4. Most of them are found in the male sex only; this is especially true of the mouthpart and leg modifications. The behavior studies of Spieth (1966b, 1968b) have demonstrated that most of these structural peculiarities are directly associated with courtship and mating. Various parts of the body are modified; examples are provided in the following paragraphs.

Head: extremely broad, showing beginning of stalk-eyedness (Fig. 4.37); elongate (Fig. 4.47); lack of frontal bristles (Fig. 4.50); excess of hairs (Fig. 4.46); pectinate antennae (Fig. 4.43,49); pointed second antennal segment (Fig. 4.48); preapical arista (Fig. 4.30); modified mouthparts (Fig. 4.34-36, 39-42, 44); bristles on palpi (Fig. 4.31-33).

Legs: "spoon-tarsi," with front basitarsus of males short, flat and concave on upper surface and modified into a spoon-like structure (Fig. 4.1-2); "forked-tarsi," males having an appendix developed at the apex of the front basitarsus and only four tarsomeres present (Fig. 4.10-15). The male of *D. freycinetiae* Hardy, however, has an appendix at the base (rather than the apex) of the front basitarsus and has the normal complement of tarsomeres (Fig. 4.3); "bristle-tarsi" have the apex of the front basitarsus of the male with a comb-like arrangement of bristles or strong setae at the apex of the front basitarsus (Fig. 4.6-7); in the *adiastola* complex the front basitarsus of the male is flat and densely setose (Fig. 4.5,16); clubbed middle tibia of male (Fig. 4.8-9); hook on tibia of male (Fig. 4.17-18); long ciliation of male forelegs (Fig. 4.19-26); knobbed tibia (Fig. 4.4).

Wings: "picture wings," characterized by rather elaborate patterns of

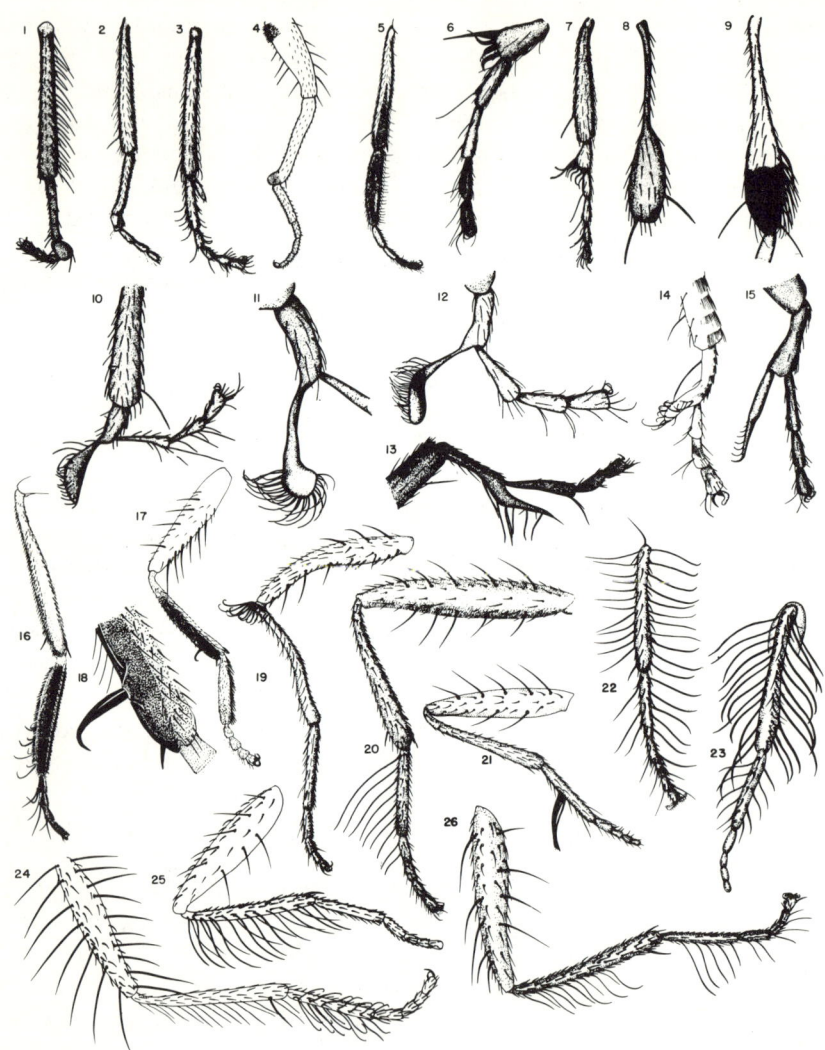

Fig. 4a. Legs of Drosophila *spp.*: *1,* dasycnemia *Hardy; 2,* percnosoma *Hardy; 3,* freycinetiae *Hardy; 4,* anomalipes *Grimshaw; 5,* adiastola *Hardy; 6,* expansa *Hardy; 7,* perissopoda *Hardy; 8,* clavitibia *Hardy; 9,* fuscoapex *Hardy; 10,* clavata *Hardy; 11,* capitata *Hardy; 12,* attenuata *Hardy; 13,* fundita *Hardy; 14,* spiethi *Hardy; 15,* enoplotarsus *Hardy; 16,* aethostoma *Hardy and Kaneshiro; 17-18,* hamifera *Hardy and Kaneshiro; 19,* flexipes *Hardy and Kaneshiro; 20,* gradata *Hardy and Kaneshiro; 21,* lineosetae *Hardy and Kaneshiro; 22,* silvarentis *Hardy and Kaneshiro; 23,* liophallus *Hardy and Kaneshiro; 24,* vesciseta *Hardy and Kaneshiro; 25,* hirtipalpus *Hardy and Kaneshiro; 26,* glabriapex *Hardy and Kaneshiro.*

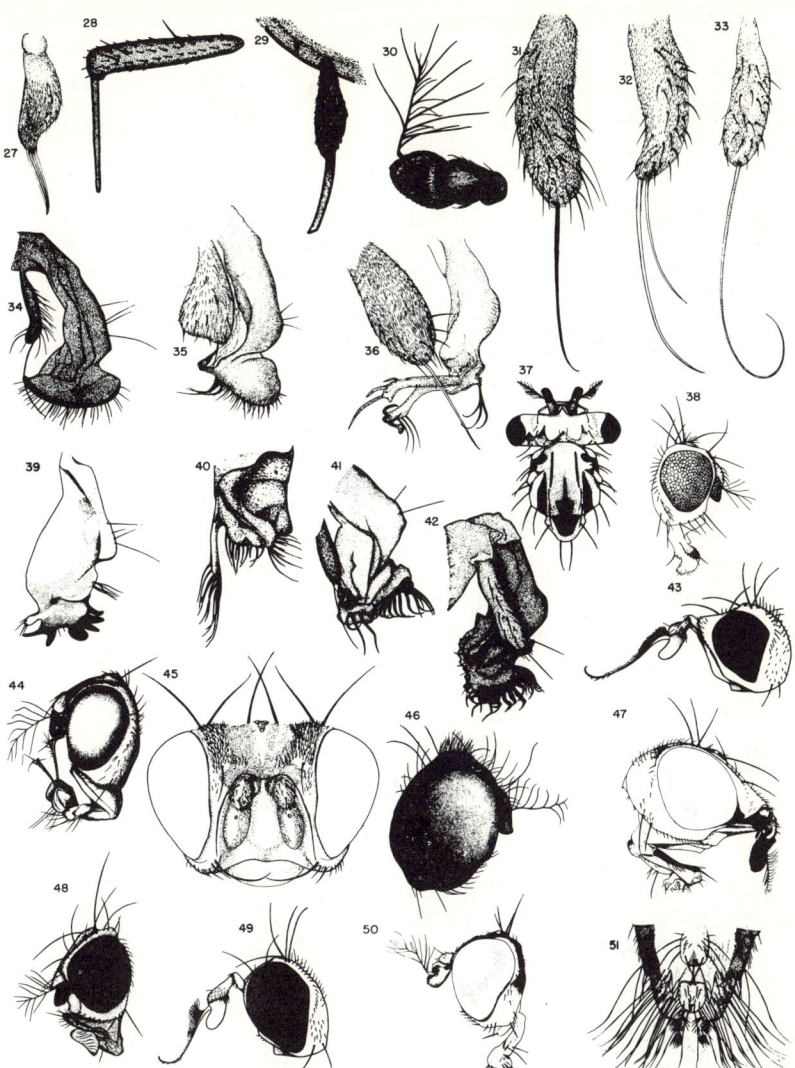

Fig. 4b. *27*, Palpus of Drosophila basisetae Hardy and Kaneshiro; *28*, palpus, Nudidrosophila gemmula Hardy; *29*, N. lepidobregma Hardy; *30*, antenna, Ateledrosophila preapicula Hardy; *31*, palpus, D. hirtipalpus Hardy and Kaneshiro; *32*, palpus, D. prostopalpis Hardy and Kaneshiro; *33*, palpus, D. macrothrix Hardy and Kaneshiro; *34*, mouthparts, D. cilifemorata Hardy; *35*, mouthparts and palpus, D. acanthostoma Hardy and Kaneshiro; *36*, mouthparts and palpus, D. aethostoma Hardy and Kaneshiro; *37*, head and thorax, D. heteroneura (Perkins); *38*, Celidosoma nigrocincta Hardy; *39*, mouthparts, D. ceratostoma Hardy; *40*, mouthparts, D. apoxyloma Hardy; *41*, mouthparts, D. artigena Hardy; *42*, mouthparts, D. scolostoma Hardy; *43*, head, Antopocerus longiseta Hardy; *44*, head, Drosophila adventitia Hardy; *45*, head, D. setosifrons Hardy and Kaneshiro; *46*, head D. (Trichotobregma) petalopeza Hardy; *47*, head, Drosophila planitibia (Hardy); *48*, head, Grimshawomyia palata Hardy; *49*, head, Antopocerus diamphidiopodus Hardy; *50*, head, Nudidrosophila aenicta Hardy; *51*, apex of male abdomen, ventral, D. clavisetae (From Hardy, Insects of Hawaii, 1965. Courtesy of University of Hawaii Press.)

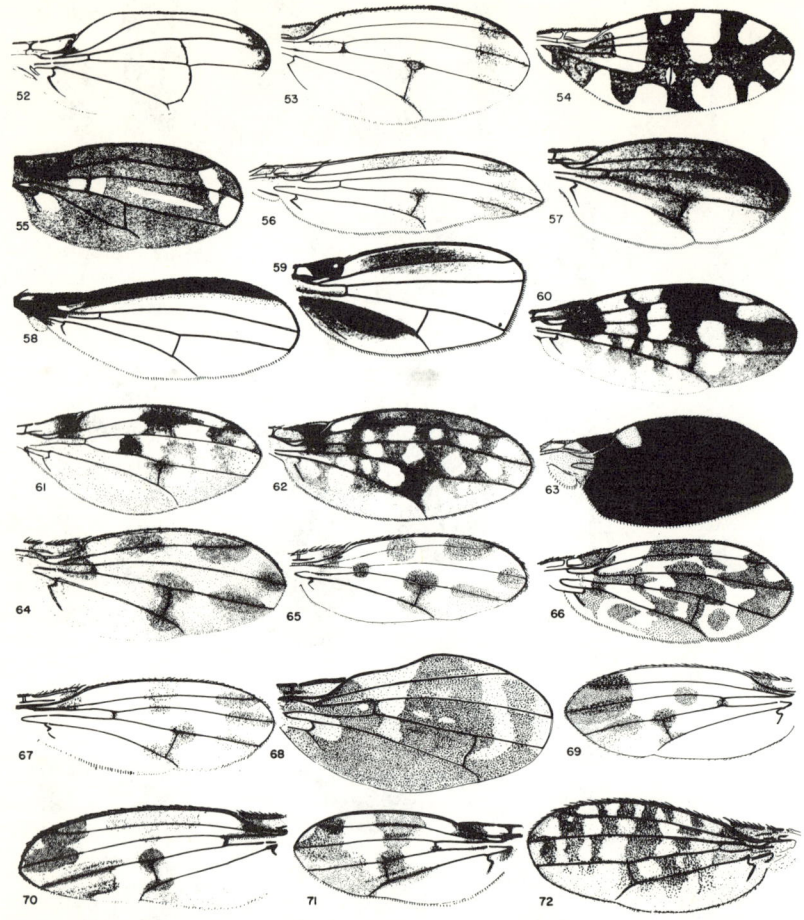

Fig. 4c. Wings: 52, Drosophila melanocephala (Hardy); 53, D. deltaneuron Bryan; 54, D. fuscoamoeba Bryan; 55, D. varipennis (Grimshaw); 56, D. lanaiensis Grimshaw; 57, D. seclusa Hardy; 58, D. eurypeza Hardy; 59, D. truncipenna Hardy; 60, D. adiastola Hardy; 61, D. clavisetae (Hardy); 62, D. spectabilis Hardy; 63, Celidosoma nigrocincta Hardy; 64, D. discreta Hardy and Kaneshiro; 65, D. ochracea Grimshaw; 66, D. crucigera Grimshaw; 67, D. flexipes Hardy and Kaneshiro; 68, D. hamifera Hardy and Kaneshiro; 69, D. liophallus Hardy and Kaneshiro; 70, D. virgulata Hardy and Kaneshiro; 71, D. silvarentis Hardy and Kaneshiro; 72, D. ochrobasis Hardy and Kaneshiro.

markings (Fig. 4.52-72); brown marks may be over one or both crossveins (Fig. 4.64,70); variously arranged maculations (Fig. 4.54-55,68); the wing may be predominantly brown with hyaline spots (Fig. 4.60,62,66); nearly all brown (Fig. 4.63); extra crossvein in cell R-5 (Fig. 4.52,61); arcuate costal margin (Fig. 4.68); long and pointed (Fig. 4.56); m crossvein split (Fig. 4.53).

Abdomen: with clavate setae on venter of male (Fig. 4.51).

The *Drosophila* may be divided into rather rough groupings of species according to morphological characteristics. In some cases, these groupings are artificial and are not subgeneric units or even species groups. As indicated in the descriptive portion above, however, some of the peculiarities are useful as group characters. Some of the more obvious species groups that can be recognized are listed in the following paragraphs.

The "picture-winged" species include most of the flies with conspicuously marked wings, although such maculations have apparently arisen independently in more than one species group. *Idiomyia* was erected as a new genus by Grimshaw (1901) based upon the presence of an extra crossvein in cell R5 (Fig. 4.52). This would appear to be a most excellent generic character, very distinctive and characteristic of only this group of Hawaiian species. As pointed out by Hardy (1965), however, the "idiomyia" are nothing more than very large *Drosophila* with a freak wing-venation. Spur-vein abnormalities are sometimes found in field-collected and laboratory strains of Hawaiian *Drosophila*, suggesting that there may be genetic variability for this character in nature. Carson et al. (1967) expressed the opinion that the genus should be abandoned because of close chromosomal similarity to other picture-winged species of the subgenus *Drosophila*. The wing-vein character is obviously not even of species-group importance. Nevertheless, this subgroup is interesting because among them are the giants of the family; some species have a wing-spread of as much as 18 to 20 mm. *D. cyrtoloma* (Fig. 5) is probably the largest species of *Drosophila* known in the world. The *eurypeza* (Fig. 4.58) and *semifuscata* (Fig. 4.57) complexes include a number of species with characteristic wing markings. Approximately half of the Hawaiian *Drosophila* have brown or fuscous markings of some sort on their wings.

Several species groups can be recognized among those flies having modified mouthparts. In addition to those which have strong bristles or processes developed from the labella or on the palpae, there is a distinct group which has a heavily sclerotized rim around the apex of each labellum (Fig. 4.34). These latter form a complex ("white-tipped scutellum") of at least 50 species, which are further characterized by being predominantly black, usually polished, having the scutellum black with a small yellow to white spot at the apex. Some of these (*polita*-complex) have the wings entirely hyaline; others (*haleakalae*-complex) have the wings clouded at the apex and over the *m* crossvein. The species of this group are all apparently associated with fleshy fungi.

As described above, a number of groups of species may be characterized by having modified leg parts in males. In the work so far, they have been conveniently referred to as the "bristle-tarsi," "spoon-tarsi," "forked-tarsi," and "clubbed-tibiae." Each are groups including numerous species.

The preliminary classification of the Hawaiian fauna (Hardy, 1965)

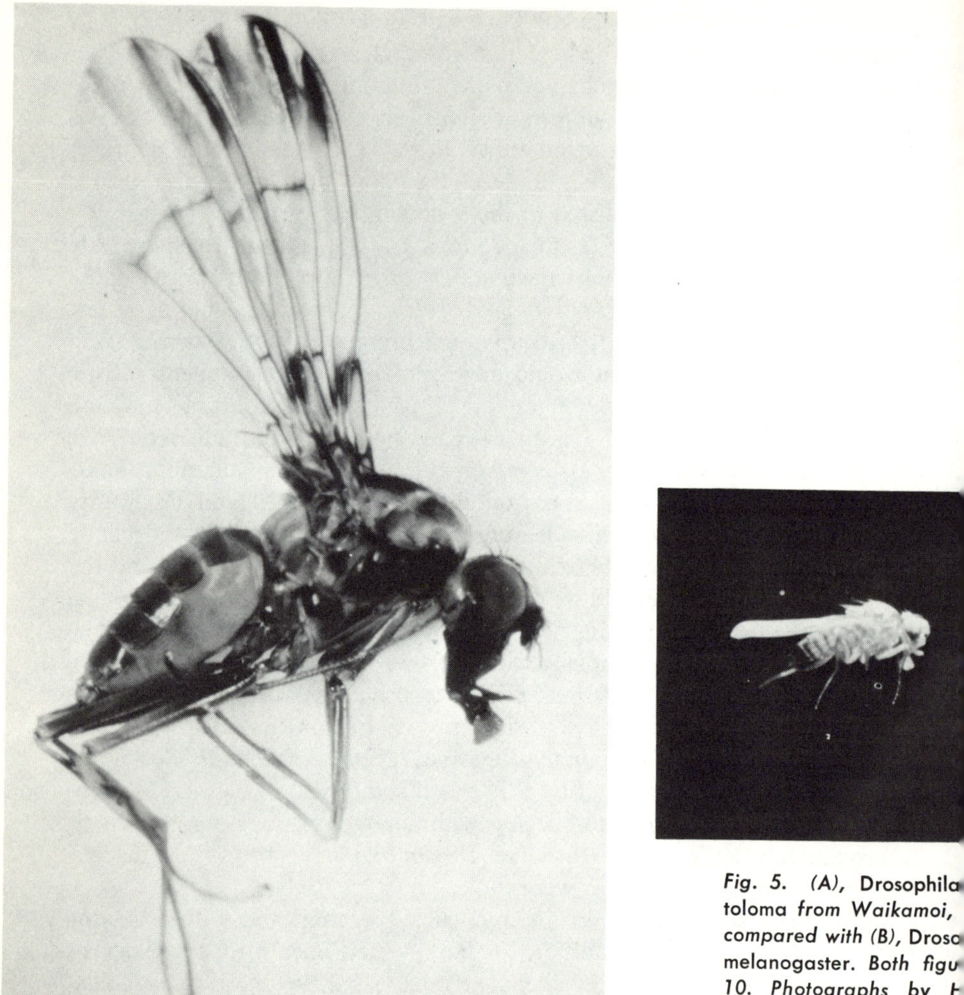

Fig. 5. (A), Drosophila toloma from Waikamoi, compared with (B), Drosophila melanogaster. Both figures 10. Photographs by H. Stalker.

divided the family into thirteen genera and ten subgenera. Two of these, *Gitonides* Knab, represented by one species, and *Pseudiastata* Coquillett, represented by three species, were purposely introduced for biological control of mealy bugs. Only *G. perspicax* Knab became established. Two were accidentally introduced, *Chymomyza* Czerny, represented by one species, and *Dettopsomyia* Lamb, represented by two species. The remaining nine "genera" are endemic, with thirteen species of *Drosophila* and one of *Scaptomyza* having been accidentally introduced from other areas.

The following introduced species have been recorded: *Gitonides perspicax*

Knab; *Chymomyza procnemis* (Williston); *Dettopsomyia formosa* Lamb; *Dettopsomyia nigrovittata* Malloch; *Drosophila (Drosophila) bizonata* Kikkawa and Peng; *D. (Drosophila) busckii* Coquillet; *D. (Drosophila) carinata* Grimshaw; *D. (D.) funebris* (Fabricius) (known from only one specimen collected at Kula Pipe Line, Maui, 4500 to 5000 feet, March 1932—this region has been collected very thoroughly, and *funebris* is obviously not established in the Islands); *D. (D.) hydei* Sturtevant; *D. (D.) immigrans* Sturtevant; *D. (D.) nasuta* Lamb; *D. (D.) polychaeta* Patterson and Wheeler; *D. (D.) repleta* Wollaston; *D. (Sophophora) ananassae* Doleschall; *D. (S.) kikkawai* Burla; *D. (S.) melanogaster* Meigen; *D. (S.) simulans* Sturtevant; and *Scaptomyza (Parascaptomyza) pallida* (Zetterstedt).

It is quite remarkable that even more species of this family have not reached Hawaii and become established here. We receive immigrant insects at an alarming rate; over the last 20 years or so an average of 16 additional species of insects from various parts of the world have become established each year.

Subsequent studies have demonstrated that the earlier generic classification was not entirely accurate. The amazing array and diversity of the morphological characters displayed by the Hawaiian species and the obviously high degree of convergence and/or parallelism that has occurred require that concepts distinctly different from those employed for other world areas be used in dealing with this fauna. Characters that would normally be considered of generic importance in other parts of the world are often found to be only species-group characters, and in some cases are of no importance in grouping the species and represent cases of convergence or parallelism. Hence, it becomes quite impossible to determine phylogenies based on morphological characters alone.

The synonymity of *Idiomyia* with *Drosophila*, being based on a character not even of species-group importance, was noted above.

Nudidrosophila Hardy (1965) is of very questionable status. It would appear to be strikingly distinct from any known genus in the family by the males lacking the major head bristles; the reclinates, the proclinates, and the ocellar bristles are completely lacking; the front is entirely bare except for microscopic pubescence, or recumbent scale-like setae (Fig. 4.50). The ocellar triangle has a series of laterally directed, recumbent, very fine and inconspicuous hairs, and a thick, rather flattened spine-like development is present at the apex of each palpus (Fig. 4.28, 29). All of the above are male characters, and the females have not been associated with some of the species. As discussed by Hardy (1966), those females of *Nudidrosophila* which have been associated with the males exhibit typical *Drosophila* characteristics; and in the case of *N. aenicta* Hardy the females show remarkable resemblance to *Drosophila hirtitibia* Hardy. Based upon the females, these species would appear to be very closely related. In most charactersitics these species appear to be identical, and they probably occupy the same or similar habitats. It is necessary that this complex be studied in detail before definite decisions can be made, but I believe it is evident that *Nudidrosophila* should not be retained as a genus. It may be necessary to synonymize it with *Drosophila,* even though on the basis of the males the striking differences would seem to be of generic importance: the male characters depart radically from those of *Drosophila,* according to present concepts. Five species have been described as *Nudidrosophila.*

Antopocerus Hardy consists of nine known endemic species, characterized by modifications in the antennae of males: first segment large, extending well beyond the margin of the front, concave on ventral surface, subequal in length to the third antennal segment, and the antennae strongly porrect with arista densely short-haired dorsally and bare ventrally (Fig. 4.43,49) except near the apex in some species. This genus is an offshoot of *Drosophila.*

Ateledrosophila Hardy, with two included species, is distinguished from *Drosophila* by having the arista preapical in position (Fig. 4.30), by having no discernible anterior reclinate or ocellar bristles, and no preapical dorsal bristles on the front tibiae. The group is poorly known. We have no information on the biology, behavior, or genetics. It is obviously close to *Drosophila.*

Celidosoma Hardy has one known species, which approximates *Chaetodrosophila* Duda in Wheeler's key to the genera of Drosophilidae of the Pacific islands (1952), but differs by lacking a comb of short spines on the front femur; by having a very weakly carinate face rather than a prominent face with a well developed carina; by having two strong sternopleural bristles rather than one strong bristle and two hairs present on upper portion of each sternopleuron; and by having the second antennal segment rather strongly produced apically extending over the basal portion of the third segment, as well as by other details. It is distinguished from other Hawaiian drosophilids by having four or five pairs of dorsocentral bristles and only four rows of acrostichal setae; by having the anterior reclinate and the proclinate bristles situated near the anterior margin of the front; and by the pointed second antennal segment. In the latter regard, it is like *Grimshawomyia* Hardy (Fig. 4.48). These two genera somewhat resemble one another in other respects, but the resemblance is superficial and they differ strikingly as follows: *Celidosoma* has only four rather than six distinct rows of acristichal setae; four or five pairs of dorsocentral bristles rather than two, with two pairs presutural in position; two strong humeral bristles; face flat medianly on lower half and slightly carinate above; head very differently shaped; anterior reclinate and proclinate bristles situated near the anterior margin of the front instead of the usual position; each tibia with two black bands; wings predominantly dark brown, hyaline only on the basal portion. This would appear to be a direct offshoot of *Scaptomyza.*

Further studies are necessary to determine its exact status. We have no knowledge of its biology.

Grimshawomyia Hardy is represented by two species. This genus is distinguished from *Drosophila* by having the second antennal segment sharply pointed at the apex, extending over the base of the third; sides of the vertex swollen, and vertical and upper ocellar bristles situated on the gibbose portion (Fig. 4.48); costal fringe elongate, extending nearly to apex of vein R_{4+5}; the wing and genital characters are distinctive. We have no information on habits, biology, or genetics of this genus.

Drosophila Fallén presently consists of 343 described species, of which 13 are introduced. It is probable that when all of the species of the Hawaiian *Drosophila* are known, the total will be near 500. In the earlier classification, the genus was divided into four subgenera. The great bulk of our species have the external morphological characters of the subgenus *D. (Drosophila)*, as presently defined. Throckmorton (1966) stated "the characteristics of the egg filaments, ventral receptacles, and vasa strongly indicate a position on the branch leading to the subgenus *Drosophila*. All other characteristics are consistent with this interpretation." More recently, Throckmorton (in litt.) has concluded that on the basis of internal anatomy, the Hawaiian *Drosophila* appear to be more closely related phylogenetically to the subgenus *D. (Hirtodrosophila)*. It is probable that when more complete information has been obtained, it will be logical to divide the genus *Drosophila* into several subgenera and many species groups. It is obvious, however, that this cannot be based upon morphological characters alone, and must entail all possible information concerning the ecology, genetics, behavior, biochemistry, and so forth.

Subgenus "*Hypenomyia*" Grimshaw was described as a genus by Grimshaw (1901) and was reduced to a subgenus by Hardy (1965). Grimshaw based this "genus" upon the dense clumps of bristles on the lower angles of the face, which is characteristic of the type, *D. varipennis* Grimshaw. This character has been found to have little or no value in grouping Hawaiian species, but *varipennis* and related species do have rather remarkable characters that set them apart. The males are characterized by the presence of a large preapical dorsal hook on the front tibia (Fig. 4.17, 18), by having the mouthparts highly modified and the costal margin of the wing strongly arched, and the male aedeagus flat and strongly clavate, spade-like. As discussed by Hardy and Kaneshiro (1968), in light of the extraordinary range of divergence of the morphological characters found in the Hawaiian species, we feel it best to treat *Hypenomyia* as nothing more than another unusual species group of *Drosophila*. This opinion is shared by Carson, who has found chromosomal similarities with *D. adiastola*. We have thus synonymized *Hypenomyia* with *Drosophila*. Three known species are in the *varipennis* group.

Subgenus "*Trichotobregma*" Hardy (1965), with only one known species,

D. petalopeza, is probably a synonym of typical *Drosophila* and nothing more than another unusual manifestation of characters. It is obviously a bristle-tarsus *Drosophila* with unusual head hairs and modified middle tarsi. This species differs strikingly from others of *Drosophila* by lacking proclinate frontal bristles and by having numerous long bristle-like hairs on the lower sides of the front, completely obscuring the reclinate bristles in the male (Fig. 4.46). In the female, two lower reclinate bristles are present as well as a series of four to six short bristle-like hairs on the orbits between the anterior and posterior reclinates. The male is also characterized by having very tiny ocellar bristles, about equal in length to the fine hairs on the ocellar triangle.

Subgenus *D. (Sophophora)* Sturtevant contains four introduced, cosmopolitan species.

Subgenus *D. (Engiscaptomyza)* Kaneshiro (1969) contains six species of the *crassifemur* and *nasalis* complexes. This is a group that Throckmorton (1966) on the basis of internal morphology placed in his *Scaptomyza*-like grouping; he refers to these as the "scaptoids" (genera or groups related to *Scaptomyza*; it should be spelled "scaptomyzoid"). Also based on mating behavior, Spieth (1966b) placed these as *Scaptomyza*-like species. I would normally prefer not to use the terms drosophiloid and scaptomyzoid, even though the words are correctly formed, with the suffix -oid or -oides (meaning "like, resembling") because there is chance for confusion since the first could represent the adjectival form of the superfamily Drosophiloidea. In dipteran taxonomy terms like "drosophiloid," "muscoid," etc., are properly used only in the superfamily sense. However, in dealing with the Hawaiian fauna it is convenient to use such terminology and as long as our meaning is understood there should be no chance for confusion. On the basis of external characters, Mr. Kaneshiro has described *Engiscaptomyza* as a new subgenus under *Drosophila*. Moreover, the work of Takada (1966) emphasized the drosophiloid nature of the genitalia of these flies. It is not practical to differentiate genera strictly upon internal characters even though it is apparent from the excellent works of Throckmorton that the characteristics of the internal reproductive systems of both sexes are more reliable for determining phylogenies than are external characters.

In the present classification the scaptomyzoids are divided into two genera, *Scaptomyza* Hardy and *Titanochaeta* Knab. *Titanochaeta,* represented by 11 species, is very close to *Scaptomyza* although their biology is very distinctive; these species are apparently all predators upon the eggs of endemic spiders. Members of this genus are characterized by having the head equal to or narrower than the thorax, rather than distinctly broader; by having the front slanted, distinctly oblique rather than almost straight; compound eyes densely pilose, except in *S. vittiger* Hardy, rather than bare or nearly so; lower margin of head short, less than one-half as long as front (except in *vittiger*) rather than having the lower margin of the head approximately equal in length to front; and arista lacking ventral rays except in *S. contestata* Hardy, which has one ventral ray. The male genitalia are rather similar to those of some *Scaptomyza (Trogloscaptomyza)* but the female ovipositor is slender, sharp-pointed, rather needle-like.

The scaptomyzoids contain 132 presently known species, 11 *Titanochaeta* and 121 *Scaptomyza,* sensu lato; 86 in the subgenus *Trogloscaptomyza* Frey, and 35 arranged in six other subgenera. These include *S. (Alloscaptomyza)* Hackman, an endemic group containing eight known species; *S. (Bunostoma)* Malloch, described from the Marquesas Islands, but best developed in the Hawaiian Islands, where eight endemic species are known at present.

S. (Exalloscaptomyza) Hardy was erected to contain one species, *mauiensis* (Grimshaw), which is characterized by having very short rays on the arista and short anterior dorsocentral and sternopleural bristles. The genitalia of both sexes are very distinctive. When the preliminary study was made *mauiensis* was thought to occur on all of the main islands. After detailed studies (Hardy, 1966), what was originally taken as *mauiensis* has been divided into six species. The islands of Kauai, Oahu, Maui, and Molokai each have one distinct species of *Exalloscaptomyza,* and the island of Hawaii has two species. All members of this subgenus breed in the flowers of morning-glory in the semi-wet areas of the islands, usually at elevations of 1,000 to 3,000 ft.

S. (Parascaptomyza) Duda is introduced and contains a cosmopolitan species, *S. pallida* (Zetterstedt). *S. (Rosenwaldia)* Malloch was described from the Marquesas Islands and contains six Hawaiian species. *S. (Tantalia)* Malloch is an endemic subgenus containing six known species.

S. (Troglosocaptomyza) Frey is an enigma. Eighty-six species are now known from Hawaii, 87 for the entire world. The type of the subgenus, *S. brevilamellata* (Frey), was described from the island of Tristan da Cunha, in the middle of the Atlantic Ocean. The group is unknown elsewhere in the world, except for the Hawaiian islands. Dr. Walter Hackman of the University of Helsinki, has done extensive studies on the genus *Scaptomyza* (1955, 1959). He has compared species of Hawaiian *Trogloscaptomyza* with Frey's type and has decided that they are congeneric. Most *Scaptomyza* over the world are characterized by having two to four rows of acrostichal setae. Many of the species of *Trogloscaptomyza* in Hawaii have six well-developed rows of acrostichals.

Throckmorton (1966) placed *Drosophila parva* Grimshaw in the scaptomyzoids. Likewise, Spieth (1966b), using mating behavior, places *parva* as a *Scaptomyza*-like species, as does Clayton (1968) on the basis of metaphase chromosomes. I have since checked *parva* again in more detail and find that it definitely is a *Scaptomyza*. It also is apparent that this is a species-complex, and from the preliminary study it appears that each island has a distinct species of this complex. It will be necessary to assign this complex to a new subgenus under *Scaptomyza*. The genitalia are definitely *Scaptomyza*-like, as is the body form, coloration, and shape of legs. The arista also is not too atypical for some *Scaptomyza,* having three dorsal and one ventral ray in addition to the large apical fork.

The following is a summary of the present taxonomic arrangement of the genera and subgenera of the Hawaiian Drosophilidae:

Subfamily Amiotinae

Gitonides Knab, 1 introduced species
Pseudiastata Coquillett, 3 introduced species (not established)

Subfamily Drosophilinae

Drosophila-like genera (drosophiloids)
Genus *Antopocerus* Hardy, 9 endemic species
　　Ateledrosophila Hardy, 2 endemic species
　　Celidosoma Hardy, 1 endemic species
　　Chymomyza Czerny, 1 introduced species
　　Dettopsomyia Lamb, 2 introduced species
　　Drosophila Fallén
　　　Subgenus *Drosophila* Fallén, 　　305 endemic species,
　　　　　　　　　　　　　　　　　　　　10 introduced species
　　　Subgenus *Sophophora* Fallén, 　　3 introduced species
　　　Subgenus *Engyscaptomyza* Kaneshiro, 　6 endemic species
　　　　　　　　　　　　　　　　　　　　324 total

　　Grimshawomyia Hardy, 2 endemic species
　　Nudidrosophila Hardy, 5 endemic species
Scaptomyza-like genera (scaptomyzoids)
Genus *Scaptomyza* Hardy
　　Subgenus *Alloscaptomyza* Hackman, 　8 endemic species
　　　　　　Bunostoma Malloch, 　　　　8 endemic species
　　　　　　Exalloscaptomyza Hardy, 　　6 endemic species
　　　　　　Parascaptomyza Duda, 　　　1 introduced species
　　　　　　Rosenwaldia Malloch, 　　　6 endemic species
　　　　　　Tantalia Malloch, 　　　　6 endemic species
　　　　　　Trogloscaptomyza Frey, 　　86 endemic species
　　　　　　　　　　　　　　　　　　　121 total

Genus *Titanochaeta* Knab, 　　　　　　11 endemic species
　　　　　　　　　　　　Total　479 presently known species in Hawaii

One of the most remarkable features of the Hawaiian fauna is the definite intergradation that is obvious between the two major genera *Drosophila* and *Scaptomyza*. As known from other world areas, these are distinct, well-defined genera, clearly differentiated by *Scaptomyza* having only two to four rows of acrostichal setae on the mesonotum; lacking ventral rays on the arista; having more elaborately developed male genitalia; frequent occurrence of weakly sclerotized, nondentate female ovipositor; very short egg filaments; primitive courtship and mating behavior; and by the characteristics of the internal reproductive system of both sexes (Throckmorton, 1966). *Scaptomyza* also differs from *Drosophila* by being dark-colored, grey pollinose, small in size and distinctive in body form. The biologies are characteristic; for the most part, *Scaptomyza* from other parts of the world are primarily leaf-miners in fresh leaves.

In Hawaii, numerous cases occur where *Drosophila* and *Scaptomyza* intergrade to a point where they cannot be separated by the normally accepted external morphological characters and may be differentiated only on the basis of mating behavior, internal morphology, and egg structure. At least half of the species of *Scaptomyza* (*Trogloscaptomyza*) have six distinct rows of acrostichal setae. Furthermore, Heed (1968) stated "the question arises whether the scaptoid line of descent overlaps the drosophilid line in ecological habits or has taken a different route. With the information available, it appears that it has done both. The *Tantalia, Exalloscaptomzya* and *Titanochaeta* have their own niches, while the *Trogloscaptomyza* overlaps somewhat the habitat of the more aggressive drosophilids."

The scaptomyzoids have comparatively elaborate male genitalia. As has been pointed out by Spieth (1966b), *Scaptomyza* exhibits a very primitive mating behavior, and it is apparent that the elaborate genitalia act as barriers for mating between different species. In *Drosophila* and closely related genera, on the other hand, the genitalia are for the most part very similar, especially in closely related species; and the isolating mechanism preventing cross-mating is the elaborate courtship and mating behavior, as has been thoroughly discussed by Spieth (1952, 1966b).

The borderline Hawaiian forms between *Drosophila* and *Scaptomyza* are as follows:

A white-tipped-scutellum species group, consisting of a complex of species presently treated under *Drosophila,* is characterized by being slender-bodied with a low body profile, having a small white to yellow spot at apex of scutellum and each labellum of the male having a heavily sclerotized dark brown to black rim along the margin (except in the species *D. fungiperda* Hardy). The larvae are apparently all fungivores and breed in polyporous and other fleshy fungi. These forms all possess the typically accepted characteristics of *Drosophila,* i.e., several ventral rays on the arista, and numerous dorsal rays, numerous rows of acrostichal setae on the mesonotum and relatively simple genitalia. Throckmorton (1966) has placed the white-tipped species with *Drosophila,* although he has stated that they show several *Scaptomyza*-like characteristics. The egg of these species have very short filaments and are somewhat *Scaptomyza*-like. Throckmorton has indicated that the paragonia are fully *Drosophila*-like. Also the ventral receptacle is *Drosophila*-like as is the female ovipositor. Spieth (1966b) said "the members of this group are quite distinctive from all other Hawaiian species not only with respect to their anatomy but also with respect to their mating behavior. The male courtship is essentially *Scaptomyza*-like." It is evident that this group should be erected to the rank of a new subgenus under *Drosophila*.

The *"Drosophila" parva* Grimshaw complex of species, which in the preliminary classification (Hardy, 1965) was placed under genus *Drosophila* because of the presence of a ventral ray on the arista and numerous rows of acrostichal setae, has now been found to be a species complex, as has been stated above; it has probably speciated by islands. The group has not been studied in detail, but it obviously will be necessary to include it as a new subgenus under *Scaptomyza.*

The *"Drosophila" crassifemur-nasalis* complexes of species, on the basis of Throckmorton's studies, have been placed as scaptomyzoid; although he did state that the pigmentation of the vasa, the ejaculatory apodeme and the anterior spiracle of the pupa are *Drosophila*-like. Spieth (1966b) says that their mating behavior is typically scaptomyzoid; furthermore, the metaphase chromosomes resemble those of scaptomyzoids. As has been stated elsewhere, a more detailed study has demonstrated that these represent two distinct complexes of species, which form a new subgenus. On the basis of their external morphological characters, Kaneshiro (1969) has described this as *Drosophila (Engiscaptomyza).* It is obviously borderline between *Drosophila* and *Scaptomyza.* The subgenus contains four species in the *crassifemur* complex and two species in the *nasalis* complex.

It is probable that the species *D. reducta* and *D. taractica* Hardy should be placed under *Scaptomyza.* Although presently under *Drosophila,* the male genitalia are definitely *Scaptomyza*-like. These species have not been studied further since the original descriptions.

The *Drosophila anomalipes* Grimshaw complex of species, consisting of two species, *anomalipes* and *quasianomalipes,* which occur in the same habitat in the Kokee area of Kauai, obviously represent a primitive group. Throckmorton (1966) points out that the spermathecae are rather *Scaptomyza-like.* Spieth (below) discusses the mating behavior.

The *Drosophila primaeva* Hardy and Kaneshiro complex, consisting of two species (*attigua* and *primaeva*), which occur in the same habitat as *anomalipes* and *quasianomalipes* on the island of Kauai, is similar in many respects to the above complex but differs in a number of important respects pointed out by Hardy and Kaneshiro (1968). The actual breeding habitat of these species is not known. The males are characterized by having a sclerotized rim on the labellum; all other known Hawaiian *Drosophila* that possess this character are thought to be fungus breeders. From preliminary cytological studies made by Stalker (1968) it appears that *primaeva* is exceptional in that one arrangement in chromosome 5 closely resembles the sequence found in certain mainland species.

The numerous examples of intergradation between *Drosophila* and *Scaptomyza,* coupled with the fact that *Scaptomyza* has attained its greatest development in the Hawaiian Islands (presently the known number of species is two times greater in the Hawaiian Islands than for the rest of the world), lead to interesting conclusions. The circumstantial evidence presented by these data would obviously indicate that the genus *Scaptomyza* originated in the Hawaiian Islands and that the entire Hawaiian fauna could have originated from one ancestral species. Throckmorton (1966) in his summary made the following conclusions with regard to the major groupings of Hawaiian species: "phylogenetically, both of these groups are very closely related to each other and they are derived from near the base of the major branch leading to the subgenus *Drosophila* of the genus *Drosophila*. Evidence is presented indicating that the genus *Scaptomyza* originated in Hawaii and arguments for and against this interpretation are discussed. If *Scaptomyza* originated in Hawaii, then the available evidence favors the introduction of only a single individual (basically a *Drosophila*) as the progenitor of the more than 400 endemic species of Drosophilids. At most, two introductions, presumably of a single individual each, are required if the Scaptoids are thought to have originated from an introduction separate from that of the Drosophiloids. Existing evidence indicates that the Drosophiloid introduction was from east Asia, perhaps from Japan, but this problem is still under investigation."

This theory has rather startling implications, since these islands are comparatively young (one to six millions of years for the present main islands) and it seems rather inconceivable that *Scaptomyza* could have spread over much of the entire world from an isolated group of small islands in the middle of the Pacific Ocean. Much remains to be learned concerning this question and also the distribution of *Scaptomyza* over the world. Throckmorton (1966, pp. 385-386) stated "Regardless of the geographical point of origin of the Scaptoids, present evidence requires that they share a close common ancestor with the Hawaiian Drosophiloids. If the origin of the Scaptoids occurred outside of Hawaii, then two successful trans-Pacific colonizations are needed. Each one of these was, in itself, an improbable event. That a successful introduction be made twice from the same family of Diptera is even less probable, and that the two successful introductions from the same family should involve species so closely related (but presumably already generically distinct) as to produce the existing patterns of variation in Hawaii is less probable still. And if we do not postulate extremely close relationships between the two original colonizers, we must then explain the broad overlaps between the Drosophiloids and Scaptoids as due to convergent evolution. This requires that we explain why the Scaptoids should have diverged in the direction of *Drosophila* (while they were competing with them among the vacant niches in Hawaii), and why, for some character states at least, they have happened to diverge specifically in the direction of the *Hawaiian* Drosophiloids. And why do the Hawaiian Drosophiloids, of all the Drosophiloids in the world, include a group of species ("white-tip scutellum" forms) that share characters with the Scaptoids, their nearest (geographical) relatives? . . . For the present, then, Hawaii must be considered to be the only place in the world where the otherwise sharp distinctions between *Scaptomyza* and *Drosophila* tend to disappear."

The Hawaiian fauna exhibits a high degree of endemicity by islands. From the preliminary studies, it is not possible to make accurate assessments of the species that are restricted to single islands, since it has become obvious that a high percentage of those which have previously been recorded from two or more islands actually represent species groups. In almost all of those cases which have been investigated thoroughly, a distinct species has been found on each island.

For the purpose of analyzing island endemism it is necessary to treat Maui, Molokai, and Lanai as one biological island, since these three islands have been fused into a single land mass at least two times, during the Pleistocene, due to the periodic retreat of the sea level. Besides, these islands are presently separated just by narrow channels, as discussed by Spieth (1968b; see also below). Their faunas show much closer relationship, and more species-overlapping occurs than between other islands, due to the comparatively recent islation and their close proximity.

From the records published in the 1965 monograph with some corrections that have subsequently been made, it would seem that approximately 85 percent of the endemic drosophiloids have speciated by islands, treating Maui, Molokai, and Lanai as a biological island. The scaptomyzoids have not been considered in this analysis, since they are still rather poorly known. It is highly probable that the actual figure for endemism by islands should be 95 to 98 percent.

The best understood of our species to date are the large *Drosophila* that have extensive maculations on their wings (picture-winged; Fig. 4.52-72). These have received special attention (Carson and Stalker, 1968a, b, c; and Hardy and Kaneshiro, 1968) due largely to the fact that they are so ideal for refined genetic and laboratory studies. For the first time we have enough detailed information concerning a large segment of native species to provide a basis for a sound systematic arrangement of the species. The results achieved with this group mark a major milestone in our understanding of the Hawaiian drosophilids. It is however, apparent that the picture-winged species constitute a small segment of the total *Drosophila* fauna, perhaps 20 percent (Carson and Stalker, 1968).

A fairly accurate analysis of the degree of speciation by islands can be made using the picture-winged species. The following have speciated by single islands. All the members of the *pilimana-hawaiiensis* complexes (51 species, Hardy and Kaneshiro, 1968, 1969); all of the *adiastola* complex (7 species, Hardy and Kaneshiro, 1968, 1969); all of the *semifuscata* complex (4 species, Hardy and Kaneshiro, 1968); and all of the "idiomyia" complexes of species (15 species, Hardy, 1969); *varipennis* complex (3 species, Hardy and Kaneshiro, 1968 and 1969); *haleakalae* complex (17 species of fungus feeders); and miscellaneous *Drosophila* that have extensive wing markings such as *fuscamoeba* Bryan and related species; *peniculipedis* Hardy, recorded from Maui and Oahu (the latter is very probably an error); and *aglaia* Hardy, recorded from Oahu and Hawaii (the latter record based upon one headless female and is no doubt an error). Considering all of the species that have rather extensive maculations in the wing, approximately 126 species to date (including 17 species of the *haleakalae* complex), only two, *crucigera* Grimshaw and *grimshawi* Oldenberg, are found on more than one main island. From an intensive study of the picture-winged species, based upon morphology it would appear that approximately 99 percent of the species are restricted to single islands.

The greatest known concentration of species is on the island of Maui as pointed out by the works of Carson and Stalker (1968a, b, c), many species groups have apparently originated on Maui, have undergone extensive speciation there and have spread to other islands, especially Oahu and Hawaii. Also from Carson's studies, using the picture-winged complexes, it is apparent that in some cases migrations have occurred from Kauai directly to Maui (bypassing Oahu) and that at least a number of species groups have developed there and radiated out to other islands. The greatest number of of known species from any area in the Islands is in the Waikamoi section on the slopes of Haleakala on Maui (Fig. 3). This is rain forest, ca. 4000 feet elevation, with rainfall of 150 to 200 inches per year, and where all of the important host plants are available. Sixty-three described species of *Drosophila* and related genera have been recorded to date from this area.

The oldest islands of the Hawaiian chain (the Leeward Islands) are now worn down to the point where they no longer support a native fauna of Drosophilidae except possibly for a few species of *Scaptomyza*. The native species are now confined to the main islands: Kauai, Oahu, Maui, Molokai, Lanai, and Hawaii. Niihau and Kahoolawe are not at present accessible for field study but apparently support no native fauna. These are low islands,

1281 and 1477 feet, respectively; both are in the rain shadow and have been completely changed by man. The ancestors of the tremendous present day fauna that has developed were doubtlessly from Kauai, and one would expect that since this is the oldest of the main islands, it would support the greatest number of species of native plants and animals; but at least with the Drosophilidae this is not the case. By comparison with Maui, Hawaii, and Oahu, Kauai has a rather small fauna. To date, 44 species of *Drosophila* are known from this island. Five of these have also been recorded from other islands, and apparently 39 species are restricted to the island of Kauai. It is interesting to note, and also highly significant in the study of the evolution of this group in Hawaii, that of the drosophiloids only the genus *Drosophila* occurs on Kauai; *Antopocerus* Hardy, *Celidosoma* Hardy, and *Nudidrosophila* Hardy are unknown from this island. Also the picture-winged species of the genus *Drosophila* are poorly represented, and the "idiomya" and *haleakalae* complexes are unknown. It is evident that several important species complexes have probably developed on the island of Maui and have radiated out from there to the other main islands (see section, below, p. 469 ff.). Maui is very rich in species by comparison to Kauai, with 106 recorded species of *Drosophila* and 7 species belonging to genera close to *Drosophila* (*Antopocerus, Grimshawomyia,* and *Nudidrosophila*) for a total of 113 drosophiloids; 5 of these species have been recorded on other islands besides Molokai and Lanai. Molokai has 55 recorded species of drosophiloids. Of these, 26 are presently known only from the island of Molokai. Twenty-two are also found on Maui, and 13 are recorded as occurring on Hawaii, Oahu, or Kauai; these need further study and many may be erroneous records. Lanai has a very impoverished fauna, the island having been completely altered by man, and most of its endemic fauna lost. Only 13 species have been recorded from Lanai, 10 of which are also found on Maui, and three may possibly be endemic to Lanai. Two of these three, *D. lanaiensis* Grimshaw and *polita* Grimshaw, are known only from female specimens collected in 1893. The other, *D. kraussi* Hardy, was described from one male specimen from Lanai, and a series of specimens from Mt. Kaala, Oahu, were also placed here. This needs further study and may represent two species. One hundred and sixty species are found on the three islands, with 142 apparently restricted to Maui (including Molokai and Lanai). This compares with 98 species from the island of Hawaii, 91 *Drosophila* and 7 species in genera related to *Drosophila*. All but 4 of these are apparently restricted to the island of Hawaii. Oahu has 84 species, 5 belonging to genera related to *Drosophila*. All but 8 species are apparently restricted to Oahu. It should be noted that the total count of species in the above analysis is greater than the presently described number, due to the inclusion of a number of recognized species complexes that have not yet been treated taxonomically.

One factor that has had and is still having a serious effect upon the native fauna is the presence of predaceous ants in the islands. Forty-two species of introduced ants have been reported. These are widespread over the main islands and have no doubt taken a toll of native insects. Surely our leaf-breeding drosophilids would be affected where ant populations are present. It is felt that this is one of the major reasons why the island of Kauai has such a comparatively small present-day fauna of *Drosophila*. Ants are widespread over this island up to an elevation of 4000 feet, at Kokee. Fortunately they have not invaded back into the native rainforest or the Alakai Swamp area, or the elevations above 4000 feet, but collecting is very poor over this island below this elevation, probably because of the effect of the ants. This is also very dramatically demonstrated on Mt. Tantalus, Oahu, where collecting is now extremely poor and where before the turn of the century when Perkins did the collecting for the Fauna Hawaiiensis, the fauna must have been comparatively rich. Several of the species (other families) that were collected on Tantalus by Perkins are now apparently extinct.

Biology and Behavior

The Hawaiian species of the family Drosophilidae, as shown by Hardy (see above), form two interlocking groups: the drosophiloids and the scaptomyzoids, each consisting of a number of closely related genera, subgenera, and species groups. Ecological and behavioral data pertaining to the scaptomyzoids are presently more fragmentary and incomplete than are those available for the drosophiloids. The present section therefore emphasizes the drosophiloids, with references to the scaptomyzoids in those instances for which adequate information is available to illustrate differences or similarities between the two groups.

Life Cycle

The eggs display interspecific and intergroup variety of considerable extent, especially involving size, sculpturing, and number of chorionic egg filaments (Throckmorton, 1966). Scaptomyzoid eggs have zero to four short, heavy chorionic filaments. Drosophiloid eggs have two basic patterns of filaments. The eggs of the white-tipped scutellum group have either two or four short, slender filaments while the remaining species all have four. Kambysellis and Heed (1969, in ms.) studied the eggs and their development in the ovaries of 60 species. Represented in their samples were both scaptomyzoid species (*Exalloscaptomyza,* 4; *Engiscaptomyza,* 6), and drosophiloids (*Antopocerus,* 5; *Drosophila,* 45). The number of ovarioles varied from 2 per fly in some species to as many as 100 in other species.

The distribution, however, was not normal and tended to cluster in three groups. Group I consists of 10 species with 2 to 5 ovarioles per fly, the majority having four. Group II encompasses 15 species with the ovariole number varying from 8 to 20, but with most species having 10 to 15. Group III had ovarioles numbering between 20 to 100, with the most common number being 40 to 50. A number of other morphological and functional features appear to be correlated with the distribution of the ovarioles for the members of each group, i.e., the number of mature eggs found in the ovaries of each individual, the ratio of length of egg filaments to length of egg, and the ovipositional behavior of the females.

The Group I flies, regardless of the number (2 to 5) of ovarioles per individual, have only one ovariole in each ovary functional at any given time and, further, these function in an alternating manner so that only one fully developed egg can be found in a mature female. Thus these flies have the potential of laying only one egg at a time and not more than one egg per day. The individual eggs are large in comparison to the size of the fly, and lack chorionic filaments. In the laboratory the eggs are deposited individually on the sides of the vial, on the paper surface or the culture medium. They are not pressed into the food surface. The natural ovipositional sites of the four species of scaptomyzoids of the subgenus *Exalloscaptomyza* are known to be the flowers of the morning-glory, in which they deposit eggs on the anthers or stamens. The natural ovipositional sites of the *Engiscaptomyza* and *Grimshawomyia* species are presently unknown. *Engiscaptomyza crassifemur* and *nasalis,* however, can be reared under laboratory conditions and do lay a single egg per day. The eggs of Group I species can be held in the vagina of the female until a full-grown first instar larva has developed, so that the flies behave as oviparous species. It is to be noted, however, that not all of the Hawaiian scaptomyzoids conform to the pattern displayed by the species that Kambysellis and Heed studied. For example, species of the subgenus *Alloscaptomyza* produce very small eggs (Throckmorton, 1966) but the details of their ovipositional behavior are not fully known.

Most species of the Group II flies have 10 to 15 ovarioles per fly but show considerable intraspecific variability. Kambysellis and Heed found that the mean number of ovarioles per fly is positively correlated with the size of the individual. Robertson et al. (1968) suggest that the differences in size of the individuals of a species probably are due to variation in larval food supply. The alternating function of the ovarioles, which is characteristic of Group I, also pertain to Group II but is modified to the extent that more than one ovariole in each of the two ovaries is functional. Kambysellis and Heed found no ovariole contained more than one mature egg at the time of dissection. Without exception, the Group II flies oviposit only in decaying leaves of various endemic plants (Heed, 1968) and, furthermore, they release the eggs singly. The eggs have four chorionic filaments, which typically do not exceed the length of the egg itself. The females possess elongate ovipositors, which enable them to insert the eggs into a decaying leaf, but the chorionic filaments remain outside the epidermis on the surface of the leaf. The eggs are deposited soon after maturation, and thus embryonic development does not precede oviposition.

The Group III species display much more interspecific variability than do those of Groups I and II. The majority of the species have 40 to 50 ovarioles per fly. The alternating function of the ovarioles does not occur, and some species may have mature eggs in three fourths of the ovarioles, others as many mature eggs as ovarioles, and still others more mature eggs than ovarioles. The eggs themselves show much interspecific variation. All species studied, with a single exception (*Drosophila conspicua*), have four chorionic filaments but the filament/egg-length ratio varies from 0.71 in *Drosophila truncipenna* to 3.93 in *D. sejuncta*. This ratio is correlated with the number of mature eggs present in the ovaries, i.e., the smaller the ratio the fewer the number of mature eggs present. The females of all these species insert the eggs into the substrate in which the larvae develop, but those having a filament/egg-length ratio of less than 1.5 deposit their eggs individually, while those with a higher ratio often will at a single insertion of the ovipositor deposit a cluster of eggs. Those species which oviposit their eggs in clusters appear to utilize specific substrates that are nutritionally rich, such as the decaying bark of *Cheirodendron* and *Clermontia*. Such substrates can support a considerable number of larvae. Those species of Group III which deposit their eggs singly typically utilize a variety of different types of substrates, e.g., *Drosophila mimica* uses fungi, leaves of several plants, and fruits of the soapberry *Sapindus*.

Kambysellis and Heed have thus been able to show that the various species of Hawaiian flies have evolved adaptive reproductive mechanisms that allow for a maximum utilization of the various types of nutritional sources that are available, at the same time minimizing the competition for food.

As indicated above, the females of the various species with certain exceptions oviposit on or in rotting (fermenting) vegetable material substrates. Heed (1968) lists nine major ovipositional substrates (see Table V). Characteristically the leaves used for ovipositional sites are those which have abscissed from trees or shrubs and have fallen to the forest floor in such a situation that they not only retain their moisture but also have absorbed additional moisture from the rainfall and the wet substrate. Leaves that become entangled in the ferns or other lower story plants, or are exposed to sunlight, desiccate to a point that they are unattractive as ovipositional sites. The females almost invariably insert the eggs into the original lower surface of the leaves; the long egg-filaments of the drosophiloids extend on the outside so that the investigator can ascertain exactly how many eggs

TABLE V

Substrate Distribution of Reared Endemic Hawaiian Drosophilidae[a]

	Leaves	Stems	Flowers	Fruits	Fungi	Slime Flux	Frass	Ferns	Parasite	No. of Species
Drosophiloids										
Picture-wings	4	8	4	4	2	3	2			13
Modified-mouthparts	10	13	3	7	4					23
Ciliated-tarsi	10	3		4	1					17
Unclassified	6	1		1						8
Nudidrosophila		2								2
Antopocerus	11									11
Bristle-tarsi	16									16
Fork-tarsi	11									11
Spoon-tarsi	14									14
White-tipped scutellum					9					9
Number	82	27	7	16	16	3	3	2	0	
Frequency	.51	.17	.05	.10	.10	.02	.02	.01	–	
Scaptomyzoids										
Trogloscaptomyza	6	2	9	8			1			19
Parascaptomyza	1		1		1	1				1
Bunostoma					1					1
Tantalia	3									3
Exalloscaptomyza				4						4
Titanochaeta									6	6
Number	10	2	14	8	2	1	1	0	6	
Frequency	.23	.05	.32	.18	.05	.02	.02		.13	
Total Number	92	29	21	24	18	4	4	2	6	
Frequency	.46	.15	.10	.12	.09	.02	.02	.01	.03	

[a] Modified from Heed (1968).

have been oviposited in each leaf. The broad-leafed evergreen trees such as *Cheirodendron, Clermontia,* and *Ilex* shed leaves throughout the year, thus making properly fermented ovipositional sites continuously available. The broken, rotting ends of small limbs and large twigs, especially of *Clermontia,* which has a thick soft bark, are regularly used for oviposition and feeding by the adults. When a *Clermontia* shrub or tree dies, then for a long period of time it will attract an amazing number of species. To a lesser extent bark of other host trees or shrubs, especially *Cheirodendron,* will be likewise used by the flies.

Fungi (Spieth, 1966b), slime fluxes, which are relatively rare, and frass occur at various levels in the forest and are also sought out by various species of the flies. In general, however, the majority of the species oviposit in materials that are lying on the forest floor, i.e., fallen leaves, fallen fruits and flowers of *Clermontia* and other lobeliads, broken limbs, and the fungi on fallen trunks or limbs.

Some species and some species groups of both scaptomyzoids and drosophiloids are substrate specific (Heed, 1968). Thus, species of *Tantalia*, the spoon-tarsi, the fork-tarsi, the bristle-tarsi groups, and *Antopocerus* utilize only leaves; *Nudidrosophila* is restricted to stems; *Exalloscaptomyza* prefers the flowers of the morning-glory; the exceptional genus *Titanochaeta* uses the egg masses of spiders, while the white-tipped scutellum group oviposit only in fungi. In comparison, the picture-wings, modified-mouthparts, and ciliated-tarsi groups, and *Trogloscaptomyza* are more catholic in their ovipositional behavior. Significantly these latter groups are, by species, the largest in Hawaii.

The accumulated data (Heed, 1968) show that, excluding fungi and the spider egg-masses, 32 endemic and 6 introduced genera of plants belonging to 30 familes serve as ovipositional sites for the endemic Drosophilidae. Of these the genera *Cheirodendron* (Araliaceae) and *Clermontia* (Lobeliaceae) serve as ovipositional and larval food sites for more than half of all the species that have been reared from all sources. The fermenting leaves, stems, and roots of *Cheirodendron* and the leaves, stems, roots, and fruits of *Clermontia* all serve as ovipositional sites. From *Cheirodendron* 51 species have been reared, and from *Clermontia* 38. Of these 89 species only three (*D. disticha, Trogloscaptomyza hackmani,* and *Tantalia gilvivirilia*) came from plants of both genera. Interestingly, one introduced drosophiloid species, *D. immigrans,* also uses the fermenting leaves and fruits of *Clermontia* as a breeding site. The next most frequently used plant is the monotypic genus *Ilex,* from which 14 species of drosophilids have been bred to date, the majority of which do not use either *Cheirodendron* or *Clermontia*. Thus, less than 10 species of the three plant genera account for approximately two thirds of all the species that have been associated with their ovipositional-larval-developmental substrates. An understandable effect of this concentration of species depending on a limited variety of food types is that many species can be reared from an individual plant or even portions of an individual plant. Heed (1968) has reared 59 individuals representing nine different species from the leaves of a single, physically isolated *Cheirodendron* tree in Kipuka Puaulu (see Table VI). Interestingly, also, the ancestral stocks that gave rise to *Cheirodendron* and *Ilex* originated in Asia, but apparently *Clermontia* and the other lobeliads came from the Americas.

The spectacular nature of the Hawaiian drosophiloids led investigators during the first half of this century to attempt to culture the flies in the laboratory. Consistently, experienced scholars found this impossible to ac-

TABLE VI

Specimens Reared from Leaves of a Single Isolated Tree of *Cheirodendron gaudichaudii* (D.C.) Sem., Kipuka Puaulu, Hawaii[a]

	13.IV.66	3.VI.66	XI.66	10.VII.67
A. cognatus				1
A. tanythrix			7	
D. cnecopleura				1
D. disticha	1	1		1
D. neutralis		3		
D. sordidapex		3		
D. spiethi	9	21		7
D. trichaetosa	2		1	
S. hackmani		1		
Total of 59:	12	29	8	10

[a] From Heed (1968).

complish. Neither the adult flies nor the developmental stages would tolerate in the laboratory those methods that were successful for rearing species from other parts of the world. New techniques and new culture methods were clearly necessary. After a number of unsuccessful attempts, Wheeler and Clayton (1965) devised a food medium that, when supplemented in each food vial with a strip of absorbent tissue, moistened on one end with a dilute sterile solution of Karo (maize-sugar syrup) and yeast hydrolysate extract and on the opposite end with distilled water, is adequate for the rearing of a number of species. Large vials are usually used but some of the larger species will oviposit more readily if maintained in small cages (Clayton, 1969). Moreover, adults of most species (both field-captured and laboratory-reared) can be maintained in good health in "sugar" vials (Spieth, 1966a) for prolonged periods of time.

Those species which can currently be reared in the laboratory belong with few exceptions to the modified-mouthparts and picture-wing groups of the drosophiloids and to the scaptomyzoid subgenera *Bunostoma* and *Trogloscaptomyza*. Excepting *Bunostoma*, for which Heed (1968) reports rearing only one species (see Table V), the other three groups are the most polyphagous of all the Hawaiian flies. The essentially monophagous groups such as *Antopocerus*, bristle-tarsi, fork-tarsi, and spoon-tarsi are currently unculturable even though their natural foods are known. Furthermore, many species of the picture-wing and modified-mouthparts species can either not be cultured or at best can be maintained for only one or two generations.

Ever since Heed in the summer of 1963 discovered that fermenting *Cheirodendron* leaves serve as a prime developmental site for many species, attempts have been made by analyzing these leaves chemically and physically to gain leads and insights into the preparation of new media for the "leaf feeders." Robertson et al. (1968) attacked the problem by analyzing the *Cheirodendron* leaf and using *Drosophila disticha* as a test organism. *D. disticha* is a leaf-breeding spoon-tarsi species that occurs in abundant, easily collectable numbers in the Waikamoi area of Maui. Heed (1968) has reared from natural substrates many individuals, the vast majority of which came from *Cheirodendron,* and an occasional individual from *Tetraplasandra, Pittosporum,* and *Myrsine* leaves.

Robertson and his associates dissected and analyzed the contents of the digestive systems of field-captured adults and larvae. The crops of the adults carried both yeasts and bacteria. The yeast contents of the males were quantitatively lower than were those of the female. Also a spectrum of bacteria occurred in the males different from that found in the females.

The larval gut content varied from the adults in that all larvae carried at least one of several recognizably different bacteria, while yeasts were present in only 3 of 40 individuals dissected, and probably 2 of these 3 were contaminants. Thus unexpected differences were found, not only between the sexes of the adults but more importantly between the adults and the larvae.

Robertson and coworkers fractionated the *Cheirodendron* leaf into three parts: Fraction I contained the combined organic solvents, Fraction II the water soluble constituents, while Fraction III was presumed to be principally cellulose. By adding Fraction I to agar they were able to induce the females to lay eggs, whereas with plain agar the females refused to oviposit. Then with axenic techniques and Medium C (Sang, 1956) they attempted to rear the larvae. On Medium C alone the larvae died in first instar. By adding appropriate amounts of supplements of various sorts to Medium C, including the *Cheirodendron* fractions and the yeasts found in the crops of the adults, they found that in some instances the larvae were able to develop into "well-grown" instar II but never to maturity (see Table VII).

Robertson et al. (1968) concluded that they apparently were "dealing with rather extensive metabolic differences between *disticha* and the species of *Drosophila* whose nutrition had been so far examined" and further that "the nutrition of a species like *disticha* may have more in common with that of soil-living nematodes than with the yeast-feeding Drosophilidae."

Subsequent to the investigations of Robertson and coworkers, attempts have been made to devise new types of food medium. Kircher et al. (1968) prepared a medium that gave some but not complete success for the rearing

TABLE VII

Effects of Supplementing Medium C with Various Concentrates, Compounds, Leaf Fractions, and Yeasts[a]

Supplement	Growth[b]
1. None	0
2. Yeast extract	0
3. Malt	0
4. Malt and Neopeptone	+
5. Malt and phospholipid concentrate	0
6. Malt and brewer's yeast/mushroom extract	+/++
7. Neopeptone	+
8. Phospholipid concentrate	+
9. Neopeptone and phospholipid concentrate	+/++
10. Fraction I	0/+
11. Fraction II	0
12. Fraction III	0/+
13. Fractions I and II	+
14. Fractions I and III	0/+
15. Fractions II and III	+
16. Fractions I and II and III	+
17. Fraction II and *Torulopsis*	+/++
18. Fraction II and *Candida*	+/++
19. Fractions I and II and *Torulopsis*	+
20. Fractions I and II and *Candida*	+
21. Fraction leaf homogenate	0

[a]Modified from Robertson et al. (1968).
[b]Symbols: 0, No growth beyond instar I; +, Survival to small-sized instar II larvae; ++, Survival to large-sized instar II larvae.

of the leaf-breeding flies. Recently Kircher (unpublished manuscript) has analyzed both green and decomposing leaves of *Cheirondendron* for sterols, triterpenes, and fatty acids. He ascertained that the fatty-acid portion consisted mainly of palmitic, linoleic, and linolenic acids, while the nonsaponifiable fractions contained the sterols β-sistosterol and stigmasterol plus the triterpene cycloartenal. This study complements that of Robertson et al. (1968), who showed that the leaves contain the usual amino acids and sugars plus bacteria during the process of decay. Kircher found the leaves to contain adequate amounts of sterols for the needs of the developing larvae during all stages of leaf decay. The fatty acids are the usual ones that are found in plants. Thus the evidence to date indicates that the *Cheirodendron* leaves neither possess some unique biochemical constituent nor lack categories of materials that set them apart from other plants. The most feasible interpretation as to why these leaves are such favorite substrates for the developing larvae appears to be the combination of a slow rate of decay, the retention of moisture, and the lack of yeasts and moulds in the interior of the leaves. Kircher believes "the leaf breeders in Hawaii may have evolved to utilize the slow release of nutrients caused by bacteria decay of *Cheirodendron* leaves." An important corollary is the lack of moulds inside the leaves, which causes consistent troubles in the laboratory when Wheeler-Clayton medium is utilized.

The important problem of developing methods for the rearing of additional species under laboratory conditions invites the attention of insect nutritionists as well as *Drosophila* investigators, for until it is solved many aspects of the over-all investigation of the biology and evolution of the Hawaiian drosophiloids simply cannot be solved.

The natural food substrates of the adults are less well known than are those of the larval stages. A number of investigators have observed adults of various species engaging in feeding behavior in the field. The substrates involved included fermenting leaves and bark of *Cheirodendron*; the leaves, flowers, fruits, and bark of lobeliads, especially species of *Clermontia;* the leaves of *Pisonia*; various fungi, especially the bracket fungus *Polyporus*; and the exuding sap from the recently cut stumps of large tree ferns (*Cibotium*). A number of different species have been observed feeding upon each of the substrates listed, and all except the stumps of tree ferns are known to serve as ovipositional sites for one or more species of the flies. There is a general rule that applies to the great bulk of the world's drosophiloids, namely, that the adults of a given species will always feed upon the substrate into which the females oviposit, but additionally they will also feed upon other substrates that are not used for oviposition. Excepting the parasitic species of the scaptomyzoid genus *Titanochaeta,* the Hawaiian flies do appear to conform to this generalization in their feeding/ovipositing behavior.

With rare exceptions when the larvae of the drosophiloids attain maturity and are ready to pupate, they leave their food site regardless of its location, and burrow into the soil; pupae have been repeatedly recovered from the soil. This unusual habit was first reported by Wheeler and Clayton (1965). Many of the larvae exhibit the habit of skipping, and the mature larvae of fungus-feeders have been observed skipping off the fungus in which they developed and falling several feet to the surface of the soil, into which they immediately burrowed (Spieth, 1966b). In those species which are culturable in the laboratory, it is necessary to remove the cotton plugs of the food vials when the larvae are mature and to place the vials in a jar or similar container partially filled with moist sterile sand. The full grown larvae leave the food, crawl out of the vial, and crawl or skip onto the sand, into which they burrow and then pupate. When the adult emerges it forces its way out of the soil (or sand, in the case of laboratory flies) as a teneral individual. Many scaptomyzoids follow this same pattern of behavior, but at least some do not for they pupate upon the material where they spend the larval developmental period. For example, the individuals of the scaptomyzoid subgenus *Tantiala* (Heed, 1968) spend their larval life scavenging the surfaces of leaves and simultaneously cover themselves with tiny bits of dirt and debris strategically placed along their bodies. They then retain this material on the pupal case when they pupate on the leaf. *Trogloscaptomyza cyrtandrae* lays its eggs on the leaves of the native *Cyrtandra* sp., and the larvae apparently feed on the exudations from the leaf hairs found on the under surface of the leaves. Pupation also occurs on the plant, with the pupal case tightly adhering to the leaf surface (Hardy, 1965). It is also of note that Swezey (1954) reported one true drosophiloid, *Drosophila sadleria*, as a borer in the parenchyma of the rachis of *Sadleria* fronds. Pupation of this species also takes place in the fronds.

The length of the life cycle is inordinately long when compared to that of other drosophiloids. The newly emerged adults typically reach sexual maturity 10 to 12 days after emergence; the females reach their peak of

TABLE VIII

Developmental Cycle Time Span[a]

Group and Species	Age of ♀ at 1st Oviposition (days)	Days to Hatching	Development from Eggs to Adults (days)
Picture-wings			
D. adiastola	16	5	32
D. conspicua	15	5	29
D. crucigera	14	5	28
D. discreta	18	4	36
D. engyochracea	25	5	48
D. fasciculasetae	16	4	32
D. grimshawi	16	5	32
D. hawaiiensis	14	5	30
D. hemipeza	15	5	30
D. paucipuncta	15	5	32
D. picticornis	18	5	35
D. pilimana	18	5	41
Modified-mouthparts			
D. asketostoma	9	4	21
D. dissita	10	4	20
D. eurypeza	12	3	24
D. fuscamoeba	12	4	24
D. mimica	10	3	21
D. quadrisetae	10	4	20

[a]Data supplied by Kathleen Resch, Genetics Foundation, University of Texas.

receptivity 14 to 18 days after emergence. Miss Kathleen Resch (personal communication) reports that at the University of Texas larvae and pupae are maintained at 20° to 24°C, and adults and eggs at 14° to 17°C. Under these conditions (Table VIII) eggs do not hatch until 3 to 5 days after oviposition; the period of time from oviposition to adult emergence varies from 20 to 48 days. In comparison, *D. melanogaster* eggs (25°C) hatch in 20 to 22 hours, and *D. funebris* eggs in 30 to 32 hours after oviposition. *D. melanogaster* at 25°C can complete a life cycle in 9 to 10 days. The adult Hawaiian drosophiloids also have a long life span as shown by the fact that field-captured adults have been maintained under laboratory conditions for prolonged periods, e.g., Carson (personal communication) captured an adult female of *D. grimshawi* on July 21, 1964. The specimen oviposited that same day, continued to lay mostly fertile eggs until March 22, 1965, and died on April 20, 1965. Since the fly was mature and inseminated, it must have been more than two weeks old when captured; it must have had a life span in excess of nine months.

The adults, as would be expected, do not randomly distribute themselves throughout the forest habitat but, rather, select and tend to accumulate in areas, often relatively small, that meet their specific ecological requirements. The major factors that control the distribution of the individuals have been delimited grossly and are as follows: wind intensity, humidity, temperature, light intensity, adult food sources, and acceptable ovipositional sites. Each species, of course, appears to have its own specific ecological requirements, but these appear to be so similar for many species that the adults of large numbers of species tend to accumulate in small areas or "pockets" of the forest. The adults of most species avoid even moderate wind currents and light intensities, humidities below 90 percent, and temperatures above 70°F (21.1°C).

For the Hawaiian flies, and especially for the drosophiloids, the temperature requirements appear to be quite exacting. Thus, in rearing the species in the laboratory, it is mandatory that the temperature be kept below 70°F, preferably about 65° to 68°F. Heed (1968) in rearing larvae kept his field-collected cultures (substrates) at 68° to 70° or temporarily at 50° to 55°F. One of the richest collecting sites known on the islands is located at an elevation of 4000 feet along the Kula pipeline at Waikamoi, Maui (Fig. 3). During the period from 13 July to 2 August 1965, a continuous record of the temperatures taken approximately 5 feet above the ground showed the temperature to vary between an average minimum of 56.7°F and an average maximum of 66.8°F. The highest daily maximum was 69°F and the lowest minimum 53°F. During this same period the relative humidity was 95 percent or higher for 77.5 percent of the time, and less than 90 percent for only 2 percent of the time. The surface of the forest floor in such situations hovers around 60°F during all times of the day. These are typical summer temperatures, and, of course, are considerably lower during the winter season, although the flies are active the year round.

The combination of physical environmental requirements listed above is not too commonly found even within the rainforests, especially when added to these factors must be the presence of suitable feeding and ovipositional sites. The preferred sites, therefore, are usually small ravines or areas that are protected from the prevailing trade winds by a ridge or similar physical obstruction. Additionally, the area must have a high and relatively dense forest canopy, usually provided by *Metrosideros* or *Metrosideros* and *Koa* or other trees of comparable size and habit, an understory of shrubs and/or tree ferns (*Cibotium*), and a dense lower level of "bracken" ferns. If in addition ovipositional substrates are present, especially *Cheirodendron* and/or *Clermontia,* areas of less than one acre ($4000m^2$) will serve as a focal point for the adults of numerous species of drosophiloids and scaptomyzoids.

Within such an area, the greatest density both of individual adults and of number of species will be found in the "bracken" fern or lowest level of vegetation. The flies prefer to sit on the under surfaces of leaves, fern pinnae, and horizontal branches of shrubs, and the like. Almost invariably they avoid sitting in a situation where light, even the low-intensity light of such forests, falls directly upon them from above. During periods of heavy overcast, when the relative humidity approaches 100 percent and especially if a mist-type of rain is falling and the light intensity is further reduced, then the flies often move upward into the vegetation and can be found on the under surfaces of leaves, limbs, etc., 6 to 10 feet above the ground.

When cloudless, sunny, warm periods occur and the humidity drops rapidly, the flies disappear, presumably seeking out small, poorly lit areas where the humidity is high and the light intensity is low. Likewise, torrential rains (but not moderate rains) will drive them from those places where the investigator can normally expect to see adults. During moderate rainfall individuals of species that normally are hidden in the dense low fern cover (e.g., *Drosophila disticha*) will move upward in the vegetation and, if the light intensity is low due to a heavy cloud cover, may sit on the upper exposed surfaces of leaves. The fact that a droplet of rain may and often does fall upon the insects does not appear to bother them in any way. A curious behavior with relation to water is displayed by species of the idiomyia group, which under laboratory conditions actually bathe themselves by immersing their bodies in shallow water. The bathing motions are ritualized and involve depressing first the sides and then the venter of the body into the water followed by a period of careful grooming.

A careful search of any given collecting site will disclose that there are a number of preferred microsites such as a particular horizontal limb, tree fern, shrub, perhaps the leaves of one limb of a shrub that is so located that light, temperature, humidity, and spatial arrangements create a microarea that is hyperattractive to the adults. Repeated visitations to such microsites over a period of months or even years will always result in the finding of adults of one to several species occupying that site.

Although the adults do accumulate in numbers in small areas where food, ovipositional sites, and the physical environment are optimal, they also migrate through other parts of the forest. Thus, if collecting is done for adults by sweeping and baiting in windswept areas such as ridges and other exposed places, few if any adults can be captured. Often in such places there may be numerous specimens of *Cheirodendron* and *Clermontia* trees, and the fermenting leaves on the forest floor will invariably have been used as ovipositional sites containing eggs and/or larvae in various stages of development. Apparently mature females must regularly move through such areas ovipositing in the available substrates. Also, in the early morning just after dawn following a windy, rainy night, adults have been collected in open areas sitting on automobiles and buildings located at considerable distances from the nearest forest in which they normally dwell.

Finally it should be noted that those species which dwell in the dry forests, such as are found at Auwahi, Maui, are exposed to greater wind velocities and light intensities than are those which dwell in the rain forests and the mixed mesophytic forests. Experienced collectors, when they first see this forest, find it difficult to believe that a rich fauna of drosophiloids dwells there. This forest, however, is at an elevation of 3000 to 4000 feet, and the temperatures are therefore within the range that the Hawaiian drosophiloids demand and the food substrates are present.

Temperature, as noted above, seems to be a limiting factor in the distribution of the Hawaiian Drosophilidae. Consequently most of the species both of drosophiloids and scaptomyzoids are found about 1000-foot (300-m) elevations and reach their maximum at 3000 to 4000 feet. A few individuals of species such as *D. crucigera* (600 feet) and *D. flexipes* (800 feet) have been taken below 1000 feet but both species also range up to 4000 feet. The introduced scaptomyzoid *S. pallida* ranges from sea level to above 8000 feet, and a number of scaptomyzoids are found from 5000 to 9000 feet in areas that are unsuitable for almost all drosophiloids. *D. asketostoma,* which is almost invariably collected from the endemic silversword *Argyroxiphium sandwicense*) in the crater of Haleakala, Maui, at 8000 to 9000 feet, seems to be living at the highest elevation of any Hawaiian drosophiloid.

Significantly, species of *Clermontia* and *Cheirodendron* that serve as larval substrates and adult food sources for many species are rarely found below 1000 feet and also are most abundant in the 3000- to 4000-foot level. Related genera, e.g., *Reynoldsia* (Araliaceae) that do grow at lower and warmer elevations do not serve as a substrate for the larvae, although

Tetraplasandra and *Pterotropia* belonging to the same family and growing above 1000 feet, often along with *Cheirodendron,* are used by a number of drosophiloid species. The intimate relationship that exists between the Hawaiian Drosophilidae and the *Clermontia* and *Cheirodendron* plants seems to be one of long standing.

Courtship Behavior

The Hawaiian drosophiloids and scaptomyzoids exhibit contrasting courtship behavior. A sexually active scaptomyzoid male, like those of most other cyclorrhaphous species, orients upon an individual to which he has become alerted, be it male or female, then maneuvers to a position behind or at the side of the other individual and with a running lunge attempts to mount. Having mounted, the males of some species vibrate one or both wings, many do not; but the male of every species elongates and curls the tip of the abdomen down and under the other individual, seeking to make genitalic contact and intromission by repeatedly drawing the genitalia across that of the individual that he has mounted. A male or nonreceptive female reacts vigorously by kicking, by twisting the tip of the abdomen away from the seeking movements of the courting male, by shaking the entire body, and by violent wing vibrations. After a variable period of time, the courting male is either dislodged or dismounts of his own accord. Receptive females typically accept the male immediately after he has mounted. Males are persistent suitors and move rapidly about seeking out other individuals. Both males and nonreceptive females are wary of any other nearby moving individuals, typically orienting themselves so that they either face such individuals or escape from the immediate vicinity (Spieth, 1966b; 1968a). Observational data on mating behavior of the Hawaiian scaptomyzoids is relatively limited, and as yet unpublished. All species observed to date, both in laboratory and field studies, conform to the basic "assault" pattern described above. There is, however, no valid reason for assuming that such a simple-appearing courtship involves fewer stimuli than are involved in the quite different pattern displayed by the drosophiloids (Spieth, 1968a).

The Hawaiian drosophiloids, with certain exceptions noted later, do indeed display a courtship pattern that differs sharply from the scaptomyzoid type. After orienting, the male approaches and assumes a display posture close to a female typically at the rear with his head under her wing tips. A small number of species posture at the front of the female and still fewer at her side. In each instance, the position is species-specific.

Having assumed the display posture, the male then engages in a number of complex, ritualized, species-specific movements utilizing various parts of his body. Nonreceptive females respond to the male's overtures by fleeing, kicking with their hind legs, depressing the tip of their abdomen toward the substrate or simply ignoring his action. If the female is nonreceptive but does not flee, the males of some species after unsuccessfully displaying for a period of time will leave the head-under-wing position and circle to the front of the female and, facing her, will engage in further complex display, exhibiting movements that typically are quite different from those performed at the rear. After a period of frontal display, the male then returns to the rear position. Males typically court nonreceptive females for prolonged periods of time, sometimes for hours. A receptive female usually accepts the male's overtures rather quickly, indicating acceptance by spreading her wings horizontally, extruding the ovipositor and depressing her body toward the substrate. (See Spieth, 1966b, 1966c, and 1968b for details of specific courtships.)

The Hawaiian drosophiloids' courtship behavior thus conforms to the general pattern displayed by drosophiloids from other parts of the world. There are, however, consistent quantitative and qualitative differences that distinguish the Hawaiian species from those found elsewhere. A major difference is the exact site selected for courtship. The usual worldwide pattern involves the diurnal morning and evening assembling upon the feeding sites by adults of all ages. The females devote most of their time to feeding, but the males after feeding for a short time then move actively about, investigating other individuals of approximately their own size and facies. Thus, the males will orientate and initiate at least the beginning of courtship with any male or female of their own species or of other similar-appearing species. If a female of his own species has been approached, the male will then proceed to court in typical fashion. At any given feeding period, most of the females will be either too young to accept the male's overtures or else have been fecundated at a previous time.

Of the numerous courtships attempted by the males, only a very small percentage results in copulation. Further, males of most non-Hawaiian species after engaging in courtship display do try to mount nonreceptive females, e.g., *D. melanogaster* and *D. immigrans*. As a total result there is a great amount of activity at a feeding site, due especially to the males' persistent courting and the negative response by many females.

In comparison the Hawaiian species do not engage in courtship *on* the feeding site. All individuals, both males and females, are quiet, move about slowly, and show no avoidance or antagonistic reactions to each other. The males feed for only a short time and then with very rare exceptions (e.g., *D. anomalipes*) leave the food with a quick darting flight into the surrounding vegetation. The flies show little diurnality. During periods of low light intensity (e.g., on cloudy days) and high humidity, a feeding site may be continuously occupied from early morning to late afternoon.

During periods of sunlight and low humidity, the same site may be deserted except early in the morning and late in the afternoon. (See also Grossfield, 1968.)

Almost all of the Hawaiian males display extraordinary sexual dimorphism (see Figs. 4a, b; the great majority of these features are found in males only). The male structures involved in these dimorphic modifications include the antennae (all *Antopocerus* spp.); the mouthparts (all the modified-mouthparts group and most of the fungus-feeders plus several picture-wing species); the reduction and/or modification of the setae on the dorsum of the head (e.g., *D. petalopeza*); the wings (many species); the prothoracic legs (picture-wings, bristle-tarsi, ciliated-tarsi, spoon-tarsi, and split-tarsi groups); metathoracic legs (various species), and the eyes (*D. heteroneura*) or rarely the abdomen (*D. clavisetae*). A given species may exhibit only one structural modification; this is the most common situation but a number of species display two or more dimorphic characters —for instance, *D. petalopeza* has a modified setal pattern on the dorsum of the head, a bristle-tarsi type of forelegs and modified tarsal segments 2, 3, 4, and 5 on the middle legs. In all cases observed to date, whenever a male of a species possesses a dimorphic character, the modified body structure is utilized in the courtship. A small percentage of the drosophiloid species from other parts of the world possess dimorphic structures—such as the sex combs of *D. melanogaster* and *D. pseudoobscura* and their relatives, but these represent rare exceptions rather than the rule, while just the opposite pertains for the Hawaiian species.

The most typical display movements during courtship of non-Hawaiian drosophiloid males involves extending *one* wing from the resting position, then vibrating it in relatively small amplitude, the employment of the mouth parts to lick the area of the female's genitalia, and the forward extension of the forelegs followed by vibration or striking movement of the tarsi against the female's abdomen. The Hawaiian males also typically use these same structures, but those species which engage in wing movements use *both* wings, with rare exceptions, and the vibration movements may be of two or three different types that occur sequentially. Likewise, foreleg movements are complex and often consist of a compound movement of first lifting and folding the leg, followed later by extension and vibration against the female. Mouthpart movements are also often complex, a series of different movements following in repeated sequence during a courtship display. Many species of Hawaiian males also, simultaneously with other display movements, pulsate a droplet of fluid from the anal papilla, while others extrude the posterior end of the rectum as a small tubular structure. Many species simultaneously curl the abdomen excessively upward, sideways, or under, often to the extent that the tip of the curled abdomen is brought *over* or *beside* the male's head or thorax.

In sum, the Hawaiian male drosophiloid engages in complex movements usually of large magnitude and simultaneously involving various parts of his body (see Spieth, 1966b, 1968b).

The typical female acceptance response is also unique in that the female extrudes her ovipositor and depresses her entire body downward against the substrate. Occasionally a male may be unable to achieve intromission with an "accepting" female. The female then may remain in a trance-like state for several minutes after the unsuccessful male has dismounted.

During copulation non-Hawaiian drosophiloid females will readily move about carrying the male, fending with her legs against other flies, and avoiding other males that may attempt to interrupt the copulation. In comparison, a copulating pair of Hawaiian flies are with rare exceptions immobile, and other individuals may actually strike against or even clamber over the pair without eliciting any responses from the mating couple. Near the end of copulation a number of males, e.g., *D. crucigera,* completely release their grasp of the female, falling backwards in a trance-like state, in which they remain for a number of seconds after the termination of copulation. When compared to displays by non-Hawaiian species, those of the Hawaiian flies can be best described as the derivative and hypertrophied modification of the basic drosophiloid pattern overlaid with uniquely novel elements.

Courtship, as noted above, does not occur at the feeding sites but rather in some other part of the habitat. Field investigations have shown that males, presumably sexually mature, tend to assemble in considerable numbers at specific localized sites in the vegetation such as the trunk of a single tree, one or more horizontal tree limbs, the leaves of one branch of a shrub, the stems of the fronds of a particular tree fern, or even the individual pinnae of a single fern-frond. These sites are not randomly distributed in the areas where the flies live, and each species has special preferences, apparently determined by factors such as light, humidity, temperature, and spatial elements. Almost invariably they are in close proximity to the food sites. Such sites retain their attractiveness over prolonged periods of time; the males of some species occupy their particular sites for 24 hours of the day. Each male defends a limited area, e.g., a single leaf, a fern pinna, a section of a fern-frond stem, or a tree limb or small area of a tree trunk. Simultaneously the male also advertises his presence, either visually by body movements, especially specific wing movements, or by pheromones (Spieth, 1968a). Some species (e.g., *D. crucigera, grimshawi, villosipedis*) repeatedly curl the abdomen downward and drag the tip against the substrate, and in doing so deposit a thin film of liquid as they move. Many assume a ritualized stance and pulsate a droplet of liquid from the anal papilla. The males, but none of the females of many species, and all of those which engage in abdomen dragging, possess a unique anal structure, which Throckmorton (1966) has named the intra-anal organ. There is considerable interspecific variation but typically this organ consists of a pair of lobes covered with a dense pile and lying between the anal plates. Associated with the organ is a special slender sclerotized rod, which lies in the median line between the two lobes. These lobes are extruded when the male is engaged in abdomen-dragging. Following such a bout of activity the intra-anal organ is carefully cleaned by means of the hind tarsi before it is retracted into its resting position between the anal sclerites.

When another individual enters the small territorial area of an advertising male, it is immediately approached and investigated. If it is another male or an individual of another species, aggressive action against the intruder is initiated. This usually involves physical contact between the individuals; most species employ complex ritualized postures and movements (Spieth, 1966b). As a result of the encounter, one of the participants invariably is displaced from the area and apparently the vigor and size of the individual is the sole determinant of success. In a number of species the mere assumption of an aggressive posture or movement prior to physical contact is sufficient to cause one of the individuals to flee.

If a female, of either the same species as the male or of another species of approximately the same size as the male, intrudes on the male's defended area, then initiation of courtship occurs. If the female is nonreceptive or belongs to another species, then the male quickly ceases courtship and turns to aggressive behavior, which invariably causes the female to flee.

These behaviors of territoriality, aggression, and advertising on the part of the males have been observed for numerous species of various species groups both in the laboratory and in the field. It appears that the Hawaiian drosophiloids have evolved a true lek behavior, which has developed in conjunction with the spatial separation of feeding and courtship.

From field and laboratory studies on more than 60 species of drosophiloids, it is clear that each of the currently accepted species groups displays a basic, unique courtship pattern that is common to the species within the group. Each, however, exhibits a species-specific courtship display, and thus we have what can be described as variations upon a basic theme existing within each species group (Spieth, 1966b, 1968b). The basic theme involves not only the shared manner of utilizing such specialized dimorphic male structures as the antennae of the *Antopocerus* group, the specialized labellar structures of the modified-mouthparts group, and the modified tarsi of the bristle-tarsi, spoon-tarsi and fork-tarsi groups, but also the movements of wings, forelegs, abdomen, and the site of posturing, i.e., in front or at the rear of the female. The spectrum of variation upon the basic theme appears to vary between groups. Thus the amount of *interspecies* variation between members of groups such as the bristle-, fork-, and spoon-tarsi is least, and that of the picture-wings and modified-mouthparts groups is greatest. These variations in amount of *interspecific* courtship display between groups are, as Heed (1968) noted, paralleled by variations in larval feeding-behavior.

A considerable number of species are as yet still unclassified as to species group. Furthermore, the relationships between some of the species groups that have been tentatively delimited are obscure, e.g., the bristle-, spoon-, and fork-tarsi. Inability to rear the species of these groups under laboratory conditions is the major block to the elucidation of these unknowns. The relationships of other groups such as the picture-wings and modified-mouthparts can be estimated on the basis of various lines of evidence, e.g., anatomy, salivary chromosomes, larval feeding habits, and courtship. For instance, the courtship studies clearly indicate that the *adiastola* subgroup of the picture-wings constitutes a connecting link between the two groups. *Drosophila ornata* from Kauai actually has modified mouthparts whose components appear homologous to the same structures of the true modified-mouthparts species. Additionally, both this species and *D. adiastola,* which lacks the mouthpart modification, display a courtship pattern that is similar in certain respects to that of the modified-mouthparts basic theme and considerably different from that of the other picture-wings.

Amongst the species that are currently unclassified as to species grouping are several, e.g., *D. (Engiscaptomyza) amphilobus, crassifemur, nasalis,* and the *Drosophila parva* species complex, whose courtship behaviors are clearly anomalous with respect to the other Hawaiian drosophiloids. Each of these species displays typical scaptomyzoid courtship behavior even though the external anatomy indicates that they are drosophiloids. Throckmorton (1966) in his detailed study of the reproductive anatomy of the Hawaiian Drosophilidae examined *crassifemur, nasalis,* and *parva* and found that each of the "species departs from the more usual *Scaptomyza* pattern some way," e.g., *parva* has drosophiloid paragonia, *nasalis* has drosophiloid coiling of the vasa and pattern of pigmentation along the vasa, while *crassifemur* has a typical drosophiloid ejaculatory apodeme.

Two other unclassified species, *Drosophila anomalipes* and *D. quasianomalipes,* also show a queer mixture of drosophiloid and scaptomyzoid characteristics plus certain non-Hawaiian features. Both of these species are restricted to the old island of Kauai, where they occur sympatrically in the drier portions of the *Metrosideros* forests in the area of Kokee. Their ovipositional sites are unknown and they can not as yet be reared under laboratory conditions. The courtship behavior of these flies has been observed both in the laboratory and in the field, and are basically similar, the differences being restricted to wing displays by the males and repelling responses by the females.

The basic courtship pattern is as follows: The male orients on another individual, approaches and with a distinct movement taps with one of his forelegs, then postures in front of the female, engages in wing display, employing both wings, circles to the rear, assumes a head-under-wing posture, crouches slightly and curls the tip of his abdomen under, and lunges onto the female in a typical scaptomyzoid fashion. He then seeks to achieve intromission, not in the scaptomyzoid fashion of drawing his genitalia across those of the female but rather in the drosophiloid manner of pressing the genitalia against those of the female. Furthermore, he dismounts quickly when he fails to achieve intromission. The nonreceptive female extrudes her ovipositor either upward (*quasianomalipes*) or downward (*anomalipes*). When perchance the male is not directly positioned behind the female as he crouches preparatory to his lunge, she will curl the tip of her abdomen toward the face of the male. Despite many hours of observation in the laboratory and in the field, involving scores of courtships, no copulation has been seen. Clearly, however, the courtship behavior is a mixture of scaptomyzoid and drosophiloid elements. Highly significant also is the clear and distinct tapping movement of the males, the extrusion of the ovipositor, and the curling motions of the tip of the abdomen of the nonreceptive females. All of these are *non*-Hawaiian in nature and are not found in any of the Hawaiian species studied to date but are characteristic of drosophiloids from other parts of the world.

Many of the Hawaiian species are not attracted to fermenting baits such as are so commonly used to collect the various species in other parts of the world, but the majority of the picture-wings will come to banana bait especially if it has been inoculated with yeast derived from the lobeliad *Clermontia*. The standard method of baiting is to smear a modest amount of food on the smooth trunk of a tree, on the under surface of a limb or on other similar surfaces, selecting a site that has environmental conditions that the flies will accept.

Both *D. anomalipes* and *quasianomalipes* are readily attracted to bait and therefore can be observed feeding in the field. The males, after feeding for a short time, do not fly away as do the other Hawaiian males but move away just past the edge of the bait area. Usually they position themselves a few centimeters away, facing toward and watching the bait area. Whenever another individual of either *anomalipes* or *quasianomalipes* arrives on the bait, the male immediately approaches, taps, and investigates the newcomer. If it is a female of his own species, he will posture in front, circle to the rear, and assume the slightly crouched position behind her. As long as she is feeding or standing on the food mass, he will go no further with his courtship; that is, he will not lunge onto the female. The female for her part then typically extrudes her ovipositor as she feeds and the male will stand posturing behind her for prolonged periods. If and when she completes her feeding, she may walk off of the bait covered area onto the surrounding substrate. The male then follows and, as soon as the pair is clear of the bait, the male will lunge onto the female.

Thus males of this species, unlike all other Hawaiian males, do not fly away from the food area into the surrounding vegetation to establish a separate mating site that they defend; rather, they merely move off the food, await the arrival of the females and do not defend the space about them. Also they will engage in courting while on the food, a typically non-Hawaiian behavior, but will not complete the courtship unless the female is spatially a short distance away from the food proper, a "primitive" expression of the typical Hawaiian behavior.

Further, the semicosmopolitan species *D. immigrans* has been introduced into Hawaii and has established itself in the same area as *anomalipes* and *quasianomalipes*. Specimens of *immigrans* also come in numbers to the baits. The males of the three species often compete with each other in investigation of "newcomers" as they arrive on the bait, and *anomalipes* has been observed courting large *immigrans* females. Picture-wings such as *D. villosipedis* and *picticornis* simultaneously come to the bait and none of the males of *immigrans, anomalipes,* or *quasianomalipes* pays any attention to these picture-wings, but the picture-wings are often disturbed and flee from the bait as a result of the commotion caused by the males of *anomalipes, quasianomalipes,* and *immigrans* vigorously competing for the privilege of investigating other females.

Throckmorton (1966) studied *anomalipes* (but not *quasianomalipes*), and found that the dorsal spermatheca is scaptomyzoid in construction and that the ventral receptacles are coiled, a non-Hawaiian drosophiloid type, while external anatomy and the remainder of the reproductive system and of the egg filaments are typically Hawaiian. Thus both structure and behavior exhibit a mixture that could well indicate that these two species arose from a primitive stock that still possessed many of the derivative characters of both the Hawaiian scaptomyzoids and drosophiloids as well as elements of the ancestral non-Hawaiian stock that colonized Hawaii.

It is interesting that the fungus-feeders, which form a distinct evolutionary branch of the drosophiloids and apparently evolved from the base of the ancestral Hawaiian stock, display a primitive lek behavior. The males of these species, after feeding, fly into the nearby vegetation only a meter or so from the feeding-ovipositional fungus and take up station on the dorsal surface of leaves. Here they orient themselves facing toward and watching the feeding site, vigorously defending their small territories. Periodically the males fly to another leaf on which they see another fly sitting. Usually they encounter another male; but females, as they leave the feeding site, also land on these leaves, and eventually a male encounters a female, whom he immediately courts.

This behavioral pattern is thus in a number of respects only quantitatively different from that displayed by *anomalipes,* and lacks the specialized features of apparent pheromone production seen in the picture-wings and modified-mouthparts species. Throckmorton (1966) has shown that this fungus-feeding group is most closely related to ciliated-tarsi species such as *D. imparisetae.* These ciliated-tarsi species breed in rotting bark of *Clermontia* and *Cheirodendron.* Behavioral data indicate that they could well represent the stock from which many of the other more specialized species groups have radiated.

Evolutionary Implications of Courtship Behavior

The major evolutionary development in the courtship of the Hawaiian drosophiloids, but not of the scaptomyzoids, was the innovation of a true lek behavior involving the spatial separation of courtship from feeding and oviposition. This intensified sexual selection, since the males under lek conditions must attract the receptive females to the lek or mating site instead of moving about on the feeding-mating-ovipositional site opportuning all available females, both receptive and nonreceptive. In other organisms the result of intense sexual selection appears typically to result in increased sexual dimorphism, and this has clearly happened to an extraordinary degree in the males of the Hawaiian drosophiloids but not in the scaptomyzoids, which do not exhibit lek behavior. Sibley (1957) observes in his discussion of bird sexual dimorphism that "the genetic basis for such characters may, and probably does, involve but a few genes and these may control only relatively superficial characters" of male structure and display movements. Bits of evidence indicate that the Hawaiian drosophiloids conform to this generalization:

(1) Many species have females that can not be separated even though the respective males differ sharply in both their structural characteristics and their courtships.

(2) The males of a number of species differing distinctly in both structure and courtship when placed with foreign females under no-choice conditions will often achieve insemination. In a considerable percentage of instances, viable F_1 hybrid adults are produced, and a number of these are fertile (Yang and Wheeler, 1969; see next section).

(3) Carson (see below, p. 492, ff.) has discovered several groups of homosequential species, i.e., groups of species in which banding-sequences of each polytene chromosome are identical. Each of these groups forms the nucleus for a species cluster involving other species that differ from the homosequential members by only one or two inversions. In those species studied to date the homosequential species are distinguishable by the male's courtship pattern and sexual dimorphic characters.

Although the factors that created the selection pressure leading to the evolution of lek behavior cannot now be specifically determined, certain inferences can be drawn. The drosophiloids congregate and feed in exactly those microhabitats and on those plants which are attractive to many of the endemic Hawaiian birds, especially the insectivorous elepaio (*Chasiempsis sandwichensis*) and the creepers (*Paroreomyza* spp.). Furthermore, many other honeycreepers catch insects and other arthropods. Both the honeycreepers and the flies are attracted to the lobeliads, especially species of the genus *Clermontia* (Spieth, 1966c). Warner (1967) placed six iiwis (*Vestaria coccinea*) in cages of approximately 60 cubic feet (1.7 m^3) and supplied the birds with a standard nectarivorus bird diet plus 50 live specimens each of *Bunostoma* sp., *Drosophila mimica,* and *D. grimshawi.* The birds avidly pursued the flies, captured and then displayed a ritualized method of eating the insects. The scaptomyzoid *Bunostoma* is small, *D. mimica* medium-sized, and *grimshawi* large. The first two were readily eaten but when a *grimshawi* specimen was captured, it buzzed violently upon being grasped by the bird's bill and was then quickly released, thus appearing to be immune to iiwi predation. Before the decimation that occurred during the 19th century, the native bird population was quite large. Today the numbers of individuals are drastically reduced and a number of species are now extinct. Individuals of some of the extinct species were larger in size than the iiwi, but many were not.

All of the known areas that are prime collecting sites for drosophiloids also have an abundant bird fauna; e.g., Kipuka Puaulu is commonly known also as Bird Park. Assuming the presence of food and ovipositional sites, then the quality of any previously uncollected area can usually be roughly gauged by the number of birds present.

The flies are also preyed upon by species of the dipteran genus *Lispocephala*. Various species of lispocephalids are abundant in the area where the drosophilids are found, and the adults lay their eggs on the same fermenting substrate as do the drosophilids; the *Lispocephala* larvae are voracious predators on the drosophilid larvae. The *Lispocephala* adults also capture and eat many of the drosophilid species, but like the birds, are unable, at least in experiments with caged individuals, successfully to attack the larger species such as the picture-wings.

Not only courtship behavior but also all other aspects of the drosophiloid's behavior indicate high adaptation to the avoidance of predators, and in the impoverished oceanic-type native Hawaiian fauna only the birds and the lispocephalids are plausible candidates for the role of major predators. Furthermore, it seems possible that the large size displayed by many Hawaiian drosophiloid species may be an evolutionary response to the selective pressure of bird and lispocephalid predation.

Finally, it is of interest that a number of the sexually dimorphic structural features of the anatomy of the males must have evolved subsequent to the behavioral characteristics associated with these structures. Thus, the male of the *Antopocerus* species (Spieth, 1968b) during courtship lunges forward and upward over the abdomen of the female and forcibly spreads her wings apart about 45° and upward 15° to 20°. In doing so, the male's antennae are thrust against and slide across the under surface of the female's wing-vane. The male antennae are adapted for this function: (1) the first antennal segment is enlarged and elongated; (2) the arista is elongated, enlarged, and whiplike; and (3) the aristal setae are extremely short but numerous and restricted to one surface (i.e., lacking on the side that slides across the female wing-vane). Clearly the normal drosophiloid antennae would be poorly suited for such vigorous physical contacts, and significantly the *Antopocerus* females do possess essentially "normal" antennae. Thus, *Antopocerus* displays not only an example of a unique sexually dimorphic character but also one that must have evolved as a result of selection pressure created by a behavioral feature that must have evolved at an earlier date than did the physical dimorphism. Various other dimorphic characters must have also followed a similar evolutionary sequence, e.g., the dorsal cranial setal pattern of *D. petalopeza,* the armored mouthparts of the modified-mouthparts group, the modified labellar lobes of *D. crucigera,* which are uniquely adapted for clamping onto the anal papillae of the female, the sclerotized oral rim of the fungus-feeders, and the specialized tarsal armaments of the bristle-, split- and spoon-tarsi groups (Spieth, 1966b). In all instances these are sexual dimorphic structures that are utilized by the males for courtship functions.

When compared to drosophiloids from other parts of the world, the courtship behavior of the Hawaiian species is clearly hypertrophied in character, and the individual elements are obviously derivative. Spieth (1952) noted that species of the subgenus *Sophophora* utilized distance stimuli, whereas those of the subgenus *Drosophila* are characterized by using contact stimuli involving prolonged licking and leg-caressing movements on the part of the males. The basic pattern of Hawaiian species when analyzed indicates clear relationship to that displayed by the primitive species groups of the subgenus *Drosophila,* rather than to the more specialized pattern of the subgenus *Sophophora.*

Chromosomal and Genetic Characteristics

Metaphase Karyotypes

Data on endemic Hawaiian species of Drosophilidae will be found in Clayton (1966, 1968, 1969) and Carson, Clayton, and Stalker (1967). Approximately 17 species of the family apparently have been introduced into the islands by man, either by the Polynesians or by later peoples. These

TABLE IX
Metaphase Configurations in Four Genera of Hawaiian Drosophilidae

Metaphase	Drosophila species	Antopocerus species	Scaptomyza species[a]	Titanochaeta species	Total species
5R 1D	80	1	–	–	81
6R	7	3	–	–	10
1V,3R,1D	2	–	12	1	15
1J,4R,1D	1	–	–	–	1
2V,1R,1D	2	–	5	1	8
2V,2R,1D	–	–	1	–	1
5V 1J	1	–	–	–	1
Total	93	4	18	2	117

[a] Includes five *Drosophila* species with strong affinities to *Scaptomyza* (see text).

are omitted from this discussion. Squash preparations have been made either from the the brains of larvae reared from eggs produced by females isolated from nature or from the testes of adult males captured in nature (see Clayton, 1960). In some instances, adult-male and brain material from laboratory stocks was used.

Table IX gives data taken from Clayton (1968) on the karyotypes observed in 117 species of 4 genera. (These summaries do not include the information given in Clayton, 1969.) The genus *Antopocerus* is closely related to *Drosophila* (subgenus *Drosophila*), and *Titanochaeta* has been characterized as scaptomyzoid by Hardy (1966), Takada (1966), and Throckmorton (1966). Indeed, *Titanochaeta* metaphases are consistent with two karyotypes found among 17 *Scaptomyza* species from Hawaii (Table IX). One of these karyotypes is illustrated in Figure 6,d. Furthermore, these configurations are the same as two types described from non-endemic species of this genus. Thus, *Scaptomyza adusta*, like the endemic species *argentifrons*, has one V, three rods, and one dot; and *pallida* has two V's, one rod, and one dot. Although in all these cases the V's appear to represent "fusions" of originally separate rod-shaped elements, there is no assurance that the same elements are involved in each case, so that the resemblance of some karyotypes may be a superficial one. The enumeration of *Scaptomyza* species from Hawaii made in Table IX includes five (including *D. nasalis*: 2V, 2R, 1D) that have been assigned to the subgenus *Drosophila* (*Engiscaptomyza*) Kaneshiro (1969). Recognition of the intermediate position of these flies is certainly necessary.

Karyotypes of 93 species of the subgenus *Drosophila* (*Drosophila*) are

Fig. 6. Metaphase karyotypes among Hawaiian Drosophilidae (diagrammatic). a, D. grimshawi; b, D. mimica; c, D. cyrtoloma; d, S. argentifrons.

entered on Table IX. The great majority of these (80) have a metaphase karyotype of five rods and one dot (Fig. 6). The other karyotypes having $n=6$ (6R; 1J, 4R, 1D; 5V, 1J), furthermore, are clearly due merely to the addition of extra heterochromatin to one or more chromosome arms. In the case of 6 rods, the heterochromatin appears to have been added to the dot chromosome. Where J-shaped and V-shaped elements are found without change in chromosome number, the evidence is strong that this is likewise due to the addition of heterochromatin (Clayton, 1968). The most extreme case of this is the condition found in the giant species, *D. cyrtoloma* from Waikamoi, Maui (see Fig. 6). Although most of the other species so far described in its subgroup have five rods and one dot, this species has acquired a full-length heterochromatic arm on each of five of the six chromosome-pairs. The remaining chromosome has became J-shaped also by acquisition of heterochromatin. This condition is not accompanied by any alteration of the usual salivary gland complement of five long and one dot-like polytene elements. The same is true of those forms which have six pairs of rods at metaphase, as well as a number of conditions, considered as basically "5R 1D" by Clayton, but which manifest, in some cases, an unusually long pair of rods or somewhat enlarged microchromosomes. No pericentric inversions have as yet been recognized in any Hawaiian species.

Accordingly, of the 93 species of the subgenus *D. (Drosophila)* indicated in Table IX, only four have had a reduction in chromosome number. This has apparently occurred by the process of whole-arm fusion and centromere loss. Three of these species belong to the modified-mouthparts group and one belongs to the bristle-tarsi group. These species are fairly close, however, and phylogenetically this may represent a single fusion, although this point remains uncertain. It is clear, however, that among the large group of picture-winged species (see Fig. 8) and the genus *Antopocerus*, there have been no fundamental changes from $n = 6$.

The above is signficant in view of the conclusion reached by Patterson and Stone (1952) that the five-rods, one-dot karyotype is a primitive one. Based on the study of 215 species of the genus *Drosophila* from areas other than Hawaii, these authors conclude that alterations of this primitive karyotype are the result of either: (1) fusion of whole arms with loss of centromere; (2) pericentric inversion; or (3) added heterochromatin. Clayton (1968) has compared the data on Hawaiian members of the subgenus *D. (Drosophila)* with the data on 150 non-Hawaiian species of the same subgenus taken from Patterson and Stone (1952) (Table X).

The data in this table underscore a most remarkable feature of the Hawaiian fauna, namely, its stability of chromosome number. Whereas there have been a minimum of 32 fusions in 150 continental species, only 4 of 93 Hawaiian species have had the karyotype so altered. On the other hand, changes via heterochromatin appear not to show very much difference from the continental situation (9/93 or 9.7 percent among insular compared with 19/150, or 12.8 percent among continental species). The count of nine Hawaiian species that have added heterochromatin, moreover, does not include those species which have a long pair of rods or which manifest rod-shaped microchromosomes. These variations also appear to be due to added heterochromatin, and inclusion of them would bring the island data even closer to that of the mainland with respect to this character.

In summary, then, it may be pointed out that endemic species of the Hawaiian Drosophilidae fall into two groups chromosomally: these groups

TABLE X

Comparison of Chromosome Numbers and Shapes from Hawaiian and Non Hawaiian Members of the Subgenus *Drosophila* (*Drosophila*)[a]

Haploid chromosome number	Non-Hawaiian species		Hawaiian species	
	No.	Percent	No.	P.
seven	1	0.7	0	
six	81	54.0	89	95.
five	36	24.0	2	2.
four	29	19.3	2	2.
three	3	2.0	0	0
Number of species	150		93	
Chromosome shapes				
5R 1D	53	35.3	80	86
6R	7	4.7	7	7
3R 1V 1D	11	7.3	2	2
1R 2V 1D	15	10.0	2	2
other	64	42.7	2	2
Number of species	150		93	

[a]After Clayton (1968).

correspond with the genus *Scaptomyza* and the subgenus *Drosophila* (*Drosophila*). Flies of the genus *Titanochaeta* appear to relate to *Scaptomyza*, and members of *Antopocerus* to the subgenus *Drosophila*.

Polytene-Chromosome Sequences

Interspecific comparisons of giant salivary gland chromosomes have been published for 53 large endemic Hawaiian species belonging to the subgenus *Drosophila* (Carson, Clayton, and Stalker, 1967; Carson and Stalker, 1968a, b, c). This work has been aided by the fact that many of these species have exceptionally large and favorable salivary gland chromosomes (Fig. 7). This circumstance, together with use of a technique that permits table-level matching of unknown sequences under the microscope with photographic maps of known sequences, has permitted the description of all five major chromosome elements of all 53 species in terms of a single arbitrarily chosen set of Standard sequences.

In making comparisons between the species, it is apparent that virtually

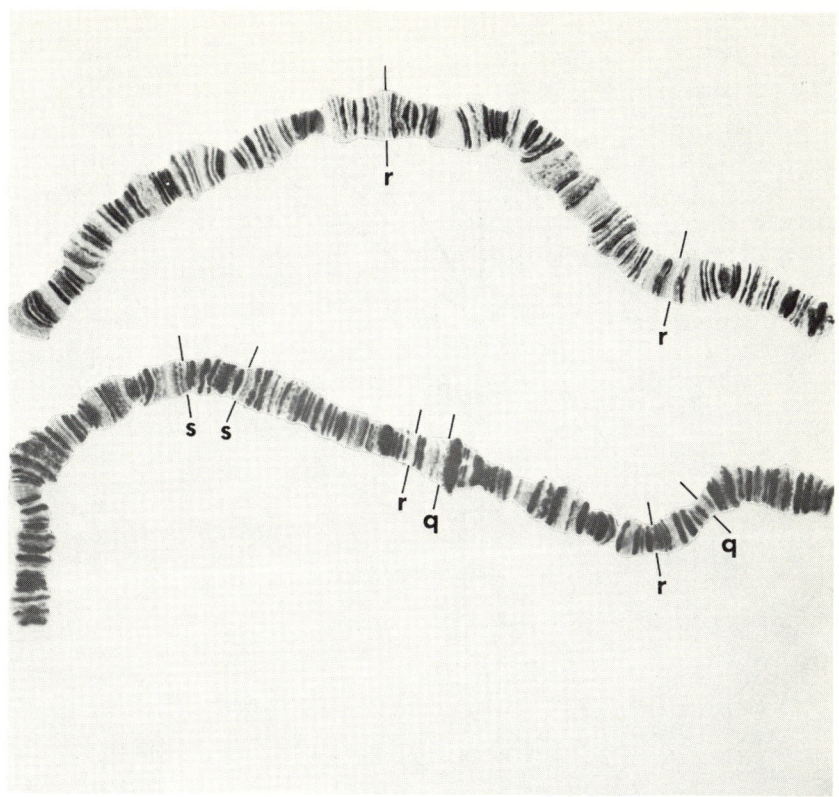

Fig. 7. Photographic chromosome maps of chromosomes 2 (upper) and 5 (lower) of Drosophila crucigera. The ordering of the bands shown is identical to the Standard D. grimshawi. The break-points of four previously unpublished inversions are entered; these are: 2r of D. inedita, 5q of D. lineosetae, 5s of D. setosifrons, and 5r of D. claytonae. The latter is based on the 5a arrangement (see Carson and Stalker, 1968a, Fig. 2).

all variation in sequences between species can be explained by paracentric inversions. Each inversion discovered has been designated by a lowercase letter, and the break-points determined. These break-points have been mapped and their exact positions entered on photographic chromosome maps. Despite the fact that the sequences of all species can be completely described in terms of the Standard *Drosophila grimshawi,* it has been found convenient to prepare five sets of chromosome maps made from each of the five species for which the subgroups are named (see Table XII, and Carson and Stalker, 1968a, b, c and 1969).

Figure 8 summarizes in a single master diagram the polytene and metaphase chromosome relationships for a total of 69 species. No species has

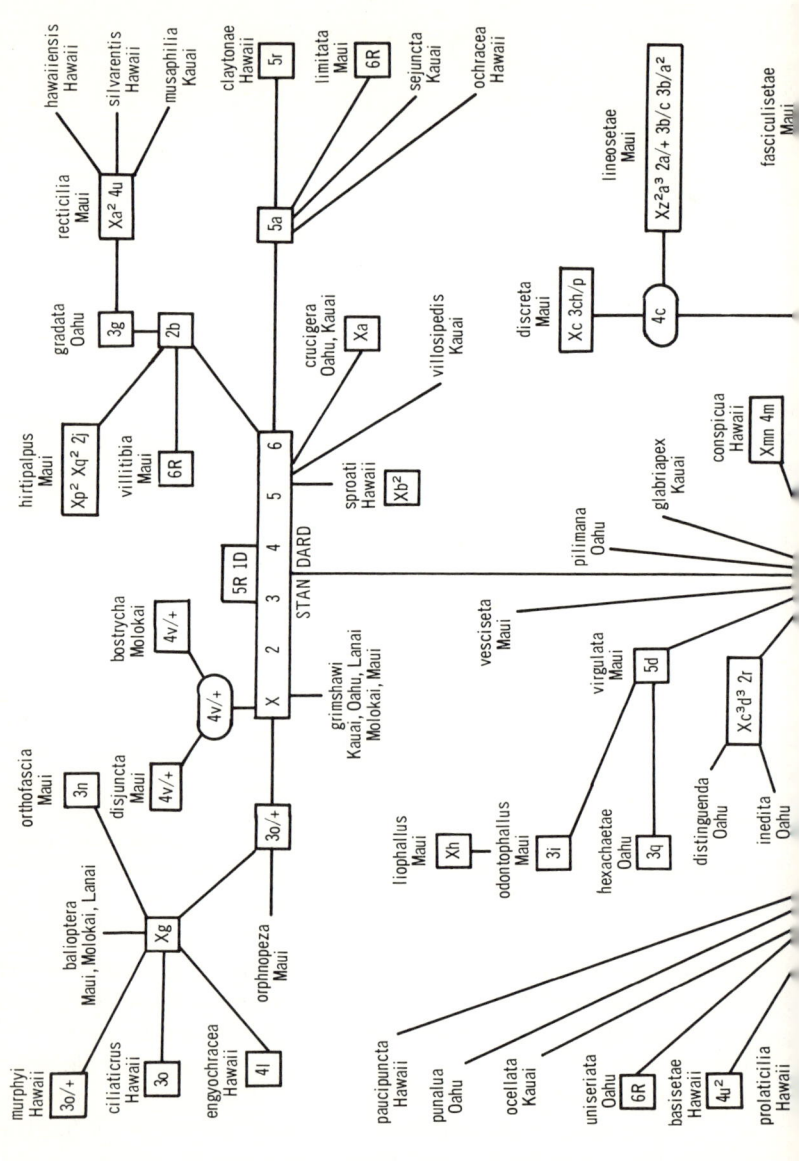

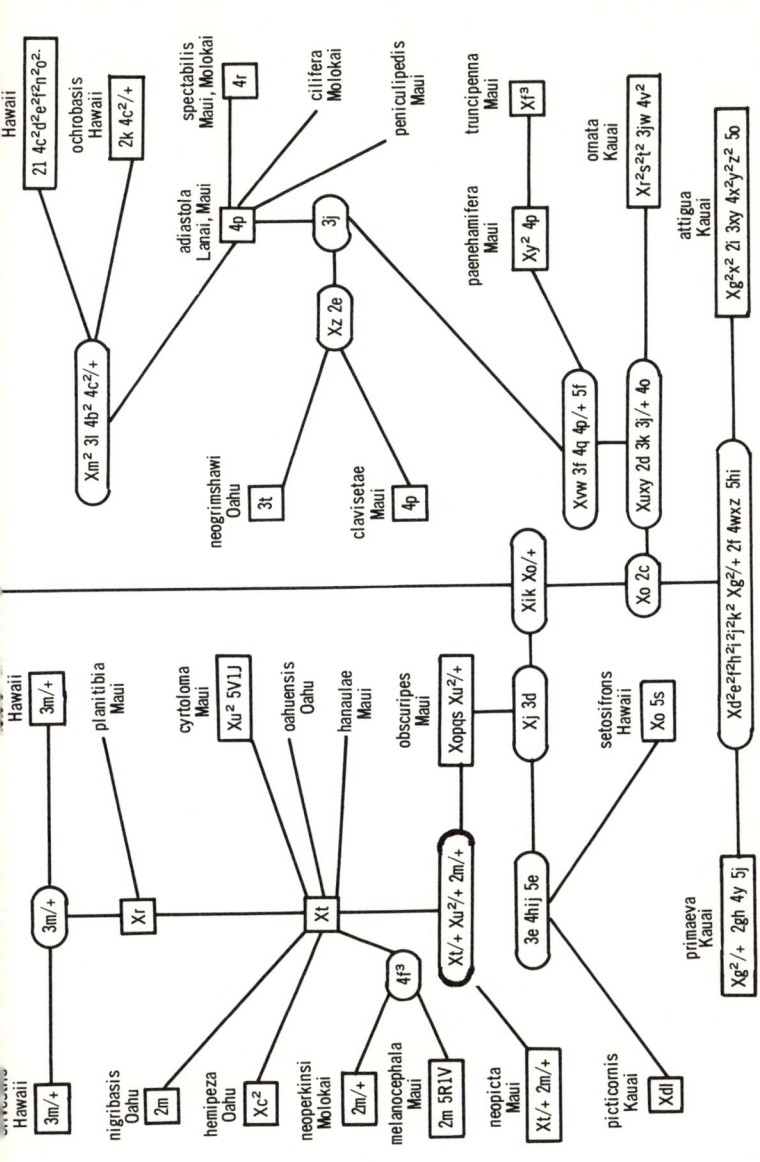

Fig. 8. Chromosomal relationship among 69 species of Hawaiian Drosophila. Read inversion formula for each species additively by following the line from the Standard chromosome sequences for the five chromosomes (box, upper center). For example, D. liophallus has the formula Xh 2 3i 4b 5d. Letters appearing singly represent fixed inversions. For details, see text.

been included in this diagram unless the sequential order of each of its five major chromosome arms can be precisely and completely described in terms of the arbitrary Standard. The figure updates the information given on similar diagrams published in Carson, Clayton, and Stalker (1967) and Carson and Stalker (1968a, b, c). Sixteen species have been added and one removed. New names are provided for some species and a few corrections have been made. Details will be found in Table XI. In order that this review may include as much recent data as possible and not be overburdened with chromosome maps, documentation of the break-points of most of the new inversions will be deferred until later publications. This also refers to those chromosomal polymorphisms which are confined to a single species; the letter designations for these, if not entered on Figure 8, will be found in Table XI.

Most of the 69 species belong to what Hardy and Kaneshiro (1968) have called "picture-winged" species. These species are mostly large and among them are probably the largest *Drosophila* species in the world. Carson and Stalker (1968a, b, c) have divided these species into four subgroups on the basis of the similarity of polytene gene orders of their members. Each subgroup is named for the one species from which full chromosome maps have been made. In the present summation, a fifth subgroup has been recognized, centering on *D. primaeva;* this will be the subject of a forthcoming paper (Carson and Stalker, 1969). The five subgroups are listed in Table XII, which also shows the minimum number of inversions fixed in each subgroup relative to the others. From these facts it may be seen that the *D. grimshawi, planitibia,* and *punalua* subgroups are quite close to one another. Among these closely knit subgroups, the greatest "distance" is 14 fixed inversions. This occurs between the *punalua* and *planitibia* subgroups. The *primaeva* subgroup is not only the most removed from the Standard but is quite far from the other subgroups as well.

Figure 8 may be used to read the basic chromosome formula (both metaphase and polytene) of each species in terms of the Standard *D. grimshawi* chromosome sequences (X 2 3 4 5 6: center, upper portion of the figure). For example, the formula of *D. liophallus* may be read additively by following the lines leading from the Standard box, e.g.: Xh 2 3i 4b 5d; 5 rods 1 dot. Existing species are given within rectangular boxes, whereas formulas for ancestral populations inferred from the data are given within boxes with rounded ends. It is noteworthy as one studies Figure 8, that for every species listed each major chromosome has been completely characterized, from one end to the other. Each and every band of the Standard has been located and accounted for in each of the 69 species. This has been made possible only by the very great favorability of these species for chrosome study (see Fig. 7). Except for a 3-band deletion (Xd of *D. picticornis*), all aberrations given are inversions. The microchromosome (6), however, has not been systematically studied.

TABLE XI

Strains and Species of Hawaiian *Drosophila* for Which New or Corrected Data are Presented in Fig. 8.

Species	Locality and strain numbers	Number of wild chromosomes observed		Remarks
		auto-somes	X chromo-somes	
D. attigua[a]	Kahili, Kauai (2500') L41C12	4	3	Described by Hardy and Kaneshiro (1969, in press); intraspecific polymorphisms: $4c^3/+$; $4b^3+$
D. claytonae[a]	Upper Olaa For. Res. Hawaii L89L1	4	3	Described by Hardy and Kaneshiro (1969, in press); intraspecific polymorphism: $4e^3/+$
D. cyrtoloma	1. Kipahulu Valley, Maui L13G9-11 (mass) (Carson and Stalker 1968 b: as *D. neoperkinsi*) 2. Waikamoi, Maui L18P4; L47C3; L86Q3	6	5	Described by Hardy (1969 in press); metaphase 5V 1J, see Clayton 1968 as "*perkinsi*?"
D. crucigera	–	–	–	See Tab. 15
D. distinguenda[a]	Makaleha Valley, Oahu (1800') L92G1-4	12	10	Described by Hardy (1965)
D. grimshawi	–	–	–	See Tab. 16
D. hanaulae[a]	Hanaula, W. Maui (4000') L91Q10	4	3	Described by Hardy (1969 in press)
D. inedita[a]	Makaleha Valley, Oahu L88G1	4	3	Described by Hardy (1965)
D. lineosetae[a]	Hanaula, W. Maui L61C1-5 (mass); G1-3 (mass); L91B2	8	6	Described by Hardy and Kaneshiro (1968; intraspecific polymorphism: $5q/+$
D. melanocephala[b]	Waikamoi, Maui M12L2	4	3	Described by Hardy (1966; 5R 1D metaphase reported by Clayton 1968 may be a different species
D. murphyi[a]	Pololu Stream, N. Kohala Hawaii (3300') L79G1,2 Upper Olaa For. Res. Haw. L82B1,2; G1; L89G5,7,10,14,18, 19,21,23,25; C19,24,26, 27,33	68	53	Described by Hardy and Kaneshiro (1969 in press); intraspecific polymorphism: $3ob^2/+$

TABLE XI (Continued)

Species	Locality and strain numbers	Number of wild chromosomes observed		Remarks
		auto-somes	X chromo-somes	
D. musaphilia[a]	Alexander Reservoir, Kauai (1700') L42G1	4	3	Described by Hardy (1965)
D. neoperkinsi[a]	Nawaihulili Stream, Molokai (2500') L98G6; L98B4-12 (mass); L98G7,9,13 (mass)	8	7	See Hardy and Kaneshiro (1968) and redescription in Hardy (1969 in press)
D. nigribasis	Mt. Kaala, Oahu L87G3,4; L87G7-11 (mass); L87B5-9 (mass)	16	12	*New name*, in Hardy (1969 in press); metaphase 5R 1D, see Clayton (1968) as: *"brunneipennis"*
D. ocellata[a]	Mt. Kualapa, Kauai L45B1	4	3	Described by Hardy and Kaneshiro (1969 in press); 5 strains reported in Carson and Stalker (1968c) as "new species 'B' "
D. ornata[a]	Pouli Stream, Kauai Mt. Kahili, Kauai			Described by Hardy and Kaneshiro (1969 in press); 3 strains reported in Carson and Stalker 1968c as "new species 'A' "
D. paenehamifera[a]	Hanaula, W. Maui L61C13,L61B11 (mass)	4	3	Described by Hardy and Kaneshiro (1969 in press)
D. peniculipedis[a]	Hanaula, W. Maui L61G9-13,35	24	18	Aberration "4s/+" (Carson, Clayton and Stalker, 1967; Carson and Stalker, 1968c) appears to be a "kink" formed by constrictions in the homozygous state; it is not a heterozygous inversion
D. perkinsi	Waikamoi, Maui J24L10. See Carson, Clayton and Stalker (1967) Table 1, p. 1283 and Carson and Stalker 1968b as *D. neoperkinsi*, Waikamoi, Maui.			This record may be in error and is being withdrawn pending reconfirmation
D. primaeva	Mohihi, Kauai (3500') G20.3B. Kokee, Kauai J81C1. Pouli Stream, Kauai L37G2,3,7-9; B6 Mt. Kahili, Kauai L41C11,20,21; G8,11-14, 16,19,21,25,26; P4-6.	2 4 24 60	2 3 18 42	Described by Hardy and Kaneshiro (1968); intraspecific polymorphisms: $Xg^2/+$; $Xg^21^2/+$ $2q/+$; $4a^2/+$; metaphase 5R 1D (Clayton, 1968)

TABLE XI (Continued)

Species	Locality and strain numbers	Number of wild chromosomes observed		Remarks
		auto-somes	X chromo-somes	
D. punalua	All strains: see Carson, Clayton and Stalker (1967); Carson and Stalker (1968c).			Inversion previously considered identical with 3f of the D. adiastola subgroup is now considered to be different and is designated 3z
D. sejuncta[a]	Mt. Kualapa, Kauai L45B3,6,10,11	16	12	Described by Hardy and Kaneshiro (1968)
D. setosifrons[a]	Upper Olaa For. Res., Hawaii L89C11-13; G1,2; L100P16	20	15	Described by Hardy and Kaneshiro (1968)
D. silvestris	Hawaii	–	–	New combination: Hardy and Kaneshiro (1968); this name replaces the combination D. nigrifacies used by Carson et al. (1967) and Carson and Stalker (1968b)
D. truncipenna[a]	Waikamoi, Maui L91C1-3 (mass)	2	2	Described by Hardy 1965; 6R metaphase reported by Clayton (1968) may be from a different species

[a] Metaphase reported as 5R 1D in Clayton (1969)
[b] Metaphase reported as 5R 1V in Clayton (1969)

TABLE XII

Cytological Subgroups of the Species of Hawaiian *Drosphila* Diagrammed in Figure 8.

Number	Subgroup Name	Number of species in subgroup	Number of species having at least one intraspecific polymorphism	Minimum number of inversions fixed relative to subgroup number				
				I	II	IV	III	V
I	D. grimshawi	36	15	X				
II	D. planitibia	14	5	9	X			
IV	D. punalua	6	1	7	14	X		
III	D. adiastola	11	2	17	17	21	X	
V	D. primaeva	2	2	22	23	27	28	X
	Total	69	25					

Intraspecific Chromosomal Polymorphism

Immediately after its origin, a new inversion is of course polymorphic within the species in which it has arisen. This condition may persist or the new arrangement may become fixed in one or more descendent populations. The number of species showing at least one intraspecific polymorphism is given in Table XII. There are only 25 such species, that is, 36 percent of the total. Although polymorphism seems to be less in the *punalua* and *adiastola* subgroups, all subgroups show at least one polymorphic species. Certain species, which are known so far from only a few strains, may of course be erroneously judged to be monomorphic. On the other hand, the reality, in fact the preponderance, of monomorphic species can hardly be doubted (especially in view of the large number of natural samples that have been examined) for certain species, without giving any indication of the presence of chromosomal polymorphism (Carson and Stalker, 1968a, b, c). Documentation of intraspecific polymorphism can be gleaned from these papers as well as Table XI of the present review.

Fixed versus Polymorphic Inversions

Further information on the distribution of inversion fixation and polymorphism will be found in Table XIII. Among a total of 178 inversions, 115 are wholly fixed. As several inversions are heterozygous in one species but fixed in one or more others, the number of fixations may be taken as somewhat greater than 115. In any event, the mean number of inversion fixations per species is slightly less than 2.

The circumstance that some inversions are polymorphic in some species and fixed in others has important phylogenetic implications. For example, *D. neopicta* of Maui is polymorphic for two inversions (Xt and 2m), which have subsequently become fixed in other lines of descent stemming from a common ancestral population (see Fig. 8). In such an instance, the polymorphic population retains a state resembling more closely the ancestral population than the other species do. As will be pointed out below (p. 508), the retention of such heterozygosity within existing species permits precise inference of ancient colonization routes for species-founder populations. Such cases, however, where an existing species actually retains the heterozygosity (boxes with square ends), should be distinguished from those wherein an ancestral population is merely inferred to have been heterozygous for a given inversion (boxes with rounded ends, Fig. 8).

Referring again to Table XIII, it may be seen that the chromosomes X and 4 have the greatest number of fixed inversions; this appears to be true in all subgroups. Conversely, chromosomes 2 and 5 (illustrated in Fig. 7) appear to be less often involved in rearrangements. In general,

TABLE XIII

Number of Fixed and Polymorphic Inversions Among Hawaiian Species of *Drosophila*. Inversions Common to More Than One Subgroup Have Been Entered Only Once. The Base Arrangements are Standard *D. grimshawi*

Chromosome	Subgroup Number											Total
	I. *grimshawi*		II. *planitibia*		III. *adiastola*		IV. *punalua*		V. *primaeva*		I-V		
	fixed	poly-morphic	fixed	poly-morphic	fixed	poly-morphic	fixed	poly-morphic	fixed	poly-morphic	fixed	poly-morphic	
X	15	0	12	4	12	3	2	0	8	2	49	9	58
2	3	1	0	3	5	1	0	0	4	1	12	6	18
3	4	10	2	2	6	1	1	0	2	0	15	13	28
4	4	12	4	4	8	9	4	0	7	3	27	28	55
5	3	4	2	0	2	1	1	2	4	0	12	7	19
Total	29	27	20	13	33	15	8	2	25	6	115	63	178

both fixed and polymorphic inversions seem to follow the same pattern, except that the number of X-chromosome polymorphisms is proportionally far less than that of the other chromosomes.

Parallel Chromosomal Polymorphisms in Different Species

Carson (1968, 1969) has made a study of intraspecific chromosomal polymorphisms in the central region of chromosome 4. This region is polymorphic in a number of closely related species of the *D. grimshawi* subgroup. Although in each instance the break-points of the inversions are closely similar, giving rise to strikingly parallel inversion configurations, these break-points are not identical. Thus, 4a of *D. grimshawi,* 4k and 4v of *D. disjuncta,* and 4c of *D. fasciculisetae* are closely similar. It was suggested that this region of the chromosome produces sectional heterozygotes yielding high fitness and that natural selection has selectively preserved similar, but not identical, inversions within different species in each case.

Chromosomal Polymorphisms That Are Identical in Two Different Species

Carson (1959) has pointed out the great rarity, among mainland faunas, of the condition where two descendent species have retained a polymorphism present in an ancestral population. This condition was ascribed to the

TABLE XIV

Pairs of Closely Related Species that Have a Chromosomal Polymorphism in Common

Case no.	Species	Chromosomal subgroup	Chromosomal polymorphism in common
1	*D. bostrycha*	*D. grimshawi* (I)	4v/+
	D. disjuncta		
2	*D. murphyi*	*D. grimshawi* (I)	3o/+
	D. orphnopeza		
3	*D. fasciculisetae*	*D. grimshawi* (I)	2a/+
	D. lineosetae		
4	*D. heteroneura*	*D. planitibia* (II)	3m/+
	D. silvestris		
5	*D. neoperkinsi*	*D. planitibia* (II)	2m/+
	D. neopicta		

tendency for species to arise from exceedingly small populations, which would be likely to be chromosomally monomorphic.

A significant number of exceptions to this rule have now been uncovered among Hawaiian species (Table XIV). In each of the five cases listed in the table the break-points of the inversion concerned have been determined as identical. In each case, there is strong evidence that the populations concerned are indeed species and are not merely subspecies. In each case, the two species appear to have inherited the polymorphism from a common ancestor.

TABLE XV

Frequency (in %) of X- and Third Chromosome Gene Arrangements in Natural Populations of *Drosophila crucigera* from Kauai and Oahu

Locality	No. of X chromosomes examined	Gene arrangement		No. of 3rd chromosomes examined	Gene arrangement	
		Xa	Xab		3	3a
Kauai						
1. Kokee	11	100.0	0.0	14	71.5	28.5
2. Halemanu	250	100.0	0.0	332	71.1	28.9
3. Kumwela Ridge	10	100.0	0.0	14	64.3	35.7
4. Alexander Reservoir	3	100.0	0.0	4	100.0	0.0
5. Kilohana Crater	3	100.0	0.0	4	100.0	0.0
6. Iliiliula River	7	71.4	28.6	10	90.0	10.0
7. Wailua River	9	33.3	66.7	12	100.0	0.0
8. Mt. Kualapa	27	81.5	18.5	36	100.0	0.0
Waianae Range, Oahu						
9. Peacock Flat	36	100.0	0.0	48	72.8	27.2
10. Makaleha Valley	33	100.0	0.0	44	59.1	40.9
11. Mt. Kaala	3	100.0	0.0	4	75.0	25.0
12. Palikea	9	100.0	0.0	12	66.7	33.3
13. Mauna Kapu	80	100.0	0.0	106	46.2	53.8
Koolau Range, Oahu						
14. Pupukea	15	0.0	100.0	23	95.7	4.3
15. Puu Kapu	38	0.0	100.0	49	98.0	2.0
16. Waimano Stream	3	0.0	100.0	4	100.0	0.0
17. Lulumahu Falls	42	12.0	88.0	62	88.7	11.3
18. Mt. Tantalus	20	0.0	100.0	26	88.5	11.5
19. Aihualama Stream	8	0.0	100.0	10	100.0	0.0
20. Manoa Falls	6	0.0	100.0	8	100.0	0.0
21. Pukele Stream	6	0.0	100.0	8	100.0	0.0
22. Kaau Crater	15	0.0	100.0	20	100.0	0.0
23. Wiliwilinui Ridge	9	0.0	100.0	12	100.0	0.0
24. E. Wailupe Gulch	20	0.0	100.0	28	100.0	0.0
25. Kului Gulch	9	0.0	100.0	12	100.0	0.0
26. Pia Valley	6	0.0	100.0	8	100.0	0.0
27. Kupaua Valley	27	3.7	96.3	36	91.6	8.4

Quantitative Studies of Chromosomal Polymorphism in Selected Species

From the inception of the Hawaiian project, data on chromosomal polymorphism have been obtained in such a way that reliable estimates of the frequencies of various gene arrangements within species could eventually be made. Because of the great diversity of species that has been uncovered, and because most species are rather rare, quantitative data have been accumulating very slowly. For several species, however, suggestive preliminary data will be discussed here.

DROSOPHILA CRUCIGERA. This species occurs on both Kauai and Oahu. Carson (1966) has presented preliminary data on populations from the Koolau Range of Oahu and the Kokee region of Kauai. In the present publication, these data have been supplemented by studies of new collections from both islands. The methods used are the same as those described in Carson (1966).

Table XV presents all data, both new and old, for a total of 27 localities, 8 from Kauai, 5 from the Waianae Range, Oahu, and 14 from the Koolau Range, Oahu. Each locality is numbered and its location is given on Figures 9 and 10.

As will be documented below (p. 511), of the two X-chromosome arrangements, Xa is more ancestral, being only one inversion step away from the Standard X. Xb overlaps Xa and is inseparable from it, giving the derived arrangement Xab. Populations from the Kokee-Halemanu area of Kauai and the Waianae Range of Oahu are characterized by very high frequencies of Xa; in fact in all but three localities on the windward side of Kauai, which are polymorphic Xa/Xab, Xa is fixed. In contrast, populations from all but two localities in the Koolau Range, Oahu, show fixation for Xab, and where polymorphic populations occur (e.g., Lulumahu Falls) the frequency of Xab is very high. With regard to the polymorphism in chromosome 3, there is again a close resemblance between the Kokee-Halemanu area of Kauai and the Waianae Range of Oahu. Similarly, windward Kauai (low frequency of 3a) resembles the Koolau Range populations. Evidence will be presented below pointing to the origin of *crucigera* from a *grimshawi*-like ancestor on Kauai. The oldest population of *crucigera* is considered to be that from the Kokee-Halemanu area. It is suggested that the Waianae, Oahu population is descended from a simple intraspecific colonizer, which reached there directly from Kauai, with the modern Waianae populations being little changed from the ancestral.

Apparently, the Xb inversion arose in populations of windward Kauai, wherein the frequency of the 3a inversion was also reduced somewhat. It is proposed that the Koolau populations are descended from a second and separate simple colonization of Oahu from Kauai by this species. Sub-

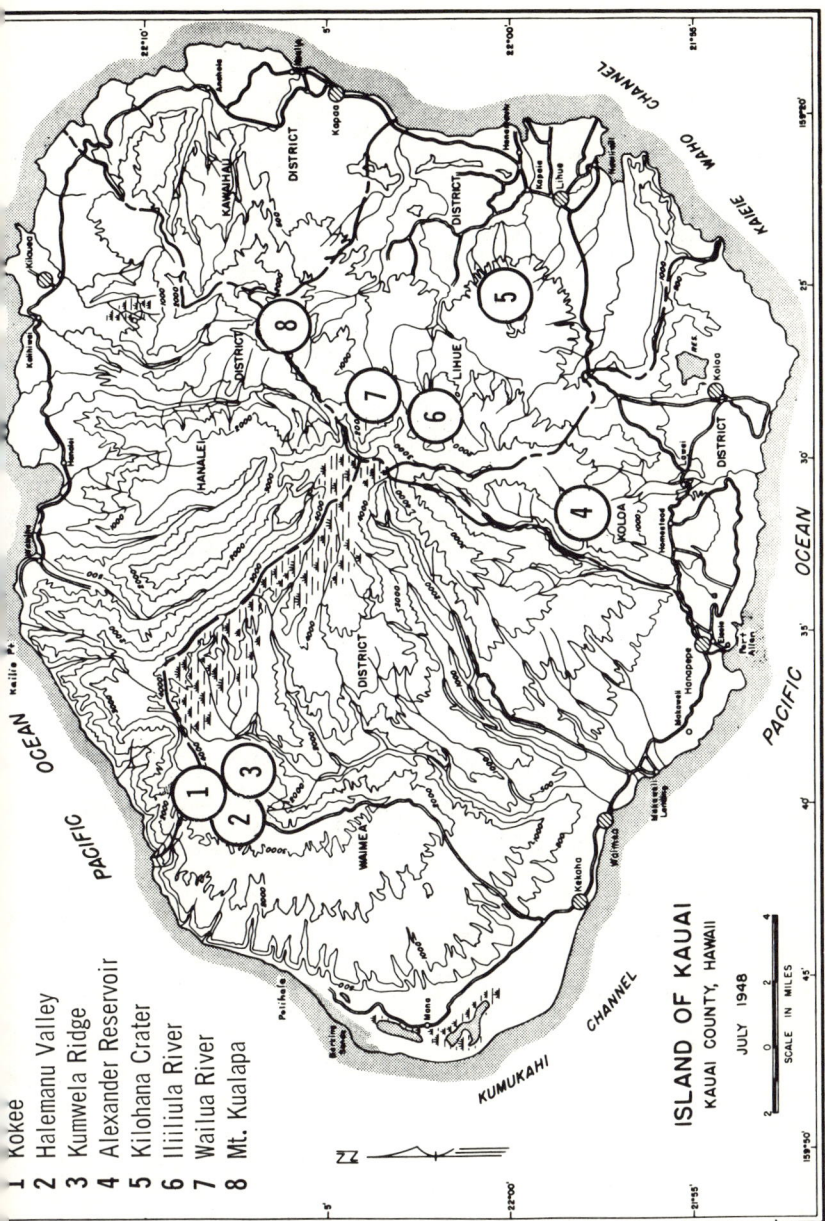

Fig. 9. Localities on Kauai from which Drosophila crucigera has been sampled cytologically.

1 Kokee
2 Halemanu Valley
3 Kumwela Ridge
4 Alexander Reservoir
5 Kilohana Crater
6 Iliiliula River
7 Wailua River
8 Mt. Kualapa

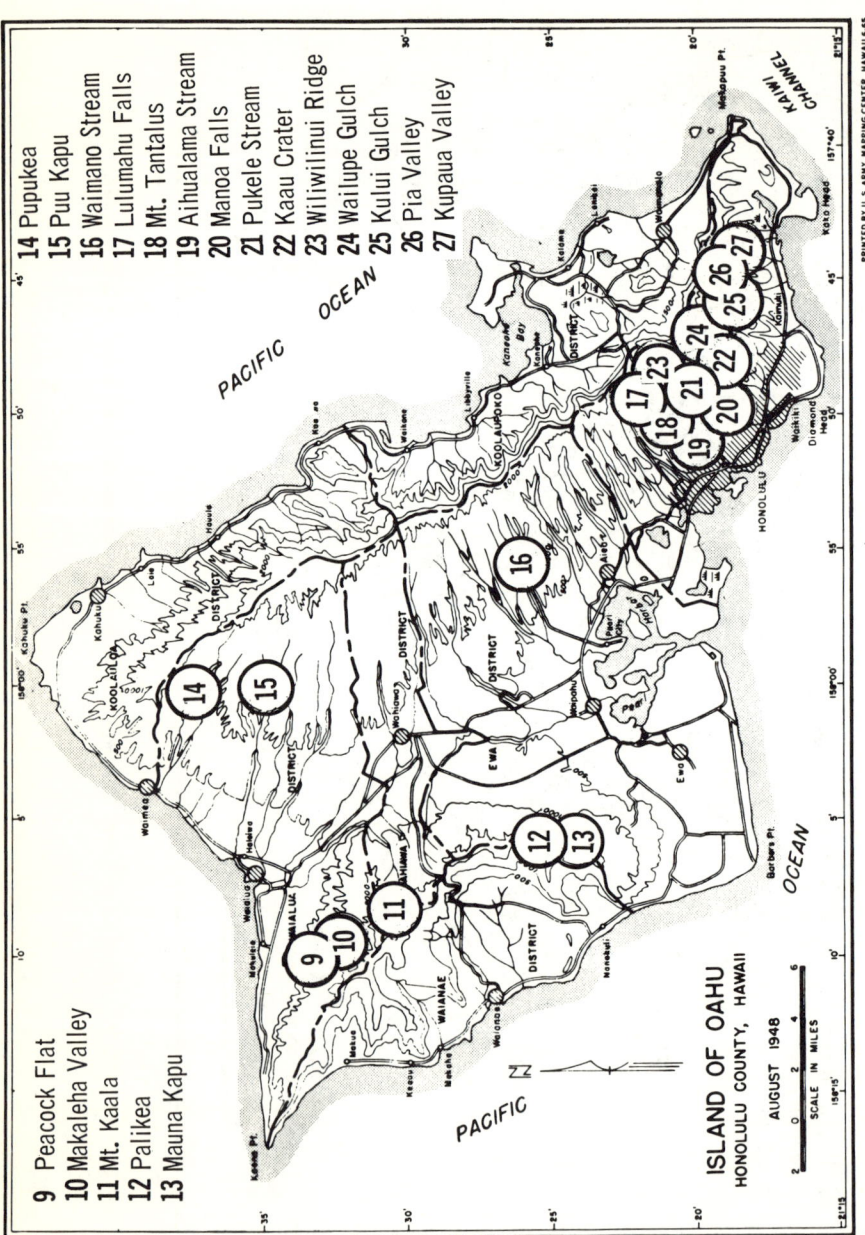

Fig. 10. Localities on Oahu from which Drosophila crucigera has been sampled cytologically. Localities 9-13 are in the Waianae Range, and 14-27 in the Koolau Range.

sequently the Xab arrangement and the Standard 3 arrangement have moved close to fixation in modern Koolau populations. Further discoveries may indeed invalidate these tentative conclusions although the writer feels that various alternate schemes lead to more difficult assumptions than the two-colonization hypothesis presented here.

The conclusion that *D. crucigera* populations of Kauai and Oahu indeed belong to the same biological species requires defense because of the exceedingly high degree of single-island endemism that is manifested in this group (see Table XVIII). In order to test the conclusion of Hardy (1965) that there is no morphological basis on which separate species could be constructed, the following genetic tests were made. Reciprocal crosses were made in the laboratory between an Oahu stock derived from a single female caught on Mt. Tantalus in the Koolau Range [C53.11] and a similar stock collected at Halemanu, Kauai [W(H)31C.2]. Each cross was made in duplicate and 10 virgin females and 10 males were used in each instance. Control crosses of the parental strains were carried out simultaneously. All cultures gave an abundant F_1 without delay, and the sex ratios were normal. All F_1 cultures, furthermore, produced vigorous and fertile F_2 generations, with the interisland hybrids producing equally well if not better than the control strains. Male interisland hybrids (either Kauai × Oahu or Oahu × Kauai) were fully fertile in backcrosses to either Kauai or Oahu females. Crosses between wild males captured in the Waianae Range and C53.11 virgin females have been routinely made as a part of the study of chromosomal polymorphism. These crosses succeed easily and give normal and vigorous F_1's. Accordingly, all crossing data support the conclusion that all these flies belong to the same biological species. Somewhat comparable data will be adduced below for *D. grimshawi* but, at the time of writing, these are the only two species of picture-wings that occur on more than one major island.

DROSOPHILA GRIMSHAWI. This is one of the easiest of all species to collect on Maui, Molokai, and Lanai. Almost all specimens from these islands are of large size and readily oviposit in the laboratory, giving vigorous F_1's and laboratory stocks. At an early time in the investigations the chromosomal arrangements found in stock G-1 from Auwahi, Maui (see Carson, Clayton, and Stalker, 1967) was chosen as the arbitrary Standard with reference to which all the other species on Figure 8 have been described chromosomally.

Since the intensity of collections on Kauai and Oahu have increased in the last few years (1966-1968), a small number of specimens closely resembling *D. grimshawi* have been collected on both islands but they can certainly be rated as extremely rare, with not more than 10 wild females having been caught so far on each island. The specimens differ from those of Maui-Molokai-Lanai in being smaller as well as having a slightly different

TABLE XVI

Frequency (in %) of 4th-Chromosome Gene Arrangements in *Drosophila grimshawi*

Locality and collection no.	No. of 4th chromosomes examined	Standard 4	Gene arrangement 4a
Kauai			
1. Pouli Stream, Hanalei District (1500 feet) L37B4 ♂ X C134.7D virgin ♀, Keanae, Maui	2	100.0	0.0
2. Kokee, M9J1; M11J2	8	100.0	0.0
Oahu			
3. Kaau Crater, Koolau Range (1200 feet) L23G5 ♂ X PK-9 virgin ♀, Puu Kolekole, Molokai	2	100.0	0.0
Molokai			
4-6	50	74.0	26.0
Lanai			
7-10	20	80.0	20.0
West Maui			
11-12	38	21.1	78.9
East Maui			
13	74	100.0	0.0
14-15	28	89.3	10.7
16-19	40	0.0	100.0

position of one of the wing-spots. Despite the fact that *grimshawi* from the southern islands breeds exceedingly well, great difficulty has been experienced in breeding Oahu and Kauai flies. To date, only two females (M9J1 and M11J2, Kokee, Kauai; Table XVI) have produced progeny in the laboratory.

In two instances, wild-caught males have been successfully crossed to virgin laboratory females from stocks derived from the southern islands (Table XVI). In each case a vigorous F_1 was obtained. The sex ratios of these progenies were normal and both wild males proved to be homozygous for the Standard 4th chromosome found in the Auwahi, Maui strain. What is even more significant in the present context, however, is the fact that the interisland F_1's in both cases proved to be fertile. F_1's from the Oahu male (L23G5) produced an abundant F_2; the one produced by the Kauai male appeared to be less so.

More data concerning the Kauai and Oahu races of *D. grimshawi* are being actively sought. It may tentatively be concluded, however, that these indeed are conspecific with *grimshawi* from the southern islands. The questions of which of these populations is ancestral and which derived will be discussed below, where it will be suggested that the Kauai and Oahu populations are primitive relicts.

Table XVI also gives data on 4th chromosome polymorphism in populations of *grimshawi* from Molokai, Maui, and Lanai. These data are in summary form; details will be presented in Carson and Sato (1970). The facts are also shown geographically in Figure 11. The Standard chromosome 4 runs 75 percent to 80 percent in most populations on Lanai and Molokai. West Maui, however, seems to be reversed, with a low frequency of Standard 4. On East Maui (Haleakala Volcano), populations from the leeward side (Auwahi) appear to have the Standard 4 in fixed condition. As one proceeds to the windward side, only the Waikamoi area (locality no. 15) shows polymorphism, and the 4a arrangement appears to be fixed in Kolea, Keanae, Hana, and Kipahulu. Thus it appears that chromosomal races can indeed be sharply manifested even over the small distances involved. Further documentation on this subject is being sought both for this species and for *D. bostrycha* and *D. disjuncta,* which are very closely related species occurring with *grimshawi* on both Maui and Molokai. *D. disjuncta,*

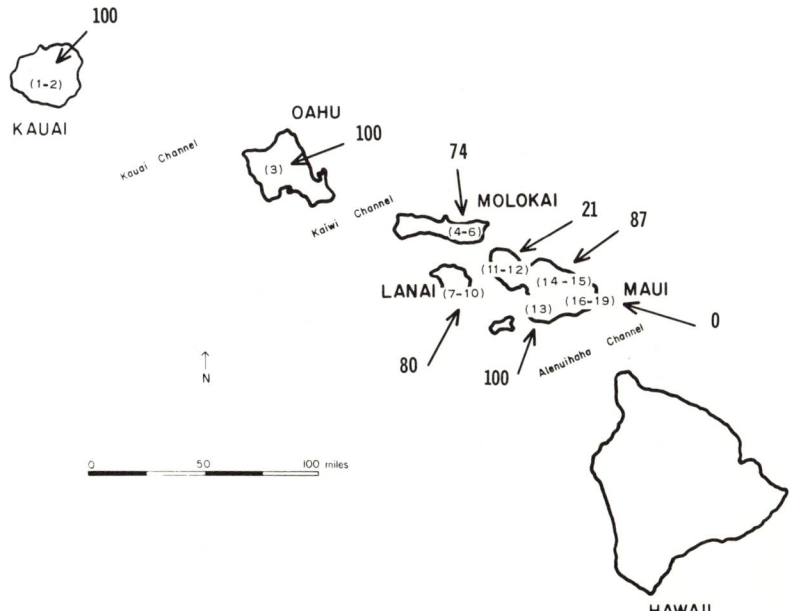

Fig. 11. Frequencies (in %) of Standard chromosome 4 of Drosophila grimshawi relative to its alternative, the derived gene arrangement 4a. Locality numbers (see Table XVI) are in parentheses.

in particular, appears to show striking chromosomal differences in different populations on the island of Maui. These will be the subject of further study in a forthcoming paper (Carson and Sato, 1970).

OTHER SPECIES. Carson and Stalker (1968a, b, c,) have recorded intraspecific polymorphism in a number of species. At least four species show considerably more polymorphism than either *D. crucigera* or *grimshawi*, but population samples are so far relatively small. These species are *D. setosimentum* and *silvestris* from the island of Hawaii, and *fasciculisetae* from Maui. Geographical data on the polymorphisms present in these species are being collected and will be presented elsewhere.

Homosequential Species

Carson, Clayton, and Stalker (1967) introduced this term to refer to two or more species that have identical polytene chromosome-banding sequences in all chromosomes. If intraspecific chromosomal polymorphism is disregarded, there are 10 groups of species among the Hawaiian fauna that can be referred to as homosequential (Table XVII). A total of 31 species fall into this category; this amounts to more than one third of the species given in Figure 8. Among the extensively studied *Drosophila* species from mainland groups, several cases are known in the *D. mulleri* and *mercatorum* subgroups (Wasserman, 1963). In these cases, however, the species concerned are essentially sibling species, whereas this term cannot be properly applied to any of the cases listed in Table XVII. Not only is each species easily distinguished from all others morphologically, but in some cases the members of a homosequential group bear little superficial resemblance and would, in some cases, hardly be grouped together by ordinary taxonomic characters. The most striking cases are *D. vesciseta* and *pilimana*, *D. punalua* and *paucipuncta*, and *D. villosipedis* and *grimshawi*. Ward et al. (1968) have described what appears to be a somewhat comparable case among mainland species.

Homosequential species are significant because their very existence stresses the importance of point mutations as a source of variability for speciation and evolution. It stresses the fact that considerable evolutionary divergence can repeatedly occur in the absence of sectional (chromosomal) mutations.

Study of Figure 8 with special attention to the distribution of homosequential species shows another interesting feature. Homosequential species, or ones that are essentially so, occur in clusters; in most cases, a relatively small number of intermediate states is found. This suggests that, at intervals, a highly stable and successful karyotype is somehow achieved through normal evolutionary processes. This karyotype may then remain essentially in a fixed condition while multiple speciation based on

TABLE XVII
Groups of Homosequential Species Among Hawaiian *Drosophila*

Homosequential group no.	Species	Island	Chromosomal subgroup	Chromosome formula			
1.	*D. bostrycha*	Molokai					
	D. disjuncta	Maui					
	D. grimshawi	Kauai, Oahu, Maui, Molokai, Lanai	I	X	2	3	4	5
	D. orphnopeza	Maui					
	D. villosipedis	Kauai					
2.	*D. glabriapex*	Kauai					
	D. pilimana	Oahu	I	X	2	3	4b	5
	D. vesciseta	Maui					
3.	*D. limitata*	Maui					
	D. sejuncta	Kauai	I	X	2	3	4	5a
	D. ochracea	Hawaii					
4.	*D. balioptera*	Maui, Molokai, Lanai	I	Xg	2	3	4	5
	D. murphyi	Hawaii					
5.	*D. hawaiiensis*	Hawaii					
	D. musaphilia	Kauai	I	Xa^2	2b	3g	4u	5
	D. recticilia	Maui					
	D. silvarentis	Hawaii					
6.	*D. heteroneura*	Hawaii					
	D. silvestris	Hawaii	II	Xijkopqrst	2	3d	4b	5
	D. planitibia	Maui					
7.	*D. neopicta*	Maui	II	Xijkopqs	2	3d	4b	5
	D. obscuripes	Maui					
8.	*D. hanaulae*	Maui	II	Xijkopqst	2	3d	4b	5
	D. oahuensis	Oahu					
9.	*D. adiastola*	Maui, Lanai					
	D. cilifera	Molokai	III	Xikouvwxy	2cd	3fjk	4bopq	5f
	D. peniculipedis	Maui					
10.	*D. ocellata*	Kauai					
	D. paucipuncta	Hawaii	IV	Xef	2	3z	4befg	5
	D. punalua	Oahu					
	D. uniseriata	Oahu					

point mutations may occur on this karyotypic base. The most striking cases of this tendency are the *D. punalua* subgroup and the cluster of species that are close to *D. grimshawi, planitibia,* and *adiastola*.

Sibling Species

This term has been applied to two or more species that resemble one another morphologically so closely that they have been frequently overlooked by the usual taxonomic procedures. Genetic tests may disclose that within what has been considered a single taxonomic entity two or more gene pools coexist without admixture. If such is the case, the separate populations concerned may be referred to as sibling species. This term is frequently retained for such entities despite the fact that in a number of instances a minor difference in genitalia, for example, may be discovered, which permits ready species recognition.

From the inception of the Hawaiian project all workers doing breeding work were sensitive to the possibility of the presence of sibling species. For over three years no evidence of such cryptic species was found: in fact the existence of morphologically divergent species that were nonetheless homosequential seemed to be the antithesis of the concept of sibling species.

Accordingly it was of great interest when, in May of 1968, a classical case of sibling species was discovered. *Drosophila primaeva* is a characteristic and widely distributed fly in the wet forests of Kauai, and since the species has a series of interesting chromosomal polymorphisms, some quantitative data have been gathered (Carson and Stalker, 1969) from various populations. In screening a large number of F_1's from Kahili, Kauai, a sibship was found, which showed numerous sequential differences from *D. primaeva*. Inspection revealed a difference of 13 fixed inversions from the latter species, which was sympatric with it. Rearing of F_1's from this culture (L41C12) was accomplished. Morphologically, these flies are virtually indistinguishable from *D. primaeva* although a small difference in male genitalia was found (Hardy and Kaneshiro, 1969). This species has been described as *Drosophila attigua*. Its difference in chromosomal makeup from *primaeva* is diagrammatically represented in Figure 8 and listed in Table XI, where it may be seen that several unique polymorphisms were also discovered in it.

Interspecific Hybridization

By and large, the species of *Drosophila* of the Hawaiian fauna are sharply distinguished from one another morphologically and behaviorally. This very discreteness of the species constitutes evidence against the widespread occurrence of interspecific hybridization in nature. In general, the cytological data regarding the picture-winged species also supports the conclusion that natural interspecific hybridization is either absent or extremely rare. Many very similar allopatric species differ by one or more fixed inversions. Hybridization between these species could be quickly and easily recognized cytologically, but no such case has been found.

On the other hand, the existence of that small number of cases where two (or in one case, three) homosequential species are sympatric requires comment. In these cases, interspecific hybridization could not be detected cytologically. On the other hand almost every one of the species concerned displays unique morphological characters, which in themselves could serve to label F_1 individuals as hybrids if such did indeed occur. Despite careful investigation of all these possibilities, no certain case of natural interspecific hybridization has been detected.

Under laboratory conditions, however, many interspecific hybridizations have been made. The most notable study is that of Yang and Wheeler (1969). These authors used 28 picture-winged species and attempted 278 interspecific crosses. In 50 instances, F_1 hybrid larvae, pupae, or imagos were produced. This number includes 11 cases in which hybrids were produced from both reciprocal crosses. F_1 imagos were obtained in crosses involving 20 species (Fig. 12). In this figure, the species are grouped according to their chromosomal similarities. Most of the successful crosses occur within the chromosomal subgroups; the rather large number of hybridizations within the *grimshawi* subgroup is not unexpected in view of the great chromosomal similarity of these species (Fig 8). Although eight inseminations have occurred involving members of different cytological subgroups, imagos were produced in only one instance, that of *picticornis* females crossed to *crucigera* males. The three other inseminations of *D. picticornis* females by members of the other two subgroups may be related to the postulated relict nature of this species (see below).

In most cases, F_1 female hybrids are fertile in backcrosses to one or both of the parental males. Of even greater interest, however, is the fact that in six instances F_1 hybrid males were at least partially fertile in backcrosses or even in F_2's (small darkened squares, Fig. 12). In each case, the strains involved are from different islands and morphological differences between the species are moderately great, and there appears to be no reason to insist that the data indicate the conspecificity of the entities involved. This is especially true because in only one case are the males fertile from both reciprocal crosses.

In the case of *crucigera* and *grimshawi,* Yang and Wheeler's data indicate fertility of F_1 males from both reciprocal crosses. These data are in disagreement with the findings of Clayton (unpublished; cited by Carson, 1966), who made the same crosses and found the F_1 males to be sterile. Such differences might be due to genetic differences between the strains used.

The *hawaiiensis-gradata* and *hemipeza-silvestris* cases are most interest-

♀♀ \ ♂♂	bostrycha	disjuncta	grimshawi	villosipedis	balioptera	engyochracea	orthofascia	crucigera	gradata	hirtipalpus	hawaiiensis	silvarentis	pilimana	discreta	paucipuncta	uniseriata	planitibia	silvestris	hemipeza	picticornis
I. *D. grimshawi* subgroup																				
bostrycha		I	N	N	N	N		L			N	N	N	N	N				N	
disjuncta	I		P	N	P	N		L			N	N	N	N	N	N			N	N
grimshawi	[I]	I		I	I	N	I	[I]	N		N	N	N	N	N	N		N	L	N
villosipedis		N	N		N	N	N	N	N		N	N	N		N	N			N	N
balioptera	P	P	I	P		N		I	L	I	N		N	P	N	N			N	N
engyochracea		L	I	I	N		N	N	N		N	N	N	N	N	N			N	N
orthofascia			L	I		N		N	N		N									
crucigera	[I]	I	[I]	P	L	N	N		N		N	N	N	L	N	L			N	N
gradata			N	N			N	N			I	N	N	N	N					
hirtipalpus																				
hawaiiensis	N	L	N	I	N	N	N	N	[I]	I		I	N	I	N	N			N	N
silvarentis		N	N	N	N	N			N		N		N	N	N	N			N	N
pilimana	N	I	I	L	N	N	N	N		N	N	N		S	N	N			N	N
discreta	N	N	N	N	N	N		N	N		N	N	N		N	N		N	N	N
II. *D. punalua* subgroup																				
pauipuncta		N	N	N	N	N		N			N	N	S	S		I			N	N
uniseriata		N	N	N		N		N			N	N	N	N	I				N	N
III. *D. planitibia* subgroup																				
planitibia																				
silvestris			N										N						N	
hemipeza	N	N	N	N	N	N		N		N	N	N	N	N	N	N	I	[I]		N
picticornis	N	N	N	N	L	N		I			N	N	N	S	N	P			N	

Fig. 12. Interspecific crosses of picture-winged species from which at least one F_1 hybrid imago was obtained (after Yang and Wheeler, 1969).

ing. In each case, one species is from Hawaii and the other from Oahu, and in both cases considerable fixed inversion and morphological differences separate the species. These cases suggest the conclusion that sexual isolation is an essentially fortuitous accompaniment of speciation. Such a conclusion is especially compelling in view of the evidence that speciation frequently has followed a single founder event (see below, p. 523). This means that complete reproductive isolation may not always accompany speciation.

The fact that the above cases involve species that are very different morphologically and chromosomally hardly tempts one to discard the previously held species criteria and declare the populations involved to be conspecific. In this connection, however, two other cases studied by Yang and Wheeler are particularly interesting since the species in both cases are morphologically and cytologically very close. *D. disjuncta* and *D. bostrycha* are allopatric species with the former confined to Maui and the latter to Molokai. They are not only homosequential but share a common polymorphism (4v/+). Thus it is impossible to distinguish the species chromosomally. Morphologically, the only difference that can easily be used is the presence of several long dorsal hairs on the middle of the tibia of *bostrycha* (Hardy, 1965). Despite these great similarities, F_1 males from both reciprocal crosses are sterile, confirming the conclusion that these two are indeed bona-fide species.

Crosses between the homosequential species *hawaiiensis* and *silvarentis* of Hawaii also resulted in sterile F_1 male hybrids, again reinforcing the judgment of the taxonomists (Hardy and Kaneshiro, 1968).

In summary, Yang and Wheeler's study of hybridization shows that the rather free crossing between species of picture-wings parallels the other data in revealing biological closeness beneath a façade of striking morphological difference. Ethological isolation is not as pronounced as might be expected from the extensive development of secondary sexual characters. Failure to mate interspecifically seems to be at least in part a fortuitous species characteristic and in part a by-product of the strong tendency for lek behavior that these species show.

The potential for natural interbreeding and gene exchange between Hawaiian species of *Drosophila* is certainly great, especially through backcrossing of F_1 hybrid females to males of parent species. Recognition of such cases should not be difficult since most species are well-marked morphologically and cytologically. The fact that no such case has as yet been recognized is noteworthy, but the search continues.

Kambysellis (1969) has studied the phylogenetic relationships of both picture-winged and other Hawaiian *Drosophila* by the method of interspecific ovarian transplantation. In contrast to what the same author found for mainland species groups of *Drosophila* (Kambysellis, 1968a, b), interspecific incompatibility among Hawaiian species is far less pronounced. This has led to the conclusion that *Drosophila* species endemic to Hawaii manifest a marked uniformity of internal environment in contrast to the high diversity observed for mainland species.

The method used by Kambysellis also leads to the possibility of the production of F_1 interspecific hybrids. This possibility is enhanced by this technique because when a donor egg laid by a host female is fertilized by a host sperm, ethological isolation is effectively bypassed. *D. grimshawi* proved to be the best host; *hawaiiensis* and *hemipeza* were also used. Among 18 picture-winged species studied, the index of oogenesis for donor ovaries developing in *grimshawi* hosts is 100 except for members of the *adiastola* subgroup, where it is somewhat less. Two modified-mouthparts species, *D. mimica* and *eurypeza*, show indices of oogenesis of 100 with the picture-winged hosts. Of all species from which transplantations were made, *nigra* and *crassifemur* have the lowest indices of oogenesis (67 and 65, respectively). This is not wholly unexpected since *D. crassifemur* has *Scaptomyza*-like properties.

In most cases, Kambysellis' data parallel what Yang and Wheeler (1969) has found. Hybrid imagos were obtained in five instances where *grimshawi* provided the male gamete (*conspicua, hexachaetae, ochracea, hemipeza,* and *silvestris*). Imagos were also obtained from the cross *gradata* female × *hawaiiensis* male. The most striking cases, of course, are the intergroup species hybrids between *grimshawi* and the two extra-vein species of the *planitibia* subgroup (*hemipeza* and *silvestris*). The production of such adults provides further evidence that these entities are congeneric. As in Yang and Wheeler's work, however, hybridization appears to be more successful within the cytological subgroups than between them.

These hybridization studies confirm in main outline the conclusions reached from the strictly cytological approach. Although they show many individual morphological and ethological peculiarities, Hawaiian *Drosophila* species show extremely close genetic affinity. One can only conclude that these startling peculiarities are, in a sense, superficial.

Evolution, Speciation, and Migration

Chromosomal Evidence for the Ultimate Origin of the Picture-Winged Species from Certain Continental Forms

As has been pointed out previously, chromosomal data on Hawaiian *Scaptomyza* are confined to metaphase studies. These, furthermore, provide no clue to the origin of these flies. On the other hand, both metaphase- and salivary-gland chromosomes have now been obtained for 69 species of the picture-winged flies, a most striking and conspicuous element of the

TABLE XVIII
Single Island Endemism Among Picture-winged Hawaiian Drosophilidae

Island	Number of species found on this island only		Species present on this island and others (cytologically studied)
	No. cytologically studied	Estimated no. not yet studied	
Kauai	9	3	*D. crucigera, D. grimshawi*
Oahu	11	17	*D. crucigera, D. grimshawi*
Maui	24	8	*D. adiastola, D. balioptera, D. grimshawi, D. spectabilis*
Molokai	3	3	*D. balioptera, D. grimshawi, D. spectabilis*
Lanai	0	1	*D. adiastola, D. balioptera D. grimshawi*
Hawaii	17	3	—
Total	64 +	35 +	5 = 104

Hawaiian Drosophilidae. The distribution of these cytologically well-known species by island of origin (Table XVIII) reveals a high degree of single-island endemism. Thus, on the six major islands, there are only 5 species out of the 69 that occur on more than one island. This tendency is all the more striking when one recalls the fact that Maui, Molokai, and Lanai have been fused into a single land mass several times because of fluctuations in sea level. If the data in Table XVIII are recalculated considering Maui-Molokai-Lanai as a single island, this unit ("the Maui complex") has a total of 30 endemic species. Thus, a total of 67 out of 69 species (all but *D. crucigera* and *D. grimshawi*) qualify as single-island endemics.

In placing value on the data from these 69 species, the quesion arises as to how representative of the entire picture-winged fauna the cytologically studied sample is. In order to approximate an answer to this, the writer has searched the collecting records, both published and unpublished, and has made a crude estimate of the number of endemic species probably belonging to this general group that are known but have not yet been studied cytologically. These numbers are given in the second column in Table XVIII.

Except for Oahu, the fauna appears to be well known, with the 69 species studied coming from all islands and being based on an apparent total of 104 species. It is clear that Oahu has had, at least until recent years, a quite substantial fauna of these large flies. Rediscovery of species taken on Oahu 10 to 20 years ago has proved difficult, and this appears to be related to the rapid and irretrievable disappearance of natural habitats under the impact of human exploitation and the inexorable inroads of exotic forest elements. The only recourse appears to be to attempt to collect in the more remote valleys, which are difficult of access.

With regard to the ceiling number of 104 species, suggested above, another note of caution should be sounded. From the beginning of *Drosophila* collections on the Hawaiian Islands, the number of species has been frequently underestimated. Accordingly, the existence of a number of as yet undescribed species remains a strong probability.

Among the eleven species studied from Kauai are represented members of all five cytologically recognized subgroups (Fig. 8 and Table XVIII). This includes two species, *D. primaeva* and *attigua,* which comprise all of the cytologically known members of subgroup V, the *D. primaeva* subgroup. Morphologically and cytologically (see Table IX) these species are distinctive in the Hawaiian fauna.

Accordingly, it is of very great interest that, in a preliminary study, Stalker (1968) has shown that certain "runs" of polytene-chromosome-banding in *D. primaeva* can be homologized with comparable sequences in certain mainland species. These homologies, moreover, relate the Hawaiian species more clearly to the *robusta* group rather than, for example, to the *virilis, repleta,* or *melanica* groups. All these groups, of course, belong to the subgenus *Drosophila*.

The *robusta* group appears to have its center of origin in Asia. Thus, on the northern Japanese island of Hokkaido alone there are six species (Okada, 1956; Takada, 1959; Kaneko et al., 1964; Kaneko and Takada, 1966), whereas in the whole of continental North America only two species, including *D. robusta* (*sensu stricto*), are found.

Although detailed comparisons are as yet unpublished, the precision of some of the banding comparisons may be exemplified by the following case (Stalker, personal communication). The basic sequence of bands in chromosome 5 in the *primaeva* subgroup is described within the Hawaiian lettering scheme as 5h (Fig. 7). It so happens that the arrangement represented by 5h is closer to the mainland forms than Standard 5; when the 5h inversion is made, producing Hawaiian Standard 5, the "run" of bands is broken up. These comparisons appear to be all the more significant because chromosome 5, in particular, is the least variable in Hawaii and the same element likewise is the most conservative among the mainland forms.

Accordingly, this type of evidence provides a valuable clue to the direction of the phylogeny. Intrinsically, and without outside evidence, a chromosomal phylogeny is not directional. Any point in such a relationship diagram as Figure 8 can be considered ancestral and all others derived from it. Geological, geographical, behavioral, and morphological evidence, together with the cytological sequential evidence just cited, all converge to indicate that, among the 69 cytologically known species, the *primaeva* subgroup of Kauai is indeed the closest to a theoretical ancestor for the group.

This is not to say that further studies may not (1) disclose other mainland forms which are even closer to Hawaiian species, or (2) disclose other Hawaiian species that are even closer to mainland forms. That this dual search continues, however, does not affect the basic outlines of the scheme given above.

Migration and Speciation of the Picture-Winged Species within the Hawaiian Islands

The establishment of Kauai as the probable ancestral home of the present *Drosophila* of Hawaii makes possible the erection, using the basic data given in Figure 8, of interpretative theories of the origin of all species and species subgroups. These theories are presented in Figure 13 through 18, with supplementary data provided in Table XIX. These interpretations, especially as regards the early evolution of the five subgroups on Kauai, will almost certainly be subject to revision and alteration as more data accumulate. They should be regarded as tentative. On the other hand, the proposed species founders for the fauna of the late Pleistocene island of

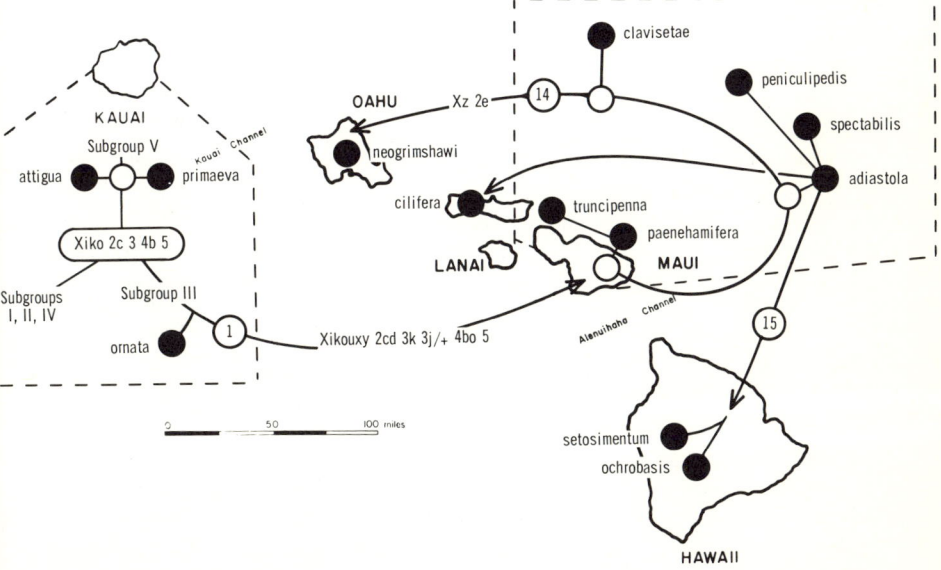

Fig. 13. Origin of D. adiastola subgroup (III) on Kauai and its subsequent history. Founders nos. 1, 14, 15, cf. Fig. 8 and Table XIX; solid symbols: existing species; open circles and brackets: hypothetical populations.

TABLE XIX
Chromosomal Composition of Interisland Founders

Founder No.	Island Donor	Island Recipient	Chromosomal formula				Derived from a population:	Gives rise to:
1	Kauai	Maui	Xikouxy	2cd	3k 3j/+	4bo	5	ancestral to *D. ornata* (subgroup III)	*D. adiastola* subgroup (III) on all eastern islands
2	Kauai	Maui	Xijko	2	3d	4b	5	descended from the above	*planitibia* subgroup (II) on all eastern islands
3	Kauai	Hawaii	Xijko	2	3de	4bhij	5e	descended from the above and ancestral to *picticornis*	*setosifrons* of Hawaii only
4	Kauai	Oahu	X	2	3	4b	5	ancestral to *glabriapex*	*pilimana* and other "4b" species of *grimshawi* subgroup (I) on Oahu
5	Kauai	Oahu	X	2	3	4	5	of *grimshawi* (Kauai)	*grimshawi* (Oahu)
6	Oahu	Maui	X	2	3	4	5	from Founder no. 5	*grimshawi* of Maui complex and other Standard 4 species of Maui
7	Kauai	Oahu	Xef	2	3z	4befg	5	ancestral to *ocellata*	*punalua* subgroup (IV) on Oahu
8	Oahu	Maui	X	2	3	4b	5	descended from Founder no. 4	"4b" species of *grimshawi* subgroup (I) on Maui
9	Oahu	Maui	X	2	3	4b	5d	descended from Founder no. 4	"5d" species on Maui (see Fig. 8)
10	Oahu	Hawaii	Xef	2	3z	4befg	5	descended from Founder no. 7	*punalua* subgroup (IV) of Hawaii
11	Maui	Oahu	X	2b	3g	4	5	descended from Founder no. 6	*gradata* (Oahu)

TABLE XIX (Continued)

Founder No.	Island Donor	Island Recipient	Chromosomal formula				Derived from a population:	Gives rise to:
12	Maui	Kauai	Xa^2	2b	3g	4u	5	similar to the above	musaphilia (Kauai)
13	Kauai	Maui	X	2	3	4	5a	ancestral to sejuncta	limitata (Maui)
14	Maui	Oahu	Xikouvxyz	2cde	3fjk	4boq	5f	descended from Founder no. 1 (see Fig. 13)	neogrimshawi (Oahu)
15	Maui	Hawaii	Xikouvxy	2cd	3fjk	4bopq	5f	descended from Founder no. 1 (see Fig. 13)	setosimentum and ochrobasis (Hawaii)
16	Maui	Oahu	Xijkopqst	2m/+ 3d		4b	5	descended from Founder no. 2 (see Fig. 14)	planitibia subgroup (II) on Oahu
17	Maui	Hawaii	Xijkopqrst	2	3d	4b	5	descended from Founder no. 2 (see Fig. 14)	heteroneura and silvestris of Hawaii
18	Maui	Hawaii	X	2	3	4b	5	descended from Founder no. 8	conspicua (Hawaii)
19	Maui	Hawaii	Xa^2	2b	3g	4u	5	similar to that for Founder no. 12	silvarentis, hawaiiensis
20	Maui	Hawaii	Xg	2	3o/+	4	5	descended from Founder no. 6	engyochracea, murphyi, ciliaticrus
21	Maui	Hawaii	X	2	3	4	5	similar to Founder no. 6	sproati
22	Maui	Hawaii	X	2	3	4	5a	similar to Founder no. 13	ochracea claytonae

Hawaii, for example, appear to be unequivocal and should be regarded as firmly established.

THE PRIMAEVA SUBGROUP (v). There are only two members of this subgroup and both have been found only on Kauai. *D. attigua* is a sympatric sibling species of *primaeva;* it has been described by Hardy and Kaneshiro (1969).

The two species differ by 13 fixed inversions; their hypothetical common ancestor differs from Standard *D. grimshawi* by 18 inversions, 14 of which are in chromosomes X and 4. These two chromosomes, it will be recalled, are the most variable among the five chromosomes (see Table XIII). In contrast, chromosome 3 of *primaeva* is identical to Standard, and 2 and 5 are not far away. If all of the inversions unique to the *primaeva* subgroup are made, the karyotype moves closer to the other subgroups, having the formula Xiko 2c 3 4b 5 (Fig. 13). The break-points of the inversions and chromosome maps are presented in Carson and Stalker (1969).

THE ADIASTOLA SUBGROUP (III). This subgroup, like the other three, apparently arose from the common ancestor just mentioned (Fig. 13). With the addition of inversions leading away from Standard (Xuxy 2d 3k 3j and 4o) a karyotype is reached which can be postulated as a common ancestor of *D. ornata* of Kauai and the rest of the *adiastola* subgroup. From this ancestor a migrant (Founder no. 1, Fig. 13) must have reached Maui directly, where, with the addition of a few more inversions (Xvw, 3f, 4p, 4q, 5f, and the fixation of 3j), the condition observed in present-day *D. adiastola* could be reached.

Most of the Maui members of this group are chromosomally very close, despite the extraordinary curious morphologies found in *D. truncipenna, paenehamifera,* and *clavisetae.* The latter, for example, not only has a remarkable set of clavate hairs at the end of the abdomen of the male, but the species is characterized by the presence of an extra cross-vein in the R-5 cell of the wing. It is of great significance that the only other member of this subgroup showing this extra cross-vein is *D. neogrimshawi* of Oahu. Since this latter species has the Xz and 2e arrangements, found also in *clavisetae,* the evidence is strong that *neogrimshawi* is derived from a specialized and peculiar section of the *adiastola* subgroup, a northward migrant from Maui (Founder no. 14, Fig. 13).

The above situation indicates clearly that the initial southward migration (Founder no. 1) must have by-passed Oahu. Only later did a member of this group reach Oahu from Maui.

As has been mentioned several times, Maui, Molokai, and Lanai have previously been joined. Accordingly, the one species unique to Molokai, *D. cilifera,* may have reached there by simple allopatric speciation over a terrestrial route. Accordingly, this is not counted as an interisland founder despite the fact that this possibility cannot be excluded. Typical *D.*

adiastola, both chromosomally and morphologically, exists on Lanai. On the Big Island (Hawaii), however, there are two distinct species having the basic arrangement of *D. adiastola* with a series of additional unique inversions (*D. setosimentum* and *ochrobasis:* see Fig. 8). These two species appear to be derived from a Pleistocene founder (no. 15) deriving from an ancestral population having the same chromosome composition as *adiastola* (Fig. 13 and Table XVIII).

THE PLANITIBIA SUBGROUP (II). This subgroup comprises 14 species. Twelve of the 14 have an extra wing-vein in cell R-5. The single Kauai member, *D. picticornis,* and *D. setosifrons* of Hawaii, do not have the extra wing-vein, yet display two inversions, Xj and 3d, which are found in all other members of the group. The absence of extra-vein flies from Kauai, yet the presence there of *picticornis,* suggests that the latter is the modern descendent of an ancestral Kauai population from which a Founder (no. 2, Fig. 14) proceeded directly to Maui and produced the extra-vein flies on that island. *D. picticornis,* however, is not itself close to the primitive population from which the founder came. In fact, it appears somewhat specialized. A number of its special inversions, furthermore, are shared by the bizarre fly *D. setosifrons* of Hawaii. Although these two have

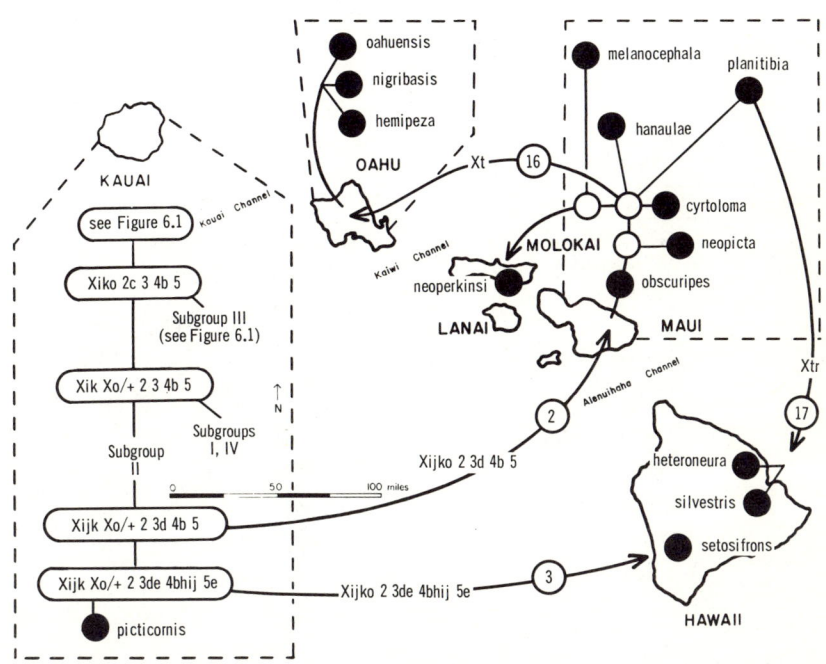

Fig. 14. Origin of D. planitibia *subgroup (II) on Kauai and its subsequent history. Founder nos. 2, 3, 16, 17.*

many inversion sequences in common, they are morphologically very different. *D. picticornis,* it will be noted, lacks Xo, and inversion that is characteristic of all of the extra-vein flies and is also present in *setosifrons*. The latter is thus just one step closer to the extra-veined flies than *D. picticornis* is.

The interpretation that has been tentatively placed on this situation is given in Figure 14. Founder no. 2 must bypass Oahu because *D. obscuripes* of Maui is the closest of the extra-veined flies to *picticornis*, with *neopicta* one step further away. This latter species is heterozygous for Xt, an inversion that is fixed in all others, including those of Oahu. This point is important because it proves that the Oahu flies are derived from a *neopicta*-like ancestor on Maui. Only three species have Xr. These are *planitibia* and the two extra-veined flies of Hawaii, *heteroneura* and *silvestris*. Clearly, the latter two are derived from a *planitibia*-like ancestor on Maui (Founder no. 17). For reasons stated earlier, *neoperkinsi* of

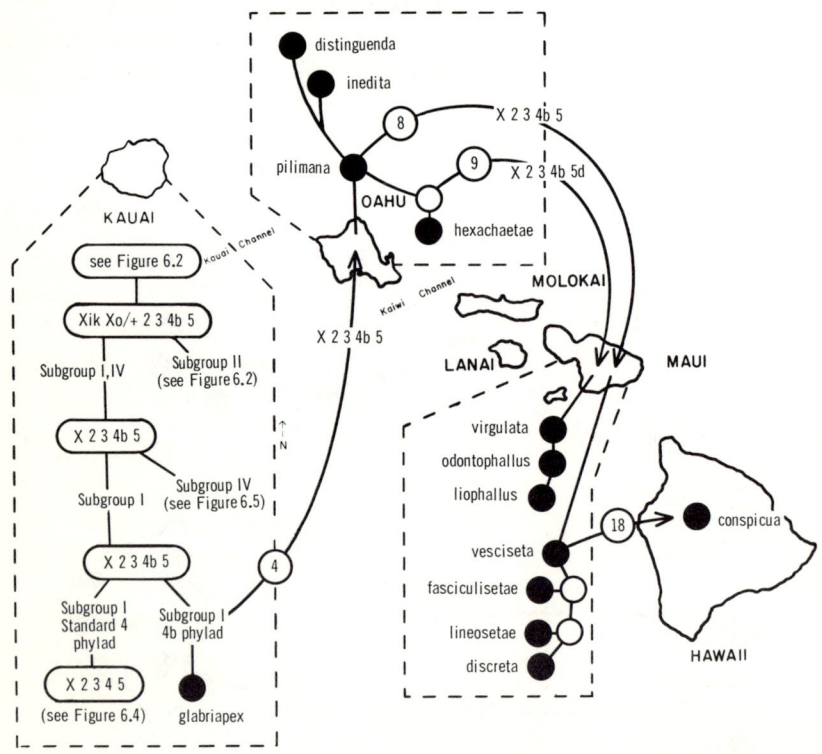

Fig. 15. Origin of the D. glabriapex section (gene arrangement "4b" phylad) of the D. grimshawi subgroup (I) on Kauai and its subsequent history. Founder nos. 4, 8, 9, 18.

Molokai is not considered a clear case of an interisland founder because of the possibility that it was formed by terrestrial allopatric speciation when the islands were joined.

THE GRIMSHAWI SUBGROUP (I). This subgroup is large and includes 36 species. Two phylads (or sections) are distinguishable on the basis of a fixed inversion difference, gene arrangement 4b. Thirteen of the species have 4b in fixed condition and 23 have the Standard Chromosome 4. These two groupings thus form separate phylads and it will be convenient to consider their history separately.

The glabriapex section (4b phylad of Subgroup I). *D. glabriapex* is the sole member of this subgroup on Kauai. It can easily be derived from the ancestor of the *planitibia* subgroup with gene arrangements Xijk and 3d becoming Standard through the occurrence of these four inversions (Fig. 15). This chromosomal condition (X 2 3 4b 5) not only can give rise directly to *glabriapex* but can produce Founder no. 4, which can give rise to *pilimana* of Oahu, as well as the other "4b" species on that island. From Oahu, the large number of "4b species" on the Maui complex can be derived from two Founders (8 and 9, Fig. 15). Despite extensive speciation on Maui, these lineages have apparently resulted in only one invasion of Hawaii (Founder no. 18, giving rise to *conspicua*).

The grimshawi section (Standard 4 phylad of Subgroup I). This group of 23 species, unlike the preceding, is well represented on Hawaii, and, furthermore, gives considerable evidence of reverse migration, with forms on Oahu and Kauai that are apparently derived from Maui.

The basic chromosomal formula of this section is easily derived by the "removal" of arrangement 4b, giving the Standard X 2 3 4 5 (see Fig. 16). This event, according to this theory, was accomplished on Kauai, with the evolution of *D. grimshawi* and *crucigera*. As has been mentioned previously, the latter 2 species are unique among the 69 species so far studied in that they are found, as apparently the same species, on more than one island. Details of the population structure of these two species were presented above (p. 511).

In Figure 16, three southward colonizations are postulated. Thus, *grimshawi* and *crucigera* of Oahu are apparently derived from migrants from Kauai. That migrant giving rise to *crucigera* is not considered a true evolutionary founder, in that the result has been apparently merely the simple colonization of Oahu from Kauai by this species. As was mentioned above (p. 529), there may have been two such colonizations of Oahu by *crucigera*.

A *D. grimshawi* migrant (Founder no. 5, Fig. 16) apparently gave rise to *D. grimshawi* of Oahu. The situation in this case, however, is different from that of *crucigera* since the Oahu populations of the former were apparently a stepping stone for the invasion of the Maui complex by *grim-*

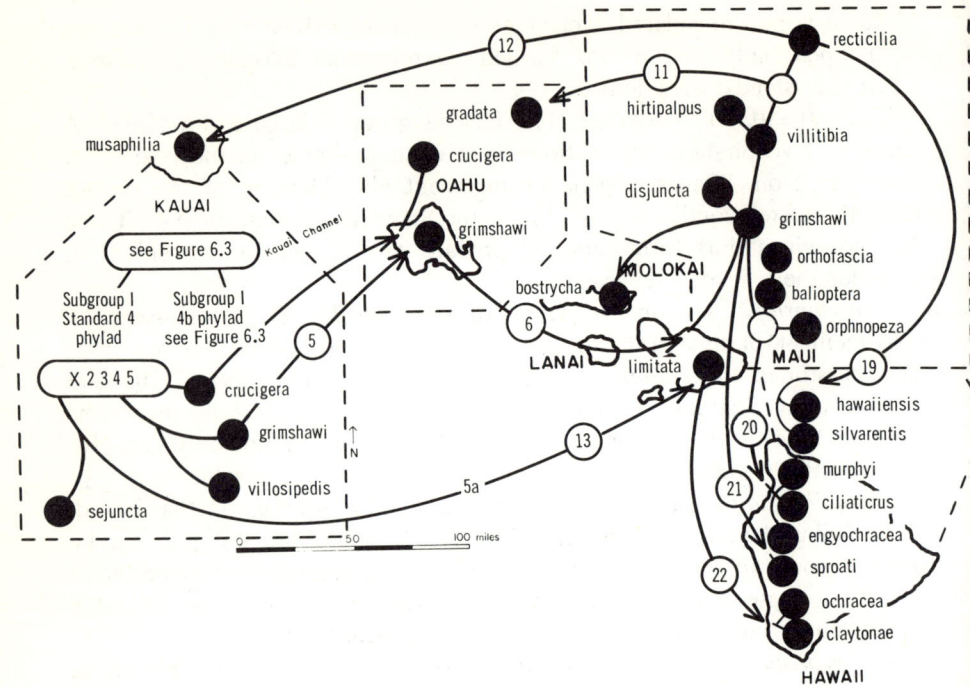

Fig. 16. Origin of the D. grimshawi section (gene arrangement Stanard 4 phylad) of the D. grimshawi subgroup (I) on Kauai and subsequent history. Founder nos. 5, 6, 11, 12, 13, 19, 20, 21, 22. The transfer of D. crucigera represents a simple colonization.

shawi. Furthermore, as indicated in Figure 16, extensive speciation in the Maui complex appears to be based on a *grimshawi*-like ancestral population, with three separate founders reaching the island of Hawaii (19, 20 and 21, Fig. 16). Reference to Figure 8 and Table XIX will show the rationale behind these interpretations.

A final set of events involves the evolution of *D. sejuncta* (X 2 3 4 5a). A founder based on this species appears to have reached Maui directly (no. 13, Fig. 16) and thence given rise to two more "5a" species on Hawaii (Founder no. 22).

The situation with regard to founder no. 20 (Fig. 16) is somewhat complex but interesting. In the first place, three Hawaiian species have the gene arrangement Xg, which is known on Maui as a fixed arrangement only in *baliopterа* and *orthofascia*. Secondly, *orphnopeza* of Maui shows polymorphism for a particular inversion (3o) and the Standard third chromosome, that is, its formula is $3o/+$. Two species found on Hawaii also show gene arrangement 3o. These are *ciliaticrus*, in which the inversion is fixed and *murphyi* which, like *orphnopeza* of Maui, is $3o/+$.

Since *D. balioptera* has not so far been found to have 3o/+, or *orphnopeza* to have Xg, it appears that Founder no. 20 must have had the formula Xg 3o/+. Such a founder could have arisen on Maui from an ancestral population having both Xg/+ and 3o/+. This same population could also have given rise to the three species *orphnopeza* (3o/+), *orthofascia* (Xg), and *balioptera* (Xg) on Maui. *D. engyochracea* of Hawaii, which has Xg but lacks 3o, must represent a case wherein this species has refixed the Standard 3+/3+ from the 3o/+ condition. These details may be gleaned from Figure 8.

THE PUNALUA SUBGROUP (IV). This rather small subgroup apparently also differentiated on Kauai from the hypothetical 4b ancestor of the *grimshawi* subgroup (Fig. 17). The sequential karyotype found in *ocellata* of Kauai occurs also in all other species of the subgroup; thus these species represent a very stable homosequential series. *D. basisetae* and *prolaticilia* of Hawaii each have an additional unique inversion, but stability characterizes the subgroup.

From an *ocellata*-like ancestor, a founder is believed to have reached Oahu (no. 7, Fig. 17) and then to have given rise to the two Oahu species that have a karyotype identical with it. It is noteworthy that no Maui species of this subgroup has been discovered. Accordingly, the facts sug-

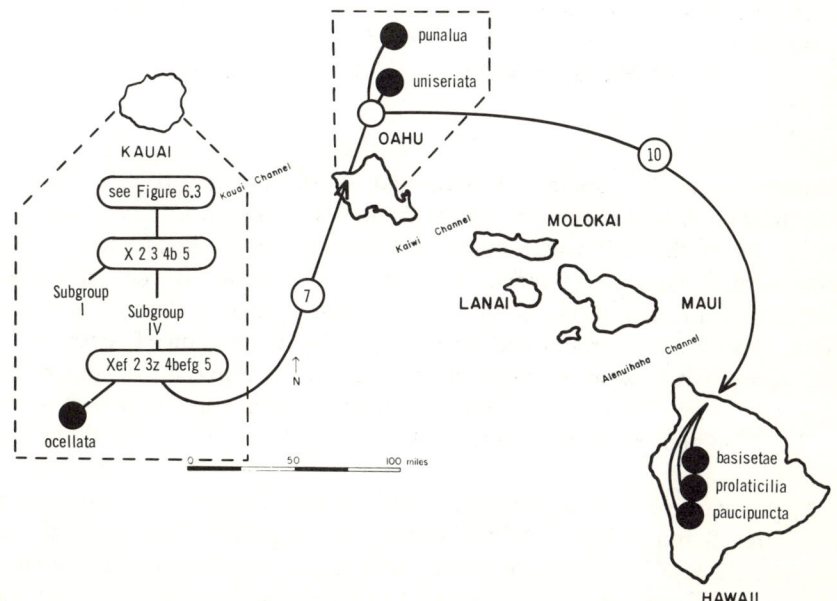

Fig. 17. Origin of the D. punalua subgroup (IV) on Kauai and its subsequent history. Founder nos. 7, 10.

gest that a founder (no. 10) went directly to Hawaii and gave rise to the three species found there.

Colonization and Speciation Patterns in Antopocerus

Among the Drosophilidae of Hawaii is a group of leaf-breeding species that have been placed in a separate genus, *Antopocerus*. The ecology and behavior of these species is well known (see above, p. 492). Spieth (1968b) recognizes three types of mating behavior among the nine species. What may be recognized as the most primitive type characterizes two species that are restricted to the Maui complex (Maui and Molokai). The three species found on Hawaii show derived types of behavior, and this is also true of the single species found on Oahu. Since no member of *Antopocerus* has ever been collected on Kauai, Spieth interprets this situation as indicating that *Antopocerus* evolved on the Maui complex, with both Hawaii and Oahu having been colonized from there. This view closely parallels that which ascribes a key role to the Maui complex in the evolution of the picture-winged species, especially the *D. adiastola* and *D. planitibia* subgroups. These also give evidence of colonization of Oahu from Maui.

So far, flies of this genus have not proved to be workable cytologically so that unfortunately it has not been possible to test Spieth's conclusions by polytene-chromosome mapping. It has likewise not been possible to obtain cytological confirmation that *A. diamphidiopodus* of the Maui complex and the island of Hawaii is indeed a single species throughout its range.

Conclusion

Reference to the papers of Carson and Stalker (1968a, b, c) will show that the *D. grimshawi* subgroup was previously supposed to have arisen on Maui. From there, it was thought to have spread southward to Hawaii, northward to Oahu, and thence to Kauai. Recent discoveries on Kauai, particularly the confirmation that *D. grimshawi* (*sensu stricto*) is found there, as well as *D. ornata* and *ocellata*, has led to revision of these ideas. Accordingly, it now seems much more likely that all five major subgroups of picture-winged flies existed on Kauai at an early time in the history of the Archipelago and spread to the other islands from there (Carson and Stalker, 1969).

The phylogenetic interpretations given in this chapter are summarized in Figure 18 and in Tables XIX and XX. In the preceding sections of this chapter, 22 interisland colonizations have been inferred from cytological information. For most of these, precise chromosomal formulas may be written (see Table XIX).

The island of Hawaii presents a case of special interest. It has been the recipient of nine founders and has apparently not been a donor in any

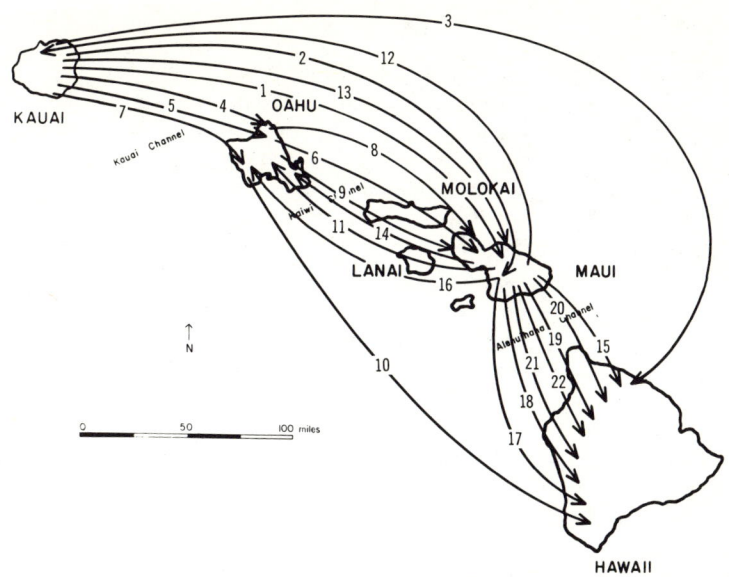

Fig. 18. Summary of the proposed minimum number (22) of interisland founders in the picture-winged Drosophila of Hawaii.

case in the present scheme. This is consistent with the fact that it is the youngest island.

Possibly because they have occurred in relatively recent geologic time, seven of the nine founder events involving the island of Hawaii can be traced with great clarity. For example, five sets of Hawaii species of the *D. grimshawi* subgroup can be traced to particular species known on Maui

TABLE XX

Minimum Number of Interisland Founders Among Hawaiian *Drosophila*

Recipient island	Donor island				Total receptions
	Kauai	Oahu	Maui complex	Hawaii	
Kauai	X	0	1	0	1
Oahu	3	X	3	0	6
Maui complex	3	3	X	0	6
Hawaii	1	1	7	X	9
Total donations	7	4	11	0	22

TABLE XXI

Five Sets of Species of the *D. grimshawi* Subgroup on Hawaii that Have Close Relatives on Maui

Set No.	Island	Species	Chromosomal formula				
1	Hawaii	*silvarentis*	Xa^2	2b	3g	4u	5
		hawaiiensis	Xa^2	2b	3g	4u	5
	Maui	*recticilia*	Xa^2	2b	3g	4u	5
2	Hawaii	*ochracea*	X	2	3	4	5a
		claytonae	X	2	3	4	5ar
	Maui	*limitata*	X	2	3	4	5a
3	Hawaii	*engyochracea*	Xg	2	3	4l	5
		ciliaticrus	Xg	2	3o	4	5
		murphyi	Xg	2	3/3o	4	5
	Maui	*orphnopeza*	X	2	3/3o	4	5
		balioptera	Xg	2	3	4	5
		orthofascia	Xg	2	3n	4	5
4	Hawaii	*sproati*	Xb^2	2	3	4	5
	Maui	*grimshawi*	X	2	3	4	5
5	Hawaii	*conspicua*	Xmm	2	3	4bm	5
	Maui	*vesciseta*	X	2	3	4b	5

(Table XXI). These facts serve as the basis for the erection of hypothetical Founders nos. 19, 22, 20, 21, and 18. In four of these five cases, the exact chromosomal formula necessary is known in an existing species on Maui. In the case of set no. 3 (Table XXI) the founder is apparently descended from a common ancestor of *orphnopeza* and *balioptera,* as discussed previously.

Two other sets of Hawaii-Maui species also show extremely close similarity. These are detailed on Table XXII and form the basis for the postulation of Founders nos. 17 and 15, discussed earlier.

Another interesting circumstance regarding the Maui-Hawaii similarities is the fact that there is as yet no known case of a "simple" colonization of the island of Hawaii by a species known from another island other than the *Antopocerus* case mentioned earlier. To put this another way, it may be said that each founding event has led to species formation and, in some cases at least, several new species have resulted on the recipient island. In fact, the rarity of simple colonizations, and, conversely, the extremely high degree of single-island endemism, seem to be the most characteristic feature of the evolution and speciation of Hawaiian Drosophilidae.

Looking at the situation with regard to the oldest island, Kauai, it will be noted that, in contrast to Hawaii, it has apparently been the recipient of only one founder (no. 12, the ancestor of *D. musaphilia*). This circumstance, however, does not take into account the hypothetical one or more original colonizations of Kauai.

The very large number of species found on Maui suggests a flowering of speciation on that island following the arrival of six founders, three from Kauai and three from Oahu (Fig. 18). As has been pointed out previously, the case for the by-passing of Oahu by founders from Kauai is well documented for two of the three cases and the facts seem unequivocal. The writer is inclined to view this as a fortuituous event perhaps no more improbable than any other interisland movement. Nothing is known as to the nature of the propagules that have accomplished interisland movements. Nevertheless, it seems likely that adult flies are involved and that populations are established directly in high-altitude locations.

In the interpretation given, only the facts available to the present-day investigator have been used, that is, the concept of extinction has not been evoked. Thus, for example, it may be argued that founders like no. 1 or no. 2 could have gone from Kauai directly to Oahu, undergone some speciation and evolution on the latter island, and then sent further founders south to Maui. If this was the case, then one must assume that the intermediates have since become extinct or remain undiscovered. This cannot, of course, be disproved but nevertheless requires many more assumptions than the rather simple explanations that make use of the existing intermediates that have actually been found on Maui. In any event, the rule

TABLE XXII

Two Additional Sets of Hawaii Species with Close Relatives on Maui

Set No.	Island	Species	Chromosomal formula				
6	Hawaii	*silvestris*	Xijkopqrst	2	3d	4b	5
		heteroneura	Xijkopqrst	2	3d	4b	5
	Maui	*planitibia*	Xijkopqrst	2	3d	4b	5
7	Hawaii	*setosimentum*	Xikouvwxym^2	2cdl	3fjkl	4bopqb^{2}c^{2}d^{2}e^{2}f^{2}n^{2}o^2	5f
		ochrobasis	Xikouvwxym^2	2cdk	3fjk	4bopqb^2	5f
	Maui	*adiastola*	Xikouvwxy	2cd	3fjk	4bopq	5f

of parsimony has been adhered to in the interpretations made so far. The writer would thus not contend that alternate schemes based on multiple extinctions could not be constructed. On the other hand, the actual existence of intermediates and of species that are polymorphic for unique inversions that ultimately become fixed in other species suggests a simpler theory. Most existing species, furthermore, appear to be widely distributed on the island to which they are endemic. More detailed distribution data, however, are needed on this point but the data so far suggest that species confined to single valleys or kipukas are fewer than originally thought.

In most continental areas it is not possible to distinguish species that are newly formed in time from those which may have existed for many millions of years. Accordingly, it is of special interest that the geological newness of the island of Hawaii and the obvious derivation of the bulk of its fauna directly from Maui suggests that Hawaii's fauna is recent. Few studies have been made of the evolutionary biology of species selected for their historical newness. Accordingly, the endemic fauna of Hawaii appears to provide unique materials for this sort of study.

Summary

The six largest islands of the Hawaiian chain are of volcanic origin and vary in age from about 5,000,000 years in the west (Kauai) to less than 1,000,000 years in the east (Hawaii), where two volcanos are still active. Maui, Molokai, and Lanai (the Maui complex) were joined and separated at least twice in recent geologic history. Thus, in the broad sense, the number of main islands with deep channels between them reduces to four: Kauai, Oahu, the Maui complex, and Hawaii. All islands receive heavy rainfall on windward slopes, due to the combination of persistent trade winds and high altitudes. These factors, coupled with the tropical latitude and rich volcanic soils support the growth of a number of forest types of which four serve as the habitat for the vast majority of drosophilids.

The oldest islands in the Hawaiian chain (the Leeward Islands: 15,000,000 years) are now worn down to the point that they probably no longer support a native fauna of Drosophilidae except possibly for a few species of *Scaptomyza*. The native species are now confined to the high main islands. Two of the main islands, however, i.e., Niihau and Kahoolawe, are low (1281 and 1415 feet, respectively); they have been greatly changed by man, and today probably support no native drosophilid fauna.

The family Drosophilidae on the islands comprises 650 to 700 species; all but 17 are endemic. A majority of species fall into two genera, *Drosophila* Fallén (324 species, 3 subgenera) and *Scaptomyza* Hardy (121 species, 7 subgenera). The remainder have been placed in endemic genera that are quite clearly either *Drosophila*-like (drosophiloid) or *Scaptomyza*-like (scaptomyzoid). Some species seem to combine certain characters of the two genera; this conclusion is based on internal anatomy, external morphology, and behavior. Structural peculiarities are extraordinarily diverse, particularly in males. Characters include protruding eyes, modified mouth parts, and elaborate foretarsi, including bristle-like, spoon-like, or forked adornments. Tibiae are sometimes knobbed; sets of bristles are sometimes missing; wings may have dark spots, an extra crossvein, or show arcuate or truncate shapes. Some *Drosophila* have extremely large body sizes; many species groups are informally recognized. The Maui complex has the greatest number of known species of drosophiloids (160), followed by Hawaii (98), Oahu (84), and Kauai (44); most are endemic to the specific islands from which they have been recorded.

Most drosophiloids are found at high altitudes in areas of high humidity and breed on fermenting vegetable materials. Nine oviposition sites on endemic plants are recognized, of which leaves of *Cheirodendron* (46 species reared) and stems of *Clermontia* (18 species reared) predominate. In all, 32 endemic genera of plants are used. The leaf-breeding species are difficult to rear in the laboratory. Ecological niches include species that breed on spider eggs, decaying flowers, and nondecaying plant hairs.

Scaptomyzoid males demonstrate a lunging, direct courtship, whereas most of the drosophiloids engage in elaborate ritualized displays, during which the secondary sexual characteristics are brought into play. Unlike drosophiloids in the rest of the world, courtship does not occur on the food, but males manifest lek behavior; they defend specific, localized territories.

Metaphase chromosomal data support the general concept of the dichotomy of scaptomyzoids and drosophiloids. The great preponderance of the latter have a karyotype that has been considered primitive for continental forms (five rods; one dot). Heterochromatin has frequently been added to this basic set. Polytene-chromosome mapping has been carried out for 69 large drosophiloids (the picture-winged species group); all can be described in terms of a single set of arbitrary Standard gene orders despite morphological diversity of the species. Except for one deletion, the chromosomes differ only by paracentric inversions, of which 115 are fixed and 63 are polymorphic; two thirds of the species lack chromosomal polymorphism. Species sometimes share a common polymorphism; within several species sharp local variations in gene arrangement frequencies occur, even within an island. This confirms the impression that population sizes are small and that inbreeding prevails. Only one case of sibling species has been recognized; on the contrary 10 series (31 species) are known in which morphological divergence has not been accompanied by any change in sequential karyotype (homosequential species). The great biological unity of this group is underscored by the ease with which morphologically diverse species may be hybridized in the laboratory; no natural hybrids have been found. Compatability in ovary transplantation is high. Cytological subgroupings generally conform to groupings by behavior and genitalia.

The Hawaiian fauna apparently developed on the older Leeward Islands and migrated down the chain as the islands weathered down. The ancestors of the tremendous present-day fauna were doubtlessly from Kauai, as the cytological and behavioral evidence indicates. On the other hand, one would expect that, since this is the oldest of the main islands, it would support the greatest number of species of native plants and animals. At least with the Drosophilidae, this is not the case. Of 44 species known from Kauai, 5 have also been recorded from other islands; thus apparently 39 species are restricted to the island. It is highly significant in the study of the evolution of this group in Hawaii that of the drosophiloids, only the genus *Drosophila* occurs on Kauai; *Antopocerus, Celidosoma,* and *Nudidrosophila* are unknown from there. The extra-vein flies ("idiomyias") are also not found there, nor is the *D. haleakalae* complex. It appears that most of these groups and species complexes have evolved on the Maui complex and radiated out from there to the other main islands. This idea is supported by the cytological data on the picture-winged species and the ethological facts on *Antopocerus*.

D. primaeva of Kauai has some polytene-banding sequences that are recognizable in certain mainland forms, especially the predominantly Palearctic *robusta* group. Since such sequences are altered by inversions in other picture-winged species, it appears that *primaeva* is chromosomally closer to the ancestral stock than any other Hawaiian species for which full information is available. With this fact and the polytene chromosome relationships, theoretical evolutionary pathways for the subgroups of picture-winged flies have been constructed. After Pliocene speciation episodes on Kauai, a minimum of 22 interisland founder events are proposed. These include two direct invasions of the Maui complex from Kauai; Oahu is bypassed and is colonized later from the east. Nine colonizations of Hawaii took place in the late Pleistocene. Five of these cases show with great clarity that the ancestors came from Maui and that speciation resulted following the founder events.

The original ancestral stock (or stocks) that invaded the Hawaiian archipelago must have found a habitat completely empty of similar flies. There is a high probability that this ancestral stock was a species that utilized rotting bark as a substrate for its larval development. Significantly, the *virilis* group, which originated from the base of the phylogenetic stem that gave rise to the subgenus *Drosophila,* is exclusively restricted to rotting bark for its developmental stages. In any case, the Hawaiian habitat offered only limited suitable substrate for larval development and adult nutrition. The native forests have an extreme paucity of fleshy fruits that are so important for many species of *Drosophila* in other parts of the world. Today the lobeliads of the genus *Clermontia* are the only segment of native flora that possesses such fruits, and even these are of modest size. Furthermore, the substrates that were available for the new immigrants must have been spartan in food value, i.e., the available food resources per given volume of substrate.

If only one immigrant was introduced, then this stock early split into the scaptomyzoids and drosophiloids. If there were separate introductions for each of these major groups, then they were constrained by the poverty of food sources to partially overlap in their use of substrates. Both groups radiated into various ecological niches, the scaptomyzoids evolving as physically small species, able to tolerate exposure to higher light intensities and thus able to use substrates such as the flowers of morning-glories and certain nonrotting vegetation as well as the same fermenting materials that the drosophiloids use. The drosophiloid stock evolved species that are in general larger than the scaptomyzoids and more restricted to the areas of low-light intensity in the dense forest.

All the species groups of drosophiloids with the exception of *anomalipes* and *quasianomalipes* appear to display lek behavior. It thus seems reasonable to assume that the ancestors of the major substrates that were used, i.e., the lobeliads, *Cheirodendron* and its relatives, were established on the islands before the drosophiloid immigrants arrived. Likewise, the agents responsible for the selection pressure that resulted in the evolution of the lek behavior must have also preceded the drosophiloids.

A number of factors appear to be responsible for the evolution of extraordinary numbers of species found today in the Hawaiian Islands. The major factors would appear to be the spartan nature of the food supply, which resulted in the evolution of a low reproductive rate and consequently small population sizes; the infrequent but repeated migrations from each island to adjacent islands, which resulted in effective isolating barriers; the added effects of volcanic and meteorological action, which further isolated small areas such as the kipukas; the evolution of lek behavior; and invasion of specialized food sources such as the leaves of a number of plants and the eggs of spiders.

Hawaii appears to be the only place in the world where the otherwise sharp distinctions between *Scaptomyza* and *Drosophila* tend to disappear. The number of species of *Scaptomyza* in Hawaii is twice that for the rest of the world. These facts provide evidence that the entire family in Hawaii could have stemmed from one ancestral introduction, with the genus *Scaptomyza* originating in Hawaii. From data presently available, however, definite espousal of the "one-introduction" over the "two-introduction" hypothesis seems premature.

Acknowledgments

This project has been supported by National Institutes of Health Grants GM 10640, GM 11609, and National Science Foundation Grants GB-711 and GF-152. Additionally the senior investigators have been supported in part by various individual grants which have contributed to their investigations.

The individuals who have been involved in the Hawaiian *Drosophila* Project are as follows: Senior Colleagues—R. Malcolm Brown, Hampton L. Carson, Frances E. Clayton, Th. Dobzhansky, Joseph Grossfield, D. Elmo Hardy, William B. Heed, Michael P. Kambysellis, Carmen G. Kanapi, Henry W. Kircher, Puliyampetta S. Nair, Toyohi Okada, Forbes W. Robertson, Herman T. Spieth, Harrison D. Stalker, Wilson S. Stone, Haruo Takada, Lynn H. Throckmorton, Marshall R. Wheeler; Graduate Students and Assistants—Eunice Au, Barry Brennan, John Ellison, Alexander S. Farm, Jr., Jack K. Fujii, Henry Davis Gaines, H. A. Navvab Gojrati, G. Gerstenberg, K. C. Goodnight, Duane Gubler, Romany Huck, Peter In, Kenneth Y. Kaneshiro, Noreen Naughton, John A. Niederkorn, Jr., Geraldine Oda, Rogene Radner, Rollin C. Richmond, E. Susan Rockwood, Karl G. Rosenstein, Joyce E. Sato, Leonard Soon, Gerald H. Takei, Joaquin A. Tenorio, Joanne Tenorio, Elizabeth Thomas, Rosemany Wong, Hei Yang; Undergraduate Students—Susan R. Aihara, Judith K. Asato, Wayne Batungbacal, Rodney E. J. Chang, Michael Conant, Jean M. Coughlin, Anne Cowan, David Cox, Linda Erickson, Vaughn T. Gammel, Jacqueline Hall, Howard Hamada, Linda Hiranaga, Mahealani (Joy) Huber, Geraldine K. Irinaka, Robert Iwamoto, Katherine Kameda, Francis Kamiya, Glenn Kawanishi, Arlene Kinoshita, Gregory Kobayashi, Audrey T. Kojima, Andrew Kuniyuki, Nathan C. S. Lee, Aileen M. Matsuyama, Eizi Momman, Larry Muramatsu, Michael Muraoka, Janice Nakama, George Nakamura, Gilbert Naong, Ronald Nitta, Stephen Ochikubo, Thomas A. Ohta, Sharon Oshiro, Gary Ota, Sharon Pyun, Travis Richardson, Shirley Sarae, Sharon Shiinoki, Michael Shook, Lynne C. Stelzer, Vernon Tam, Rae Tanabe, Muriel Toma, Fumiko Yamasato, Marvin Yoshinaga, Doreen E. Yoshizumi, Calvin Yokote, Patricia Zane. Other persons who helped were Meredith S. Carson, June Grossfield, and Marion L. Stalker. Special thanks are due to John P. Murphy and Kathleen M. Resch who have served both as laboratory administrators and active investigators, and to Mmes. Ora Barber, Anne Bruch, and Glen Seaholm whose tact, understanding, and attention to administrative details facilitated the efforts of all the other personnel involved in the Project.

References

CARSON, H. L. 1959. Genetic conditions which promote or retard the formation of species. Sympos. Quant. Biol., 24:87-105.

——— 1966. Chromosomal Races of *Drosophila crucigera* from the Islands of Oahu and Kauai, State of Hawaii. Univ. Texas Publ., 6615:405-412.

——— 1968. Parallel inversion polymorphisms in different species of Hawaiian *Drosophila*. Proc. XII Int. Congr. Genet., 1:321.

——— 1969. Parallel polymorphisms in different species of Hawaiian *Drosophila*. Amer. Natural, 103:323-329.
——— F. E. CLAYTON, and H. D. STALKER. 1967. Karyotypic stability and speciation in Hawaiian *Drosophila*. Proc. Nat. Acad. Sci. U.S.A., 57:1280-1285.
——— and J. E. SATO. 1970. Microevolution within three species of Hawaiian *Drosophila*. Evolution, (in press).
——— and H. D. STALKER. 1968a. Polytene chromosome relationships in Hawaiian species of *Drosophila*. I. The *D. grimshawi* subgroup. Univ. Texas Publ., 6818: 335-354.
——— and H. D. STALKER. 1968b. Polytene chromosome relationships in Hawaiian species of *Drosophila*. II. The *D. planitibia* subgroup. Univ. Texas Publ., 6818: 355-365.
——— and H. D. STALKER. 1968c. Polytene chromosome relationships in Hawaiian species of *Drosophila* III. The *D. adiastola* and *D. punalua* subgroups. Univ. Texas Publ., 6818:367-380.
——— and H. D. STALKER. 1969. Polytene chromosome relationships in Hawaiian species of *Drosophila*. IV. The *D. primaeva* subgroup. Univ. Texas Publ. (in press).
CLAYTON, F. E. 1960. Determination of *Drosophila* karyotypes from adult males. Evolution, 14:134-135.
——— 1966. Preliminary report on the karotypes of Hawaiian Drosophilidae. Univ. Texas Publ., 6615:397-404.
——— 1968. Metaphase configurations in species of the Hawaiian Drosophilidae. Univ. Texas Publ., 6818:263-278.
——— 1969. Variations in metaphase chromosomes of Hawaiian Drosophilidae. Univ. Texas Publ. (in press).
FOSBERG, F. T. 1961. Guide to excursion III, Tenth Pacific Science Congress. Honolulu, University of Hawaii. 207 pp.
GRIMSHAW, P. H. 1901. Fauna Hawaiiensis, 3(1):51-73.
——— 1902. Fauna Hawaiiensis, 3(2):86.
GROSSFIELD, J. 1968. Visual stimuli in the biology of the Hawaiian *Drosophila*. Univ. Texas Publ., 6818:301-317.
HACKMAN, W. 1955. On the genera *Scaptomyza* Hardy and *Parascaptomyza* Duda (Dipt. Drosophilidae). Notulae Entom., 35:74-91.
——— 1959. On the Genus *Scaptomyza* Hardy (Dipt. Drosophilidae) with Descriptions of new species from various parts of the world. Acta Fennica, 97:3-73.
——— 1962. On Hawaiian *Scaptomyza* species (*Dipt. Drosophilidae*). Notulae Entom., 42:33-42.
HARDY, D. E. 1965. Insects of Hawaii, Vol. 12. Diptera: Cyclorrhapha II, Series Schizophora, Section Acalypterae I. Family Drosophilidae. Honolulu, University of Hawaii Press. 814 pp.
——— 1966. Descriptions and notes on Hawaiian Drosophilidae (Diptera). Univ. Texas Publ., 6615:195-244.
——— 1969. Notes on Hawaiian "idiomyia" (*Drosophila*). Univ. Texas Publ. (in press).
——— and K. Y. KANESHIRO. 1968. New picture-winged *Drosophila* from Hawaii. Univ. Texas Publ., 6818:171-262.
——— and K. Y. KANESHIRO. 1969. Descriptions of New Hawaiian *Drosophila*. Univ. Texas Publ. (in press).
HEED, WILLIAM B. 1968. Ecology of the Hawaiian Drosophilidae. Univ. Texas Publ., 6818:387-419.
HILLEBRAND, W. 1888. Flora of the Hawaiian Islands. New York, B. Westerman. 673 pp.
HOSAKA, E. Y. 1937. Ecological and floristic studies in Kipapa Gulch, Oahu. Occas. Papers Bishop Mus., 13:175-232.
KAMBYSELLIS, M. P. 1968a. Comparative studies of oogenesis and egg morphology among species of the genus *Drosophila*. Univ. Texas Publ., 6818:71-92.

―――― 1968b. Studies on interspecific ovarian transplantations among species of the Genus *Drosophila*. Univ. Texas Publ., 6818:93-134.
―――― 1969. Compatibility in insect tissue transplantations. I. Ovarian transplantation between *Drosophila* species endemic to Hawaii. (in ms.).
―――― and W. B. HEED. 1969. Studies of oögenesis in natural populations of Drosophilidae. I. Correlation of ovarian development and ecological needs of the Hawaiian species. (in ms.).
KANEKO, A. and H. TAKADA. 1966. *Drosophila* survey of Hokkaido XXI. Description of a new species *Drosophila neokadai* sp. nov. (Diptera, Drosophilidae). Ann. Zool., 39:55-59.
―――― T. TOKUMITSU, and H. TAKADA. 1964. *Drosophila* survey of Hokkaido XX. Description of a new species, *Drosophila pseudosordidula*. J. Fac. Sci. Hokkaido Univ., 15:374-394.
KANESHIRO, K. Y. 1968. A study of the relationships of Hawaiian *Drosophila* species based on external male genitalia. Honolulu, University of Hawaii, Master's Thesis.
KIRCHER, H. W. 1969. *Engyoscaptomyza*, a new subgenus of *Drosophila* endemic to Hawaii. Univ. Texas Publ. (in press).
――――, K. G. GOODNIGHT, and R. W. JENSEN. 1968. A medium for *Drosophila* that are difficult to rear in the laboratory. Drosophila Inform. Serv., 43:191.
―――― 1969. Sterols in the leaves of the *Cheirodendron gaudichaudii* tree and their relationship to Hawaiian *Drosophila* ecology (in ms.).
KRAJINA, V. J. 1963. Biogeoclimatic zones on the Hawaiian Islands. Newsletter Hawaiian Bot. Soc., 2:93-98.
OKADA, T. 1953. Comparative morphology of the Drosophilid flies. III. The "Phallosomal Index" and its relation with systematics. Zool. Mag., 62:278-283.
―――― 1954. Comparative morphology of the Drosophilid flies. I. Phallic organs of the *melanogaster* species group. Kontyû, 22:36-46.
―――― 1955. Comparative morphology of the Drosophilid flies. II. Phallic organs of the subgenus *Drosophila*. Kontyû, 23:97-104.
―――― 1956. Systematic study of Drosophilidae and allied families of Japan. Tokyo, Gihodo 183 pp.
PATTERSON, J. T., and W. S. STONE. 1952. Evolution in the genus *Drosophila*. New York, The Macmillan Company. 610 pp.
PERKINS, R. C. L. 1910. Fauna Hawaiiensis, 2 (6) Suppl. to Diptera, pp. 697-700.
―――― 1913. Fauna Hawaiiensis, Intro. 1 (6): CLXXX-CLXXXIX.
RIPPERTON, J. C., and E. Y. HOSAKA. 1942. Vegetation zones of Hawaii. Hawaii Agric. Exp. Sta. Bull., 89:1-60.
ROBERTSON, F. W., M. SHOOK, G. TAKEI, and H. GAINES. 1968. Observations on the biology and nutrition of *Drosophila disticha* Hardy, an indigenous Hawaiian species. Univ. Texas Publ., 6818:279-299.
ROBYNS, W., and S. H. LAMB. 1939. Preliminary ecological survey of the Island of Hawaii. Bull Jard. Bot. Bruxelles, 9:241-293.
ROCK, J. F. 1913. The indigenous trees of the Hawaiian Islands. Honolulu, 518 pp.
SANG, J. H. 1956. The quantitative nutritional requirements of *Drosophila melanogaster*. J. Exp. Biol., 33:45-72.
SELLING, O. 1948. Studies on Hawaiian pollen statistics. Part III. On the late Quaternary history of the Hawaiian vegetation. Bishop Mus. Spec. Publ., 39:1-154.
SIBLEY, C. G. 1957. The evolutionary and taxonomic significance of sexual dimorphism and hybridization in birds. Condor, 59:166-191.
SPIETH, H. T. 1952. Mating behavior within the genus *Drosophila* (Diptera). Bull. Amer. Mus. Natural. Hist., 99:399-474.
―――― 1966a. A method for transporting adult *Drosophila*. Drosophila Inform. Serv., 44:196-197.
―――― 1966b. Courtship behavior of Hawaiian Drosophilidae. Univ. Texas Publ., 6615:245-313.
―――― 1966c. Hawaiian honeycreeper, *Vestaria coccinea* (Forster), feeding on lobeliad flowers, *Clermontia arborescens* (Mann) Hillebrand. Amer. Natural., 100:470-473.

———— 1968a. Evolutionary implications of sexual behavior in *Drosophila*. Evolutionary Biology, New York, Appleton-Century-Crofts. pp. 157-193.

———— 1968b. Evolutionary implications of the mating behavior of the species of *Antopocerus* (Drosophilidae) in Hawaii. Univ. Texas Publ., 6818:319-333.

STALKER, H. D. 1968. The phylogenetic relationships of *Drosophila* species groups as determined by the analysis of photographic chromosome maps. Proc. XII Int. Congr. Genet., 1:194.

STEARNS, H. T. 1966. Geology of the State of Hawaii. Palo Alto, Calif., Pacific House, Inc., Publishers, 266 pp.

SWEZEY, O. H. 1954. Forest entomology in Hawaii. Bishop Mus. Spec. Publ., 44: 1-266.

TAKADA, H. 1959. *Drosophila* survey of Hokkaido IX. On *Drosophila okadai* sp. nov. with supplementary notes on the female of *Scaptomyza polygonia* Okada. Ann. Zool. Jap., 32:152-155.

———— 1966. Male genitalia of some Hawaiian Drosophilidae. Univ. Texas Publ., 6615:315-333.

TALIAFERRO, W. J. 1959. Rainfall of the Hawaiian Islands. Honolulu, Hawaii Water Authority. 394 pp.

THROCKMORTON, L. H. 1966. The relationships of the endemic Hawaiian Drosophilidae. Univ. Texas Publ., 6615:335-396.

WARD, B. L., W. B. HEED, and J. S. RUSSELL. 1968. Salivary gland chromosome analyses of *Drosophila pachea* and related species. Genetics, 60:235.

WARNER, R. E. 1967. [Unpublished observations, cited by permission, from the Scientific Report of the Kipahulu Valley Expedition, 2 August-31 August 1967.]

WASSERMAN, M. 1963. Cytology and phylogeny of *Drosophila*. Amer. Natural., 97:333-352.

WHEELER, M. R. 1952. A key to the genera of Drosophilidae of the Pacific Islands (Diptera). Proc. Hawaiian Entom. Soc., 14:421-423.

———— and F. E. CLAYTON. 1965. A new *Drosophila* culture technique. Drosophila Inform. Serv., 40:98.

YANG, H., and M. R. WHEELER. 1969. Studies on interspecific hybridization within the picture-winged group of endemic Hawaiian *Drosophila*. Univ. Texas Publ. (in press).

ZIMMERMAN, E. C. 1948. Insects of Hawaii, Vol. 1, Introduction. Honolulu, University of Hawaii Press. 206 pp.

———— 1958. Three hundred species of *Drosophila* in Hawaii? A challenge to geneticists and evolutionists. Evolution, 12:557-558.

16

Human Genetic Adaptation

C. C. LI

*Graduate School of Public Health, University of Pittsburgh,
Pittsburgh, Pennsylvania*

Introduction	545
Heritability	546
Estimation of Heritability	550
Random Mating	552
The Markov Property	557
Shall We Count the Living or the Dead?	560
Simplest Selection Model	561
Correlated Responses	564
Genetic Improvement of Mankind	567
Genetic Deterioration	568
Decline of Intelligence	570
Control of Human Evolution: Reproductive Specialization	574
Genotype-Environment Interaction	575
Summary and Conclusions	576
References	577

Introduction

Although we use the same word "adaptation" for both physiological readjustment and genetic change, a sharp distinction must be made between the two processes, especially in view of the fact that they both involve reactions to environmental agents. A physiological adaptation refers to the regulatory mechanisms of an individual who, as a rule, does not undergo an immediate and corresponding change in his genetic makeup. A true genetic change, although it must originate in certain individuals, is more appropriately applied to future generations rather than to immediate individual responses to current environmental conditions. A physiological adaptation is temporary; a genetic change is more lasting. They are two different kinds (or levels) of reactions to environmental changes. This paper discusses exclusively the nature of genetic change, with special reference to human populations.

Genetic changes may be affected by various means: gene mutations, migrations among population groups, selection through differential reproduction of various genotypes, to mention a few. If the reproducing population is of limited size, there will be an additional factor—random sampling of parents—which may cause considerable change in the genetic makeup of the subsequent generations. Gene mutations occur "spontaneously" and "at random" and hence have usually no immediate significance in adaptation to prevailing environment. Methods of inducing specific and directional mutations are yet to be found. Therefore, a large population (such as a human population) usually has a great variety of apparently useless or even harmful mutant genes of low frequency in storage. This phenomenon must be accepted as part of Nature, a consequence of the almost (but not quite) perfect duplication of the long and complicated DNA molecule during meiosis. Perfection means a permanent standstill; an occasional imperfection opens up the great possibility of genetic diversity. Mutations provide the ultimate source of new genes in a population but could not by themselves explain adaptation to environment.

Migrations among various population groups, like mutations, are a constant source of genetic change for a particular population or a force to maintain genetic equilibrium. Very few populations are completely isolated. However, migrations merely affect a new shuffling or intermixture of existing genes in the various groups from different local environments and with different genetic compositions. It is then clear that the crux of the problem of genetic change and adaptation is selection through differential reproduction of different genotypes under given environmental conditions. It is selection which molds the genetic makeup of a population. Selection is the tool for genetic adaptation to environment. The following discussions, in essence, deal with genetic selection and its effectiveness and possible consequences.

Heritability

In order that selection be effective in causing a genetic change in future generations, the characteristics or traits being selected for or against must have a genetic basis. If the traits being selected are not related to certain genotypes, the selection will have no genetic effect on future generations. Thus, the effectiveness of selection depends upon the degree to which the trait is associated with genotypes, technically known as the "heritability" of the trait under selection. An analytical definition will be developed

TABLE I

Identical Distribution of Family Size y = Number of Children per Family for All Three Genotypes. In Such a Case, the Trait y Has no Heritability

Genotype	Number of children per family (y)									Total number of families
	0	1	2	3	4	5	6	7	8+	
AA	7	10	9	7	5	4	3	2	3	50
Aa	14	20	18	14	10	8	6	4	6	100
aa	7	10	9	7	5	4	3	2	3	50

later. It is somewhat alarming to find that many discussions on genetic selection in man have either ignored the problem of heritability or have assumed complete heritability. The fact is that the heritability of most common traits and diseases in man has not been seriously studied and remains unknown. This one single area of our ignorance has undoubtedly diminished the validity and importance of many current essays on human selection, including the present one, of course. In fact, one of the primary purposes of this communication is to point out our great ignorance and to plead for more basic research and less philosophical predictions on the genetic future of man.

In order to understand the positive meaning of heritability, it will be helpful to examine a hypothetical example in which there is no heritability. The number of children (y) varies from family to family and is obviously an important biological characteristic that determines the next generation. However, suppose that the number of children per family is distributed (roughly as a negative binomial) with the same mean (3) and variance (7.20) for all the three genotypes of mother with respect to one locus as indicated in Table I. From the table, it is clear that selection for either large families or small families will be ineffective, for large and small families consist of genotypes in the same proportion ($1AA:2Aa:1aa$). In such a case we say that the trait (y) has no heritability, despite the apparently very large variation in the number of offspring produced by various families. This example brings out the sharp difference between the "variability" and the "heritability" of a trait. There is no necessary relationship between these two phenomena.

To measure the variability of a trait is comparatively simple and is usually the first thing done by statisticians. A convenient measure is known as the standard deviation. However, in order to assess the degree of variability independent of physical units of measurement, statisticians often employ the ratio of the standard deviation to the mean value of the trait, which is known as the "coefficient of variability" (C.V.). The standard deviation of body weight of elephants is of course much greater than that of mice in absolute units, but not necessarily so relative to their mean body weight. Some geneticists borrow this idea and apply the method to the number of offspring (y) discussed above. Thus,

$$\text{C.V.} = \frac{\sigma_y}{\bar{y}} \qquad (\text{C.V.})^2 = \frac{\sigma_y^2}{\bar{y}^2}$$

the latter being called the "index of total selection" (Crow, 1958), setting an upper limit to the rate of change by selection. Without any knowledge whatsoever about the heritability of trait y in man, the writer fails to see any connection between genetic selection and the value of (C.V.)^2 (or any other function of y). As it stands, it is merely a measurement of the relative variability of the number of offspring per family.

Now we shall show how heritability is defined, and, later, how it may be estimated from appropriate data. We shall continue to use y = number of children in a family as an example. For each given genotype there is a

TABLE II

Mean Number (y) of Offspring for Each Genotype in a Random Mating Population and Its Heritability, Assuming the Environmental Variance to Be $\sigma_E^2 = 7.20$ in Both Examples

Genotype	Frequency f	Mean no. children y	Example (1) f	y	Example (2) f	y
AA	p^2	y_2	.25	3.00	.25	2.00
Aa	$2pq$	y_1	.50	2.50	.50	3.00
aa	q^2	y_0	.25	2.00	.25	2.00
Total	1		1.00		1.00	
mean		$\bar{y}$		$\bar{y} = 2.50$		$\bar{y} = 2.50$
genotype var.	$\sigma_G^2 = \Sigma fy^2 - \bar{y}^2$			$\sigma_G^2 = 0.125$		$\sigma_G^2 = 0.250$
L-component	$\sigma_L^2 = 2pq\beta^2$			$\sigma_L^2 = 0.125$		$\sigma_L^2 = 0$
heritability	$h^2 = \sigma_L^2/(\sigma_G^2 + \sigma_E^2)$			$.125/7.325 = .017$		$h^2 = 0$

probability distribution of y; its variance is nongenetical and be denoted by σ_E^2. Suppose that the form of distribution is the same for all genotypes but the mean value of y differs from genotype to genotype; this variance is genetical and be denoted by σ_G^2. In the setup of Table II, random mating is assumed, but its full implications in a human population will not be discussed until a later section. Let p be the frequency of gene A and $q = 1-p$ be that of gene a. The symbols y_2, y_1, y_0 denote the *mean* number of children produced by the three genotypes of mother (average over father's genotype). The number of children for each genotype is distributed somewhat like that indicated in Table I and its variance is assumed to be $\sigma_E^2 = 7.20$ for all three genotypes. All nongenetical influences are defined to be environmental.

With these preliminary explanations in mind, we may now proceed to make certain calculations that would lead to the definition of heritability. The mean of the three genotypes is

$$\bar{y} = p^2 y_2 + 2pq\ y_1 + q^2 y_0$$

and the variance of the genotype means,

$$\sigma_G^2 = p^2 y_2^2 + 2pq\ y_1^2 + q^2 y_0^2 - \bar{y}^2$$

is the genotype variance. The calculations up to this point are all too familiar. Two numerical examples are given in Table II for easy verification, taking $p = p = 1/2$. The subsequent steps are genetical and we shall merely give the procedure without proof (but see Li, 1961b). The linear regression coefficient of the genotype value (y_2, y_1, y_0) on the genotypes (AA, Aa, aa regarded as 2, 1, 0) is

$$\beta = p(y_2 - y_1) + q(y_1 - y_0)$$

In example (1), $\beta = 0.5(0.5) + 0.5(0.5) = 0.50$, and in example (2), $\beta = 0.5(-1) + 0.5(1) = 0$. Once the linear regression β is found, the linear component (L-component) of the genotype variance σ_G^2 is easily calculated by the formula $\sigma_L^2 = 2pq\beta^2$. This value is always smaller than the genotype variance except when the genotype means themselves are linear (such as 3.0, 2.5, 2.0), in which case the linear component is the same as the genotype variance, as shown in example (1), Table II. Then the heritability (in the broad sense) of the trait y is defined as the ratio of the linear component to the total variance of the variable, viz.,

$$h^2 = \frac{\sigma_L^2}{\sigma_G^2 + \sigma_E^2}$$

In example (1), we see that although the three genotypes do differ appreciably in their reproductive performance, the heritability is only 1.7 percent on account of the comparatively large nongenetical variance. (The conception rate in cattle has a heritability of about 1 percent.) The heritability in example (2) is 0, despite the genotype differences. This means that selection for either large families or small families is totally ineffective; and the genetic composition of the offspring population will remain the same as the parental. This may sound unreasonable to nongeneticists but it may easily be seen by shifting the distribution in Table I. Suppose that the heterozygotes produce on the average one offspring more than the others. Then the distribution for the heterozygotes should be shifted to the right by one step. Now, if we select for families of size 6, the selected group will consist of, not $3\ AA + 6\ Aa + 3\ aa$, but of $3\ AA + 8\ Aa + 3\ aa$. On random mating, however, this group will produce the offspring population $1\ AA : 2\ Aa : 1\ aa$ again, as if no selection had occurred. This example emphasizes the fact (entirely too familiar to breeders but not so familiar to other scientists) that *genotypic difference does not always mean the existence of heritability.* (See Falconer, 1960, for further discussion on heritability.)

One may object to the assumption of the large nongenetical variance ($\sigma_E^2 = 7.20$) in the examples above. This number is not entirely arbitrary. The reader may consult the United States census data on the distribution of family size and see that the variance is of the same order of magnitude as that examplified here (United States Census, 1950 and 1960).

It is important to realize that the heritability is not an intrinsic and fixed property of the trait under consideration; it varies with environmental conditions as well as the gene frequencies involved in determining the trait. Hence, it varies from population to population and from time to time for the same trait. However, three categories of traits have been found by breeders to have usually low heritability, viz.: traits that are easily influenced by nongenetical factors, traits that are directly related with reproduction, and traits that have long been regulated by selection. In view of these principles, it is not surprising that the number of offspring per family should have low heritability. With the advent of the many artificial means of birth control as well as birth induction, its heritability may be expected to go down even further, eventually reaching a point that the number of children would have no more heritability than the household furniture of the family.

Estimation of Heritability

In practice the genotypes are generally indistinguishable and unknown, and the heritability as defined above cannot be directly observed. We must resort to indirect means of estimation. Breeders, enjoying the freedom of having controlled mating among the experimental plants and animals, have devised a number of methods by which the heritability of various traits may be estimated. Furthermore, the breeders can actually exercise artificial selection of specified intensity to measure the response to selection over a number of generations. Human geneticists, of course, can do none of these things. One of the few things the human geneticist can do is study the correlations (or regressions) among the family members. Theoretically, the heritability can be estimated from the parent-child, full-sib, and half-sib correlations. Genetic information on other relatives decreases rapidly with remoteness. Of these, the full-sib correlation is the easiest to obtain in a human population but is also the least reliable one in estimating heritability. The full-sib correlation tends to overestimate heritability for two reasons: one is that it contains, in addition to the linear component, a "dominance component" of the genotype variance, and the other is that it contains a large nongenetical component due to the similarity of their environments. The correlation between half-sibs, raised in different families, may provide a better estimate of heritability but they are comparatively rare and difficult to identify and study in a human population. So, it seems that our best approach is to study the correlation between parent and child (or the linear regression of child on parent). The reason that the parent-child correlation provides an estimate of heritability is that it is directly proportional to the linear component of the genotype variance. We state without proof:

$$r \text{ (parent-child)} = \tfrac{1}{2} h^2,$$

so that the study of the parent-child correlation is almost a direct study of heritability within limits of sampling errors (see Li, 1961b). If information on both parents is known, we may take the average (or the sum) of the two parents, known as the mid-parent, and find the regression of child on the mid-parent. The regression coefficient thus found is the heritability. Then, it is seen that the heritability of human traits can be studied after all. The painful fact is that very few serious studies of this kind have been made.

The study of parent-offspring correlation, though theoretically simple, is not without pitfalls, especially in a human population. The difficulties in obtaining a random sample of parent-child pairs cannot be overemphasized. Those who intend to make such a study are strongly advised to consult a statistician with experience in sampling human populations. Most of the biases that occurred in sampling would tend to overestimate the correlation.

In addition to the technical problem of sampling, the study of parent-offspring correlation is plagued with other biologically complicating factors. Probably the most obvious one is the influence of the home environment.

In studying plants and small animals, the geneticists assume that the environment of parents and offspring are "uncorrelated" and does not affect the correlation coefficient. In a human family, the assumption of uncorrelated environments is probably an oversimplification. If the home environment increases the correlation, it would lead to an overestimate of heritability. Another complicating factor is assortative mating in human marriages. That is, like tends to marry like. This practice would lead to a higher correlation between parents and offspring than random mating. Still another factor is a certain amount of inbreeding (marriages between relatives), which also tends to increase the parent-offspring correlation. Some biases are difficult to overcome, while others may be readily remedied. For instance, the bias due to inbreeding may be eliminated simply by excluding the inbreeding families from calculation. On the other hand, the effect of assortative mating may be quite difficult to assess and to remedy. As to the influence of home environment and possible maternal effect, it is believed that father-child correlation should be more reliable than the mother-child correlation for the purpose of estimating heritability.

Now if we wish to study the heritability of $y=$ number of children in a family, we probably have to study the correlation between mother's y and daughter's y. For this particular trait, however, there is an additional complication. A childless mother can not have a daughter, still less grandchildren, so that a whole class of mother-daughter combinations will be missing from the correlation table. If mother and daughter are uncorrelated with respect to trait y, the omission has no effect. If, on the other hand, mother and daughter are correlated, the omission of an extreme class of combinations will impede our effort in estimating the heritability of y.

The whole purpose of discussing the problem of heritability is to remind us of our vast ignorance in this area. Without knowing the heritability, we have no way to predict the result of selection. This should cast some doubt on the validity of predictions as to the genetic future of man, or of controlling the course of human evolution.

Random Mating

Random mating is not an absolute term; it is relative to the criterion under consideration. We have mentioned assortative mating in the foregoing section. For instance, if we study the height or education level of husband and wife, we would find the correlation appreciable—maybe of the order of 0.15 and 0.30. If the *MN* blood types of the same husband-wife pairs were also to be studied, no association would be found. Thus, the mating is at random with respect to the *MN* traits but assortative with respect to height or education. It is seen immediately that, when taken as a whole, the mating system in a human population is a highly complicated phenomenon. The following discussions apply only to those genetic factors with respect to which the mating is essentially at random.

The implications of random mating (with respect to genetic factors) in a human society are truly profound. Only a few of them that can be readily appreciated will be discussed here. The mates in a random mating society are "strangers" (genetically speaking); they are unrelated by "blood." The genes of Mr. Roth are not correlated with those of Mrs. Roth. Let us consider a pair of genes at a certain locus, and tag those of Mr. Roth as (1, 2) and those of Mrs. Roth (formerly Miss Abt) as (3, 4). These numbered tags will help us to trace the whereabouts of these genes in subsequent generations. One of the many possible outcomes of continued random mating is illustrated in Figure 1, where only the male descendants of Mr. Roth are shown. Since half of the son's genes are derived from his father and half from mother, his genes have been designated by (1, 3), which represents one of the four possibilities. This son (Roth I) marries an unrelated woman, formerly Miss Bey, whose genes are labeled as (5, 6), uncorrelated with the genes 1, 2, 3, 4. Their son Roth II (3, 6) marries

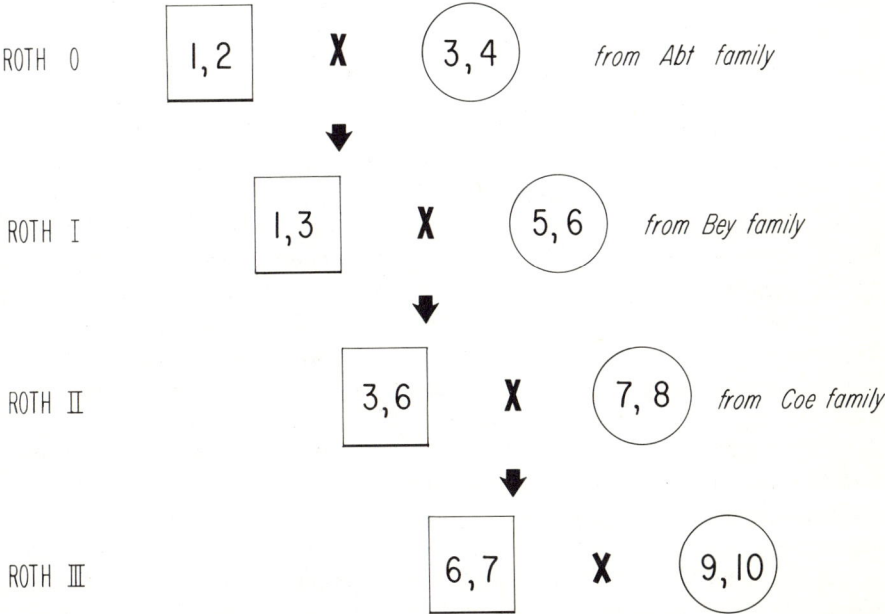

Fig. 1. Gene shuffling among families in a random mating population. Square indicates male; circle female. This particular diagram shows that grandfather Roth 0 and grandson Roth II have no genes in common, while great-grandson Roth III has no genes in common with either great-grandparent.

the former Miss Coe, whose random genes are labeled as (7, 8). The last Mr. Roth (III) shown in Figure 1 has the genes (6, 7). It is seen from the diagram that Roth III (6, 7), although bearing the Roth family name and only three generations away from the original Roth under consideration, has inherited none of the original four genes (1, 2, 3, 4) of the family. As a matter of fact, half of his genes are derived from the Bey family and half from the Coe family, both unrelated with Roth. Furthermore, if the original Roth is a highly successful man in some sense despite the interference from the Abt family, and we wish to consider his genes (1, 2) only, then we see that his grandson, Roth II (3, 6), already has no genes in common with him. This could happen in merely two generations. What has been said about the descendants of Roth equally apply to the ancestors of Roth. From the genetic viewpoint, one sees how utterly meaningless a family name is in a random-mating population.

If many pairs of genes are considered, then Roth 0 and Roth I have one half of their genes in common; Roth 0 and Roth II have one quarter of their genes in common; Roth 0 and Roth III have one eighth of their genes in common, and so on.

In the foregoing it should be carefully noted that I said "genetically meaningless," not "socially valueless." Quite on the contrary, we recognize the great social and economic value of a name under the present socioeconomic system. A large portion of human energy is directed toward gaining an economically profitable name. Those who are engaged in such an endeavor should perhaps be reminded that the "wealth may be still there, but the genes are gone." Any further pursuit of the subject would be beyond the scope of this chapter. Suffice it to say that the usual notion of a good family "breed" has no genetic basis beyond a few generations. If there is such a thing as family "tradition," it is due to strong artificial and environmental factors, including early training, forced learning, and continued indoctrination. This analysis has also weakened the belief that the human race may be immediately improved by allowing only the "good" families to reproduce. This leads us to a closer examination of the genetic relationship between two successive generations.

"Like begets like." Sayings of this type are applicable to separate species and certainly not to individuals within a population, on account of the segregation of genes in meiosis. In a random mating population, the relationship among the various genotypes between two successive generations may be easily found for one pair of genes. To illustrate, we may use the *MN* blood system as an example without going through the mathematical derivations. In Figure 2, it has been assumed that the frequencies of the genes *M* and *N* are each equal to ½ for easy calculation. In many human populations, they do not deviate from ½ too much. Then the parental generation consists of ¼ *MM* + ½ *MN* + ¼ *NN* individuals, and so does

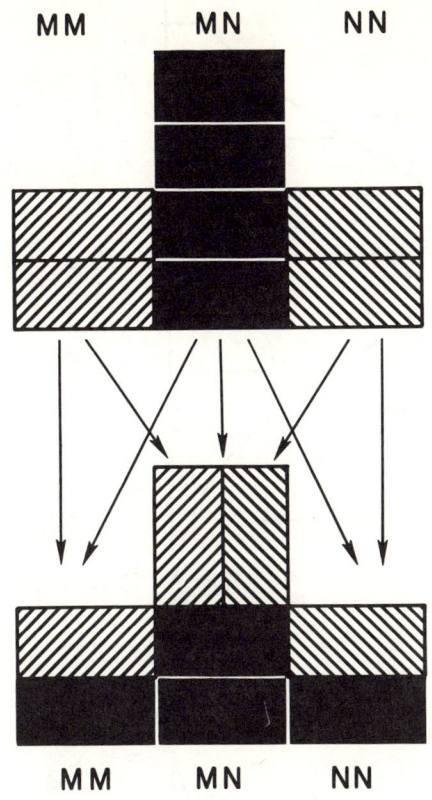

Fig. 2. Parent-offspring relationship for one pair of genes in a random-mating population. The MN parent produces all three types of offspring. Conversely, the MN offspring is derived from all three types of parents.

the offspring generation, assuming a stationary population. The important feature of the relationship is that the ½ *MN* individuals in the offspring generation is *not* the descent of the ½ *MN* individuals in the parent generation. They are derived from various sources. Similarly, the ¼ *MM* offspring is *not* the direct descendant of the ¼ *MM* parents but also derived from various sources.

For the sake of simplicity in presentation, Figure 2 shows only the relationship of one parent (say, mother, with father averaged out, or vice versa) with child. The diagram may be viewed in two different ways. From parent to child, we see that the *MN* parents contribute to all three blood types in the offspring generation. Conversely, we may also say that the *MN* children are derived from all three types of parents. The different patterns of shading in the diagram help the reader to trace the quantitative relationship of the contributions from one generation to the next. This diagram is the simplest of its kind and it gives us some inkling as to the intricacy of the genetic relationship between successive generations.

Extending from a trait controlled by one pair of genes, we may venture

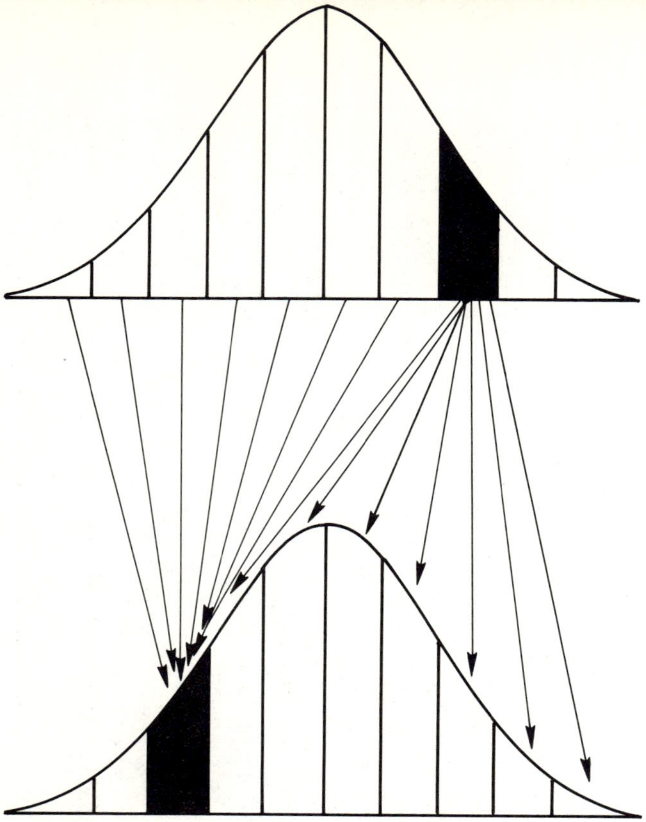

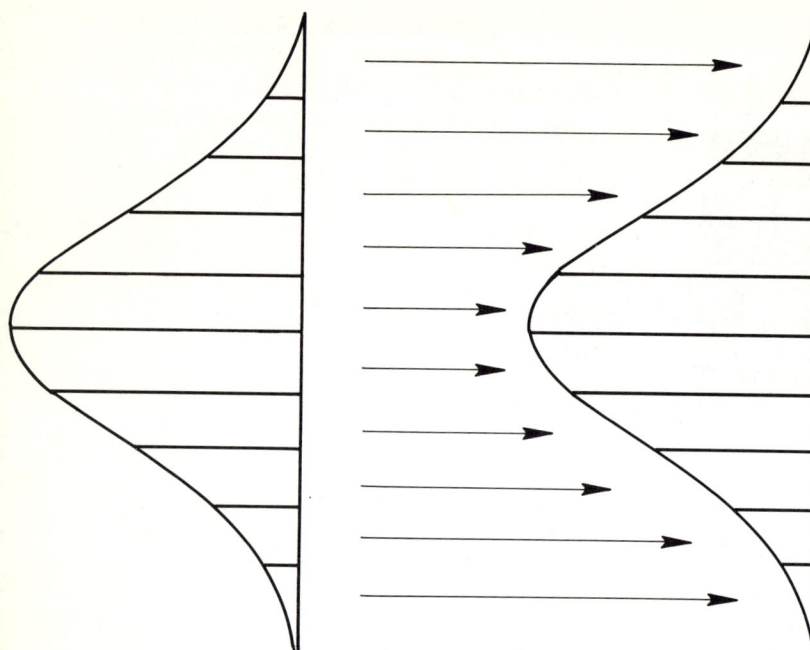

Fig. 3. *Left:* Parent-offspring relationship for many pairs of genes in a random mating population. Any given class of parents (shaded) produces many different classes of offspring. Conversely, any given class of offspring (shaded) is derived from many different classes of parents.

Right: Wrong concept of the parent-offspring relationship in a bisexual random-mating population. The one-to-one class relationship is applicable only to pure lines or asexually reproducing populations.

to visualize the situation for a trait partially controlled by many pairs of genes, such as intelligence, compounded by environmental influences. An attempt to diagram the situation is shown in the left portion of Figure 3, however inadequate it may be. The diagram shows a hypothetical distribution of the trait values. Actually, the distribution need not be symmetrical or normal; the shape depends on the type of gene effects and gene frequencies as well as environmental influences. The diagram is merely intended to show that any given class of parents will contribute to a wide range of classes of offspring. Conversely, any given class of offspring is derived from many different classes of parents. Such is the intricate network of genetic relationship between two successive generations. This, however, is not to say that selection will be ineffective with respect to traits controlled by many genes, but it does say that it will be far less effective than the case depicted in the right portion of Figure 3, where the classes of parents and offspring have a one-to-one correspondence. The latter situation is true only for "pure lines" and asexual reproduction; it is not applicable to bisexual random-mating populations.

The Markov Property

As long as we are on the subject of parent-offspring relationship, we may as well discuss briefly a very basic principle in the field of genetics. For two pairs of genes (A, a and B, b) there are nine different genotypes and 81 different types of mating if we distinguish the sex. Now let us consider the genotype $AaBb$ in the offspring generation. Where do they come from? Who are their parents? Table III shows that no less than 49 of the 81 possible types of families are capable of producing such an offspring (with different probabilities). The table also shows that the offspring genotype $AaBb$ is derived from all genotypes of the parental generation. The new point we wish to emphasize is that the offspring genotype $AaBb$ is exactly the same genotype no matter whether it is produced by the family,

(I) $AABb \times AaBB$ or   (II) $aaBb \times Aabb$,

or by any other of the possible families. The offspring $AaBb$ from family (I) and the offspring $AaBb$ from family (II) will have the same genetic potentialities and will have the same performance (phenotype) *under the same environment,* within the limits of accidental effects. In practice, the home environments of these two $AaBb$'s will be different, including different brothers and sisters. Nevertheless, their genetic makeup and genetic properties are exactly the same.

To clarify the concept, we shall restate the principle in a more general

TABLE III

Parental Genotype Combinations That Are Capable (indicated by x) of Producing an AaBb Offspring

Parental genotypes	AA BB	AA Bb	AA bb	Aa BB	Aa Bb	Aa bb	aa BB	aa Bb	aa bb
AABB	0	0	0	0	x	x	0	x	x
AABb	0	0	0	x	x	x	x	x	x
AAbb	0	0	0	x	x	0	x	x	0
AaBB	0	x	x	0	x	x	0	x	x
AaBb	x	x	x	x	x	x	x	x	x
Aabb	x	x	0	x	x	0	x	x	0
aaBB	0	x	x	0	x	x	0	0	0
aaBb	x	x	x	x	x	x	0	0	0
aabb	x	x	0	x	x	0	0	0	0

language. A genotype may be thought of as a "state" or a condition of existence of an object or of an affair. There may exist a number of states. An object may change from one state to another at each unit-time interval. If the future characteristics or behaviors of an object depend only on the state in which it finds itself and not on the states from which it has been transferred, then we say that the object has the (simplest) Markov property, named after the Russian mathematician who studied the consequences of such systems (see Li and Sacks, 1954, for detailed discussion of implications for genetics). The Markov property is essentially a property that depends solely on the present state, and is independent of history. In homely words, the Markov property is: "It depends on where you are, not where you are from." The laws of heredity provide an excellent and clear-cut example of the Markov property; and this is why the mathematical method of "Markov chains" is readily applicable to certain problems in genetics. Thus, an *AaBb* individual has certain definite genetic properties no matter from what family he is derived.

The Markov property also applies to genetic populations. Two populations of identical genetic composition will have identical genetic behaviors, regardless of how these two populations are derived. Two examples in the field of population genetics should suffice to clarify the concept. I have occasionally heard the statement, allegedly supported by mathematics, that an initial population of heterozygotes, $P_0 = (0\ AA,\ 1\ Aa,\ 0\ aa)$, behaves *differently* (in some respect that does not concern us here) from an identical population derived from an initial cross AA × aa, which yields the F_1 population (0 *AA*, 1 *Aa*, 0 *aa*). This is obviously impossible; a population of heterozygotes is a population of heterozygotes, whether we call it P_0 or F_1. The properties—the future properties, not the past—of such populations should be the same, regardless of how they were obtained by the breeders.

As a second example, let us consider an equilibrium population (p^2 *AA*, $2pq$ *Aa*, q^2 *aa*). The value of q may be determined by a large number of different factors, some of which may become known upon study; but most will remain obscure to us at the present state of our knowledge. Two populations with the same value of q do not mean that the same factors are playing the same role with the same importance in the two populations. To investigate the genetic composition of a population is to study the status quo; it reveals very little of what had happened in the past. The histories of these two populations may be very different. Our point is that the two populations with the same value of q will have the same future properties with respect to q, regardless of history. Suppose that $p = 0.96$ and $q = 0.04$ in both populations; and we wish to study the effect of inbreeding by observing the children of first-cousin marriages from such populations. We will find that the frequency of homozygotes *AA* and *aa* will be each increased by $pq/16 = (0.96)(0.04)/16 = 0.0024$, and that of the heterozygotes decreased by 0.0048 as shown in the following:

Genotype	AA	Aa	aa	Total
Random mating:	.9216	.0768	.0016	1.00
Children of 1st cousins:	.9240	.0720	.0040	1.00
If completely inbred:	.9600	0	.0400	1.00

This will be so for all populations with $q = 0.04$, regardless of how and why the value $q = 0.04$ was reached historically. If the recessive genotype *aa* invariably develops a genetic disease, then all we can see from the inbreeding data is that the disease incidence among the children of first-cousin parents is $40/16 = 2.50$ times that among the random mating offspring. These inbreeding properties solely depend on the existing value of q, not its history. But cf. Crow (1958), and the claim that the inbreeding results can distinguish certain two types of equilibrium populations with different patterns of natural selection. Needless to say, this is impossible from observing inbreeding properties alone. The claim was apparently based on a series of artifacts—one built upon another.

The purpose of citing these two examples, however, is not to show the fallibility of scientists (we are all fallible) but to help crystallize the concept of Markov's property in genetics, which has profound implications in the theory of evolution. Owing to this remarkable property, each generation may be regarded as a new starting point, from which the population may evolve, regardless of its past. It implies that the existing state of a population is all that counts, with respect to its future. Evolution is a series of such step-by-step conditional developments.

It should be reiterated that the discussions above refer to biological inheritance and do not apply to what has been called "social inheritance," a set of man-made laws and conventions, which is sometimes diagonally opposite to laws of Nature. (For a detailed discussion of the interplay between the biological inheritance and the social inheritance, the reader may consult Dozhansky's *Mankind Evolving,* Yale University Press, 1962.) 1962.)

Shall We Count the Living or the Dead?

Before we discuss the specific example of natural selection in the next section, we have to clarify the definition of fitness first. In all models of selection problems, we have to introduce a parameter known as the "fitness" value or selection value for each genotype, to measure the genetic contribution of that genotype to the next generation. Hence, the fitness value of a genotype is made proportional to the number of living offspring produced by that genotype. For instance, consider the following total reproduction results of genotypes AA and Aa in a population:

Parent genotype	Number	Total number of births	Number dead	Mortality rate	Number living	Average number of living children per parent
AA	100	125	5	4%	120	1.20
Aa	200	500	100	20%	400	2.00

It is seen that the children of genotype Aa have a much higher mortality rate ($100/500 = 20$ percent) than those of AA. But this should not be taken to mean that genotype Aa would contribute less than AA to the next generation. In the numerical example above, the contrary is true. In spite of the differential mortality rates among the two groups of children, the final average number of *living* children for parents AA and Aa are 1.20 and 2.00, respectively. We say that the fitness of the genotypes AA and Aa have the relative values $12:20 = 3:5$. The genotype Aa contributes more to the next generation in spite of the high mortality rate among their children.

In the usual study of toxicity of drugs on animals, employing experimental groups of equal size (say, 30 mice in each treatment group), the question of "Shall we count the living or the dead?" does not arise, since a high mortality rate is equivalent to a low survival rate. But the size of human families is not a constant. The genetic contribution of a genotype to the offspring generation is determined by the number of living children produced by that genotype, not the number of deaths that occurred previously. Hence, in evaluating the fitness value of a genotype we always count the living children, not the dead.

The numerical example cited above is not entirely unrealistic. In fact, its general features are found to be true in almost all human populations. Families with a larger number of births accompanied by a higher mortality rate, nevertheless, contribute more living children than those having small number of births accompanied by low mortality rate. This reproductive pattern may, however, change according to cultural as well as scientific advances.

The example also points out that in studying selection values of man, the mortality data have very little direct bearing on the problem. It is the positive contribution as measured by living children which counts. This leads us to doubt the validity of the operational aspects of the theory of "genetic load" which are based primarily on mortality data. For further discussion, see Li (1963).

Simplest Selection Model

We really know very little as to how natural selection operates, especially in man. It is much easier to observe the results of selection than to study the mechanism and procedure by which the selection is achieved. There are undoubtedly many different types of selection operating in Nature; and hence many different models may be constructed to describe the situation. Most of the selection models so far studied by geneticists are limited to one pair of genes. In fact, a general solution for selection with respect to two pair of genes has not yet been obtained and is still under active investigation, despite the liberal use of high-speed computers. Then one may imagine how little we actually know about natural selection with respect to complex traits partly controlled by heredity and partly influenced by environment. For a review, see Li (1967).

In order to gain some idea about the model and effects of selection, we present the simplest case of its kind for discussion (Table IV). Let us assume that the gentoype AA and Aa are the normals with average reproductive capability. For purpose of calculation, we assign to these genotypes a "fitness" of unity. The recessive homozygote (aa) is assumed to be lethal or unable to have any offspring, and is given a "fitness" of 0. If the initial

TABLE IV

A. Complete Elimination of Recessives in Every Generation of A Random Mating Population

Genotype	Frequency at birth	Fitness value	After selection	New gene freq. after selection	Next generation after random mating
AA	p_0^2	1	p_0^2	$p_1 = \dfrac{1}{1+q_0}$	p_1^2
Aa	$2p_0 q_0$	1	$2p_0 q_0$	$q_1 = \dfrac{q_0}{1+q_0}$	$2p_1 q_1$
aa	q_0^2	0	0		q_1^2
Total	1		$p_0^2 + 2p_0 q_0$	1.00	1.00

B. Number of Generations (n) Required to Reduce an Existing Gene Frequency (q_0) to Half of Its Value ($q_n = \tfrac{1}{2} q_0$).

Initial q_0	to	Final q_n	Generations $n = \dfrac{1}{q_0}$	Initial q_0	to	Final q_n	Generations $n = \dfrac{1}{q_0}$
.5000	to	.2500	2	.0156	to	.0078	64
.2500		.1250	4	.0078		.0039	128
.1250		.0625	8	.0039		.0020	256
.0625		.0312	16	.0020		.0010	512
.0312		.0156	32	.0010		.0005	1024

population before selection is $(p_0^2, 2p_0 q_0, q_0^2)$, then the actual mating population (parents) after the operation of selection will be $(p_0^2, 2p_0 q_0, 0)$, the genotype *aa* being completely eliminated from contributing to the next generation. The frequency of the recessive gene in this selected parental group is (see Li, 1961b):

$$q_1 = \frac{p_0 q_0}{p_0^2 + 2p_0 q_0} = \frac{q_0}{1 + q_0}$$

which is the recurrence relation for the gene frequency of two successive generations. It is seen that $q_1 < q_0$; the recessive gene is decreasing in frequency due to selection against the recessives. After random mating among these parents, the offspring generation will be $(p_1^2, 2p_1 q_1, q_1^2)$ as shown in the last column of Table IV. It is the new initial population before selection, thus completing a generation-cycle. This is the usual type of model adopted by geneticists for selection in random-mating populations. This model implicitly assumes the Markov property, since the relative fitness is directly assigned to individual genotypes, regardless of the families from which they are produced.

If the same selection force continues to operate in all the succeeding generations, we will find by the same recurrence relations that the gene frequencies after 2, 3, ... generations are:

$$q_2 = \frac{q_1}{1+q_1} = \frac{q_0}{1+2q_0}, q_3 = \frac{q_2}{1+q_2} = \frac{q_0}{1+3q_0}$$

and after n generations of continued selection,

$$q_n = \frac{q_0}{1+nq_0}$$

The last expression gives the relationship among the three quantities:

initial, q_0 ... number of generations, n, ... final, q_n.

When any two of the three quantities are given, we may calculate the third. In particular, when $nq_0 = 1$, we have $q_n = \frac{1}{2}q_0$. That is to say, in order to reduce the existing gene frequency to half of its value, it is necessary to continue the selection for $n = 1/q_0$ generations. The lower part of Table IV gives some examples. Thus, to reduce q value of 0.0156 to 0.0078 requires 64 generations.

Several remarks may be made with respect to the above example. One might think that complete elimination of recessives in every generation would quickly get rid of the harmful gene from the population. This is not so, especially when the recessive gene has a low frequency. In terms of human history, the selection effect is practically nil for rare recessives. We will appreciate what 64 generations mean when we recall that from Jesus Christ to us is merely a period of 60 generations. Even if we had enforced a program of complete elimination for about 2,000 years and had succeeded in reducing $q = 0.0156$ to 0.0078, we would still have an incidence of $(0.0078)^2 = 6/100,000$ among the new births, instead of the original $(0.0156)^2 = 24/100,000$.

We should also realize that calculations of this nature are purely schematic and may bear little resemblance to reality. Environmental conditions change over time and may change drastically in a short time. The assumption that a selection pattern remains constant for hundreds of generations is unsupportable. Finally, mutation effects have been ignored in the above calculations. Suppose that the mutation rate from the normal allele to the recessive gene is $\mu = 4 \times 10^{-6}$ per generation. Under the assumption of complete elimination of the recessives in every generation, it may be shown that the equilibrium value of the gene frequency is $q = \sqrt{\mu} = 0.0020$. This means that we may never be able to get the value of q below 0.0020 no matter how long the selection continues.

The ineffectiveness of selection against rare recessives may also be seen from the heritability viewpoint. Letting $y = 1, 1, 0$, we find that the fitness variance is $\sigma_y^2 = q^2(1-q^2)$, the regression coefficient is $\beta = q$, the linear component is $\sigma_L^2 = 2pq\beta^2 = 2pq^3$ and the heritability is

$$h^2 = \frac{\sigma_L^2}{\sigma_y^2} = \frac{2pq^3}{q^2(1-q^2)} = \frac{2q}{1+q}$$

and

$$r = \text{correlation (parent-offspring)} = \tfrac{1}{2} h^2 = \frac{q}{1+q}$$

assuming no environmental variance (heritability in the narrow sense). It is seen that when q is very small, the heritability is correspondingly low. As simple as this example is, it does illustrate the general principle that a trait with a long history of selection usually has low heritability.

If the deleterious effect on fitness of the "recessive" gene is not so completely recessive, the heterozygote may have a slightly lower fitness than that of the normal homozygote. In such a case the selection effect would be much greater than that against the true recessives. On the other hand, if the gene has a heterotic effect so that the heterozygote (carrier) has a higher fitness than both homozygotes, an equilibrium known as genetic polymorphism will be reached where there is no heritability in fitness at all ($\sigma_L^2 = 0$.). From our present knowledge, no general conclusion may be made. No doubt, however, both types of heterozygotes exist in Nature, with respect to different loci as well as for the same locus in different environments.

Correlated Responses

The selection example discussed in the foregoing section is too much oversimplified in one important aspect: it is limited to the case in which the trait under consideration is determined by only one pair of genes and independent of all other genes. This is almost never the case with respect to quantitative traits. Selection, with respect to quantitative traits, is much too complicated a subject to be discussed here. Nevertheless, we shall mention just one outstanding feature as briefly as possible. One of the consistent phenomena discovered by selection experiments with animals and plants is the correlated responses to selection, for which there is still no established or unique explanation. Suppose that we select for trait A (greater number of bristles, large or small body size, high protein content, and the like), which has a certain degree of heritability. After a number of generations of continued selection, we will obtain a high-A stock, but this is *not* all. It will be noticed that many other traits ($B, C, D, \ldots$) of the population have also changed, although they have not been selected for or against. The "responses" in these traits ($B, C, D, \ldots$) are called the *correlated responses*.

Selection experiments also show that the severer the selection, the greater the correlated responses. When selection is relaxed or suspended, the correlated responses become milder. Very frequently, such correlated responses are in the undesirable direction. One of the commonest is sterility; the reason for this is unknown. Sometimes the experimenter has found it necessary to stop his selection experiments for a few generations in order that the experimental animals may recover their fertility to some extent, or he will lose his experimental stock. The correlated responses can be more drastic than the main responses we are breeding for.

Why the correlated responses? As mentioned before, there is no satisfactory explanation yet, and the selection experiments are still going on. However, some general conjectures may be made. One is the limited size of the selection experiments, even with small animals (e.g., vinegar flies). Since only a small number of parents are selected to produce the next generation, there is bound to be a large sampling fluctuation accumulating over the generations. The correlated response is then the result of random sampling fixation. No doubt, sampling "error" is omnipresent in all selection experiments, but it can hardly explain the consistent appearance of certain correlated responses (e.g., sterility) in many experiments.

Another explanation is based on the two well-known genetic facts: (1) a quantitative trait is determined by many genes (multiple loci), and (2) a gene has manifold effects (pleiotropism). Suppose that trait A is influenced by genes (loci) $G_1, G_2, G_3, G_4, G_5, G_6, G_7, G_8$; trait B by loci $G_3, G_4, G_7, G_8, G_9, G_{10}$; and trait C by $G_2, G_3, G_5, G_6, G_{11}, G_{12}$; and so on. Selection for A would automatically change trait B because these two traits share the common genes G_3, G_4, G_7, G_8. Likewise, traits A and C share the genes G_2, G_3, G_5, G_6. The correlated responses are due to change in their common genes. Selection for one trait would affect other traits simultaneously. Again, there is no doubt a great deal of truth in this explanation. However, it is known that a given amount of change in trait A accomplished by slow selection over a long period of time induces far less correlated responses than the same change accomplished by severe selection within a short period. The common-gene theory cannot explain the difference between slow and drastic selection.

Still another explanation is that the genes under natural selection are in general not combined at random but are associated. To illustrate, if p, q are the frequencies of the A,a genes determining one trait and u, v are the frequencies of the B,b genes determining another trait ($p+q=1$, $u+v=1$), the situations for the random combination and associated distribution in the gametes (and thus also in genotypes) produced by the population are as follows:

Random combination

	B	b	
A	pu	pv	p
a	qu	qv	q
	u	v	1

Associated genes

	B	b	
A	x_1	x_2	p
a	x_3	x_4	q
	u	v	1

The gamete distributions are not achieved in one generation of random mating but are gradually approached in a number of generations. In case of random combination, artificial selection with respect to the trait affected by one locus does not change the gene frequencies of the other locus. In case of associated distribution, however, any abrupt change in one locus will cause a change in the other. Hence, the correlated response.

Since a quantitative trait is controlled by many genes, the association pattern among the loci must be manifold and very complicated. There are the primary associations for pairs of loci; the secondary associations for triplets of loci; the tertiary associations for quadruplets of loci; and so on. It is even difficult to devise a meaningful single overall measure for the degree of association involving so many factors. Yet, the principle is clear: a change in one trait will cause changes in others; any abrupt change in one trait will upset the entire system of association among the genes established by natural selection over a long period of time.

The system of association of genes does not only explain the correlated responses to selection but also explains the difference between the slow selection and abrupt selection. A very slow selection, the effect of which is hardly discernible in any one generation, allows the various loci to undertake new combinations (which takes time, especially when there is genetic linkage), establishing a new system of association with minimum correlated responses. A drastic selection, disturbing the old system and allowing no new system of gene combination to establish itself, imposes upon the population an *arbitrary distribution* of genes with a host of correlated responses, most of which are expected to be unfavorable to the genotypes. Sterility is one example.

There may be other explanations for correlated responses to selection. It should be noted that the three possibilities mentioned above are not mutually exclusive of each other. Indeed, all three causes may be working simultaneously. Experimental evidences favoring one hypothesis are not to be construed as contradictory to the others. In most cases so far, the experimenter regards the correlated responses as incidental side-effects of selection, and pays cursory attention to them only when they interfere with his further experiments. I feel that the correlated response should be a major topic of genetic research on its own merit.

Genetic Improvement of Mankind

There are apparently many reasons for urging the improvement of the human race genetically. Only a few of them can be discussed here. One reason is based on the spectacular success of our plant and animal breeders in producing "improved" varieties. If we can accomplish that much with the chickens and cows, why cannot we accomplish something with man? Those who reason this way have apparently misunderstood the function of the experimental stations and the meaning of such "improvements." At the outset it must be emphatically stated that these improvements refer to the *commercial value* of the traits, not the biological well-being of the plants or animals. If the breeders accomplished both, it would be purely incidental. More frequently, however, they do the opposite, enhancing the commercial value at the expense of the biological well-being. Thus, we (I was a plant breeder) breed a virus-infected variety of certain flowers because it has a much greater commercial value. This diseased plant is then called an "improved" variety. We also hybridize watermelons to produce a sterile variety (triploidy)—the seedless watermelon—which are regarded almost as a treasure because they will bring the poor country foreign currency. This "improved" variety cannot reproduce. Agricultural experimental stations are much like the laboratories of private industry, engaging essentially in commercial research to meet the demands of the customers. When large turkeys are demanded by large family reunions, they breed for large size. When small turkeys are demanded to fit into the oven of efficiency apartments, they breed for small size. The "improved" variety at one time becomes undesirable at another.

One may point out that the breeders have been able to change some basic biological characteristics such as the growth rate of chickens and the milk yield of cows. Again, these "improved" traits do the animals themselves no good at all. The faster a chicken grows, the sooner she gets slaughtered and consumed by man. We have greatly improved our poultry business, but not the welfare of the chickens. Since we are not allowed to add water to milk, the next best thing to do is find a cow who is willing to add water for us—legally. My agricultural friends point out to me that the cow also yields more total dry matter, not just water. Of course, the cow cannot help it; she does not know how to add pure water.

There is no need to labor this point any further. To summarize, the breeders have a definite goal to achieve; the goal is commercial value to man, not the welfare of the plants and animals. Experimental stations frequently have to shift the goals to cope with new economic opportunities. They are not institutions trying to save the biological world or concerned with the well-being of this or that species. Without the economic value as a yardstick, the word "improvement" would instantly become controversial, if not meaningless.

Genetic Deterioration

Another reason for urging the genetic improvement of man is the fear or belief that man is now genetically deteriorating because of the soft environment and the advances made in the medical science. The argument is that modern medicine is saving too many lives of those who have hereditary diseases and would have died in the old days, thus now preserving too many harmful genes in the population. According to this theory, medical science will harm people in the long run by helping them in the short run. Realizing that the medical profession needs no protection from anybody, I shall limit my discussions in the following paragraphs to the genetic aspects of the problem.

It is true that modern medicine has saved some people who might have died before, as it has always been true with medicine since ancient times. It is then clear that the genetic pollution must have started from the very dawn of human civilization. As our knowledge accumulates, we will be able to save more and more lives. How could it be otherwise? We have exhibited the ineffectiveness of selection against rare deleterious genes in a preceding section. When the selection pressure relaxes, the frequency of the deleterious gene will rise to a new equilibrium level, but it does so very, very slowly.

In order to obtain a quantitative idea of this type of change in gene frequency, let us continue with the example given in Table IV, still assuming the mutation rate from the normal allele to the harmful form to be $\mu = 4 \times 10^{-6}$. When the relative fitness of the genotypes are 1, 1, 0, the equilibrium frequency of the gene is $q = \sqrt{\mu} = 0.0020$. Now let us assume that modern medicine and surgery are able to save half of those with diseased recessives so that the new relative fitness of the three genotypes becomes 1, 1, ½. Then the new equilibrium frequency of the deleterious gene may be shown to be

$$q = \sqrt{2\mu} = 0.002828$$

as compared with the previous 0.0020. The incidence of the hereditary disease in the new equilibrium population will be $q^2 = 8 \times 10^{-6}$ as compared with the previous 4×10^{-6}. This is far from the doomsday of mankind. Furthermore, since we save half of the eight recessives per million, there will be still four deaths per million from the hereditary disease, exactly the same as before the advent of the modern medical practice. Note that in the paragraph above we have been careful to say that the new *equilibrium* frequency is $\sqrt{2\mu} = 0.002828$; it is not the frequency in the immediate next generation as soon as the recessives are saved. It will take a large number of generations for the gene frequency to rise from 0.0020 to 0.0028. The calculations for the change in the immediate next generation yield the following result:

(1) given initial gene frequency $\quad q_0 = 0.002,000$
(2) decreased due to selection $\quad q' = 0.001,998$
(3) contribution by mutation $\quad p\mu \doteq \mu = 0.000,004$
(4) new gene frequency $\quad q' + \mu = q_1 = 0.002,002$
(5) net amount of gain $\quad q_1 - q_0 = \triangle q = 0.000,002$

It is seen that the net amount of increase in the next generation, $\triangle q$, is approximately equal to one half of the mutation rate. And this is the largest of increases per generation, since $\triangle q$ decreases as the gene frequency approaches the equilibrium value. Even on the basis of $\triangle q = 0.000,002$, it will take more than 400 generations (probably close to 800 generations) to reach the new equilibrium level of $q = 0.002,828$. This is the genetic deterioration that some of our scientists are worrying about!

The increase of fitness from 0 to ½ for the recessives, as assumed in the foregoing discussions, is an exaggeration for two reasons. First, every time the newspapers or journals announce a new success in medical treatment or surgery, we know that there are perhaps several dozens of similar patients without the benefit of the new practice. The medical profession, as practiced under the present system, has saved only a small fraction of those who could be saved. Second, the recessives we did save do not necessarily reproduce, or reproduce as many as the normals. The net fitness of the recessive genotype as a whole could be far below the 50 percent mark. (It depends on the particular disease under consideration.) With the help of an "heredity clinic" or "genetic counseling," most of these patients may be made to understand the nature of the problem and would voluntarily refrain from reproducing, not with a view to improve the human race, but purely as a matter of family economics, health, and happiness. To summarize, I do not see the day when the harmful genes would spread all over the population. Only the "good" genes can increase in frequency.

Decline of Intelligence

A third reason for urging the genetic improvement of man is based on the "fact" of differential reproduction of the various social classes. To discuss this problem in any detail would require a book-length article; we can only make a few remarks on the subject here. The reproductive pattern probably has never been the same for all social classes in human history. There is no reason to think that it is a new phenomenon of the twentieth century. One may argue that the differential reproduction pattern, though not entirely new, becomes more pronounced in our time than before. A problem is always more acute and pronounced to the contemporary observer than to someone centuries away.

The basic "fact" on which the theory of the decline of human intelligence relies is that the low economic-social-intelligence class reproduces more offspring than the upper economic-social-intelligence class, so that the former contributes proportionally more to the next generation than the latter. Since it is doubtful whether the economic social status has any heritability at all, we shall in the following discuss only the matter of intelligence, which is known to have high heritability.

It is an irrefutable evolutionary fact that man's intelligence has been increasing over the entire history of man, from the primitive to the modern. This, even those who believe that human intelligence is declining with genetic deterioration, would not deny. They merely assert that human intelligence is *now* beginning to decline. The key word is "now," the time. In other words, during the period before us, there has been progress; during the period after us, there will be deterioration. It follows that the present generation is at the maximum level of intelligence in the entire evolutionary event of man, as shown in the following rough sketch:

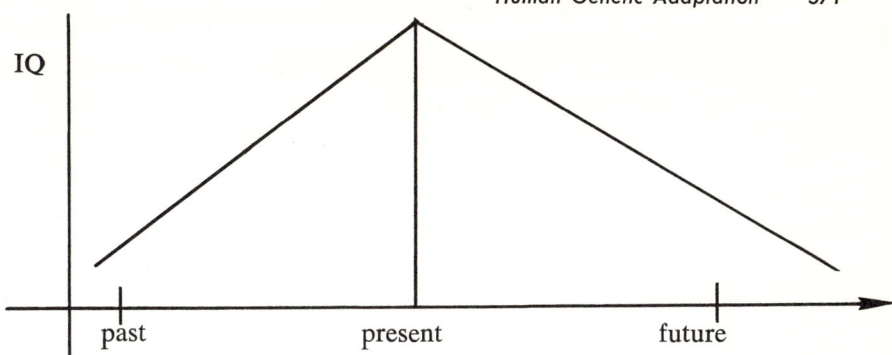

If so, we would be in a truly unique position in the natural history of man, witnessing the turning point along the time axis which is so long that it has no beginning and has no end! Of all the human efforts trying to make ourselves important, this one must rank with the most sophisticated. We cannot bear the thought that we too might be regarded as primitive by those living five thousand generations later.

The unlikeliness of an event does not necessarily mean it is untrue. So we proceed to examine the evidence that human intelligence is declining. There are really no studies designed to compare the intelligence of two generations. This is perhaps fortunate from the technical viewpoint, on account of the difficulties in obtaining random samples from human populations and the minute amount of change in intelligence expected in any one generation. Should there be a gain or loss of 10 points on the IQ scale in 60 generations, it would be a substantial evolutionary change but it could not be detected by any kind of study at all, especially in view of the environmental influences on the development of intelligence. The closest thing to an attempt to detect any change in intelligence level within a man's life time is the Scottish Survey of all school children aged 11 years in 1947 compared with a similar survey of 11-year-olds in 1932 (Scottish Council for Research in Education, 1949, 1953). A period of 15 years is approximately equal to half of a human generation. The survey shows a *gain* of a few points in average intelligence-test score over the 15 years. The apparent increase in intelligence could easily be attributed to better environment, better development, and perhaps also better education. At the outset, surveys of half-generation intervals could not possibly detect any change due to genetic selection. The gain in intelligence cannot be attributed to real genetic gain. By the same token, the survey result would be still less useful in substantiating the theory of the decline of human intelligence. Yet, those who "believe" in the decline have a ready interpretation: there is actually a decline in intelligence due to genetic factors but it has been overshadowed by the gain due to environmental improvements. The truth is unknown to us. This interpretation (or any other interpretation at all) could be true.

However, the nature of the argument has violated almost every rule in drawing inferences from observations. The theory of decline cannot be disproved either way the survey result turns out. Then, why the survey in the first place? An hypothesis that cannot be subjected to possible disproof serves no scientific purpose, since the basic procedure of scientific inquiry is to set up an hypothesis and then try to disprove it.

The belief that the human intelligence is declining is actually based not on the comparison of two generations but on the phenomenon that the low intelligence group has larger families than the normals. This seems to be the crux of the whole problem. It has long been observed by social scientists that there is a negative correlation between intelligence of the children and the size of the family in which they belong. It is impossible to review the voluminous literature on the subject, but, with a brief digression, an exposition of the source and nature of biases in an hypothetical study would be helpful to clarify the situation.

Suppose that we wish to find out the average size of family in a certain community and for this purpose conduct a survey of the high-school seniors, asking each of them: "How many children does your mother have?" or "How many brothers or sisters do you have?" If the student is the only child of the family, we count the size of family as $s=1$. If he has one brother or sister, we enter $s=2$, and so on. The answers from 100

TABLE V

An Hypothetical Study of the Family Size by Surveying 100 High-school Seniors

Size of family s	I Direct results from survey	II Corrected survey results	III "True" distribution in population
0	–	22	0.18
1	19	19	.20
2	18	18	.19
3	18	18	.15
4	15	15	.11
5	13	13	.08
6	10	10	.05
7	5	5	.03
8	2	2	.01
Total	100	122	1.00
mean $\bar{s}$	3.45	2.83	2.46

hypothetical seniors are listed in Table V, column I, where the average size is found to be $s=3.45$. This average size, however, is a gross overestimate of the true size in the community, for the student could not possibly have answered: "My mother never had any children." In other words, a childless family has no representative in the classroom, and therefore has not been included in the survey. Now, suppose that the investigator is told that 18 percent of all families in the community are childless. With this information, the investigator may "correct" the bias by adding a number (n_0) of childless families to his survey result and recalculate the average size. Thus, $n_0:100=18:82$; $n_0=22$ approximately. When this number is added to the result, the new average size is reduced to $\bar{s}=2.83$ as shown in column II. One would think that the corrected result should be free of bias; unfortunately, this is not so. There is a second source of bias which is even more difficult to overcome. That is: a family with eight children is more likely to have a child in the classroom than a family with a single child. In general, the large families are proportionally overrepresented than the small families in a survey of this type ("from-bottom-to-top"). Hence, the corrected mean, $\bar{s}=2.83$, is still too large. An ideal method of finding out the true distribution and average of family size is to survey, say, the 50-year-old couples and record the number of children they have had. The result of this "from-top-to-botom" survey is listed in column III, which yields an average size $\bar{s}=2.46$. The second source of bias is difficult to remedy because we do not know to what extent the large families have been overrepresented in any particular survey, although we are sure they have been to some extent. There are other possible sources of bias, which we need not consider here.

From the hypothetical example above, we may generalize that there are two main sources of bias in population studies of the from-bottom-up type. One is the exclusion of a certain extreme class and one is the differential representation of the various classes. In studying the reproductive performance of the low-intelligence group or the correlation between the intelligence of children and the size of family, the usual procedure is of the from-bottom-up type (starting with propositi) and hence both sources of bias have been at work at one stage or another. As a result, proportionally more large low-intelligence families have been included in the usual studies. Recently, Higgens, Reed, and Reed (1962: see Reed and Reed, 1965, *Mental Retardation: A Family Study,* Saunders, Philadelphia and London) tried to correct at least the first source of bias. Many very-low-intelligence persons remain unmarried and have no offspring; these persons have never been included in previous studies. When these childless low-intelligence people are included, it was found that the low-intelligence group reproduces less than the normal or the high-intelligence group. If the second source of bias could also be remedied, it would be found that the low-intelligence group reproduces even less than indicated. Hence, Reed and Reed conclude that human intelligence is not decreasing, but may be increasing—slowly and inconspicuously.

Control of Human Evolution: Reproduction Specialization

Another urge for the improvement of mankind is even more academic than those discussed so far; that is, improvement for the sake of improvement, whatever that means. Without the commercial value as a criterion (as is the case with plants and animals) it is difficult to judge what constitutes an improvement except by an arbitrary definition. Yet, assuming that we know what we want and assuming we will obtain what we want, some scientists are proposing the control of human evolution, a long-term project of positive eugenics or human breeding. How could we be sure that what we want now is also what our descendants want later? One may argue that there are certain intrinsic human properties that are desirable—for instance, high intelligence. It is doubtful that even such an obviously admirable trait could by itself be a blessing to mankind. Social scientists find that the intelligence of the leaders of the big city criminals rank with that of our elite physicists. Reed and Reed (1965), after following up the details of more than 82,000 persons' intelligence and behavior, made this observation: "Social performance can be hopelessly inept in persons with high intelligence."

Honesty has been mentioned as another desirable trait in a human society. It is much easier to devise a social system that encourages honesty than to breed a stock that is genetically honest, even if honesty has heritability. Disease resistance is also highly desirable property. Again, it is easier to control the diseases by other means than to breed a disease-resistant human race. The latter has been proved difficult even for plants and animals, because the pathogens are also constantly changing. This brings us back to the problem of genetic deterioration discussed before, and boils down to the choice between (1) being unhealthy people with healthy genes or (2) being healthy people with unhealthy genes.

Another argument for the control of human evolution is that man has now conquered his environment, and natural selection has ceased to operate and hence we must substitute human guidance. Apparently this is the way we compliment ourselves nowadays, and it may not be unique in human history. A caveman, with weapons to hunt, with fire to cook, furs to keep him warm, and a cave for comfortable living, might also have thought he has at last conquered nature. In perspective, every living species has conquered nature in one way or another.

If the aims of controlling evolution are difficult to formulate, the means by which it might be accomplished is even more so. To achieve the drastic selection required for controlling evolution, one proposal is "reproductive specialization," allowing only a chosen few for propagation. The sperms of the chosen sire may be preserved and distributed to inseminate a large number of women. One of the many problems is: whose sperms should be preserved and distributed? Consider a man of "success" and suppose that he has four sons: the first became a War Hero, the second the Chief Executive of a large nation, the third an Attorney General, and the fourth a Senator. Would not we wish to preserve the sperms of such a man? Yes, if you also wish to risk a mentally retarded daughter! Briefly, the basic defect of the proposal is to apply a social standard to a biological problem.

Reproductive specialization would also drastically diminish the "effective size" of the reproducing population. Suppose there are $N_1 = 20$ chosen sires from a population and $N_2 = 20,000$ women to be inseminated, the effective size of the breeding population, N_e, may be shown to be:

$$\frac{1}{N_e} = \frac{1}{4N_1} + \frac{1}{4N_2} \doteq \frac{1}{4N_1}, \text{ when } N_2 \gg N_1$$

In our example, $N_e = 4N_1 = 80$ very nearly. In other words, instead of having a large random-mating population, we have reduced it to an equivalent one having only 40 males and 40 females. The evolutionary consequences of having a small breeding population are far-reaching indeed (Wright, 1931, 1938; Li, 1955). The offspring, or at least the long-term result, of a small breeding population may vary widely and are more unpredictable than the correlated responses mentioned earlier. The random variation of small populations is so large that it may nullify the expected effects of selection. Most of the offspring from the matings of 20 males × 20,000 females are half-brothers and sisters, and continuation of the practice in subsequent generations amounts to close inbreeding. But, the resulting genotypes actually obtained are not necessarily the ones selected for! In this aspect, the reduction of effective population size by selecting a few sires will take the human species back to the ancient times when the breeding populations were small.

The chief advantage of having a large random-mating population are its stability (negligible random variation) and its capability to respond to selection (e.g., adaptation) through genetic diversity. Homozygosis is probably the worst long-run prospect (genetically) that could happen to any species.

Genotype-Environment Interaction

In closing, we shall mention briefly just one more important aspect of human evolution, and that is the genotype-environment interaction. There is no absolute measure of biological fitness; that fitness is always relative to

TABLE VI

Strong Genotype-environment Interaction in Fitness

	Environment				
Genotype	E_1	E_2	E_3	E_4	Total
G_1	8	3	5	4	20
G_2	4	5	3	8	20
G_3	3	7	7	3	20
Total	15	15	15	15	60

a particular environment. Different genotypes react differently to different environments. To exaggerate the point, an hypothetical situation is shown in Table VI, in which the numbers represent the relative fitness of the genotypes. It is seen that G_1 fits E_1 best, G_2 fits E_4 best, and G_3 fits E_2 and E_3 equally well. Taken as a whole, there is no environment that is better than others, nor is there a genotype that is better than others. The entire variation in fitness is due to interaction between the genotype and its environment. Since there exist many different genotypes in our population, this shows the importance of providing different environments for the maximum welfare of all individuals. The environment in a modern complicated society is more varied, not more uniform, than a primitive society. Hence, it seems that the immediate task of our society is to find the right environment for the right genotype.

Summary and Conclusions

In view of the low heritability of the human social-economic traits, the random shuffling of genes from family to family in each generation, the ineffectiveness of selection against rare deleterious genes, and the possible correlated responses to selection, we believe it is prudent to go slow on any improvement plan for man. There is no danger that human heredity is deteriorating; and there is no evidence that human intelligence is declining. Drastic selection on a human population would greatly diminish its effective size and would lead to unpredictable and probably undesirable results. Even without our good intentions, human evolution will go on, and on, and on: and man will continue to adapt as he did before. The immediate task of our society is to provide or find the right environment for the right genotype to fully develop its potentiality and usefulness.

References

CROW, J. F. 1958. Some possibilities for measuring selection intensities in man. Hum. Biol., 30:1-13.

DOBZHANSKY, TH. 1962. Mankind Evolving. New Haven, Yale University Press.

FALCONER, D. S. 1960. Introduction to Quantitative Genetics. New York, The Ronald Press Company.

HIGGENS, J. V., E. W. REED, and S. C. REED. 1962. Intelligence and family size: A paradox resolved. Eugen. Quart., 9:84-90.

LI, C. C. 1955. Population Genetics. Chicago, University of Chicago Press.

———— 1961a. The diminishing jaw of civilized people. Amer. J. Hum. Genet., 13:1-8.

———— 1961b. Human Genetics: Principles and Methods. New York, McGraw-Hill Book Company.

———— 1963. The way the load ratio works. Amer. J. Hum. Genet., 15:316-321.

———— 1967. LI, C.C. 1967. Genetic equilibrium under selection. Biometrics, 23:397-484.

———— and L. SACKS. 1954. The derivation of joint distribution and correlation between relatives by the use of stochastic matrices. Biometrics, 10:347-360.

REED, E. W., and S. C. REED. 1965. Mental Retardation: A Family Study. Philadelphia, W. B. Saunders Company.

Scottish Council for Research in Education. 1949. The Trend of Scottish Intelligence. London, University of London Press.

Scottish Council for Research in Education. 1953. Social Implications of the 1947 Scottish Mental Survey. London, University of London Press.

United States Census. 1950 and 1960. Washington, D.C., U.S. Government Printing Office.

WRIGHT, S. 1931. Evolution in Mendelian populations. Genetics, 16:97-159.

———— 1938. Size of population and breeding structure in relation to evolution. Science, 87:430-431.

Author Index

Abel, O., 56, 91
Abramoff, P., 241-242, 259-261
Adams, F. D., 45, 91
Agassiz, L., 51, 55
Albritton, C. C., Jr., 91
Allard, R. W., 182, 187, 203
Allen, G., 31
Alston, R. E., 177, 203
Ambrose, L., 40
Anderson, E., 203
Anderson, W. W., 37-40, 150, 155
Andrewartha, H. G., 129, 131-132, 155, 257, 259
Arnold, C. A., 76, 91
Asimov, I., 38
Atwood, K. C., 298
Avery, A. G., 180, 203
Avi-Dor, Y., 171
Ayala, F. J., 121, 131, 139, 141-142, 146-147, 150, 152, 155
Axelrod, D. I., 188, 202-204

Bacon, F., 64
Badash, L., 74, 91
Baetcke, K. P., 198, 203
Bagenal, T. B., 128, 138, 155
Bailey, D. W., 184, 203, 268, 286
Bailey, E. E., 45, 91
Baker, H. G., 37, 186, 203-205
Bakker, K., 125-126, 155
Ball, F. M., 204
Balsano, J. S., 259
Barbier, M., 170-171
Bar-Geev, N., 171
Barghoorn, E., 71, 95, 202, 207
Barker, J. S. F., 359, 375
Barlow (Lady), 53, 92
Barnett, S. A., 32
Basden, E. B., 408, 432
Basrur, V. R., 252-253, 260
Bastock, M., 319, 323-325, 327-328, 357, 375, 423, 432
Bateman, A. J., 316, 324, 339, 375, 398
Baumrim, B., 36
Bazinet, M., 172
Beadle, G. W., 22, 81, 297-298

Beardmore, J. A., 32, 34, 299-314, 396-397
Beaufoy, Mr., Mrs., 270, 274, 276, 280, 286
Beck, L. W., 65, 91
Becker, G. C., 376
Beckman, L., 176, 203
Beckner, M., 66, 86, 91
Bennet-Clark, H. C., 362, 375
Benson, L., 179, 203
Berg, L. S., 41
Berkeley, G., 111, 118
Berry, H. K., 31
Bertalanffy, L. von, 104, 118
Birch, L. C., 36-37, 124, 127, 129, 132, 155, 257, 259, 416, 432
Bishop Usher, 47
Björkman, E., 212, 234
Black, E. D., 172
Black, G. A., 28-29
Black, M., 65, 91
Blackwell, D. L., 286
Blum, M. S., 169, 171
Blyth, E., 33
Boam, T. B., 284, 287, 300, 314
Boche, R. D., 21
Bock, W. J., 80, 85, 91
Bodmer, W. F., 197, 203, 426, 432
Bolton, E. T., 176, 205
Born, M., 99, 118
Bösiger, E., 375, 424, 432
Bossert, W. H., 169, 172
Bradley, W. H., 44, 91
Bradshaw, A. D., 178, 180, 203, 207, 284, 286
Brewbaker, J. L., 176, 203
Bridges, C. B., 3, 19, 21
Brncic, D., 32, 355, 375, 401-436
Brodbeck, M., 92
Brouwer, A., 56, 91
Brown, R. G. B., 362-363, 375
Brown, R. R., 298
Brown, S. W., 198, 204, 239, 260
Brown, W. L., 154, 155
Bryan, G. T., 297-298
Budnik, M., 422-423, 433

Burla, H., 27-28, 382, 398, 429, 434
Burton, D. L., 175, 205
Butler, C. G., 170-171

Cain, A. J., 184, 204, 300, 313
Callan, H. G., 199, 204
Callow, R. K., 170-171
Camin, J., H., 175, 204
Campbell, J. H., 56, 93
Cannon, W. F., 45, 53, 91
Carlisle, D. B., 168, 171
Carmody, G., 36
Carson, H. L., 197, 204, 250, 260-261, 291, 301, 303, 313, 406-407, 421, 426, 429-431, 433, 437, 451, 455, 466-467, 479, 490, 492, 496-497, 500, 504, 506, 508, 514, 516, 532, 540
Caspari, E., 316, 375
Cavalcanti, A. G. L., 28, 398
Cavill, G. W. K., 169
Chaney, R. W., 188, 204
Chapman, R. N., 158
Cheney, J., 262
Chetverikov, S. S., 3-4, 39
Chitty, D., 127, 131, 155
Cibis, C., 114, 118
Clarke, B., 150, 155, 364, 375
Clarke, C. A., 300, 313
Clausen, J., 179, 182, 187, 204
Clayton, F. E., 461, 474, 477, 492-496, 500, 514, 541, 543
Cloud, P. E., Jr., 71, 91
Coe, M. D., 2
Coe, S., 2
Cole, L. C., 129, 155
Collier, A., 167, 171
Condit, C., 188, 204
Conybeare, W. D., 43, 49-53, 67, 69, 88, 92
Cook, S. A., 180, 204
Cooke, D., 146, 156
Cooper, D. M., 31, 136, 155
Cordeiro, A. R., 27, 29-30
Creed, E. R., 263, 268, 270, 286
Crombie, A. C., 128, 155
Crossley, S., 375
Crow, J. F., 548, 559
Cuvier, L. C., 48, 52, 56-57, 92
Cyrén, O., 248, 260

DaCunha, A. B., 27-31, 33, 153, 156, 300, 313, 398, 410, 416, 423, 429-430, 432-434
Daly, K., 178, 192, 207
Darevsky, I. S., 247-248, 259-260
Darlington, C. D., 24
Darlington, P. J., Jr., 76, 92, 405, 433
Darnell, R. M., 242, 259-261
Darwin, C., 32-34, 43, 47-48, 53-58, 79-80, 84, 88, 92, 105, 118, 188
Dawood, M. M., 125-126, 156
Dawson, P., 127, 156
Day, A., 202, 207
De Bach, P., 128-129, 156
De Castro, D., 198, 204
De George, F. V., 33
Delevoryas, T., 76, 92
Del Solar, E., 124, 153, 156, 409, 421, 425, 435
Demerec, M., 39
Democritos, 98, 111
Dempster, E. R., 300, 314
Dethier, V., 373, 375
DeWet, J. M. J., 196, 204
Diaz Collazo, A., 36
Doak, W., 2
Dobrzhansky, G., 1
Dobzhansky, N. P. 29. *See also* Sivertzev, N. P. and Sivertzev-Dobzhansky, N. P.
Dobzhansky, Th., 1-15, 81, 92, 97, 105, 118-119, 127, 131, 136-137, 142, 150, 155-157, 174, 184, 188, 190, 197, 203-204, 210, 236, 238, 257, 260, 264, 279, 283, 286, 289, 293, 298, 300-301, 312-314, 343, 348, 351, 358, 364, 375-376, 379, 381-382, 398-399, 401-403, 405-406, 408, 410, 416-418, 421-423, 425, 429-430, 433
bibliography, 18-41
career summary, 16-17
collecting expedition, 7-15
scientific work, 3-7
Dollo, L., 89, 92
Doty, P., 205
Dowdeswell, W. H., 263, 267-268, 286-287
Downes, A., 258, 260
Drescher, W., 35
Dreyfus, A., 24
Driesch, H., 102, 118
Dubinin, N. P., 300, 314, 382, 399, 408-409, 417, 423, 434

Dunbar, C. O., 44, 92
Duncan, F. N., 21
Dunn, L. C., 26, 32-33
Dure, L. S., 190, 204, 208

Eardley, A. J., 75, 92
Eddington, A., 113, 118
Ehrendorfer, F., 180, 196, 204
Ehrlich, P. R., 168, 171
Ehrman, L., 1, 32, 36-38, 40, 150, 157, 319, 347, 351, 357, 359, 364-365, 368, 370-371, 376, 378, 386, 399
Eibl-Eibesfeldt, I., 169, 171
Eiche, V., 209-210, 212, 214, 216, 223-224, 228, 233-234
Einstein, A., 99, 112
Eisner, T., 169, 171
Elens, A. A., 376
El-Gazzar, A., 175, 204
Elliot, P. O., 36-37
Ellis, P. E., 171
Elton, C. S., 122, 128-129, 138, 156
Emberger, L., 76, 92
Eneroth, O., 233, 235
Ephrussi, B., 296, 298
Epling, C., 4, 24-25, 27, 185, 204
Estabrook, G., 175, 206
Ewing, A. W., 319, 362, 375-376

Falconer, D. S., 550, 577
Farrand, W. R., 56-58, 92
Faul, H., 74, 83, 92
Favarger, C., 194, 204
Feigl, H., 92
Feinberg, E. H., 158
Feller, W., 132, 156
Firschein, I. L., 31
Fisher, R. A., 4, 426, 434
Fleming, H., 175, 206
Florkin, M., 159-161, 164, 171
Ford, E. B., 105, 118, 146, 156, 263-264, 268, 270, 283, 286-287
Fosberg, F. R., 36, 447-449, 541
Fraenkel, G., 168, 171
Freire-Maia, A., 434
Freire-Maia, N., 409-410, 428, 434
Fried, M. H., 32
Fuenzalida, H., 402, 434
Fulker, D. W., 340-342, 356, 376

Gaines, H., 542
Galun, K., 168, 171

Garn, S., 37
Gartler, S., 30-31
Gaul, H., 182, 205
Gause, G. F., 128, 130, 132, 156, 252, 260
Geer, B. W., 358, 376
Geerts, S. J., 36-37
Geikie, A., 49, 92
Gibson, J., 312
Gilbert, L. E., 286
Gillespie, C. C., 45, 49, 92
Gilluly, J., 74-75, 77, 92
Glaessner, M. F., 71, 92
Glass, H. B., 40
Glisin, M. V., 205
Glisin, V. R., 189, 205
Goldblatt, S. M., 243-244, 262
Goldschmidt, E., 430, 434
Goldschmidt, R., 85, 92, 191, 438
Goodman, G. T., 286
Goodman, N., 64-65, 92
Goodnight, K. G., 542
Goodspeed, T. H., 180, 205
Gould, S. J., 49-50, 58-59, 63, 65, 66, 81, 88, 92
Grant, K. A., 201, 205
Grant, V., 81, 92, 163, 171, 180, 188, 192-193, 201, 205
Green, M. M., 291, 358, 376
Gregory, R. P. G., 178, 204, 286
Grene, M., 80, 85, 92
Grimshaw, P. H., 438, 455, 541
Grinnell, J., 122, 128, 156
Grossfield, J., 358, 362, 376, 484, 541
Grünbaum, A., 92
Guest, W. C., 436
Gustafsson, Å., 209, 216, 228, 234-235

Habeler, E., 204
Hackman, W., 461, 541
Haldane, J. B. S., 4, 128, 130, 151, 156
Hall, C. P., 40
Hall, O., 176, 205
Hapgood, C. H., 56-57, 93
Hardin, G., 128-129, 156, 252, 260
Hardy, A. C., 118
Hardy, D. E., 412, 434, 437-439, 454-455, 457, 459, 464-467, 478, 493, 500, 516, 519, 541
Harland, S. C., 294, 298
Harper, J. L., 178, 183, 205
Harrington, C. R., 119

Harris, H., 311, 314
Harris, P., 146, 156
Harrison, B. J., 325-327, 377
Harrison, G., 33
Hartmann, M., 99, 118
Hartmann-Goldstein, I., 418, 434
Haskins, C. P., 242, 260
Haskins, E. F., 260
Hayes, J. T., 158
Heberer, G., 34
Hecht, M. K., 39-40, 119, 157, 203
Heed, W. B., 260, 407, 434, 463, 469-471, 473, 475, 478-479, 487, 541, 543
Heisenberg, 99, 103, 112
Heiser, C. B., Jr., 175, 195, 205
Hempel, C. G., 66, 85-86, 90, 93
Henbest, L. G., 76-77, 93
Herschel, J. F. W., 93
Heslop-Harrison, J., 177-178, 205
Hewitt, R. E., 260
Hiesey, W. M., 179, 182, 204
Higgens, J. V., 573
Hildreth, P. E., 324, 376
Hillebrand, W., 447, 541
Hinegardner, R., 199, 205
Ho, F. K., 139, 156
Hoenigsberg, H. F., 327, 376, 424, 434
Holmes, A., 75, 93
Holmgren, A., 234-235
Holz, A. M., 24
Hooykaas, R., 45, 48, 50, 58-60, 62-63, 66, 81, 88, 93
Hosaka, E. Y., 447-448, 541-542
Hosgood, S. M. W., 339, 376, 378
Hosoi, T., 168, 171
Hovanitz, W., 24
Howells, W., 32
Hoyer, B. H., 176, 205
Hubbert, M. K., 45, 50, 53, 67-68, 74, 93
Hubbs, C. L., 260
Hubbs, L. C., 260
Hubby, J. L., 176, 205-206, 311, 314
Hubel, D. H., 114, 118
Hughes-Schrader, S., 239, 260
Hull, P., 426, 435
Hunter, A. S., 36, 427, 434
Hunter, D. A., 412
Hunter, R. A., 427, 434
Husserl, E., 100, 118
Hutchinson, G. E., 122-123, 128-129, 136, 156

Hutton, J., 43, 45-51, 57, 59-61, 66-67, 73, 75, 84, 87, 93
Huxley, J. S., 32, 105, 118

Ingram, V. M., 171, 182, 205

Jacob, J., 250, 261
Jacobs, M. E., 377
Jacobson, M., 170-171
Jaeger, C. P., 434
Jaffrey, I. S., 36
Jastrow, R., 69, 93
Jayakar, S. D., 151, 156
Jeans, J., 118
Jensen, R. W., 542
Jeuniaux, Ch., 166, 171
Johnson, A. W., 194, 205
Johnson, B. L., 176, 205
Johnston, N. C., 170-171
Jordan, P., 99, 105, 118
Jowett, D., 178, 205
Jukes, T. H., 181, 205, 311, 314

Kallman, K. D., 242, 260-261
Kambysellis, M. P., 469-471, 519-520, 541
Kaneko, A., 522, 542
Kaneshiro, K. Y., 451, 459-460, 464-467, 493, 500, 516, 519, 541-542
Kant, I., 99, 111, 118
Karlson, P., 169, 171
Kastritsis, C. D., 39-40
Katz, E., 297-298
Kaul, D., 350-351, 357-361, 377-378
Kessler, S., 335-337, 342-343, 350, 377
Ketchel, M., 435
Kettlewell, H. B. D., 146, 156
Key, K. H. L., 262
Kikkawa, H., 296, 298
Kimball, S., 36
Kimura, M., 311, 314, 426, 435
King, J. L., 311, 314
Kinne, B. P., 435
Kircher, H. W., 475-476, 542
Kitts, D. B., 44, 93
Knapp, E. P., 31
Kohn, A. J., 167, 172
Kojima, K., 150, 156, 312
Koller, P. C., 23
Koref-Santibañez, S., 327, 355, 375-376, 419, 421, 424-425, 427, 432-435
Krajina, V. J., 447, 542
Krendl, F., 204

Krimbas, C. B., 34, 426, 435
Krimbas, M. G., 34
Krivshenko, J. D., 407, 435
Kruckeberg, A. R., 178, 205
Kulikova, V. N., 247, 260
Kullenberg, B., 189, 205
Kulp, J. L., 74, 93
Kurabayashi, M., 192, 205
Kvelland, I., 362, 377
Kyhos, D. W., 192, 206

Lack, D., 129, 156
Ladd, H. S., 55, 93
Laird, C. D., 295, 298
Lamarck, J. B. M. de, 69, 93, 160
Lamb, E., 242, 260
Lamb, S. H., 447, 542
Lamborot, M., 422-423
Lamotte, M., 184, 206
Lamprecht, H., 191, 206
Lance, G. N., 204
Langer, B., 336, 342, 346, 348, 350, 357, 379, 424, 436
Langlet, A., 210, 212, 233, 235
Langridge, J., 358, 376
Lantz, L. A., 248, 260
Lederer, E., 171
Lee, B. T. O., 378
Lee, J. A., 183, 206, 339
Leibniz, 116
León, J., 435
Leopold, E. B., 202, 207
Lerner, I. M., 139, 156, 303, 314
Levene, H., 1-2, 25, 27, 29-33, 150-151, 156, 300, 314, 358, 376, 423, 434
Levin, B. R., 128, 154, 157
Levin, L., 36
Levine, L., 304-305, 307, 313
Levins, R., 300, 314
Levitan, M., 417, 423, 426, 435
Lewis, H., 180, 204, 206, 381
Lewis, M. E., 191-192, 206
Lewontin, R. C., 36-37, 126, 150, 157, 176, 197, 205-206, 300-301, 303, 311, 314, 316, 343, 364, 376-377, 416, 426, 435
L'Heritier, P., 150, 157
Li, C. C., 300, 314, 364, 376, 544, 551, 558, 561-562, 575
Light, R. E., 40
Lindauer, M., 169, 172
Lippman, H. E., 56-58, 93

Locke, J., 111, 119
Long, T. C., 304, 306, 314
Lotka, A. J., 128, 130, 132, 157
Lovejoy, A. O., 52, 93
Lowe, C. H., 244-245, 260, 262
Ludwig, W., 238, 260, 300, 314
Lüscher, M., 169, 171
Lyell, C., 43, 46, 49-55, 57-59, 67, 69-70, 75, 84, 88-89, 93
Lysenko, T. D., 6, 32, 181

MacBean, I. T., 354, 377
Macgregor, H. C., 243, 260
Mägdefrau, K., 76, 93
Mainardi, D., 324, 377
Mainardi, M., 324, 377
Malogolowkin-Cohen, C., 2, 27, 29
Mangenot, G., 196, 206
Mangenot, S., 196, 206
Manning, A., 316, 319, 325, 330-336, 341-342, 347, 357-359, 361-363, 373, 375-377
Manton, I., 196, 206
Martin, J., 426, 435
Martin, P. G., 199, 208
Maslin, T. P., 244-246, 260
Maslow, A., 33
Mather, K., 284, 287, 300, 314, 325-327, 342, 377
Mather, W. B., 35-36
Matsuo, Y., 126, 157, 416, 435
Matthey, R., 240, 260
Maynard Smith, J., 424, 435
Mayr, E., 81, 86, 93, 95, 105, 119, 178, 185, 188, 206, 274, 402, 406, 435
McCarthy, B. J., 176, 205, 295, 298
McCoy, C. J., 244, 246, 260
McCrea, W. H., 58, 63, 65, 94
McGrath, K. H., 176, 208
McIntyre, D. B., 94
McNeilly, T. S., 178, 204, 286
McWhirter, K. G., 263-264, 267-268, 284, 286-287
Mead, M., 40
Medawar, P. B., 65, 85, 94
Mehra, P. N., 196, 206
Meinwald, J., 169, 171
Mendel, G., 39, 102-103, 106, 108
Merrell, D. J., 153, 157, 324-325, 327-328, 359, 377
Merrill, G. P., 94
Merrit, C., 172

Meyer, F., 67, 72, 94
Michelbacher, A. E., 397
Middlekauff, W. W., 397, 399
Mikulska, I., 248, 261
Miller, R. S., 122, 125, 128-129, 136, 148, 156-157
Milstead, W. W., 246, 261
Miner, J. R., 131, 157
Minton, S. A., Jr., 244, 261
Montagu, M. F. Ashley, 26
Moore, B. P., 250, 261
Moore, J. A., 151, 153, 157
Moore, R. C., 76, 94
Moorehouse, J. E., 171
Morgan, L. R., 297-298
Morgan, T. H., 2-3, 16, 438
Morgenbesser, S., 94
Morton, J. K., 196, 206
Muller, H. J., 5, 187, 299, 314
Mulligan, G. A., 119, 194, 206, 208
Munday, K. A., 171
Murca Pires, J., 29-30
Murdy, W. H., 250, 261

Narbel-Hofstetter, M., 240, 261
Narise, T., 150, 157
Nauman, C. H., 203
Navashin, M., 294, 298
Nemirov, Russia, 1, 16
Newell, N. D., 48, 77-78, 80, 93-94
Nicholson, A. J., 126, 146, 157
Nordenskiöld, H., 198, 206
Nothdurft, H., 118
Nur, U., 239, 261

Obrebski, S., 36
O'Donald, P., 316, 343, 364, 375, 378
Okada, T., 451, 522, 542
Omodeo, P., 241, 261
Oparin, A. L., 172
Oppenheim, P., 66, 85, 93
Osborn, F., 40
Osborne, D. J., 171
Osborne, R. H., 31, 33
Oshima, C., 40
Oster, I., 297
Ostwald, W., 113, 119
Owen, R., 55

Packer, J. G., 194-195, 205
Paik, Y. K., 398-399
Palomino, H., 124, 156

Pap, A., 65, 94
Paparello, F. N., 33
Park, T., 125, 128, 138-139, 148, 157
Parker, S. L., 131, 157
Parsons, P. A., 197, 203, 319, 337-339, 341, 350-352, 354, 357-361, 376-378, 426, 432
Patten, B. C., 129, 157
Patterson, J. T., 25, 403, 405-407, 431, 435, 438-439, 495, 542
Pavan, C., 24, 27-28, 398
Pavlovsky, O., 29-40, 150, 157, 188, 203-204, 301, 313, 343, 351, 364, 376, 423, 425, 434
Peacock, A. D., 239, 261
Pearl, R., 130-131, 157
Peckham, M., 48, 54, 79
Pennock, L. A., 244, 261
Penrose, L. S., 30
Perje, A. M., 408, 435
Perkins, R. C. L., 438, 542
Petit, C., 150, 157, 319, 357, 364-365, 376, 378, 424, 435
Pfeiffer, W., 169, 172
Phaff, H. J., 31
Philipchenko, J., 16
Phillips, E. A., 203
Pimentel, D., 127, 131, 150, 151-153, 157, 416, 435
Pipkin, S. B., 435
Piveteau, J., 76, 94
Planck, M., 99, 119
Plato, 111
Playfair, J., 49, 94
Pokrovskaia, I. M., 202, 206
Pontecorvo, G., 294, 298
Popper, K., 66, 85, 94
Porsild, A., 101, 119
Poulson, D. F., 21
Prakash, S., 355-356, 378
Price, J. M., 298
Prout, T., 151, 158

Queal, M. L., 22-23

Rall, J. E., 34
Rao, P., 297-298
Rasch, E. R., 242, 261
Raven, P. H., 168, 171, 196, 206
Reed, E. W., 378, 573-574
Reed, L. J., 130, 157
Reed, S. C., 378, 573-574

Reese, G., 194, 205-206
Reiche, K., 402, 435
Reichstein, T., 171
Rembold, H., 170, 172
Rendel, J. M., 358-359, 378, 424, 435
Rensch, B., 33, 97, 105-106, 110, 114, 119, 402, 435
Rhoades, M. M., 23
Richmond, R., 1, 41
Rietsema, J., 180, 203
Riley, R., 193, 206
Riordan, D. F., 146, 156
Ripperton, J. C., 447, 542
Ritossa, F. M., 294, 298
Rizki, R. M., 289-298
Rizki, T. M., 289-298, 397
Robertson, A., 296, 298
Robertson, F. W., 330, 358, 378, 470, 475-476, 542
Robertson, P. L., 169, 171
Robyns, W., 447, 542
Roche, J., 166, 172
Rock, J. F., 447, 449, 542
Rodgers, J., 44, 92
Rodriguez, R. L., 435
Rogers, D. J., 175, 206
Rolansky, I., 41
Romer, A. S., 76, 94
Ross, G. N., 169, 171
Ross, H., 136, 158
Rothblatt, B., 40
Rothfels, H. K., 252-253, 260
Rudwick, M. J. S., 45, 50, 52, 94
Russell, J. S., 543
Russell, L. S., 65, 75, 94
Rutherford, E., 73, 94

Sacks, L., 558
Sagan, C., 69, 95
Sakai, A., 224, 235
Sakai, Kan-ichi, 183, 206
Salzano, F. M., 433-434
Sánchez, P., 408, 433
Sanderson, A. R., 250, 261
Sanderson, I., 93
Sang, J. H., 131, 158, 475, 542
Sasamoto, K., 184, 206
Satina, S., 180, 203
Sato, J. E., 514, 541
Sauer, W., 204
Scali, V., 286
Scandalios, J. G., 176, 203, 207

Schaffner, K. F., 87, 94
Scheibe, A., 184, 207
Schindewolf, O. H., 48, 76-80, 85, 94
Schindler, O., 171
Schlick, M., 113, 119
Schmalhausen, I. I., 297-298
Schoffeniels, E., 159, 161, 170-172
Scholl, H., 250, 261
Schopf, J. W., 71, 95
Schreiber, K., 184, 207
Schrödinger, E., 104, 112, 119
Schuchert, C., 67, 95
Schultz, J., 20, 21, 293, 298
Schultz, R. J., 242, 261
Schutz, F. 169, 172
Schwanitz, F., 34
Schwartz, D., 176, 207
Schwemmer, S. S., 203
Scott, R. A., 202, 207
Scott, W. B., 70, 95
Scriven, M., 82, 86, 95
Sears, E. R., 193
Seiler, J., 248, 258, 261
Selling, O., 443, 447, 542
Selye, H., 37
Semenov-Tian-Shanskij, A., 19
Shapley, H., 34
Shaw, C. R., 311, 314
Sheppard, P. M., 184, 204, 300, 313
Shklovskii, I. S., 69, 95
Shook, M., 542
Shorey, H. H., 362, 378
Sibley, C. G., 490, 542
Siever, R., 86-87, 95
Silagi, S., 36
Simak, M., 234-235
Simmons, A. S., 2
Simpson, G. G., 37, 43, 105, 119, 164, 172
Singh, R., 298
Sinko, M., 233
Sinnott, E. W., 33
Sivertzev, N. P., 2
Sivertzev-Dobzhansky, N. P., 21. See also Sivertzev, N. P., and Dobzansky, N. P.
Slobodkin, L. B., 128-129, 133, 138, 146, 158
Smith, H. H., 192, 207
Smith, J. M., 316, 326-327, 378
Smith, M. A., 262
Smith, V. E., 38

Snaydon, R. W., 178, 207
Sneath, P. H. A., 174-175, 207
Sokal, R. R., 174-175, 204, 207
Sokolov, D., 23, 29, 156, 381, 434
Soria, J., 175, 205
Spackman, P., 40
Sparrow, A. H., 203
Spassky, B., 23, 24-26, 28-31, 33-41, 138, 156, 158, 313, 351, 364, 376, 416, 434
Spassky, N., 29-31, 33-35
Sperlich, D., 355, 378, 418, 421, 423, 430, 434-435
Spiegelman, S., 298
Spiess, E. B., 315-379, 423-424, 435
Spieth, H. T., 319, 362-363, 379, 437, 451, 459-461, 463-464, 466, 472, 474, 477, 482-483, 485-486, 491-492, 532, 542
Spiro, M. E., 38
Spuhler, J. N., 39
Spurway, H., 424, 435
Stalker, H. D., 250, 262, 426, 436, 465-467, 492, 496-497, 500, 504, 514, 516, 532, 541, 543
Stearns, H. T., 441, 444-445, 543
Stebbins, G. L., 38, 173-208, 294, 298
Steere, W. C., 39-40, 119, 157, 203
Stefani, R., 256, 262
Stefansson, E., 233, 235
Steiner, E., 198, 207
Stepanov, D. L., 80, 95
Stephens, S. G., 184, 208
Stern, C., 296, 298, 438
Stone, W. S., 403, 405-407, 411, 431, 435-436, 437-439, 495, 542
Streisinger, G., 25
Strickberger, M. W., 125-126, 156
Sturtevant, A. H., 4, 20, 22, 291-293, 295-296, 298, 318-319, 343, 376, 379
Subba, V., 298
Sued, J., 41
Sumner, F. B., 283, 287
Suomalainen, E., 248-249, 259, 262
Swanson, C. P., 193, 208
Swezey, O. H., 478, 543
Sylvest, V., 298

Takada, H., 436, 451, 460, 522, 542-543
Takei, G., 542
Takenouchi, Y., 248, 262

Taliaferro, W. J., 543
Tan, C. C., 22, 324, 379
Taylor, R. L., 194, 208
Tax, S., 34, 207
Tebb, G., 379
Teilhard de Chardin, P., 40, 69, 95
Teissier, G., 150, 157
Thoday, J. M., 284, 287, 300, 314, 379
Thomas, R., 248, 262
Thompson, H. J., 196, 206
Thomson, W. (Lord Kelvin), 47-48, 67, 73, 93
Throckmorton, L. H., 459-461, 464-465, 469-470, 486, 489-490, 493, 543
Tidwell, T., 36
Timofeeff-Ressovsky, H. A., 399
Timofeeff-Ressovsky, N. W., 304, 314, 382, 397-399
Tiniakov, G. G., 300, 382, 399, 409, 417, 423, 434
Tinkle, D. W., 244, 262
Tobach, E., 40
Tokumitsu, T., 542
Toulmin, G. H., 95
Tremblay, E., 239, 260
Tshetverikov, S. S., (see Chetverikov, S. S.)
Tucker, J. M., 188, 208
Turesson, G., 234-235
Turner, B. L., 177, 203

Ullrich, R., 36
Ullyett, G. C., 125-126, 158
Umbgrove, J. H. F., 75, 95
Utida, S., 131, 158
Uzzell, T. M. Jr., 243-244, 260, 262

Vaidyanathan, C. S., 297-298
Valentine, D. H., 189, 208
Van Delden, W., 301, 305, 313
Van der Pijl, L., 201, 208
Van Valen, L., 129, 158, 358, 379
Velikovsky, I., 56, 95
Verhulst, P. F., 130, 158
Voinarsky, S., 1
Vojtiazky, B. P., 19
Volterra, V., 128, 130, 132, 158
Von Wahlert, G., 80, 85, 91
Vysotskii, B. P., 58, 62, 66, 88, 95

Waddington, C. H., 103, 119, 181, 208
Wagner, M., 402, 436

Waldron, I., 362, 379
Walker, I., 416, 436
Wall, J. R., 191, 208
Wallace, B., 5, 26, 29-30, 33, 35-36, 344, 346, 379, 381-399, 420, 436
Walsch, J. T., 172
Walters, L., 204
Wangenheim, K. H. von, 189-190, 208
Ward, B. L., 514, 543
Warner, R. E., 491, 543
Warner, R. M., 397, 399
Warters, M., 407, 436
Washburn, S. L., 36
Wasserman, M., 407, 436, 514, 543
Watanabe, T., 40
Waterbolk, H. T., 202, 208
Waters, L. C., 190, 208
Watson, J. D., 81, 96
Watson, L., 204
Watson, R. A., 86-87, 96
Wattiaux, J. M., 361, 376, 379
Webb, G. C., 255, 262
Wedel, M., 29
Weidmann, U., 239, 261
Weimorts, D., 298
Weisbrot, D. R., 125-126, 158
Weismann, A., 181
Weiss, M. C., 296, 298
Weissbach, H., 297-298
Wharton, D. R. A., 170, 172
Wharton, M. L., 172
Wheeler, M. R., 260, 406, 411-412, 436, 474, 477, 490, 517-520, 543
Whewell, W., 49, 93, 96
Whitaker, T. O., 195, 205

White, M. J. D., 150, 157, 236, 253, 255, 262
Wiesel, T. N., 114, 118
Wilder, P. A., 203
Williams, C. B., 128, 138, 158
Williams, G. C., 160, 172
Williams, W. T., 204
Wills, C. J., 37
Wilmarth, M. G., 74, 96
Wilson, E. O., 154-155, 169, 172
Wilson, F. D., 436
Wilson, L. G., 45, 49, 96
Witt, A. A., 128, 156
Witthöft, W., 110, 119
Wolfe, S. L., 199, 208
Wood, P. W., 158
Woodroffe, G. E., 261
Wright, J. W., 244-245, 260, 262, 313
Wright, S., 4, 23-24, 26, 117, 119, 382, 398-399, 417, 422, 436, 575, 577
Wright, T. R. F., 308, 311-312, 314

Yang, H., 490, 517-520, 543
Yarbrough, K., 150, 156, 312, 314
Yerington, A. P., 397, 399
York, T. L., 191, 208

Zanardini, I. F., 434
Zeuner, F. E., 83, 96
Ziehen, Th., 99-100, 112, 114, 116, 119
Zimmering, S., 28
Zimmerman, E. C., 438, 449, 543
Zimmerman, W., 201, 208
Zweifel, R. G., 245-246, 248, 262

Subject Index

Acacia
  loderi, 254
  Wilhelmiana, 254
Adaptation at molecular level, 159-163
Age of Mammals, 52
Aleurodidae, 256
Ambystoma
  jeffersonianum, 243-244
  laterale, 243-244
  platineum, 243-244
  tremblayi, 243-244
Animal Species and Evolution, 178
Annonaceae, 196
Antopocerus, 458, 462, 468-469, 472, 474, 484, 486, 492-493, 495-496, 532, 534, 538
  Colonization, speciation, 532
Aphelandra micans, 412
Araliaceae, 481
Argyroxiphium sandwicense, 481
Atelodrosophila, 458, 462

Barynotus, 248
Bergmann's rule, 107-108
Biological Basis of Human Freedom, The, 7
Biology, behavior Hawaiian Drosophilidae, 469
  courtship, 482-490
  evolutionary implications, 490-492
  life cycle, 469-482
Biology of Ultimate Concern, The, 7
Bothriochloa, 196
Branchiostomata (*Amphioxus lanceolatum*), 165-166
Bufo
  bufo, 169
  calamita, 169
Bufonidae, 169
Bupalus piniarius, 283

Calathea allouia, 412
Cambrian, 70-71, 75-77
Camelina, 183
Carex, 198
Catapionus, 248

Celidosoma, 458, 462, 468, 538
Cenozoic, 74-76, 78
Cepaea, 300
Cephalochordata, 165
Cestrum, 413
  euanthes, 412
  parqui, 411-412, 415
  tomentosum, 412
Chaenactis, 192
Chaetodrosophila, 458
Charpentiera, 449
Chasiempsis sandwichensis, 491
Cheirodendron, 443, 448-449, 471-473, 475-477, 480-482, 490, 537, 539
Cheyletus eruditus, 239
Chironomus intertinctus, 426
Chromosomal polymorphism, mating propensity, *Drosophila pseudoobscura*, 343-346
  sex ratio, 344
  third chromosome arrangements, 346-356
  in other species, 354-356
Chrysomya
  albiceps, 125-126
  chloropyga, 125
Chusquea, 411
Chymomyza, 456, 462
  procnemis, 457
Cibotium, 447, 477, 480
Clarkia, 180, 191-192
Clermontia, 448, 471-473, 477, 480-482, 488, 490, 491, 537, 539
Chemidophorus, 244, 246, 248
  exsanguis, 246
  inornatus, 244-246
  neomexicanus, 244-245
  perplexus, 245
  sacki, 246
  septemvittatus, 244-246
  sexlineatus, 245
  tesselatus, 245, 255
  tigris, 244-246
Cnephia
  mutata, 252-253, 257
Coactones, 167

Coccidae, 256
Coccinellidae, 3
Colombia, 7-15
Competitive coexistence
  frequency-dependent selection, 148-150
  in laboratory, 138-146
  niche heterogeneity and temporal variations, 150-151
  related species in nature, 136-138
  selection and life cycles, 146-147
Competitive exclusion, 128-130
  Volterra's equation, 132-136, 154
Compositae, 200
Cope's law, 107
Cosmology, 57-58
Crepis, 186, 199, 294
Cretaceous, 89, 202
Crustacea, 169
Curculionidae, 248, 250
Cyanea, 448
Cyperaceae, 198
Cyrtandra, 478
Cysticetes, 101

Danthonia, 197
Daphnia, 146
Datura, 180
Deoxyribonucleic acid (DNA), 176, 180, 189, 198-199, 250, 295, 546
Dettopsomyia, 456, 462
  formosa, 457
  nigrovittata, 457
Devonian, 75, 77
Diaspididae, 239
Dichanthium, 196
Diplocystis clerci, 256
Dipsacaceae, 180
Diptera, 124, 282, 465
Directionalism
  in evolutionism, 69-71
Dollo's Law, 70-71
DNA (see Deoxyribonucleic acid)
Drosophila, 3, 7, 9, 12-13, 82, 124, 131, 289-379, 437-543
  adiostola, 451-452, 487, 504, 516, 520, 523, 526-527
  ananassae, 290, 405-406, 422, 429, 457
  anomalepis, 464, 483, 487-490
  busckii, 126, 405-406, 429, 457
  crucigera, 481, 485, 492, 508-511, 514, 517, 521
  cyrtoloma, 455-456, 494, 520

Drosophila (continued)
  fitness in populations, 301-312
  flavopilosa, 401, 411-415, 418, 422, 428, 430-432
  funebris, 139, 300, 381, 401, 405, 408-411, 417, 422-423, 429, 432, 457, 479
  grimshawi, 467, 479, 485, 491, 497, 500, 506, 508, 511-514, 516-517, 520-521, 526
  hydei, 405, 407, 418, 429, 457
  immigrans, 401, 405, 408-411, 418, 428, 473, 483, 489
  melanogaster, 124-126, 139, 147-148, 151-152, 289-296, 302, 304-305, 312-313, 318, 326-328, 335, 337, 339-340, 343, 358-359, 362, 364-365, 373, 381-399, 403, 405-406, 410, 416, 418, 422, 424, 429, 438, 456, 483-484
  nebulosa, 139-141, 147, 152-153
  paulistorum, 5, 9, 14, 137, 284, 364, 425
  pavani, 355-356, 401, 418, 427, 429, 432
  persimilis, 136, 290, 346, 351-352, 356-360, 364, 369-372, 417, 422, 424
  primaeva, 464, 500, 516, 522-523, 526
  pseudoobscura, 4, 124, 127, 136, 141-147, 152, 283, 290, 300, 309, 328, 335, 342-344, 348, 352, 356-360, 364, 368-369, 371, 382, 398, 416-417, 422-423, 429-430
  repleta, 405, 407, 418, 457, 522
  robusta, 355-356, 417, 422-423, 426, 429-430, 522
  serrata, 125, 139-147, 152-153
  simulans, 125, 148, 151, 289-291, 293-295, 335, 358, 362, 405-406, 457
  subobscura, 253, 296, 355, 358, 361, 397, 418, 422, 424, 426, 430
  willisfoni, 2, 7, 128, 137, 253, 290, 300, 364, 382, 422, 430
Drosophila subgenera
  Engyscaptomyza, 460, 462, 464, 469-470, 487
    amphilobus, 487
    crassifemur, 460, 462, 464, 469-470, 487
    nasalis, 460, 464, 470, 487, 493
  Hirtodrosophila, 459
  Hypenomia, 459

Drosophila subgenera (*continued*)
  Sophophora, 328, 362, 460, 462, 492
  Trichotobregma, 459
Drosophilidae, Evolutionary biology in Hawaiian, 437-543
  characteristics, chromosomal and genetic, 492-520
  fauna, 450-569
  habitat, 439-450
  speciation and migration, 520-532

Ecological Factors, *Drosophila*, 299-314
Ecological niche, 122-124
Ecomones, 167
Elephants, 56, 57
Elymus, 192
Endocoactones, 168
Epilobium, 189
Erythrina, 196
Erythroneura, 136
  bella, 136
  lawsoni, 136
  morgani, 136
  torella, 136
Euphausiids, 169
Eusomus, 248
Evolution
  of matter, 97-119
  orthogenesis, specialization, and differentiation, 201
  paleobotanical evidence, 202
  patterns of distribution, localities of origin, and rates, 202
  principle of irreversibility, 201
  recapitulation and embryonic similarity, 201
Evolution, Chilean species, *Drosophila*, 401-436
  Isolation and chromosomal structure, 428-432
  Restricted spp., 411-418
  Versatile spp., 418-428
  Widespread spp., 405-411
Evolutionary Biology, 2, 110

Factors and laws, organismic evolution, 104-109
Factors modifying mating propensity, 357
  extrinsic factors, 357-361
    age of parents, 361
    density, 359
    food volume, 358

Factors, extrinsic factors (*continued*)
    plight, 358
    sex ratio, 359
    temperature, 357
  frequency dependency, 364-373
  intrinsic factors, 361-364
Festuca, 197
"founder effect," 6
Freycinetia, 447

Ganaspis, 415
Gause's principle, 128
Genetic adaptation
  correlated responses, 564-567
  deterioration, 568-570
  improvement, 567-568
  intelligence, 570-574
Genetic systems, factors in evolution, 185-187
Genetics and the Origin of Species, 4, 17, 174, 289
Genetics of the Evolutionary Process, The, 4, 6
Genotype-Environment interaction, 575-576
Gilia, 180, 192
  malior × *modocensis*, 193
Gitonides, 456, 462
  perspicax, 456
Godetia, 191
Grimshawomyia, 458, 462, 468, 470
Gueriniella serratulae, 239
Gymnopais, sp. 252, 258
Gymnophthalmus underwoodi, 248

Halimeda, 175
Haploembia solieri, 255-256
Helianthus, 178
Heliconia spp., 412
Hemidactylus garneti, 248
Heritability, 546-550
  Estimation of, 550-552
Heterozygosity, genetic polymorphism in parthenogenetic animals, 236-262
Homoptera, 239, 256
Homo sapiens, 72, 86
Human evolution, 574-575
Huttonian-Wernerian, 46
Hybrid autofluorescent patterns, 291-297
Hybridity, structural and chromosomal evolution, 197-200

Hybridization and its effects, 192-193
  interspecific, 516-520
Hymenoptera, 282
Hypenomyia, 459

Identism, panpsychistic, polynomistic, realistic, 117-118
Idiomyia, 455, 457
Ilex, 448, 472
Inversions, fixed versus polymorphic, 504-506
Irreversibility
  in evolutionism, 69-71
Isolation, origin of species, 187-188
  isolating mechanisms, 188-191
  origin of, 191-192

Jesup Lectures, 4
Jurassic, 52, 77

Karotypes
  metaphase, 492-496
  Drosophilidae, 492-496
Kiev, University of, 1, 16
Koa, 480
Kynurenine distribution in larval fat body, 290

Lacerta
  armeniaca, 247
  dahli, 247
  rostombekovi, 247
  saxicola, 247-248, 259
    bithynica, 248
    defilippi, 248
    mixta, 248
    unisexualis, 247-248
Leguminosae, 196
Leningrad, University of, 1, 16
Life processes, causal analysis, 101-104
Linanthus parryae, 185
Liophloeus, 248
Lispocephala, 491
Listroderes, 248
Lobeliaceae, 473
Logistic theory of population growth, 130-132
Lonchoptera dubia, 247, 250-251, 257
  furcata, 251
Lucilia
  cuprina, 146
  sericata, 125-126
Luzula, 198

Mammoths, 56
Mammuthus primigenius, 56
Maniola jurtina, 263-287
Mankind Evolving, 6-7, 17, 560
Markov property, 557-560
Mating propensity as a genetic character, *Drosophila,* 315-379
  associations with mutants, 318-324
  genetic analysis by selection techniques, 328-337
  quantitative genetic analysis, wild type strains, 337-343
  strain differences without genetic markers, 324-328
Matter, physical analysis, 112-113
Megaceros, 106
Mesozoic, 74-76
Metrosideros, 447-450, 480, 487
Microdispersion, *Drosophila melanogaster*
  Bryant Park Experiments, 382-386
  Experimental populations, 386-393
  "Tropical rainforest" greenhouse experiments, 393-395
  Venetian experiments, 395-396
Microtatobiotes, 101
Molecular adaptation
  and phylogeny, 163-166
  at population level, 166-170
Molliensia formosa, 241
Moraba virgo, 241, 243-244, 253-254, 256, 258
Morella, 175
Miocene, 188
Morpho, 12
Musca
  domestica, 152, 416
Mycoplasmatales, 101
Myrsine, 448-449, 475

Nasonia vitripennis, 416
Natural selection, 151-154
  dynamics, random variation, 184-185
  experimental evidence, 182-184
Neptunist-Cuvierian model, 74
Nicotiana, 180, 192
Nudidrosophila, 457-458, 462, 468, 472, 538
  aenicta, 457

Oenothera, 189, 198
Opius trimaculatus, 415
Opuntia, 431

Ordovician, 72, 77
Oryza, 198
Osmanthus, 449
Otiorrhynchus, 248
  chrysocomus, 249
  salicus, 249
  scaber, 249
  singularis, 249

Paleocene, 89
Paleoecology, 55-56
Paleozoic, 68, 71-72, 74-76
Panicum, 198
Paramecium, 130
Paroreomyza, 491
Parthenogenesis
  Insects, 248-256
    beetles, 248-250
    dipterous flies, 250-253
    embiid, 255-256
    Moraba virgo, 253-255
  vertebrates, 241-248
    ambystomas, 243-244
    lizards, 244-248
    poeciliid fishes, 241-243
  Nature of, 238-241
Partula, 268
  taeniata, 283
Patterns, variation
  examples of, 177-180
    ecotype concept, 177-180
    species, genus level, 180
  exploring, charting patterns, 174-175
  population, concept, 177
Peritelus, 248
  hirticornis, 249
Permian, 89
Phacidium infestans, 212
Phaenicia
  sericata, 152
Phaseolus
  coccineus, 191
  vulgaris, 191
Pheromones, 169
Phormia, 373
Phylogeny
  and molecular adaptation, 163-166
Picea abies, 210
Pinus sylvestris, 209-235
Pisum
  arvense, 184
Pittosporum, 475

Plantago, 183, 202
  lanceolata, 251
Pleistocene, 51, 77, 106-107, 194, 443 447 466, 527, 538
Pliocene, 188, 538
Poa, 187, 197
Poecilia
  formosa, 241-243, 257
  latipinna, 241-242
  mexicana, 241-242
  sphenops, 242
Poecilopsis
  *latiden*s, 243
  lucida, 242-243
Polydrusus, 248
Polymorphism, Drosophilidae
  chromosomal, 504
    identical in two species, 506-507
    intraspecific, 504
    parallel, 506
    quantitative studies, 508-514
Polyploidy and apomixis, 193-197
Polyporus, 477
Population research, *Pinus sylvestris,* 209-235
  climatic conditions, experimental and natural populations, 212-214
  cumulative mortality and survival, 214-216
  in selected experimental plantations, 216-223
  height growth, 228-231
  single-year mortality, 223-228
Potentilla glandulosa, 179
Precambrian, 71-72, 75, 77
Primula, 189
  elatior × *veris,* 190
Prosimulium
  macropyga, 252
  ursinum, 252
Pseudiastata, 456, 462
Pseudocneothinus, 248
Pseudocoila bochei, 416
Psychogenesis and epistemological analysis, 109-112
Pterotropia, 482
Ptinus
  clavipes, 250
  mobilis, 250
Pulvinaria hydrangeae, 239

Quaternary, 258

Quercus, 179,
  douglasii, 188
  dumosa, 188
  turbinella, 188

Radiation, Genes and Man, 5
Ranales, 196
Random mating, 552-557
Ranunculus
  flammula, 180
Reynoldsia, 481
Revolutionism, in geohistory, 74-75
Ribonucleic acid (DNA), 189-190
RNA, see Ribonucleic acid
Rosaceae, 200
Rumex, 183

Sadleria, 478
Saga pedo, 240
Salvia, 175
Sapindus, 449, 471
Satyridae, 263-285
Scaptomyza, 450, 456, 458-459, 461-465, 467, 487, 493, 496, 520, 536-537, 539
  adusta, 493
  argentifrons, 493
  contesta, 460
  pallida, 481, 493
  parva, 461, 463, 487
  vittiger, 460-461
Scaptomyza subgenera
  Alloscaptomyza, 461-462, 470, 478
  Bunostoma, 461-462, 474, 491
  Exalloscaptomyza, 461-463, 469-470, 473
  Parascaptomyza, 461-462
  Rosenwaldia, 461-462
  Tantalia, 461-463, 472, 473
  Trogloscaptomyza, 461-463, 472-473, 478
Scepticus, 248
Sciaphilus, 248
Secale, 198
Senecio, 197
Silene, 197
Silliman Lectures, 6
Solanum, 175, 189
  rubidum, 412
Solenobia, 259

Species,
  homosequential, 514-516
  sibling, 4, 516
Strophosomus, 248
Struggle for Existence, The, 128, 156

Taeniodonta, 78
Taraxacum, 186
Tertiary floras, 202
Tetraplasandra, 447, 449, 475, 482
Thrips tabaci, 415
Titanochaeta, 460, 462-463, 473, 477, 493, 496
Trachyphlorus, 248
Triassic, 77, 89
Tribolium, 148
  castaneum, 127, 138-139
  confusum, 127, 138-139
Trifolium
  repens, 251
Trillium, 192, 199
Triticinae, 176
Triticum, 193
Tropidophorus, 248
Tropiphorus, 248
Tunicata, 165

Uniformitarianism, 43 ff
  Actualism, 61-66
  Evolutionism, 68-72
  Gradualism, 72-81
  Historical Inferences, 81-87
  Historicism, 66-68
  Immanence and Configuration, 58-60
  Naturalism, 61

Variability of gene-pools, and ecological factors, *Drosophila,* 299-314
Variation, individual, 180-182
  modification, environmental, 180-181
  mutations
    genetic effects, 182
    rates, 182
    types, 181-182
Vestaria coccinea, 491
Vicia, 199
  faba, 199
Vulcanist-Neptunist, 46-47, 60